National Center for Construction Education and Research

Welding Level Two

AWS Entry Level Welder – Phase Two

PEARSON
Prentice Hall

Upper Saddle River, New Jersey
Columbus, Ohio

contren™
Learning Series

National Center for Construction Education and Research

President: Don Whyte
Director of Curriculum Revision and Development: Daniele Dixon
Welding Project Manager: Daniele Dixon
Production Manager: Debie Ness
Quality Assurance Coordinator: Jessica Martin
Editor: Bethany Harvey
Desktop Publishers: Laura Parker, Rachel Ivines

The NCCER would like to acknowledge the contract service provider for this curriculum:
Topaz Publications, Liverpool, New York.

This information is general in nature and intended for training purposes only. Actual performance of activities described in this manual requires compliance with all applicable operating, service, maintenance, and safety procedures under the direction of qualified personnel. References in this manual to patented or proprietary devices do not constitute a recommendation of their use.

---

10 9 8 7 6 5 4 3 2
ISBN 0-13-102581-3

Preface

This volume was developed by the National Center for Construction Education and Research (NCCER) in response to the training needs of the construction, maintenance, and pipeline industries. It is one of many in NCCER's *Contren™ Learning Series*. The program, covering training for close to 40 construction and maintenance areas, and including skills assessments, safety training, and management education, was developed over a period of years by industry and education specialists.

NCCER also maintains a National Registry that provides transcripts, certificates, and wallet cards to individuals who have successfully completed modules of NCCER's *Contren™ Learning Series*, when the training program is delivered by an NCCER Accredited training Sponsor.

The NCCER is a not-for-profit 501(c)(3) education foundation established in 1995 by the world's largest and most progressive construction companies and national construction associations. It was founded to address the severe workforce shortage facing the industry and to develop a standardized training process and curricula. Today, NCCER is supported by hundreds of leading construction and maintenance companies, manufacturers, and national associations, including the following partnering organizations:

PARTNERING ASSOCIATIONS

- American Fire Sprinkler Association
- American Petroleum Institute
- American Society for Training & Development
- American Welding Society
- Associated Builders & Contractors, Inc.
- Association for Career and Technical Education
- Associated General Contractors of America
- Carolinas AGC, Inc.
- Carolinas Electrical Contractors Association
- Citizens Democracy Corps
- Construction Industry Institute
- Construction Users Roundtable
- Design-Build Institute of America
- Merit Contractors Association of Canada
- Metal Building Manufacturers Association
- National Association of Minority Contractors
- National Association of State Supervisors for Trade and Industrial Education
- National Association of Women in Construction
- National Insulation Association
- National Ready Mixed Concrete Association
- National Utility Contractors Association
- National Vocational Technical Honor Society
- North American Crane Bureau
- Painting & Decorating Contractors of America
- Plumbing-Heating-Cooling Contractors National Association
- Portland Cement Association
- SkillsUSA
- Steel Erectors Association of America
- Texas Gulf Coast Chapter ABC
- U.S. Army Corps of Engineers
- University of Florida
- Women Construction Owners & Executives, USA

Some features of NCCER's *Contren™ Learning Series* are:

- An industry-proven record of success
- Curricula developed by the industry for the industry
- National standardization providing portability of learned job skills and educational credits
- Credentials for individuals through NCCER's National Registry
- Compliance with Apprenticeship, Training, Employer, and Labor Services (ATELS) requirements for related classroom training (CFR 29:29)
- Well-illustrated, up-to-date, and practical information

FEATURES OF THIS BOOK

Capitalizing on a well-received campaign to redesign our textbooks, NCCER is publishing select textbooks in a two-column format. *Welding Level Two* incorporates the design and layout of our full-color books along with special pedagogical features. The features augment the technical material to maintain the trainees' interest and foster a deeper appreciation of the trade.

Think About It uses "What If?" questions to help trainees apply theory to real-world experiences and put ideas into action.

Hot Tip provides a head start for those entering the trade by highlighting important safety requirements and presenting tricks of the trade from experienced welders.

We're excited to be able to offer you these improvements and hope they lead to a more rewarding learning experience.

As always, your feedback is welcome! Please let us know how we are doing by visiting NCCER at www.nccer.org or e-mail us at info@nccer.org.

UNIQUE ASPECTS OF THE NEW AWS ENTRY LEVEL WELDER PROGRAM

In this new edition of the Contren™ Welding series, you'll note features and icons which relate to the American Welding Society's (AWS) *AWS EG2.0-95, Guide for the Training and Qualification of Welding Personnel: Entry Level.* These features correlate with specific guidelines of AWS S.E.N.S.E. (Schools Excelling through National Skills Education) programs. But they can also be used as helpful tools in delivering and evaluating NCCER performance examinations.

This unique version of NCCER/Prentice Hall's welding program allows dual credentialing through both NCCER and AWS, provided all applicable requirements are met by the participating organizations.

For more information on AWS's Entry Level Welder program, contact AWS at (305) 443-9353 or visit www.aws.org. For information on NCCER's Accreditation and National Registry, contact NCCER Customer Service at 1-888-622-3720 or visit www.nccer.org.

Acknowledgments

This curriculum was revised as a result of the farsightedness and leadership of the following sponsors:

Applied Welding Technology
Exxon Mobile
Flint Hills Resources
Fluor Daniel Inc.
Kawerak
Koch Refining
Lee College
Northern Arizona Vocational Institute of Technology
Northland Pioneer College
Spec-Weld Technologies, Inc.
Texas A & M University System—Texas Engineering Extension Service
William H. Turner Adult Education Center
Zachry Construction Corporation

This curriculum would not exist were it not for the dedication and unselfish energy of those volunteers who served on the Authoring Team. A sincere thanks is extended to:

Tom Atkinson
Rex Ball
Curtis Casey
Bill D. Cherry
John Gault
Terry A. Lowe
David McGrath
John D. Murray
Joe D. Sanders
Jerry Trainor
John R. Yochum

Contents

*Please note that Modules 29203-03 and 29210-03 are electives for *Welding Level Two* but are requirements for *AWS Entry Level Welder*.

Module 29201-03

Welding Symbols

COURSE MAP

This course map shows all of the modules in the second level of the Welding curriculum. The suggested training order begins at the bottom and proceeds up. Skill levels increase as you advance on the course map. The local Training Program Sponsor may adjust the training order.

WELDING LEVEL TWO

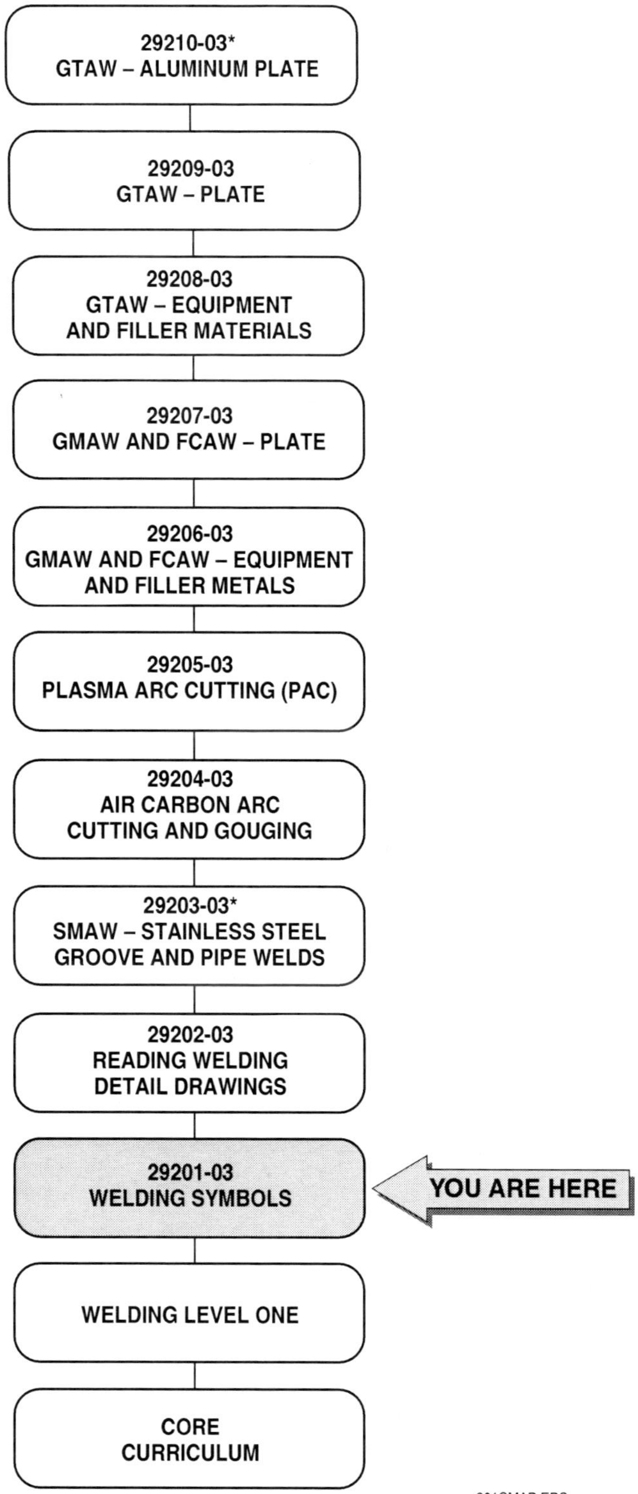

201CMAP.EPS

Please note that Modules 29203-03 and 29210-03 are electives for *Welding Level Two* but are requirements for *AWS Entry Level Welder.

MODULE 29201-03 CONTENTS

Figures

Table

Welding Symbols

Objectives

When you have completed this module, you will be able to do the following:

1. Identify and explain the various parts of a welding symbol.
2. Identify and explain fillet and groove weld symbols.
3. Read welding symbols on drawings, specifications, and welding procedure specifications (WPSs).
4. Interpret welding symbols from a print.
5. Draw welding symbols based on the observation of actual welds.

Prerequisites

Before you begin this module, it is recommended that you successfully complete the following: Core Curriculum; Welding Level One.

1.0.0 ◆ INTRODUCTION

Welding symbols are used on drawings, project specifications, and welding procedure specifications (WPSs) to convey the design specifications for welds. They use a series of symbols to indicate the joint configuration and weld type, location, size, and length required. Welders must be able to properly interpret welding symbols to ensure the welds they make will meet the design specifications.

2.0.0 ◆ STANDARD SYMBOLS

Most companies use the symbols that have been standardized by the American Welding Society (AWS) in *AWS Standard A2.4, Standard Symbols for Welding, Brazing, and Nondestructive Examination.* Most companies also have a site quality standard that provides guidelines and examples of how welding symbols are to be used. Always refer to your site quality standard when interpreting welding symbols.

3.0.0 ◆ WELDING SYMBOL BASE

The base for all welding symbols is the horizontal reference line with an arrow at one end. The arrow, which can come off either side of the reference line, points to the location to which the welding symbol applies (*Figure 1*). The welding symbol describes the type of weld, its size, and its surface finish.

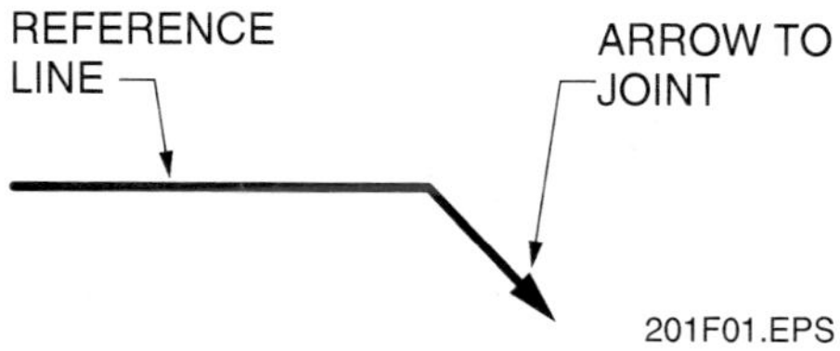

Figure 1 ◆ Horizontal reference line.

The opposite end of the arrow is used for information that aids in the making of the weld but doesn't have its own special place on the symbol (*Figure 2*). For instance, it is used for designating the welding and cutting processes, the reference

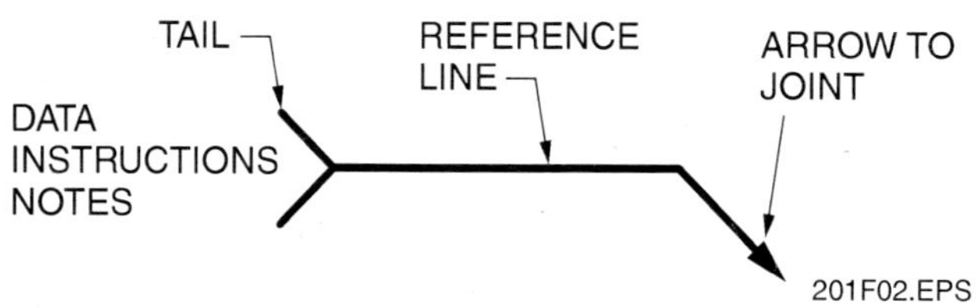

Figure 2 ◆ Tail side information.

> **Hot Tip**
>
> ### *Welding and Cutting Processes*
>
> The appendix in the back of this module details AWS abbreviations for various welding and cutting processes. *Table A-1* covers welding processes, *Table A-2* covers cutting processes, and *Table A-3* covers supplementary notations for applying welding processes. Abbreviations from all three tables can be used in tail references or in single-line or multiple-line welding symbols.

to a note, and the welding procedure or electrode type to be used. The tail is omitted when a reference is not required.

In the case of a T-joint, there are two sides to the joint to which the arrow points. Thus, welding may be accomplished on either side of the joint. The welding symbol distinguishes between the two sides of a joint by using the arrow and the spaces above and below the reference line. The side of the joint to which the arrow points is called the arrow side, and its weld is made in accordance with those instructions below the reference line. The other side of the joint is referred to as the other side, and its weld is made in accordance with those instructions found above the reference line. *Figure 3* shows examples of these symbols.

Regardless of which end of the reference line the arrow is on, information on the reference line is always read from left to right. *Figure 4* shows a welding symbol base and numerous elements of a welding symbol.

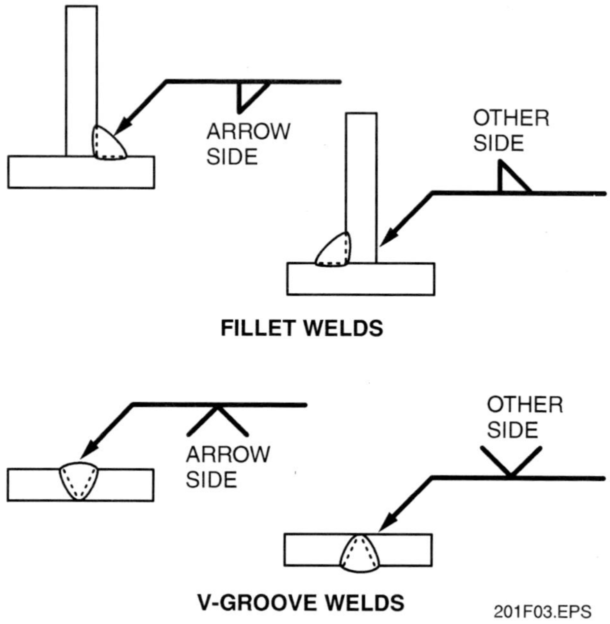

Figure 3 ◆ Arrow side and other side significance.

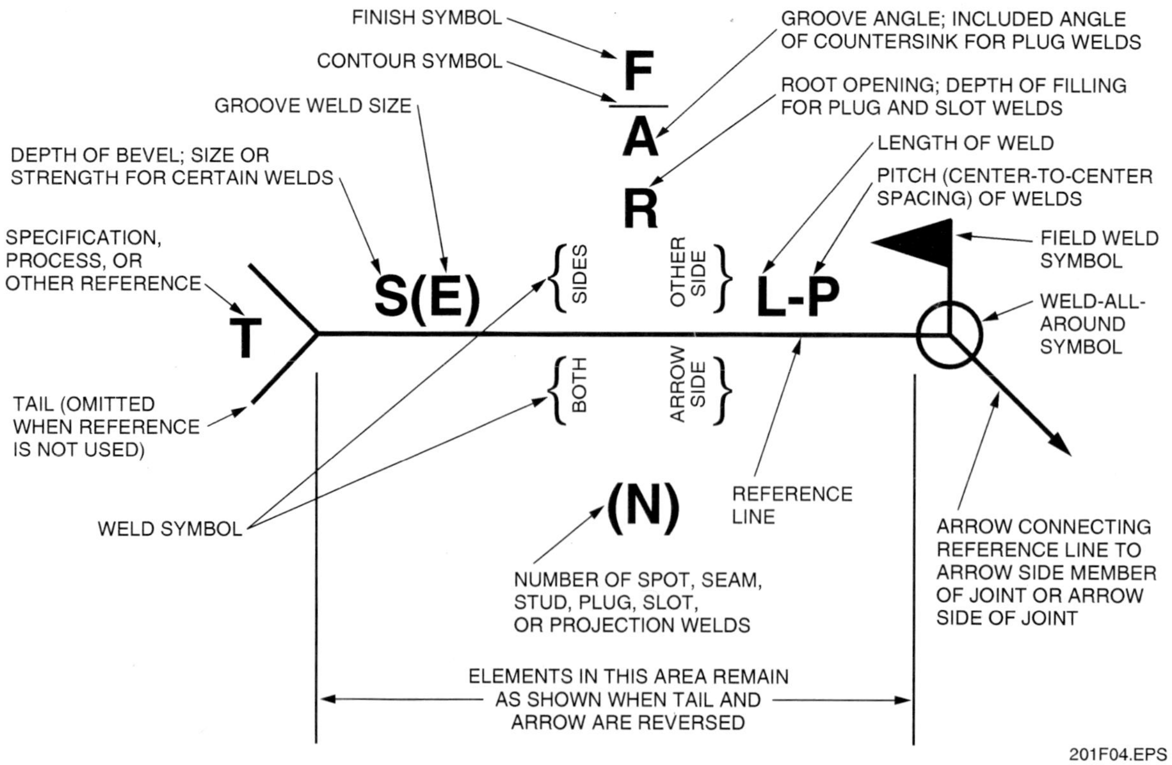

Figure 4 ◆ Standard location of elements of a welding symbol.

Hot Tip

Nondestructive Examination Symbols

These symbols may appear on the same drawings as, and look very similar to, welding symbols. They are used by inspectors to examine joints once welding is complete. The nondestructive methods of inspection are defined by letter codes on the reference line. Most of the methods involve some type of electrical, electronic, or magnetic examination.

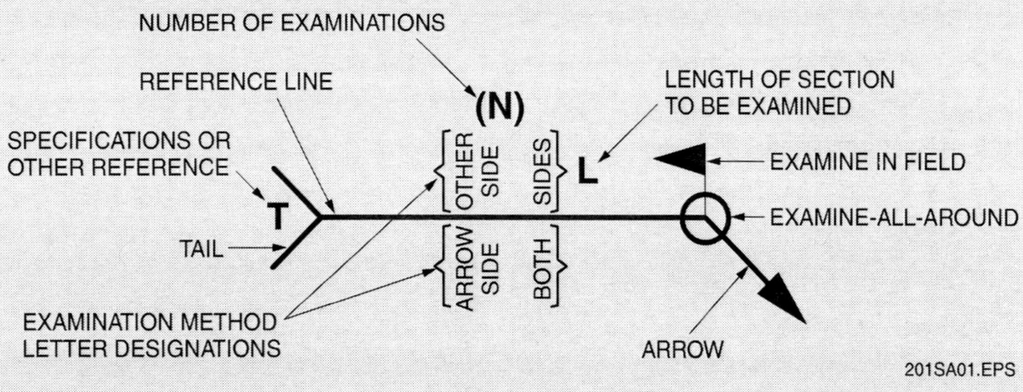

The symbols used to identify the type of weld to be made signify a basic type of weld or joint preparation. *Figure 5* shows the basic **weld symbols**.

3.1.0 Symbols for Welds

The type of weld used is determined by the way the members to be welded are positioned to form the joint. The basic types of weld joints are the butt, lap, corner, T-, edge, flanged butt, flanged corner, flanged T-, flanged lap, and flanged edge joints. The basic types of weld joints are shown in *Figure 6*.

3.2.0 Location of Weld Symbols

When the weld symbol is on both the top and bottom of the reference line, the information applies to both sides of the joint or surface to which the arrow is pointing. Symbols appear on both sides of the reference line when a double-fillet or double-groove weld is specified. *Figure 7* shows both sides for fillet and V-groove welds.

The shape of the groove weld symbol indicates how the groove is to be prepared. In the case of the bevel groove and J-groove weld symbols, only

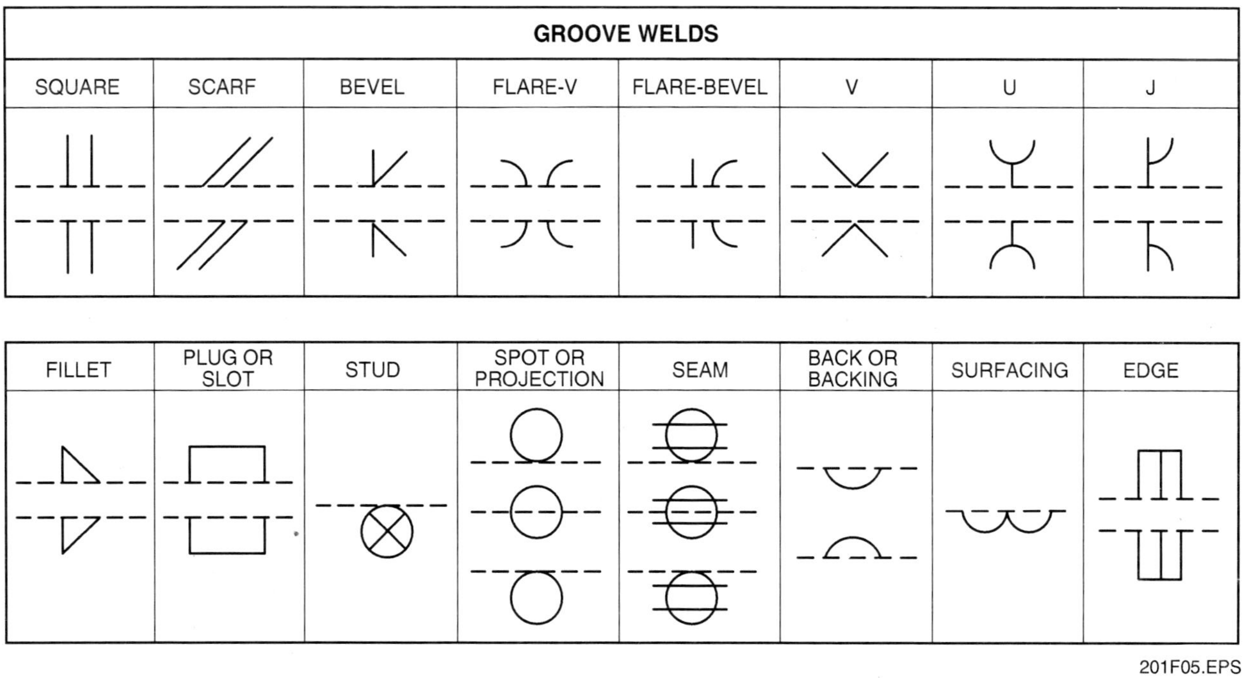

Figure 5 ◆ Basic weld symbols.

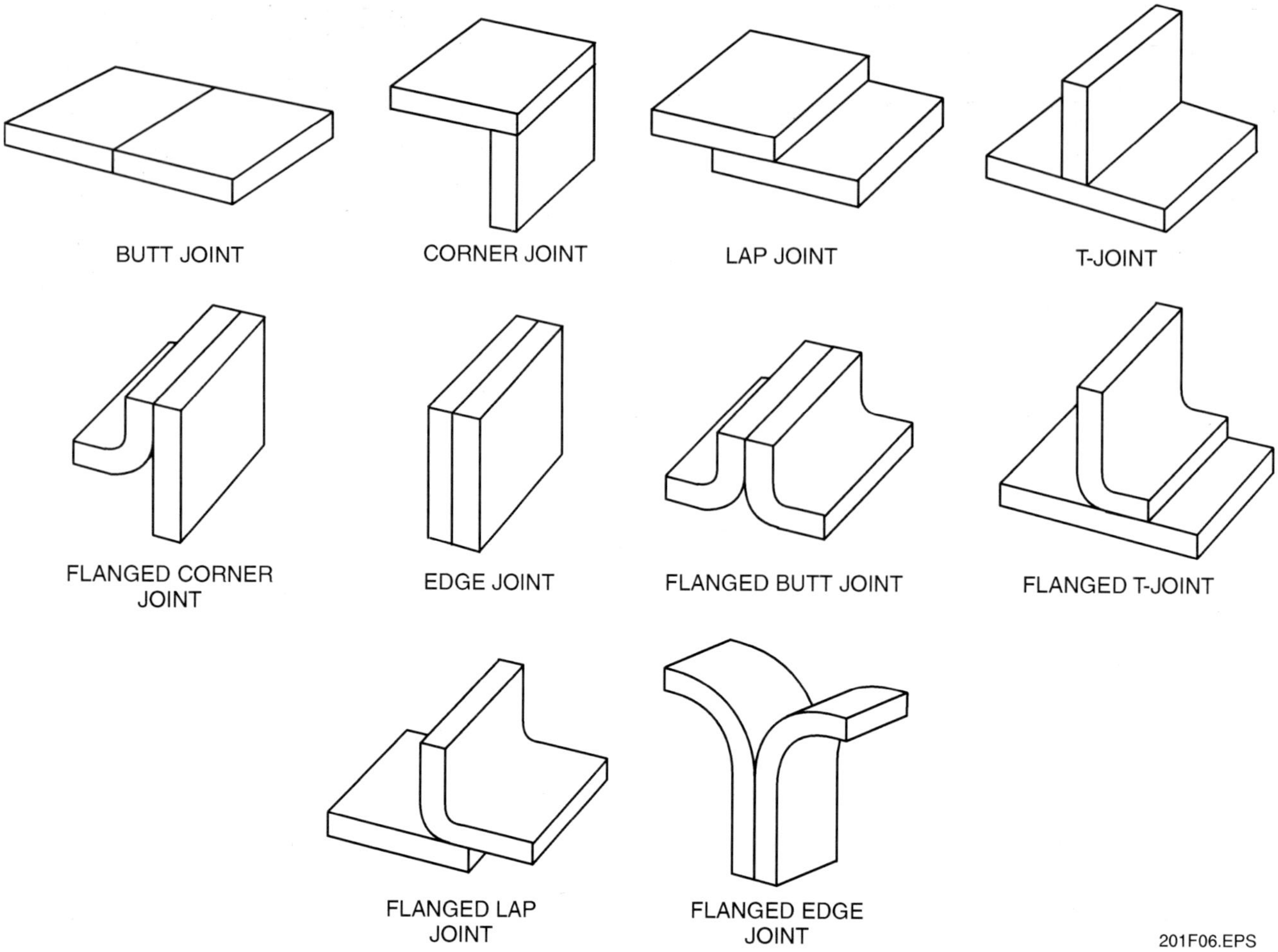

Figure 6 ◆ Basic types of weld joints.

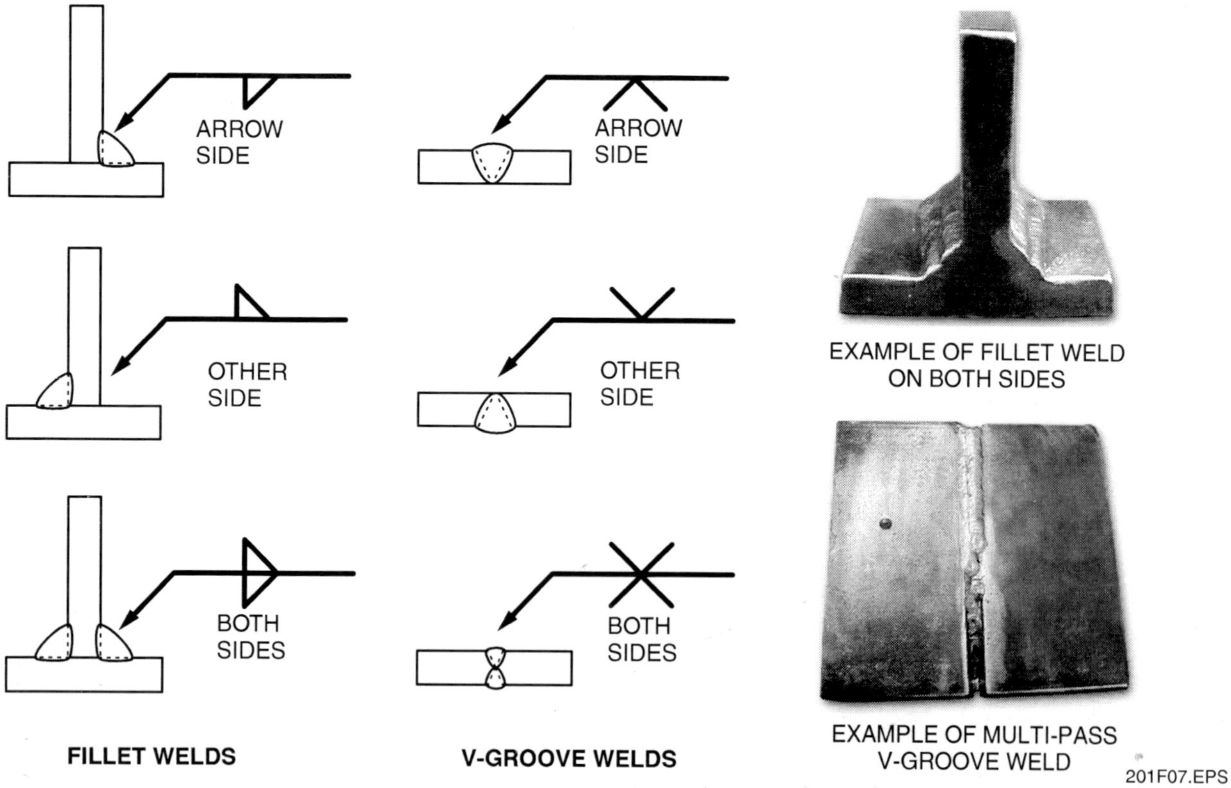

Figure 7 ◆ Fillet and V-groove welds.

Hot Tip

Weld Symbols and Welding Symbols

A weld symbol is a graphic character connected to the reference line of a welding symbol, specifying the weld type. A welding symbol graphically represents the specifications for producing a welded joint. All welding symbols include a reference line and arrow. Many also include a weld symbol.

one of the two members to be welded is prepared. The arrow indicates which member is to be prepared by breaking toward that member. *Figure 8* shows bevel and J-groove weld symbols. Notice also that the vertical leg of the weld symbol is shown drawn to the left of the slanted/curved leg(s). Regardless of whether the symbol is for a fillet, bevel groove, J-groove, or flare bevel groove weld, the vertical leg is always drawn to the left.

3.3.0 Combining Weld Symbols

When more than one type of weld is to be made on the same joint, it is necessary to combine weld symbols. One of the more common combination welds is made by adding a fillet weld to a single or double bevel groove weld in a T-joint. The fillet weld symbol is always placed on top of the groove weld symbol, just as it would be on the actual weld. *Figure 9* shows an example of a combined weld symbol.

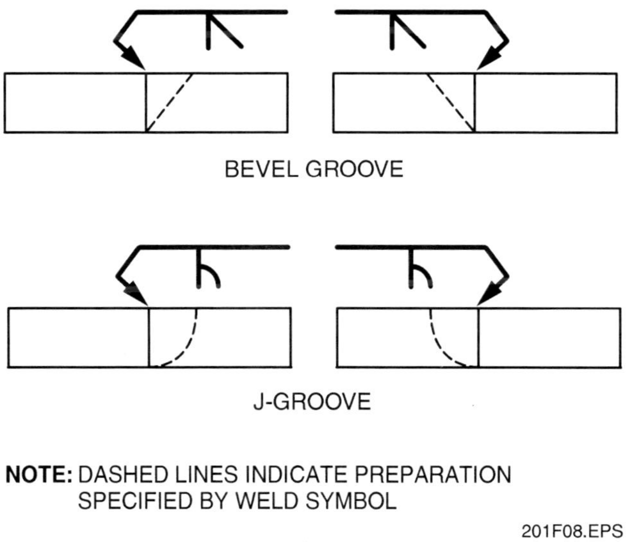

Figure 8 ◆ Bevel and J-groove weld symbols.

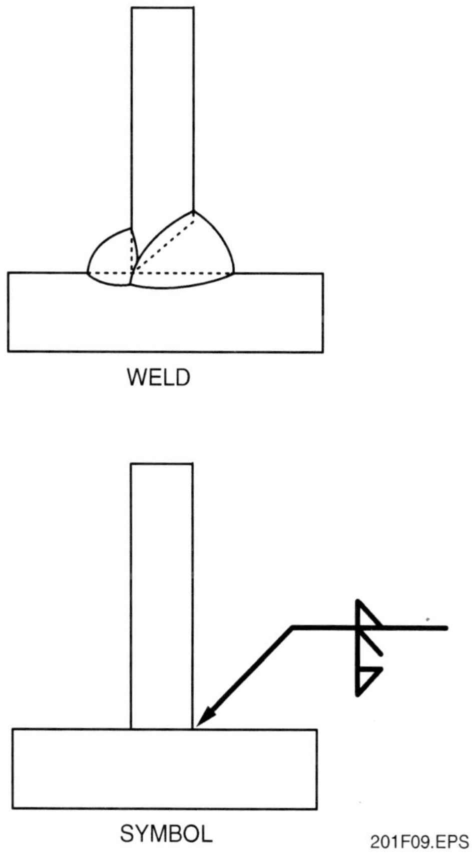

Figure 9 ◆ Combination weld symbol.

Think About It

The Breaking Arrow

Are the symbol representations shown in *Figure 8* the only ways a breaking arrow can be shown?

3.4.0 Sizing and Dimensioning Welds

Unless defined in a drawing note, the size data for a weld is always shown to the left of the symbol to which it applies. For example, in *Figure 10*, the size of the fillet weld shown is ¼". The length of the weld is always shown to the right of the symbol to which it applies. An example of this is seen in *Figure 11*. In this case, the length is 6".

Dimensions can be given as fractions, decimals, or metric measurements. The following sections explain how to size and dimension the following:

- Fillet welds
- Groove welds
- Plug welds
- Slot welds

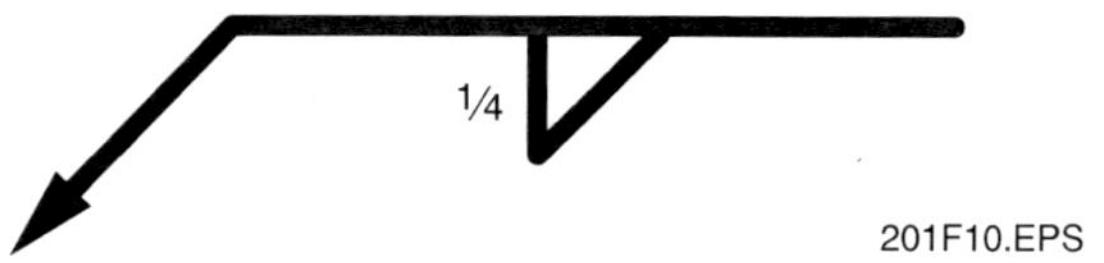

Figure 10 ◆ Sizing welds.

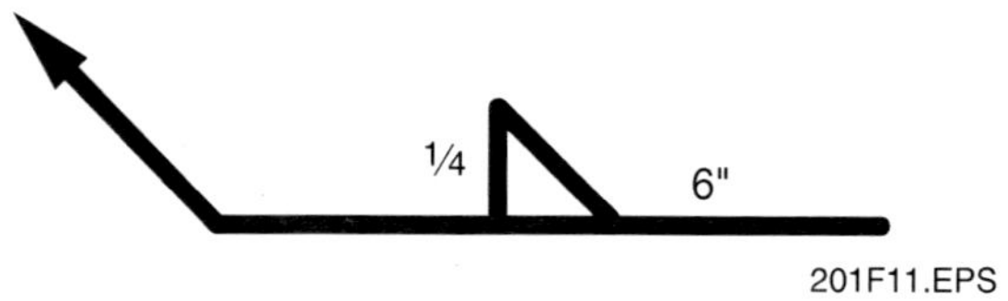

Figure 11 ◆ Dimensioning welds.

3.4.1 Sizing Fillet Welds

A fillet weld is a weld made in a corner. It is roughly triangular in shape and joins two welding surfaces that are perpendicular to each other, as in a lap, T-, or corner joint. The size of a fillet weld is the length of the legs (sides) of the largest triangle that can be inscribed within the fillet weld cross section (see *Figure 12*).

The lengths of the legs on a typical fillet weld are equal, so only one dimension is given. Equal leg fillet welds are sized by placing the size to the left of the fillet weld symbol or as defined in a drawing note.

If the fillet weld has unequal legs, the size of each leg is given in parentheses or brackets to the left of the fillet weld symbol. An example of this is shown in *Figure 13*. In this case the unequal legs are ¼" and ½".

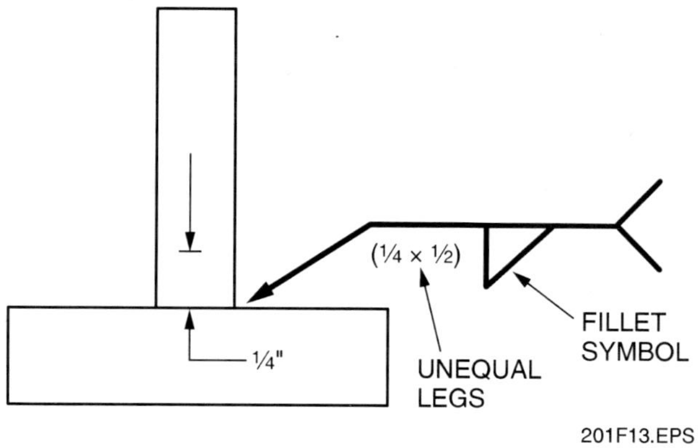

Figure 13 ◆ Sizing fillet welds.

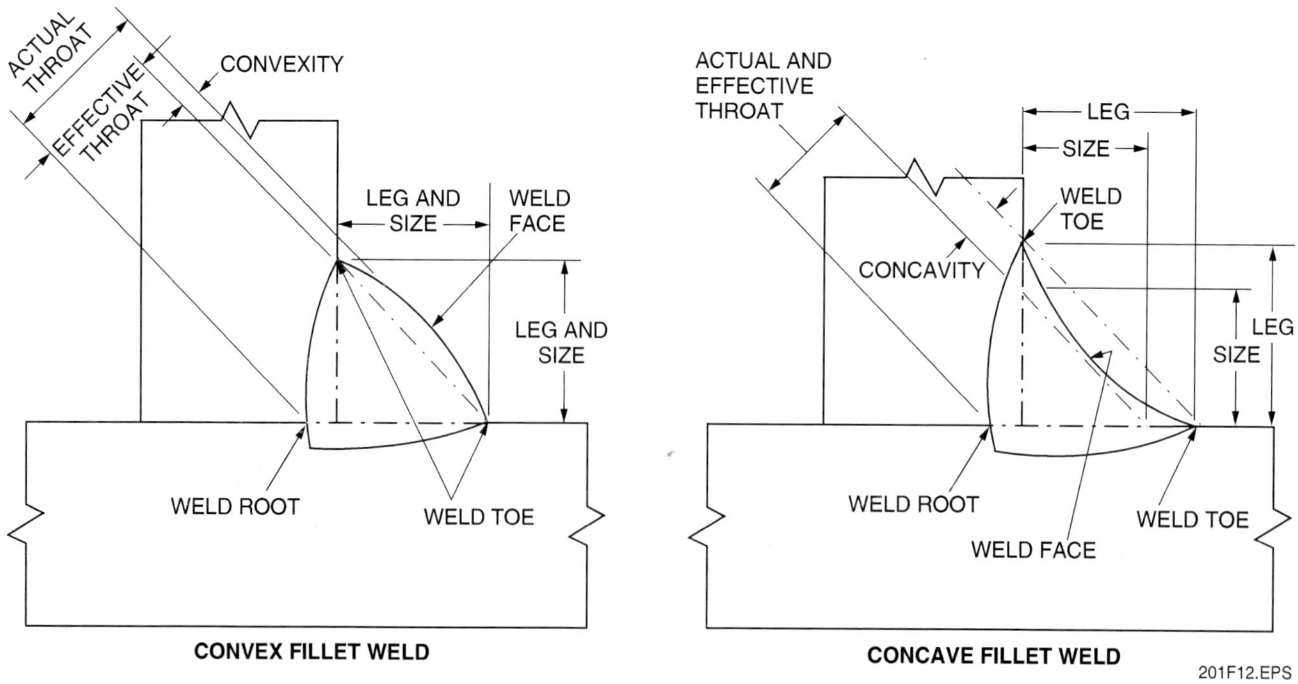

Figure 12 ◆ Profile of fillet welds.

Location of Unequal Fillet Weld Legs

How does a welder determine which side of a fillet weld has the longer leg?

3.4.2 Dimensioning Fillet Welds

The length of a fillet weld is shown to the right of the fillet weld symbol. If the fillet weld is to be an intermittent fillet weld, the length (length of each segment) and the pitch (center-to-center distance between welds) are shown with a dash. The dash appears between the length, which is always shown first, and the pitch, which is always shown second. *Figure 14* shows a length of 2" and a pitch of 6". If fillet welds are required on both sides of the joint, they can be back-to-back or staggered. If the welds are back-to-back, as shown in *Figure 14(A)*, the fillet weld symbols are lined evenly on both sides of the reference line, as shown in *Figure 14(B)*. If the welds are to be staggered, as shown in *Figure 14(C)*, the fillet weld symbols above and below the reference line are offset, as shown in *Figure 14(D)*. *Figure 14* shows dimensioning of fillet welds.

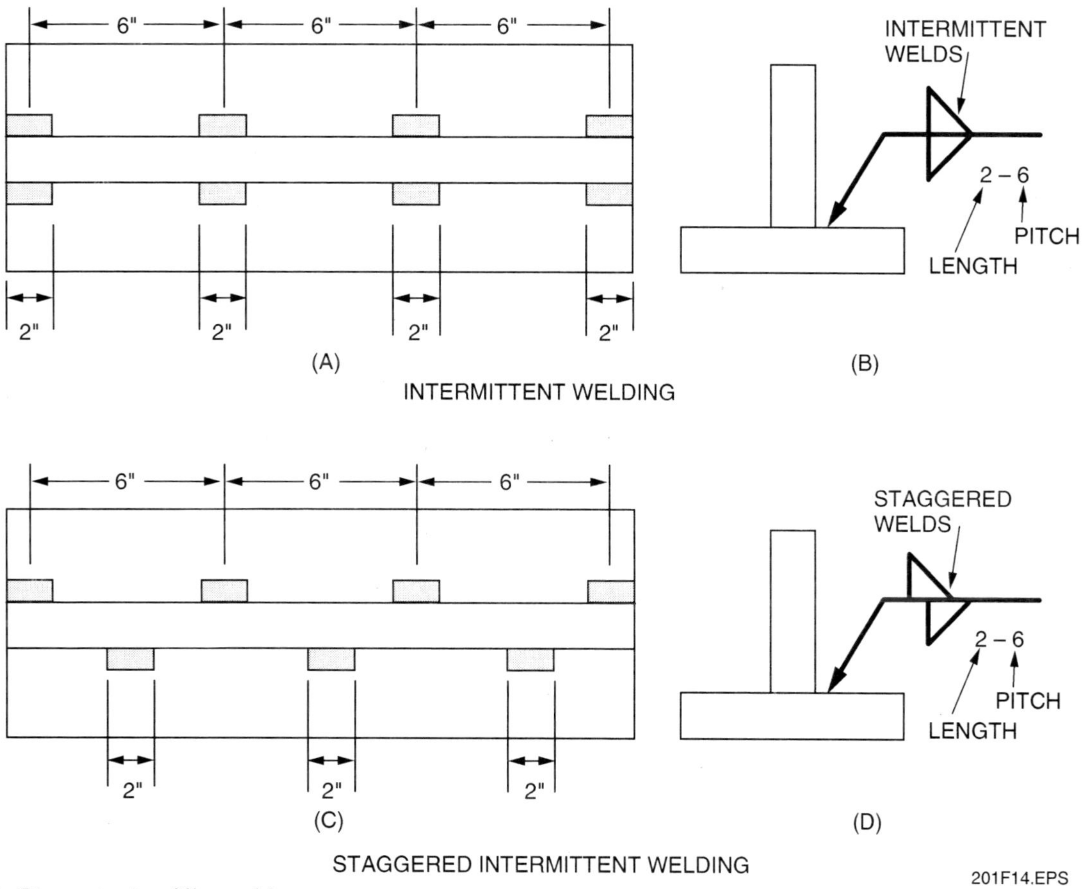

Figure 14 ◆ Dimensioning fillet welds.

Hot Tip: Unequal Fillet Weld Legs

The only practical use of unequal fillet weld legs is on metal of unequal thickness. Unequal legs do not make the weld stronger; the thinner material prevails. It helps distribute loads and eliminates lack of fusion discontinuities caused by making a small weld on heavy material without preheat.

When there is no dimension on the right of the fillet weld symbol, the weld extends the full length of the joint.

3.4.3 Sizing Groove Welds

A groove weld is a weld made in the groove formed by two beveled pieces of metal. There are several items that must be considered when sizing groove welds. These items apply to all the groove weld types. *Figure 15* shows several types of groove welds.

Groove preparation is the depth to which the groove penetrates into the base metal. When the groove preparation extends all the way through the joint, as shown in *Figure 16(A)* and *16(B)*, no preparation size needs to be shown. When the groove preparation extends only part of the way through the joint, as shown in *Figure 16(C)* and *Figure 16(D)*, the depth of the groove is shown to the left of the symbol. *Figure 16(C)* and *Figure 16(D)* show sizing a groove with a depth of ¼".

For partial joint penetration (PJP) groove welds, both the depth of groove penetration and weld size (joint penetration) are shown, as in *Figure 17(A)* and *Figure 17(B)*. The depth of the groove is shown first and the weld size is shown next in parentheses. In this example the groove depth is 0.25" and the weld size is 0.385". For com-

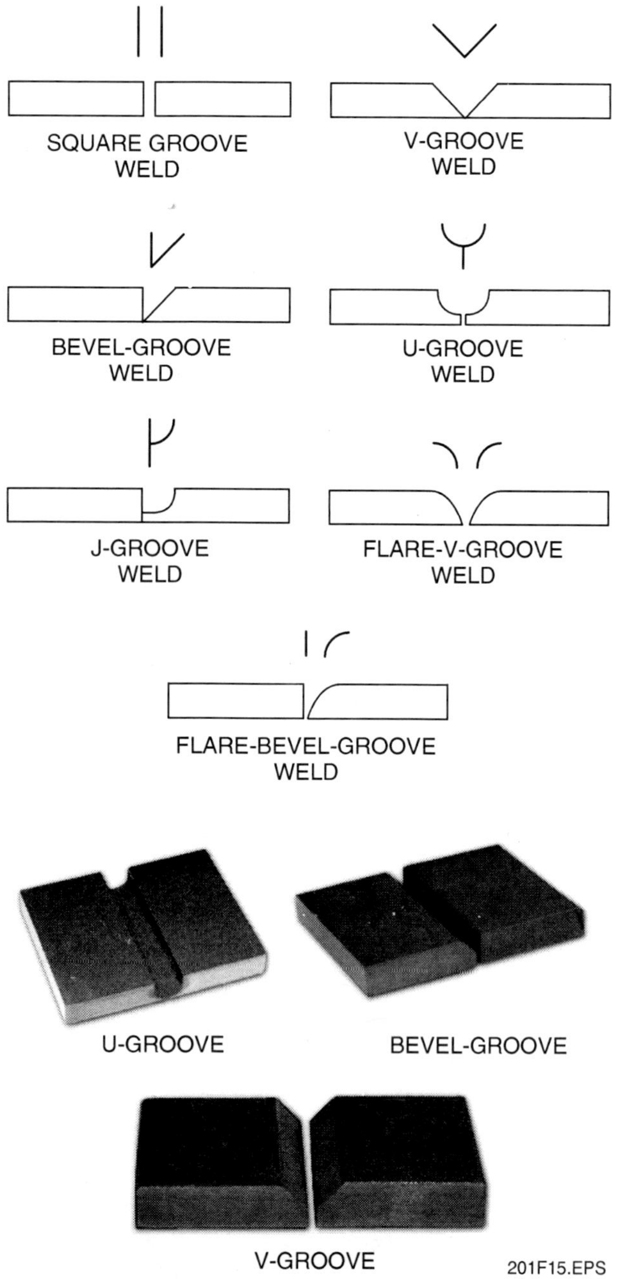

Figure 15 ◆ Types of groove welds.

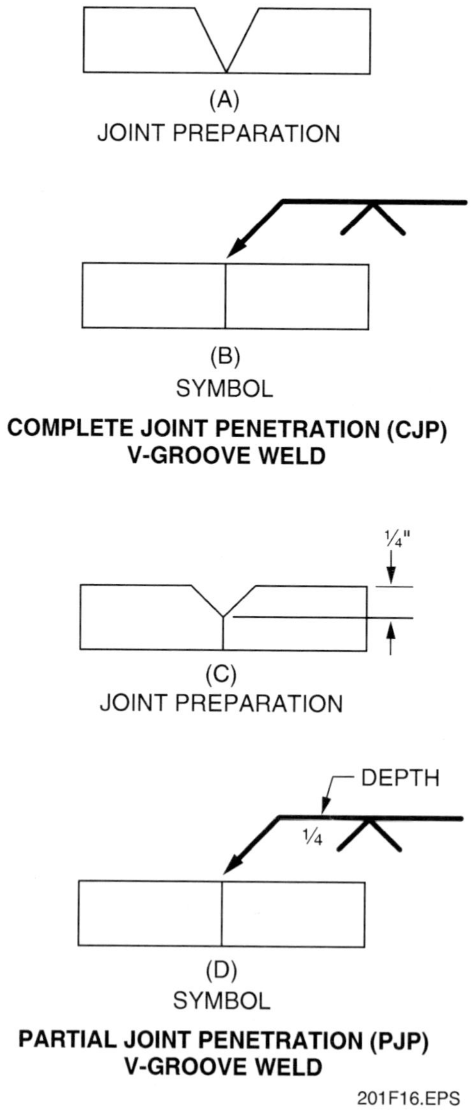

Figure 16 ◆ Sizing groove welds.

Groove Weld Penetration

When the groove preparation extends all the way through the joint, welding of the joint results in a *complete joint penetration (CJP)* weld. Welding of a partially prepared groove joint results in a *partial joint penetration (PJP)* weld.

plete joint penetration (CJP) groove welds, the weld size (joint penetration) is the thickness of the base metal, not including face and root reinforcement.

If a root opening is required, the root opening dimension of a groove weld is shown inside the symbol. In *Figure 18*, the root opening is ⅛" and is shown in only the arrow-side symbol because it is common to both grooves.

The angle for the type of groove is shown above or below the welding symbol, depending upon whether the symbol is above or below the reference line. In *Figure 19*, the groove angle is shown as 45° and is shown only by the arrow-side symbol because the groove is the same on both sides. The bevel angle for groove preparation on each workpiece is 22½°.

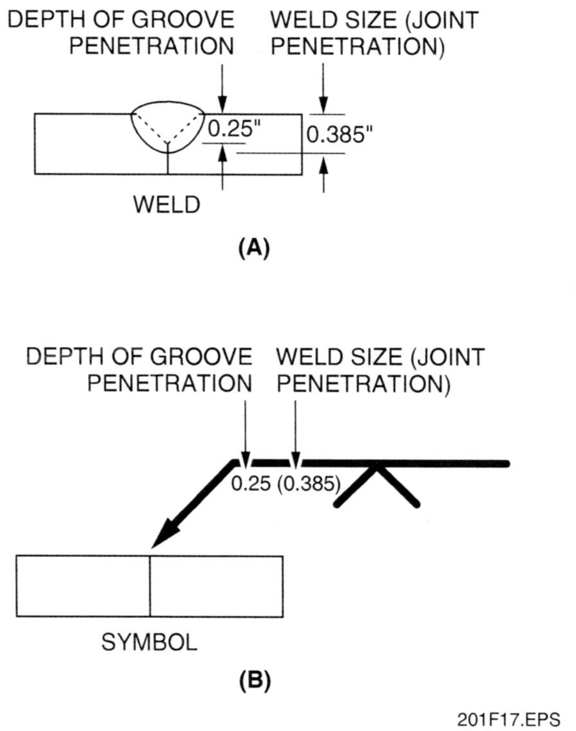

Figure 17 ◆ Sizing partial joint penetration (PJP) welds.

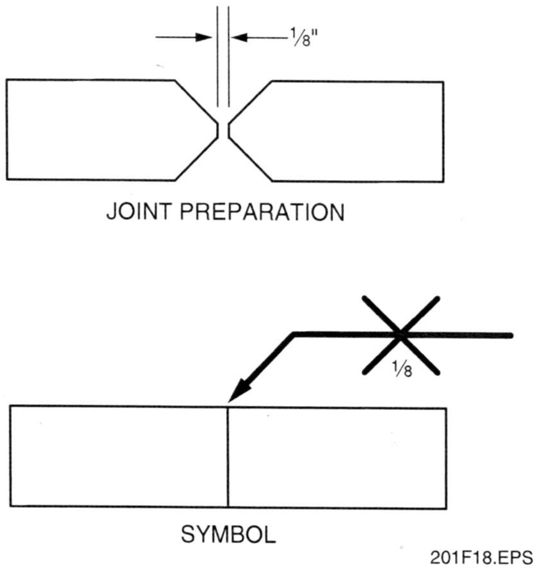

Figure 18 ◆ Sizing the root opening.

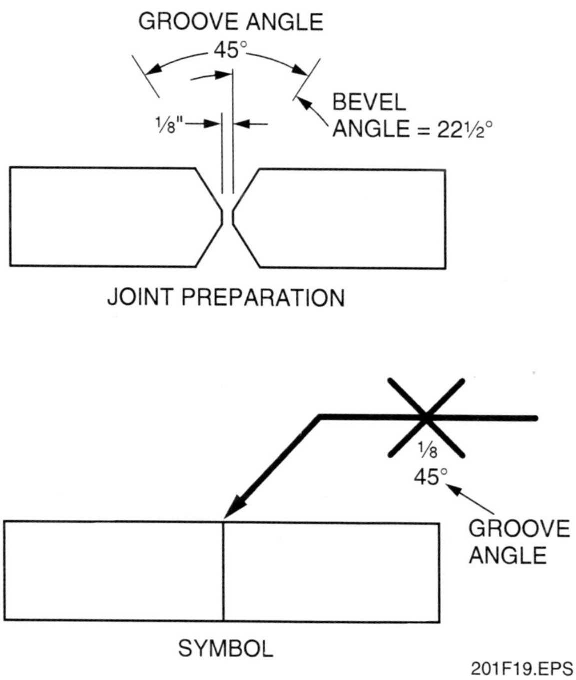

Figure 19 ◆ Specifying the groove angle.

3.4.4 Dimensioning Groove Welds

The length of a groove weld is shown to the right of the groove-weld symbol. If there is no dimension on the right of the groove-weld symbol, the weld extends the full length of the joint.

3.4.5 Sizing and Dimensioning Plug Welds

A plug weld is a weld made in a circular hole in one member of a joint, fusing that member to another member. The size of a plug weld, which is the diameter of the plug, is shown on the left of the welding symbol. If the plug is not completely filled, the depth of the fill is shown inside the plug weld symbol, as represented by *Figure 20(A)*. If the plug weld is to be **countersunk**, the angle of the countersink is shown below or above the weld symbol, depending on which side of the reference line the symbol is on, as shown in *Figure 20(B)*. The pitch, shown in *Figure 20(C)*, represents the center-to-center spacing of the plugs shown in *Figure 20(B)* on the right of the plug weld symbol. *Figure 20* shows a composite view of sizing and dimensioning plug welds.

3.4.6 Sizing and Dimensioning Slot Welds

The only time a size is used with a slot weld symbol is when the slots are partially filled. The partial depth of fill is shown in the symbol. All other information, such as length, width, spacing, angle of countersink, and location of slots, is shown in a special detail on the print. *Figure 21* shows sizing of slot welds.

3.4.7 Multiple Reference Lines

When more than one welding operation must be performed in a certain sequence, two or more reference lines may be used to indicate the sequence of operations. *Figure 22(A)* and *(B)* illustrates the typical sequence of operations. In each sequence, the first operation to be performed is closest to the arrow. The first operation must be shown on the reference line nearest the arrow. Subsequent operations may be shown sequentially on other reference lines. The last operation to be performed is the one farthest from the arrow.

Multiple reference lines may also be used to show data that are supplementary to welding symbol information included on the reference line nearest the arrow. As shown in *Figure 23(A)* and *23(B)*, test information may be shown on a second or third line away from the arrow.

When required, the weld-all-around symbol shown in *Figure 24(A)* must be placed at the junction

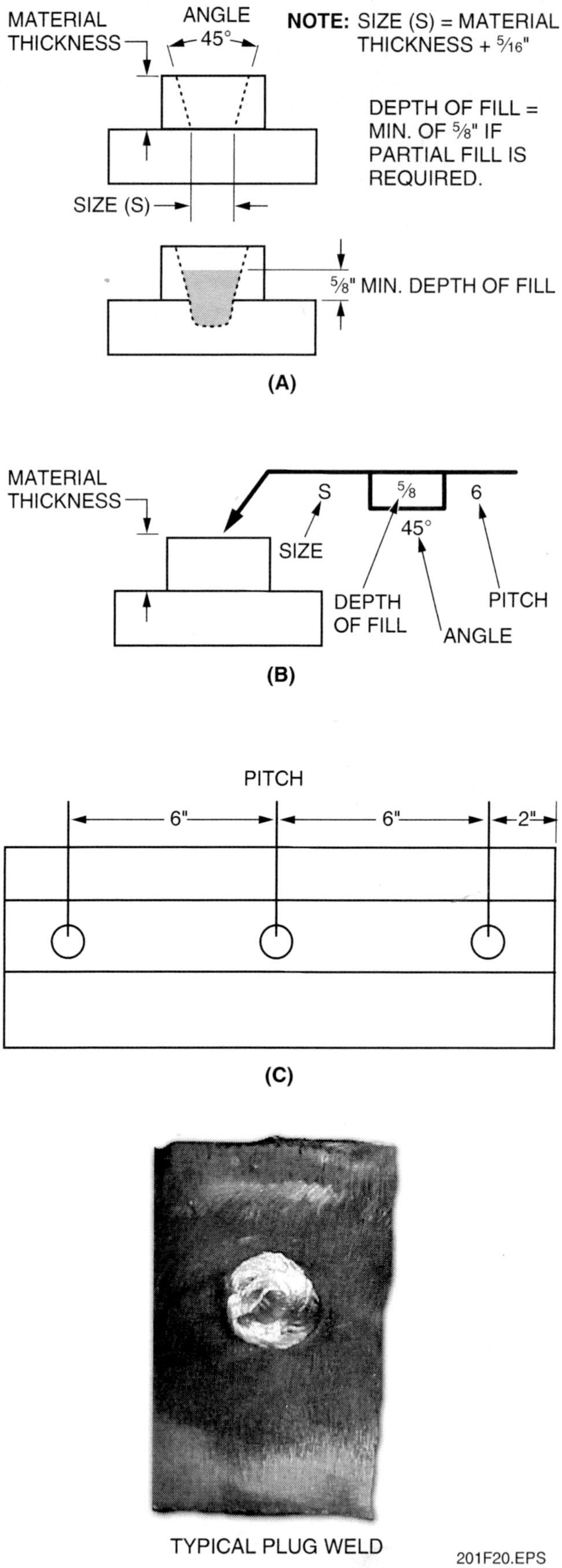

Figure 20 ◆ Sizing and dimensioning plug welds.

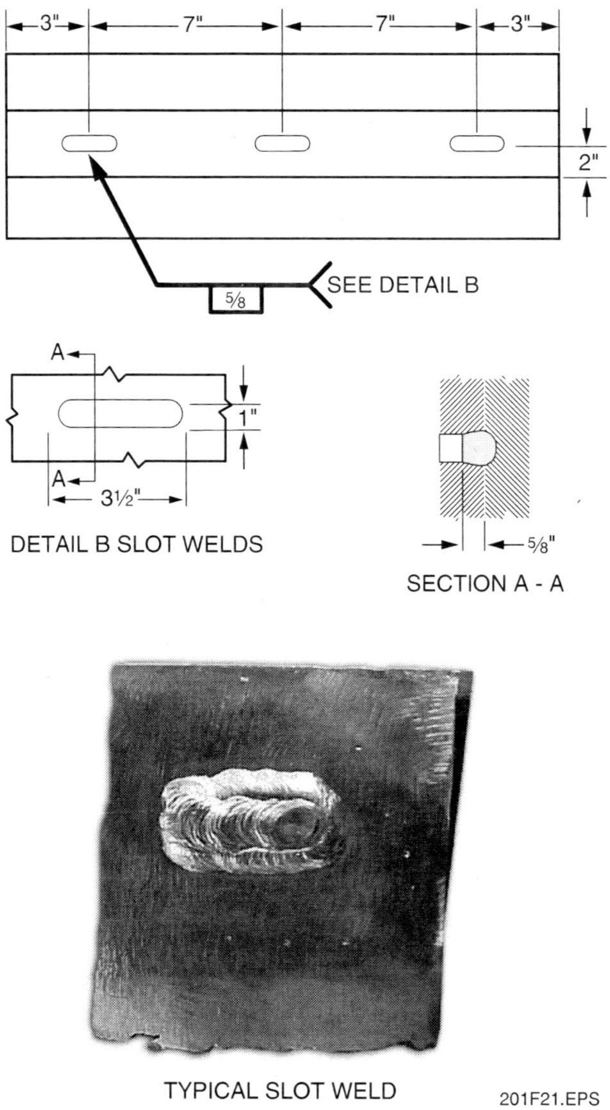

Figure 21 ◆ Sizing slot welds.

of the arrow line and reference line for each operation to which it applies. The field weld symbol shown in *Figure 24(B)* may also be used in this manner. These supplemental symbols are explained in the next section.

4.0.0 ◆ SUPPLEMENTAL SYMBOLS

Supplemental symbols are used to convey special instructions pertaining to the welding symbol. These include:

- Weld-all-around
- Field weld
- Contour finish

Figure 25 shows the supplemental symbols.

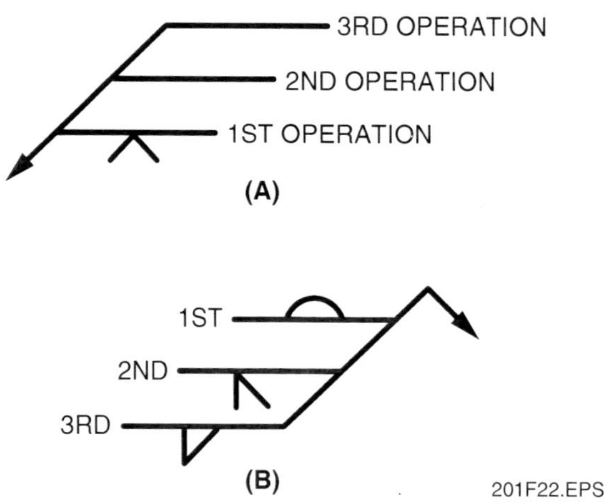

Figure 22 ◆ Multiple reference lines.

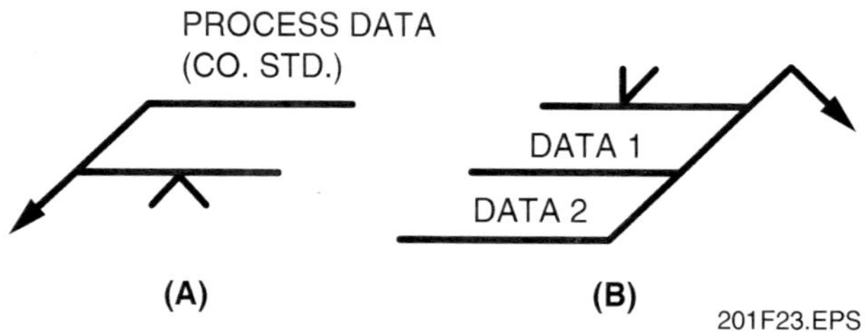

Figure 23 ◆ Supplementary data.

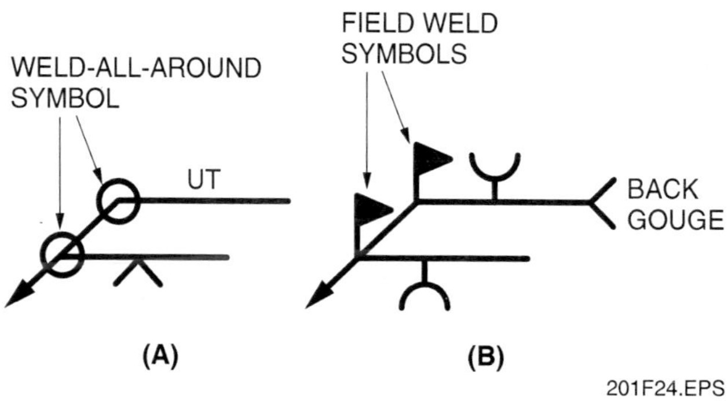

Figure 24 ◆ Typical placement of supplemental symbols.

4.1.0 Weld-All-Around

When a weld is to extend completely around the joint (such as when welding a channel to a base plate), a small circle is shown where the arrow meets the reference line. *Figures 24, 25,* and *26* show the symbol placement. *Figure 26* also shows the intended result of a typical weld-all-around symbol.

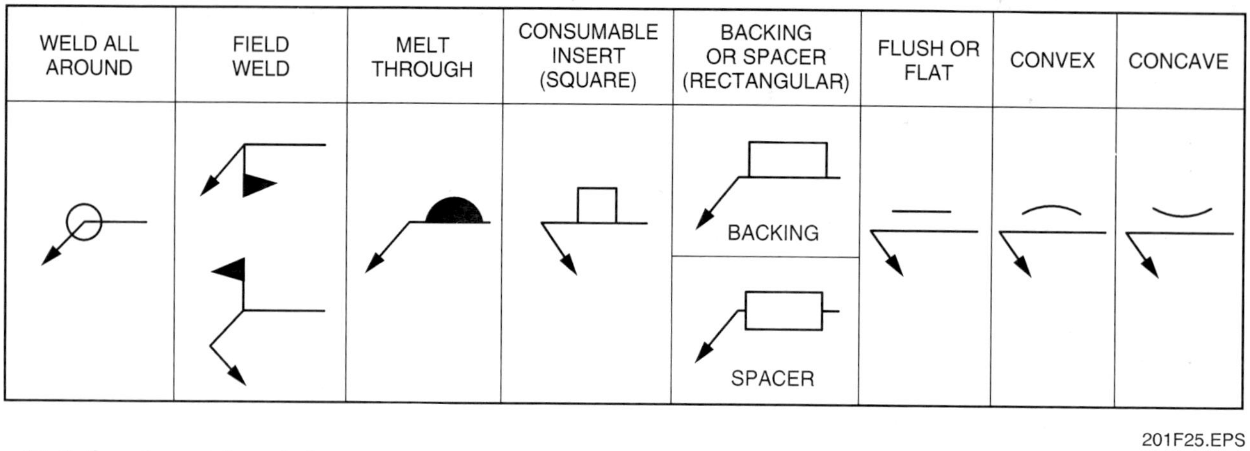

Figure 25 ◆ Supplemental symbols.

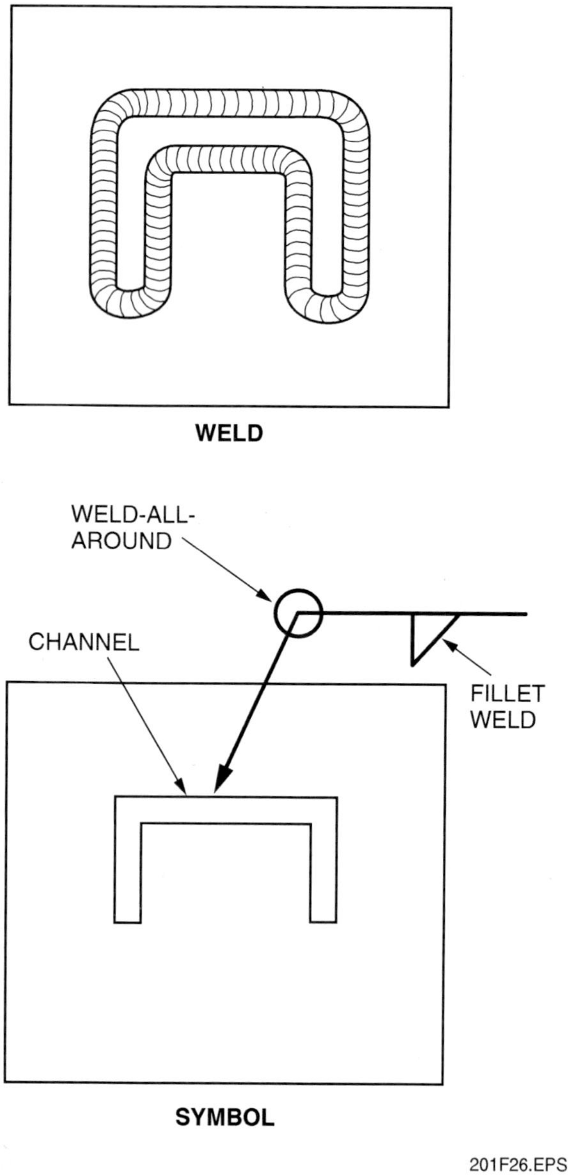

Figure 26 ◆ Weld-all-around symbol.

4.2.0 Field Weld

When the weld indicated by the welding symbol is to be made at a location other than that of the initial construction, a small flag is placed where the arrow meets the reference line. An obsolete field welding symbol may also be found on older prints. The obsolete field welding symbol is a small black dot placed where the arrow meets the reference line. *Figure 27* shows field welding symbols.

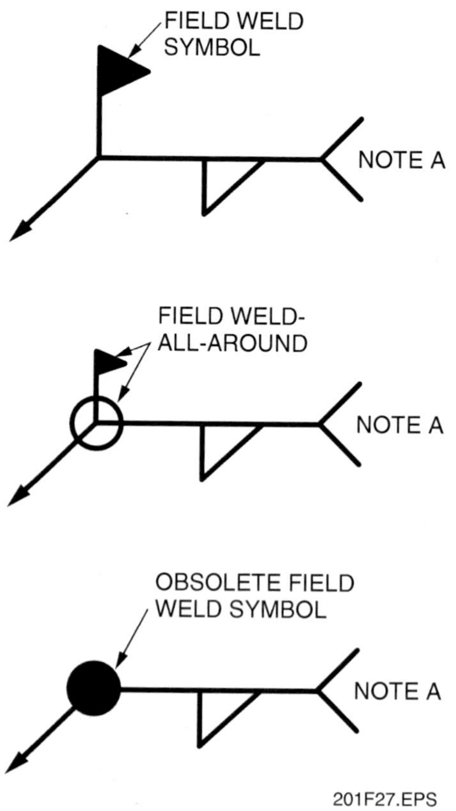

Figure 27 ◆ Field welding symbols.

4.3.0 Contour Finish

When the face of the weld must have a finished shape that is not its normal as-welded condition, a finish symbol indicating a flush, convex, or concave surface is placed adjacent to the welding symbol. The method to finish the weld face is specified by a letter indicating the method of finishing (*Figure 28*). See *Table* 1 for the finishing methods.

Table 1 Finishing Methods

Method	Letter
Chipping	C
Grinding	G
Hammering	H
Machining	M
Rolling	R
Unspecified	U

5.0.0 ◆ OTHER SYMBOLS

Other, less common symbols are:

- Backing and spacer
- Back or backing weld
- Melt-through weld
- Surfacing weld
- Edge weld
- Spot weld
- Seam weld
- Consumable insert weld

These less common symbols are explained in the following sections. *Figure 29* shows these other symbols.

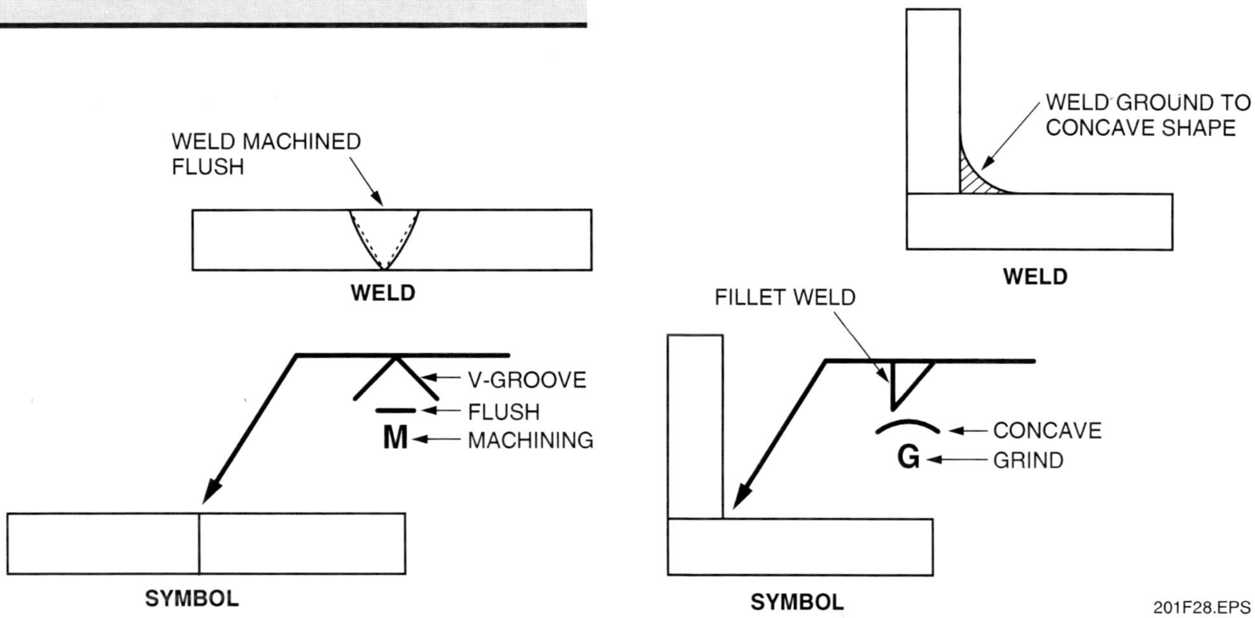

Figure 28 ◆ Examples of contour finishes and symbols.

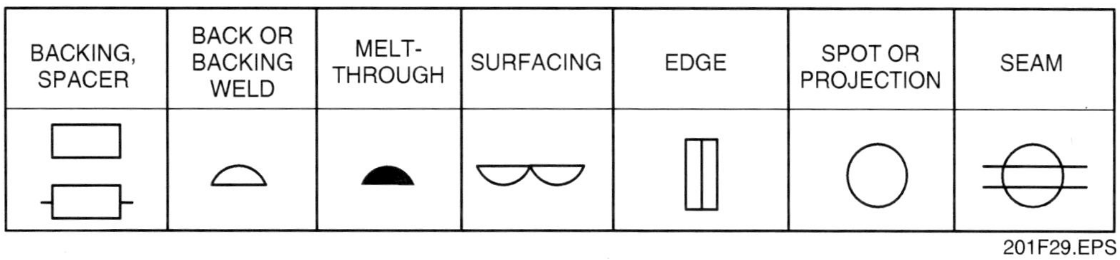

Figure 29 ◆ Other symbols.

THINK ABOUT IT

Unspecified Contour Finishes

What does an unspecified (U) contour finish mean?

Hot Tip

The Letters TYP

The letters TYP (meaning typical) are placed in the tail of the weld symbol to denote multiple joints of the same configuration. This eliminates duplication of the same weld symbol on the drawing.

5.1.0 Backing and Spacer Symbols

Backing is used at the root of the weld to control melt-through. The backing can be metal strips or shapes, flux-coated tape, fiberglass-coated tape, or ceramic tape. The backing symbol is placed opposite a groove weld symbol on the reference line. The material, size, and shape of the backing are specified in the tail of the welding symbol.

Spacers are metal strips or shapes that are placed at the root of the joint to control burn-through. They are used most often on double groove joint preparations. The spacer symbol is centered on the reference line. The joint weld symbol is modified to project from the corners of the spacer symbol. The material and size of the spacer are noted in the tail of the welding symbol. *Figure 30* shows backing and spacer symbols.

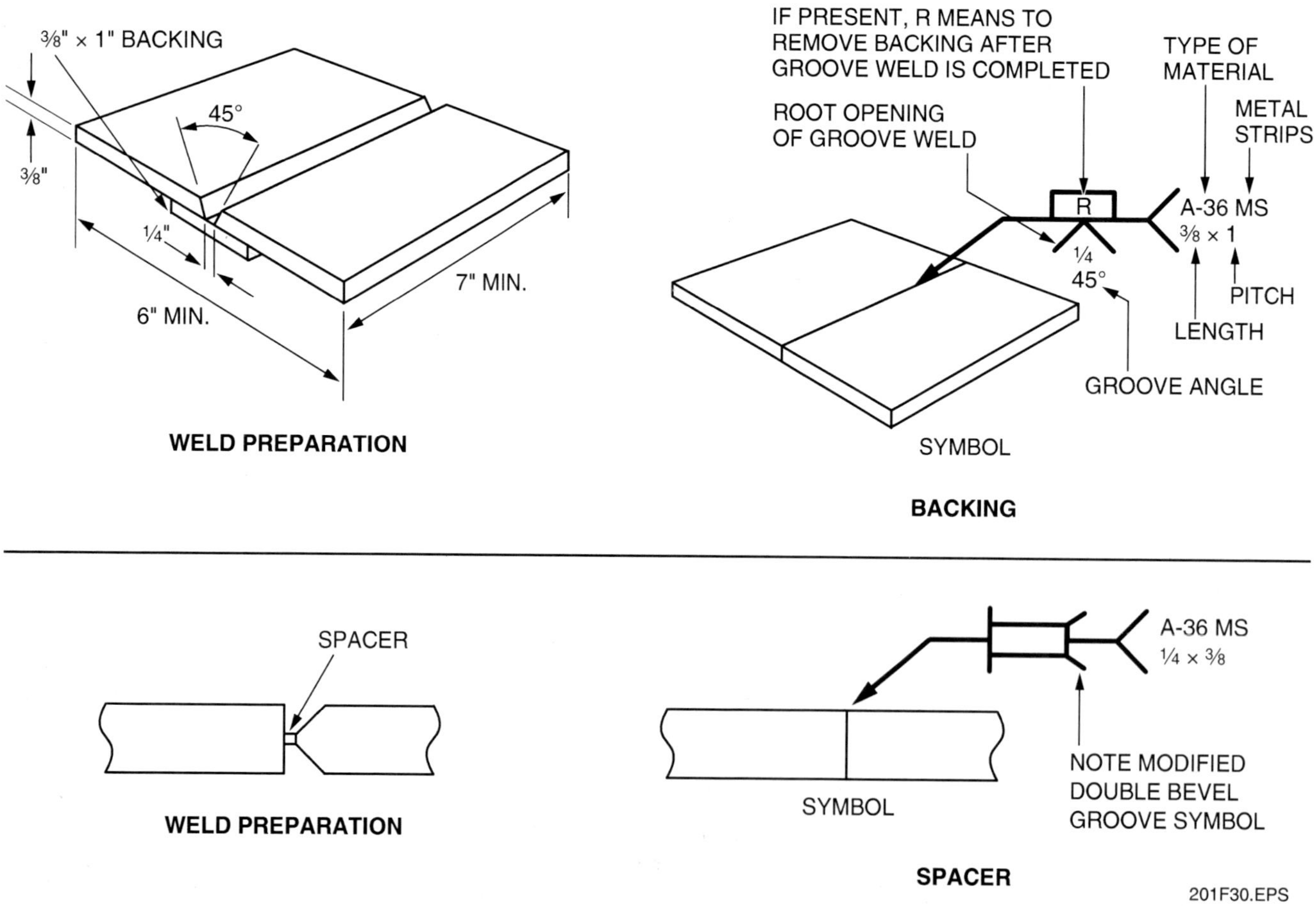

Figure 30 ◆ Backing and spacer symbols.

5.2.0 Back or Backing Weld Symbols

A back weld is made after the groove weld. A backing weld is made before the groove weld. The same symbol is used to indicate either a back or backing weld (*Figure 31*). The back or backing weld symbol is placed opposite a groove weld symbol on the reference line. A note in the tail of the welding symbol indicates which weld is to be made.

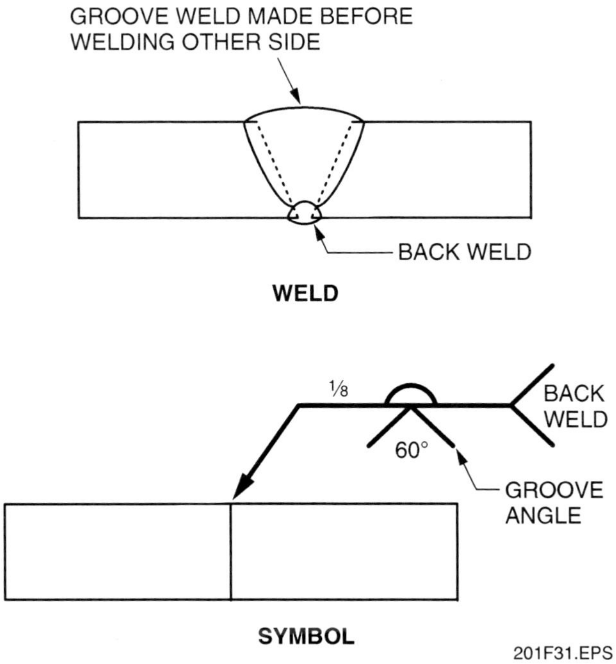

Figure 31 ◆ Back or backing weld symbol.

5.3.0 Melt-Through Symbols

When complete penetration is required from one side of a joint and root reinforcement is desired, a melt-through symbol (*Figure 32*) can be used to specify the amount of melt-through. It is placed opposite a groove weld symbol on the reference line. It is similar to a back or backing symbol, except it is filled in. The amount of melt-through required is placed to the left of the melt-through symbol.

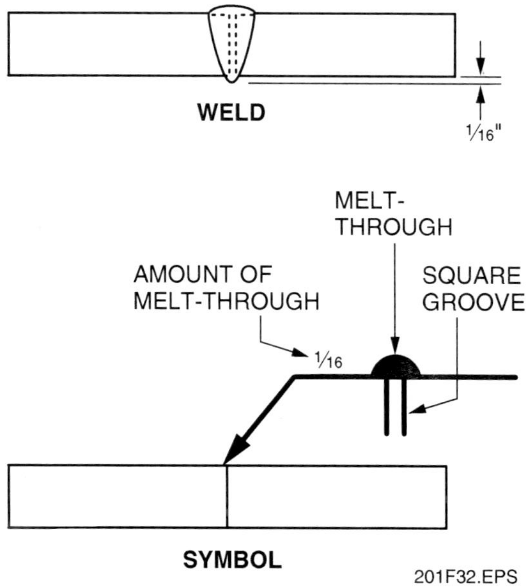

Figure 32 ◆ Melt-through symbol.

Hot Tip

An Alternate Welding Symbol

The welding symbol shown in *Figure 31* could also be presented as a multiple line symbol. By using a multiple line symbol, the tail information designating the back weld is eliminated. This is because the first operation is the groove weld and the second operation, by definition, can only be a back weld rather than a backing weld.

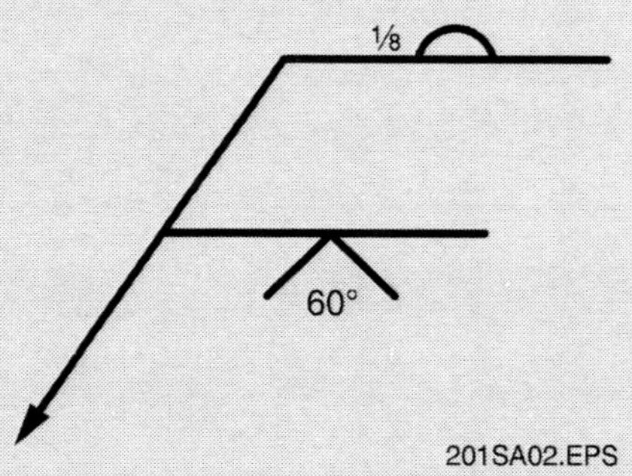

5.4.0 Surfacing Weld Symbols

The American Welding Society defines surfacing as "the application by welding, brazing, or thermal spraying of a layer of material to a surface to obtain desired properties or dimensions, as opposed to making a joint." Examples of surfacing welds are shown in *Figure 33*. A surfacing weld symbol is used when a surface is to be built up with weld. The depth of buildup is indicated to the left of the symbol.

When less than the entire surface is to be built up, the surface to be built up is dimensioned on the blueprint in a detail drawing.

5.5.0 Edge Weld Symbols

The American Welding Society defines an edge weld as "a joint between the edges of two or more parallel or nearly parallel members." Edge welds are commonly used on sheet metal and can be flanged butt joints or flanged T-joints. *Figure 34* shows edge welds, various types of symbols, and dimensioning.

5.6.0 Spot Weld Symbols

Spot welding is a process used to join sheet metal using a series of small spots. There are two types: resistance spot welding (RSW) and arc spot welding. RSW uses electrical resistance heating and clamping pressure to fuse sheet metal panels. Arc spot welding involves either gas metal arc welding (GMAW) or gas tungsten arc welding (GTAW). The process used to make the spot weld is specified in the tail. The size of a spot weld is indicated on the left of the symbol. The pitch (center-to-center spacing) is indicated on the right of the symbol. The number of spot welds is defined in parentheses beneath the symbol. For arrow-side spot welds, the circle is placed on the bottom of the reference line. When the spot weld is placed on the other side, the circle is placed on top of the reference line. When there is no side significance (such as with a resistance spot), the circle is centered on the reference line. The drawings will define the starting or ending locations for the spot welds. *Figure 35* shows examples of spot weld symbols.

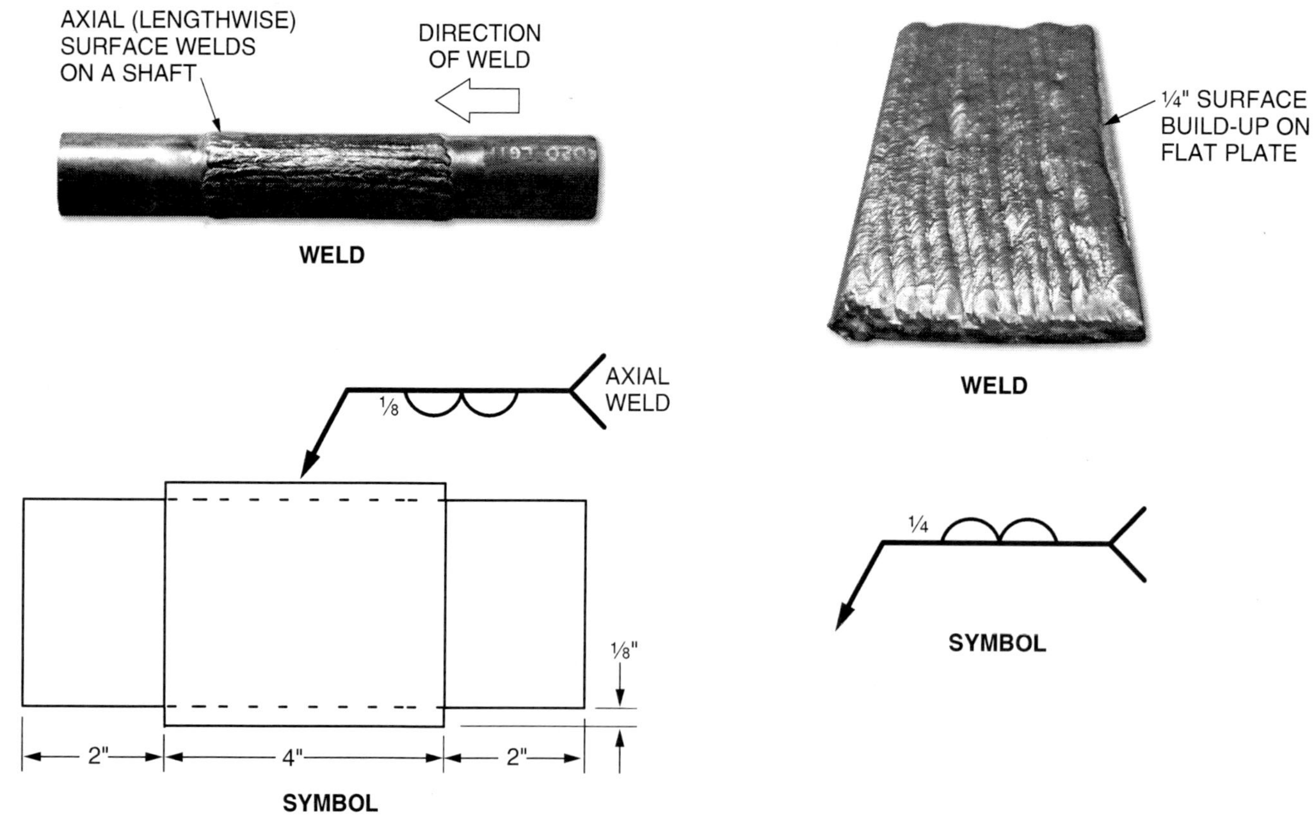

201F33.EPS

Figure 33 ◆ Surfacing welds and symbols.

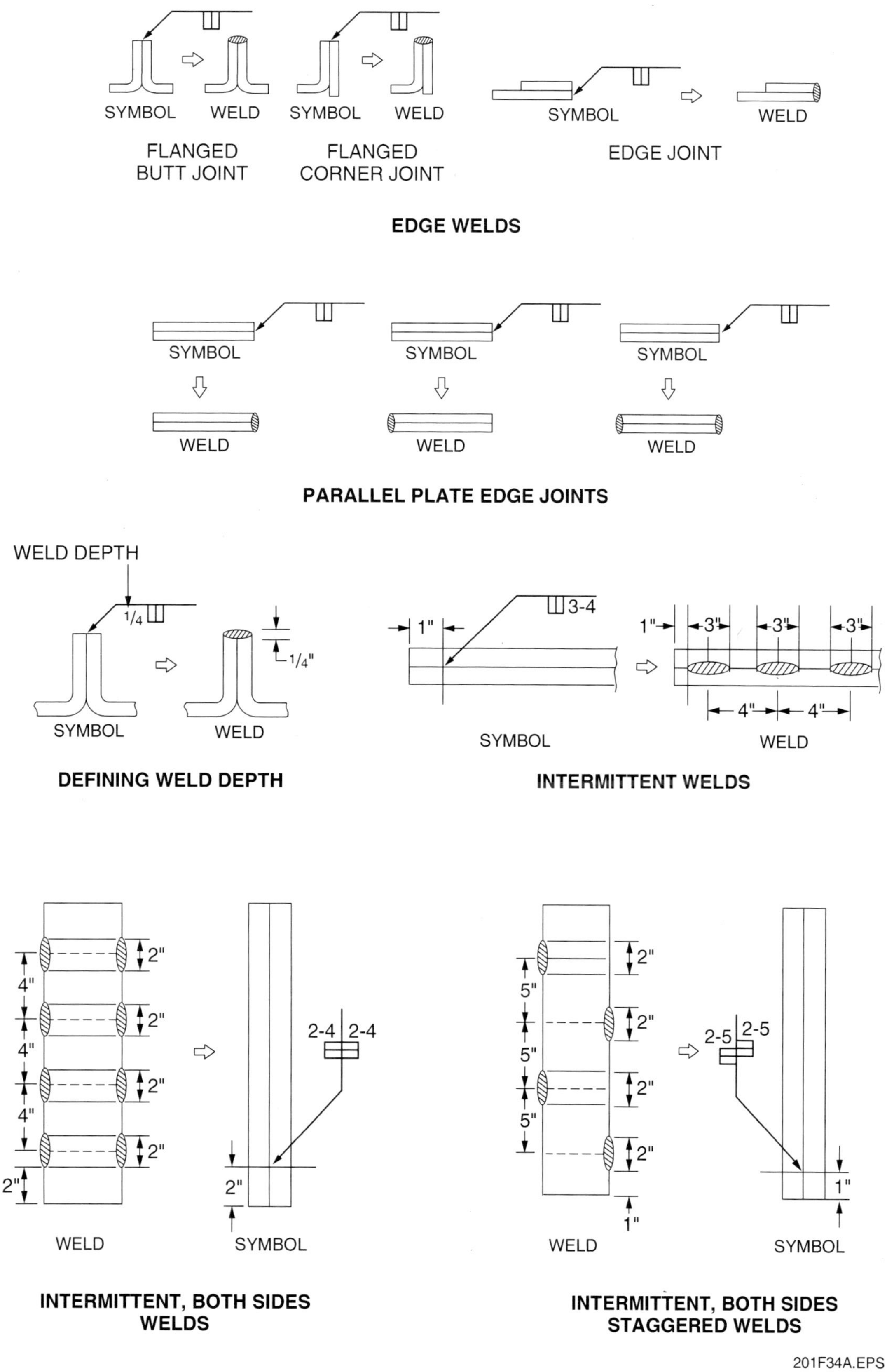

Figure 34 ◈ Edge welds and symbols (1 of 2).

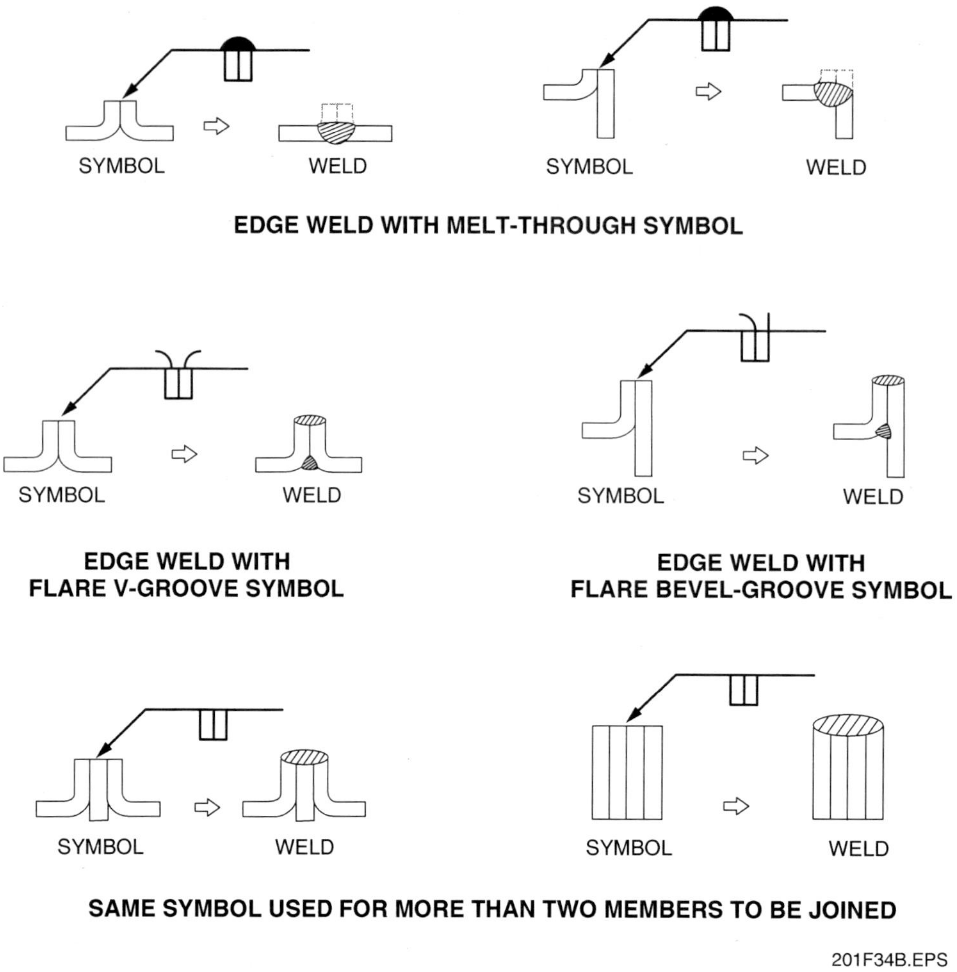

Figure 34 ◆ Edge welds and symbols (2 of 2).

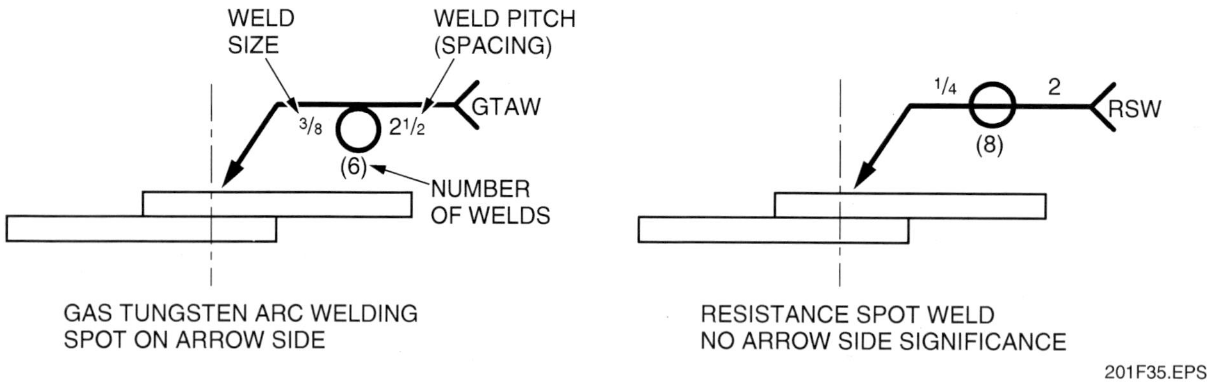

Figure 35 ◆ Spot weld symbols.

Hot Tip

Projection Welding

Projection welding (PW) is a type of spot welding in which a number of welds are performed simultaneously with one welding machine. Because of tooling costs, projection welding is primarily used in high-production applications.

Prior to welding, one of the members to be joined is passed through a machine that presses projections at specified locations on the member. This member, along with another member, is joined in a welding machine by pressing the two members together and applying current through the projections. The resistance at the point of contact for the projections causes the members to heat, melt, and fuse together at the projections while under pressure. The resultant welded members are fused closely together. The side without the projections has minimal surface marking as a result of the weld.

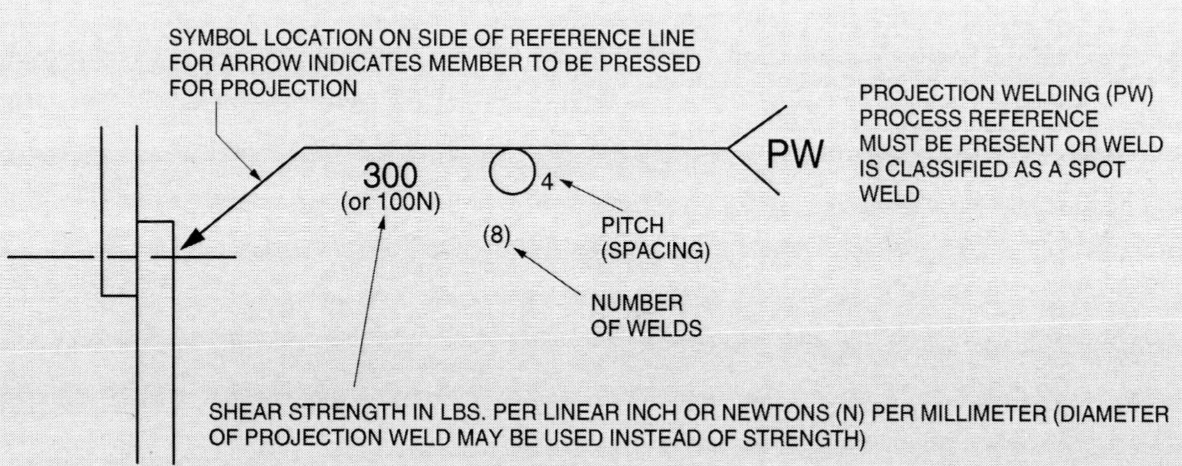

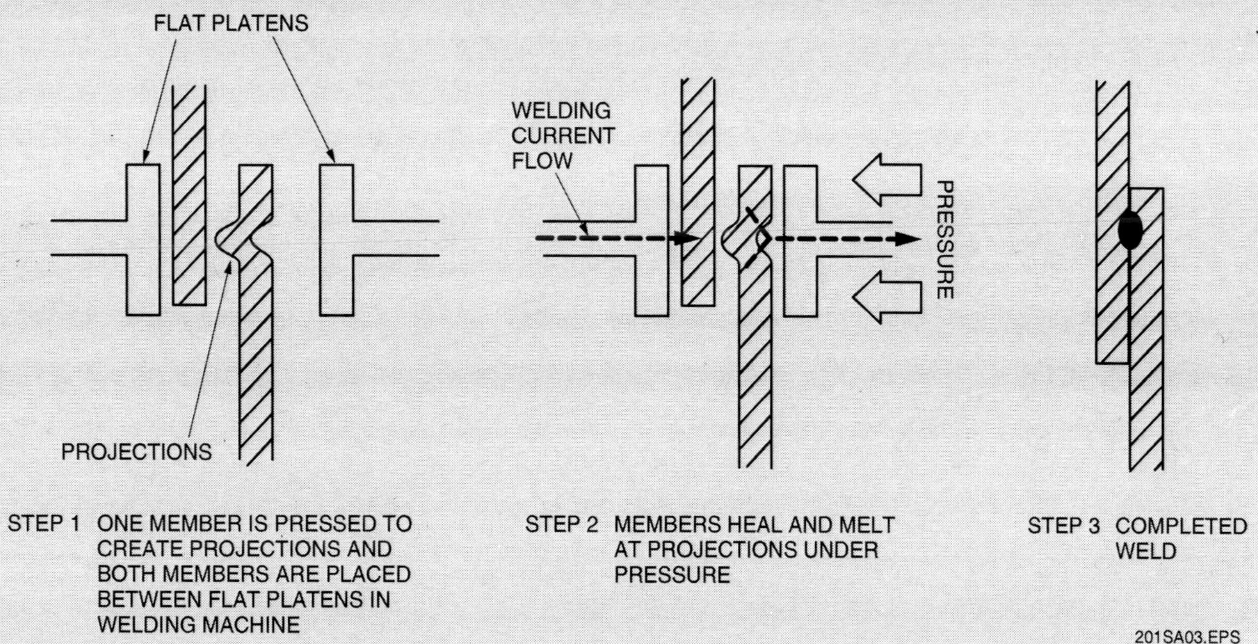

5.7.0 Seam Weld Symbols

Seam welding is used to join two pieces of sheet metal in a continuous seam or in a series of overlapping spot welds. Several processes can make a seam weld. The most common are GMAW, GTAW, and resistance seam welding (RSEW). The process used to make the seam weld is specified in the tail. The size of a seam weld (width) or strength of the weld is given to the left of the symbol. The length and pitch are given on the right of the symbol for intermittent welds. For welds that are less than the full length of the joined members, the drawing details the starting point and the length of the weld.

For arrow-side seam welds, the circle is placed on the bottom of the reference line as shown in *Figure 36(A)*. When the seam weld is placed on the other side, the circle is placed on top of the reference line. When there is no side significance, such as with a resistance weld seam, the circle is centered on the reference line, as shown in *Figure 36(B)*. Flush contour symbols can be applied to seam weld symbols for other than resistance weld seams. For parallel seams, a drawing reference detailing the spacing and number of seams is included in the tail.

Summary

Welding symbols are a language expressed in graphic format and are used to convey precise instructions about how a weld is to be made. By learning the symbols and basic rules for applying them, a welder can easily understand how to prepare precisely and make the required welds. This module has described welding symbols and weld symbols used with various types of welds. Understanding these symbols is essential when using WPSs and drawings.

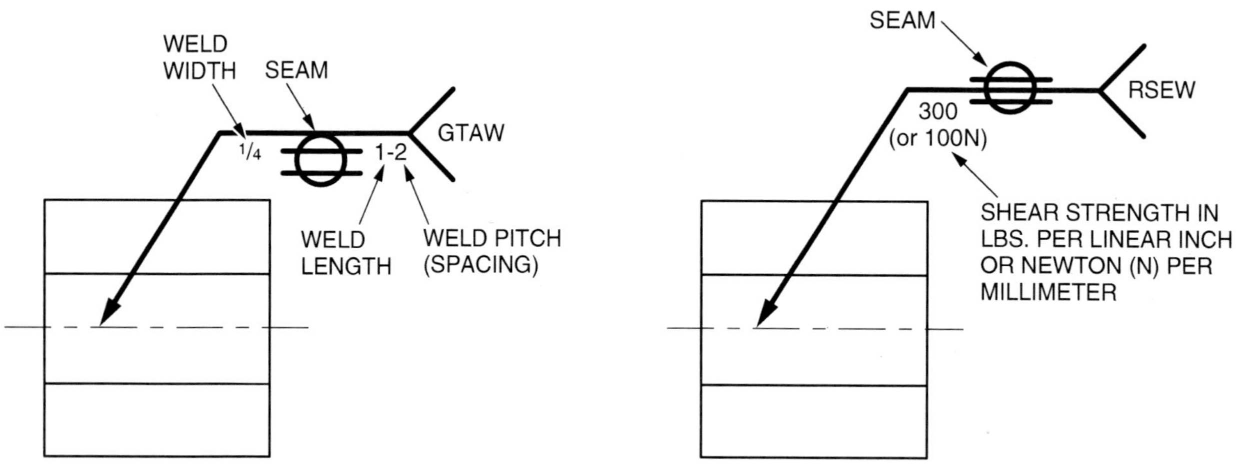

(A) GAS TUNGSTEN ARC WELD SEAM ON ARROW SIDE

(B) RESISTANCE SEAM WELD NO ARROW SIDE SIGNIFICANCE

201F36.EPS

Figure 36 ◆ Seam weld symbols.

Review Questions

1. Most companies use the welding symbols that have been standardized by *Standard A2.4, Standard Symbols for Welding, Brazing, and Non-destructive Examination*, of the _____.
   a. American Welding Society (AWS)
   b. American Petroleum Institute (API)
   c. American National Standards Institute (ANSI)
   d. American Society of Mechanical Engineers (ASME)

2. The base element for all welding symbols is the _____.
   a. arrow at the end of the reference line
   b. reference line with the arrow at one end
   c. letters above the welding symbol's reference line
   d. letters below the welding symbol's reference line

3. The side of the joint or surface that is opposite the side to which the arrow of a reference line is pointing is called the _____.
   a. bottom side
   b. arrow side
   c. other side
   d. top side

4. The fillet weld size is placed _____ the fillet weld symbol.
   a. above
   b. below
   c. to the left of
   d. to the right of

5. The pitch (center-to-center spacing) of a plug weld is shown _____ the plug weld symbol.
   a. above
   b. below
   c. to the left of
   d. to the right of

6. When a weld is to extend completely around the joint, a small circle is placed where the arrow line meets the _____.
   a. weld symbol
   b. weld location
   c. reference line
   d. tail

7. When backing is to be used at the root of the weld to control melt-through, the backing symbol is placed opposite the _____ weld symbol on the reference line.
   a. tail
   b. field
   c. spacer
   d. groove

8. When complete penetration is required from one side of a weld, a melt-through symbol may be placed opposite a _____ weld symbol on the reference line.
   a. tail
   b. field
   c. spacer
   d. groove

Figure 1

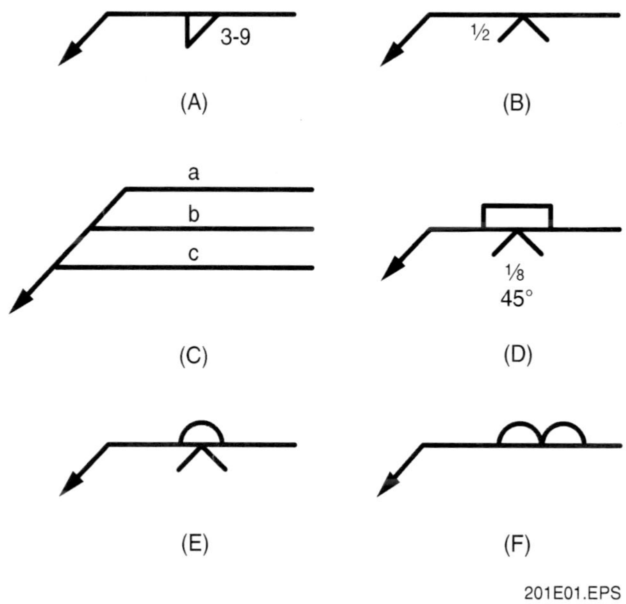

Refer to *Figure 1* to answer Questions 9 through 15.

9. The type of weld represented by the welding symbol shown in *A* is a _____ weld.
   a. fillet
   b. U-groove
   c. slot
   d. bevel groove

10. In the welding symbol shown in *A*, the 9 represents the _____ of the weld.

a. length
b. width
c. pitch
d. depth

11. In the welding symbol shown in *B*, ½" represents the _____.

a. length of the weld
b. width of the weld
c. size of the weld
d. depth of preparation

12. In the welding symbol shown in *C*, the operation to be conducted first is _____.

a. a
b. b
c. c
d. not critical

13. In the weld symbol shown in *D*, the numbers inside the symbol represent the _____.

a. pitch and angle
b. groove angle and depth
c. groove angle and root opening
d. root opening and pitch

14. The type of weld symbol shown above the reference line in *E* is a _____ weld.

a. back
b. surfacing
c. melt-through
d. spot

15. The type of weld represented by the weld symbol in *F* is a(n) _____ weld.

a. edge
b. backing
c. melt-through
d. surfacing

GLOSSARY

Trade Terms Introduced in This Module

Countersink: To chamfer a hole so that a flat head fastener will seat flush with the surface.

Weld symbol: A graphic character connected to the reference line of a welding symbol, specifying the weld type.

Welding symbol: A graphical representation of the specifications for producing a welded joint. All welding symbols include a reference line and arrow, and many also include a weld symbol.

APPENDIX

Welding Abbreviations

Letter Designations for Welding Processes

AAW	Air Acetylene Welding	GTAW	Gas Tungsten Arc Welding
AB	Adhesive Bonding	GTAW-P	Gas Tungsten Arc Welding – Pulsed Arc
AHW	Atomic Hydrogen Welding	HPW	Hot Pressure Welding
AW	Arc Welding	IB	Induction Brazing
B	Brazing	IRB	Infrared Brazing
BB	Block Brazing	IW	Induction Welding
BMAW	Bare Metal Arc Welding	LBW	Laser Beam Welding
CAW	Carbon Arc Welding	OAW	Oxyacetylene Welding
CAW-G	Gas Carbon Arc Welding	OHW	Oxyhydrogen Welding
CAW-T	Twin Carbon Arc Welding	PAW	Plasma Arc Welding
CEW	Coextrusion Welding	PEW	Percussion Welding
CW	Cold Welding	PGW	Pressure Gas Welding
DP	Dip Brazing	PW	Projection Welding
DFB	Diffusion Brazing	RB	Resistance Brazing
DFW	Diffusion Welding	ROW	Roll Welding
EBW	Electronic Beam Welding	RSEW	Resistance Seam Welding
EBW-HV	Electronic Beam Welding – High Vacuum	RSEW-HF	Resistance Seam Welding – High Frequency
EBW-MV	Electronic Beam Welding – Medium Vacuum	RSEW-I	Resistance Seam Welding – Induction
EBW-NV	Electronic Beam Welding – Nonvacuum	RSW	Resistance Spot Welding
EGW	Electrogas Welding	RW	Resistance Welding
ESW	Electroslag Welding	SAW	Submerged Arc Welding
EXW	Explosion Welding	SAW-S	Series Submerged Arc Welding
FB	Furnace Brazing	SMAW	Shielded Metal Arc Welding
FCAW	Flux Cored Arc Welding	SSW	Solid-State Welding
FLB	Flow Brazing	SW	Stud Arc Welding
FLOW	Flow Welding	TB	Torch Brazing
FOW	Forge Welding	TCAB	Twin Carbon Arc Brazing
FRW	Friction Welding	TW	Thermite Welding
FW	Flash Welding	USW	Ultrasonic Welding
GMAW	Gas Metal Arc Welding	UW	Upset Welding
GMAW-P	Gas Metal Arc Welding – Pulsed Arc	UW-HF	Upset Welding – High Frequency
GMAW-S	Gas Metal Arc Welding – Short Circuiting Arc	UW-I	Upset Welding – Induction

Letter Designations for Cutting Processes

AC	Arc Cutting	LOC	Oxygen Lance Cutting
AOC	Oxygen Arc Cutting	MAC	Metal Arc Cutting
CAC	Carbon Arc Cutting	OC	Oxygen Cutting
CAC-A	Air Carbon Arc Cutting	OC-F	Flux Cutting
EBC	Electron Beam Cutting	OFC	Oxyfuel Gas Cutting
GMAC	Gas Metal Arc Cutting	OFC-A	Oxyacetylene Cutting
GTAC	Gas Tungsten Arc Cutting	OFC-H	Oxyhydrogen Cutting
LBC	Laser Beam Cutting	OFC-N	Oxynatural Gas Cutting
LBC-A	Laser Beam Cutting – Air	OFC-P	Oxypropane Cutting
LBC-EV	Laser Beam Cutting – Evaporative	OC-P	Metal Powder Cutting
LBC-IG	Laser Beam Cutting – Inert Gas	PAC	Plasma Arc Cutting
LBC-O	Laser Beam Cutting – Oxygen	SMAC	Shielded Metal Arc Cutting
		TC Thermal Cutting	

Letter Designations for Applying Welding Processes

AU	Automatic
ME	Mechanized
MA	Manual
SA	Semi-Automatic

Additional Resources

This module is intended to be a thorough resource for task training. The following reference works are suggested for further study. These are optional materials for continued education rather than for task training.

AWS A3.0: Standard Welding Terms and Definitions, Latest Edition. Miami, FL: The American Welding Society.

Modern Welding. A.D. Althouse, C.H. Turnquist, W.A. Bowditch, K.E. Bowditch. Tinley Park, IL: The Goodheart-Willcox Company, Inc.

ANSI/AWS A2.4, Standard Symbols for Welding, Brazing, and Nondestructive Examination, Latest Edition. Miami, FL: The American Welding Society.

Welding Handbook, Latest Edition. Miami, FL: The American Welding Society.

Welding Print Reading. John R. Walker and W. Richard Polanin. Tinley Park, IL: The Goodheart-Willcox Company, Inc.

Figure Credits

Figure 4 courtesy of American Welding Society

AWS Committee on Filler Metals, 1998, *Standard Symbols for Welding, Brazing, and Nondestructive Examination*, AWS A2.4-98, Miami: American Welding Society.

CONTREN™ LEARNING SERIES — USER UPDATES

The NCCER makes every effort to keep these textbooks up-to-date and free of technical errors. We appreciate your help in this process. If you have an idea for improving this textbook, or if you find an error, a typographical mistake, or an inaccuracy in NCCER's Contren™ textbooks, please write us, using this form or a photocopy. Be sure to include the exact module number, page number, a detailed description, and the correction, if applicable. Your input will be brought to the attention of the Technical Review Committee. Thank you for your assistance.

Instructors – If you found that additional materials were necessary in order to teach this module effectively, please let us know so that we may include them in the Equipment/Materials list in the Instructor's Guide.

Write: Curriculum Revision and Development Department
National Center for Construction Education and Research
P.O. Box 141104, Gainesville, FL 32614-1104

Fax: 352-334-0932

E-mail: curriculum@nccer.org

Craft ____________________ Module Name ____________________

Copyright Date __________ Module Number __________ Page Number(s) __________

Description

____

____

____

____

(Optional) Correction

____

____

____

(Optional) Your Name and Address

____

____

____

Module 29202-03

Reading Welding Detail Drawings

COURSE MAP

This course map shows all of the modules in the second level of the Welding curriculum. The suggested training order begins at the bottom and proceeds up. Skill levels increase as you advance on the course map. The local Training Program Sponsor may adjust the training order.

WELDING LEVEL TWO

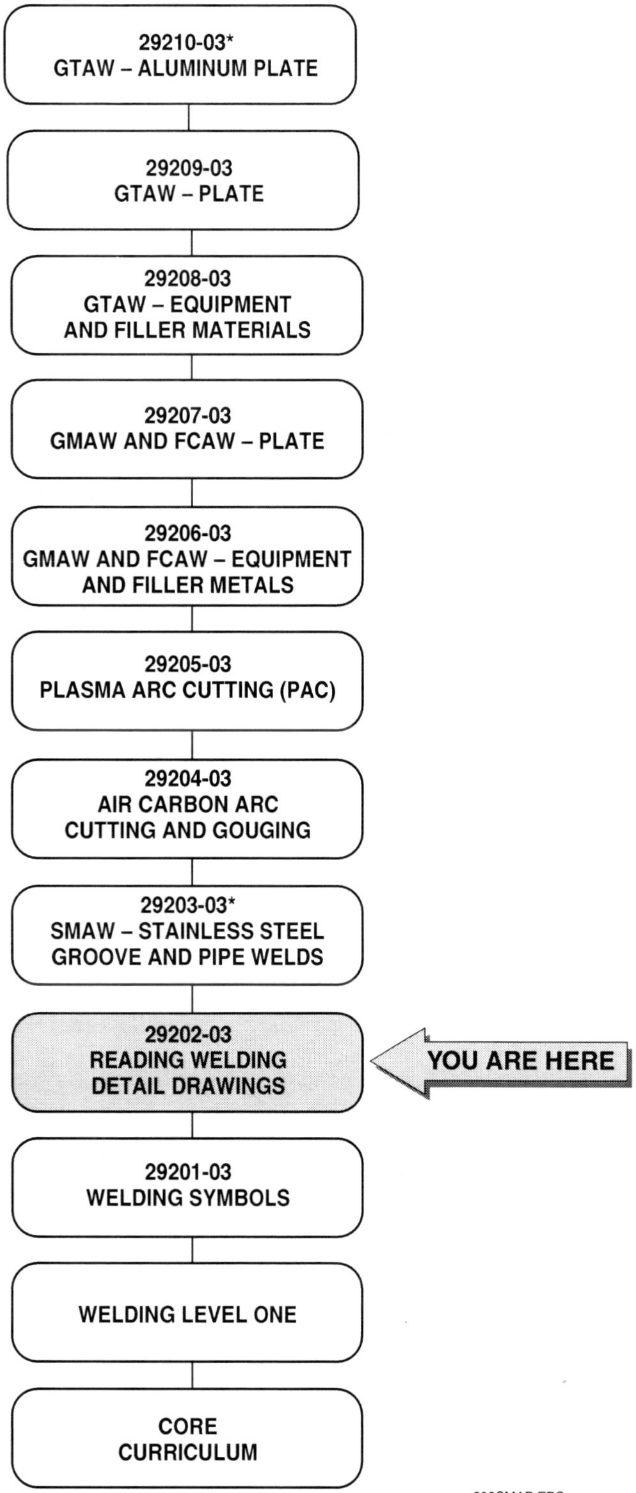

202CMAP.EPS

***Please note that Modules 29203-03 and 29210-03 are electives for *Welding Level Two* but are requirements for *AWS Entry Level Welder*.**

MODULE 29202-03 CONTENTS

Figures

MODULE 29202-03

Reading Welding Detail Drawings

Objectives

When you have complete this module, you will be able to do the following:

1. Identify and explain a welding detail drawing.
2. Identify and explain lines, material fills, and sections.
3. Identify and explain object views.
4. Identify and explain dimensioning.
5. Identify and explain notes and bill of materials.
6. Interpret basic elements of a welding detail drawing.
7. Develop basic welding drawings.

Prerequisites

Before you begin this module, it is recommended that you successfully complete the following: Core Curriculum; Welding Level One; Welding Level Two, Module 29201-03.

1.0.0 ◆ INTRODUCTION

Blueprints, also called prints or drawings, are used by engineers, designers, and drafters to convey information. Welders must be able to correctly interpret prints so the design specifications and safety considerations engineered into the drawing will be incorporated into the finished product. Altering or failing to comply with a drawing could result in a violation of federal or state standards or building codes, leading to fines and other penalties. More importantly, a deviation from the drawings could create an unsafe structure or product, which might result in the death or injury of others. To avoid these problems, welders must be able to accurately read and interpret prints. This module explains how to read and interpret assembly and detail prints.

2.0.0 ◆ WELDING DETAIL DRAWINGS

There are many categories of prints, such as civil, structural, mechanical, piping, electrical, and instrumentation. Within each category of prints there are several types of drawings. The most common drawing used by welders is the welding detail drawing, which generally depicts the weldment that must be fabricated and the pieces to be joined by welding. The parts are drawn assembled as they will appear after welding, except that the welds are not shown. Welding symbols are used to show how the pieces are to be

Hot Tip

Blueprints

Blueprints are so named because the obsolete chemical process used to develop chemically treated print paper turned paper blue. The lines produced on the paper were white against the blue background. Prints are made today using other methods.

joined. *Figure 1* shows an example of a small bracket that may have to be fabricated on the job.

Figure 2 shows how this same bracket would appear on a welding detail drawing.

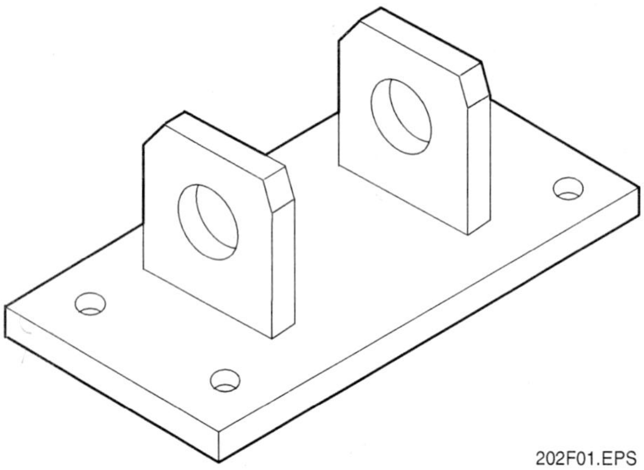

Figure 1 ◆ Bracket.

When the object to be fabricated is too complicated for one drawing, it may be shown using an assembly drawing and one or more detail drawings. The assembly drawing shows the complete and assembled object. Its main purpose is to show how the parts are assembled in relation to one another. Detail drawings of each part are then provided. The detail drawings will contain all the information necessary to make the part. *Figure 3* shows an assembly drawing and *Figure 4* shows the detail drawing for the parts.

> **NOTE**
> The parts are identified with letters and leader lines in the assembly drawing and then labeled with the same letters in the detail drawing.

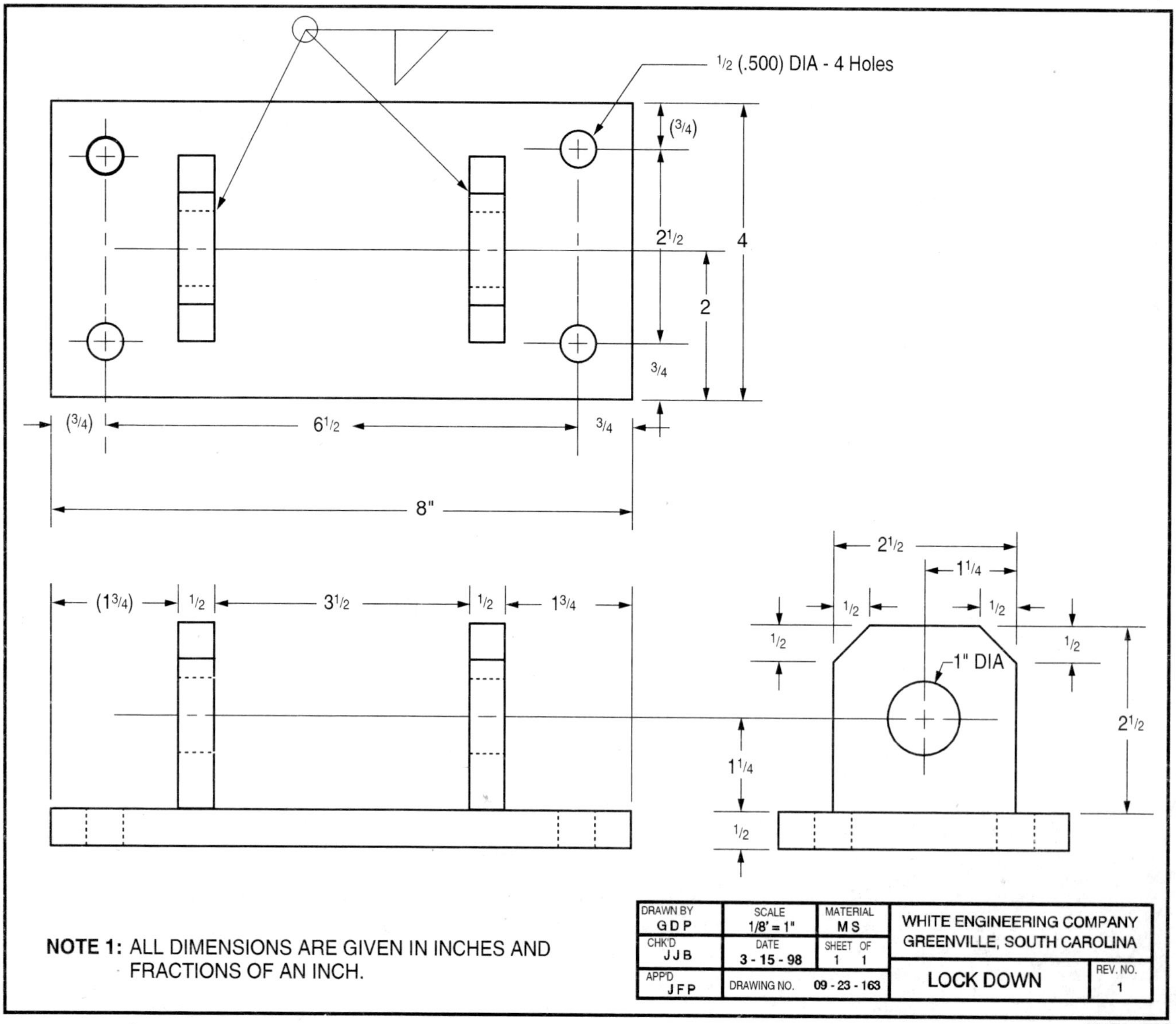

Figure 2 ◆ Welding detail drawing.

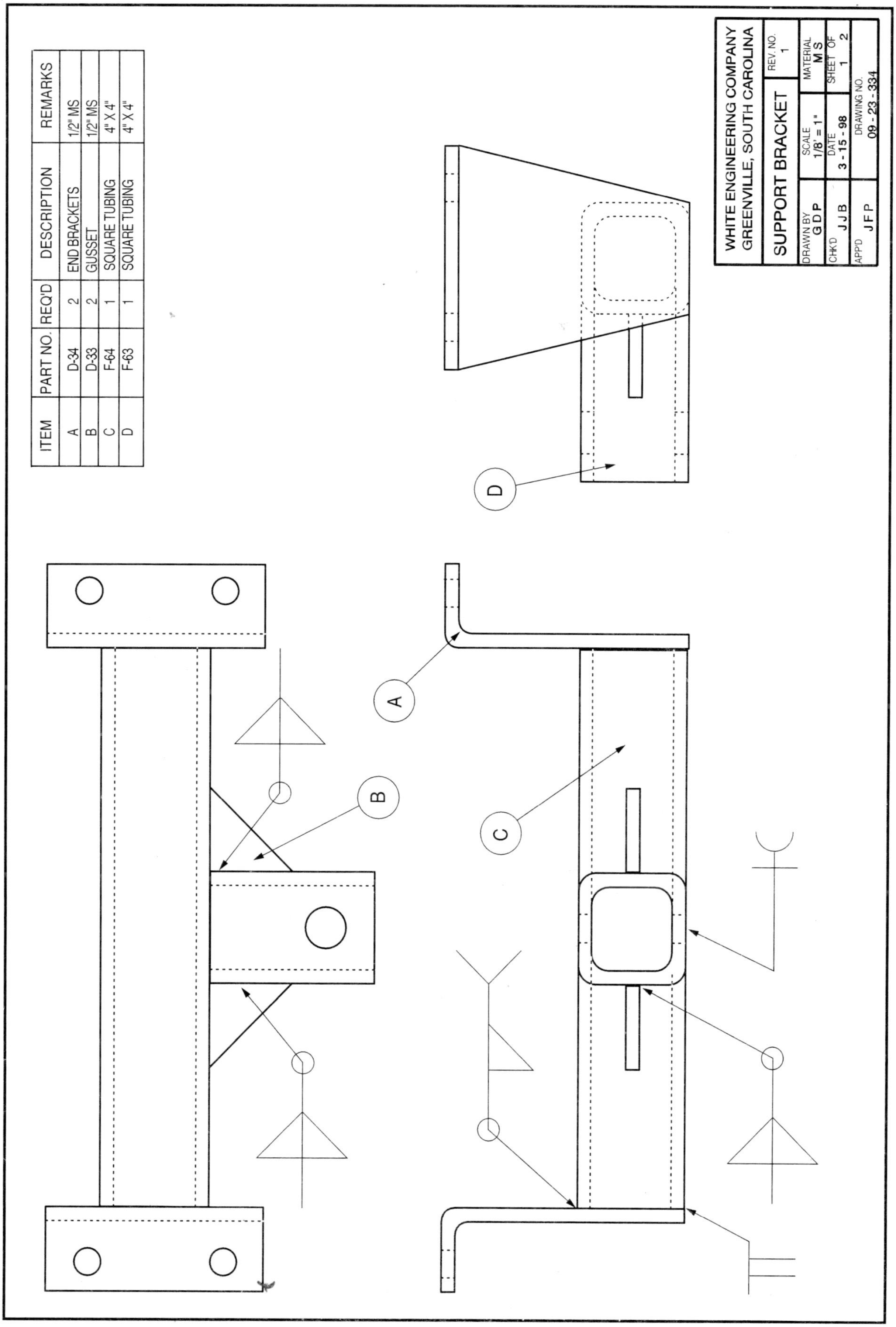

202F03.EPS

Figure 3 ◈ Assembly drawing.

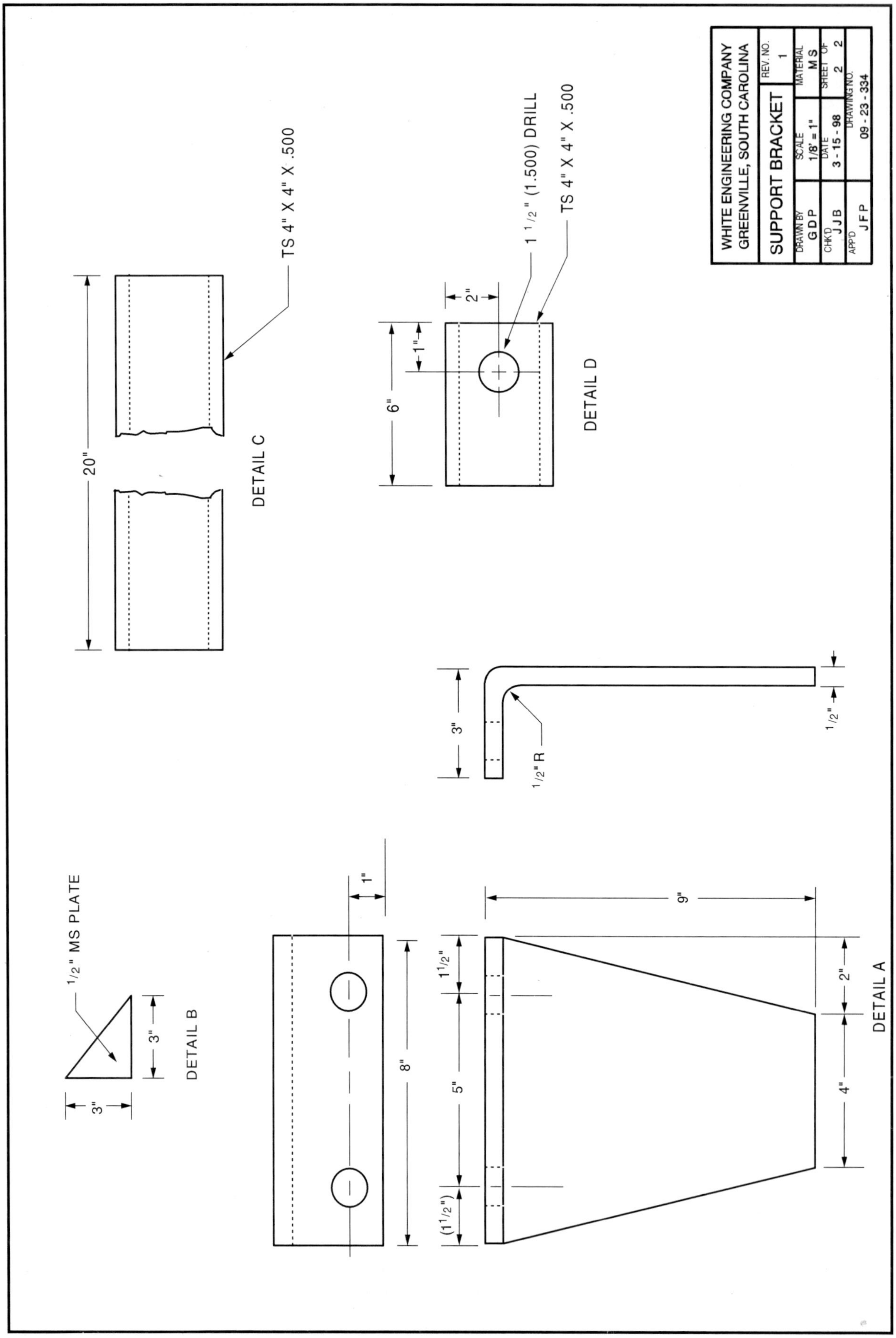

202F04.EPS

Figure 4 ◆ Detail drawing.

3.0.0 ◆ DRAWING COMPONENTS

Welding drawings have several components, including lines, material symbols, solid round, pipe, or tubing breaks, and revolved sections.

3.1.0 Lines

Many different types of lines are used on drawings, each with its own special meaning. The lines are drawn to a standard that specifies thickness (thick, medium, or thin) and the way they are interpreted or broken. The more commonly used lines are shown in *Figure 5* and follow:

- *Object line* – An object line is a thick, solid line used to show all visible edges of a part. Also called a visible line.
- *Hidden line* – A hidden line is a medium, broken line made up of evenly spaced dashes that is used to show edges that are not visible.
- *Center line* – A center line is a thin, broken line made up of short and long dashes that is used to show the center of holes or other objects.
- *Extension and dimension lines* – Extension lines are thin lines used with dimension lines. They extend from the point on the object that the dimension applies to. Dimension lines are thin lines with arrowheads that show the extent of the dimension. They are broken by the dimension number.
- *Leader line* – A leader line is a thin horizontal line with an arrowhead that angles off one end. It is used with a note or dimension. The arrowhead points directly to the surface of the note or dimension to which it applies.
- *Short break line* – A short break is a thick, solid, irregular line drawn freehand to indicate that part of the object has been removed or broken away. It is used to conserve space or show internal detail.
- *Long break line* – A long break is a thin, solid line with zigzags inserted at several places. It is used to show where a section of an object has been removed to conserve space on the drawing.
- *Cutting plane line* – A cutting plane line is used to show where an imaginary cut has been made through the object. It has a heavy line that can be made up of either a series of one long and two short dashes or a series of short dashes. Short lines with arrowheads are on each end of the line. The arrowheads point in the direction the section should be viewed.

3.2.0 Material Symbols

Material symbols are a series of thin lines, either solid or solid and broken, arranged in a pattern to represent various types of materials. The symbols are used to identify internal surfaces exposed by a cutting plane line or a short cutaway view. *Figure 6* shows common types of material symbols used to represent various metals.

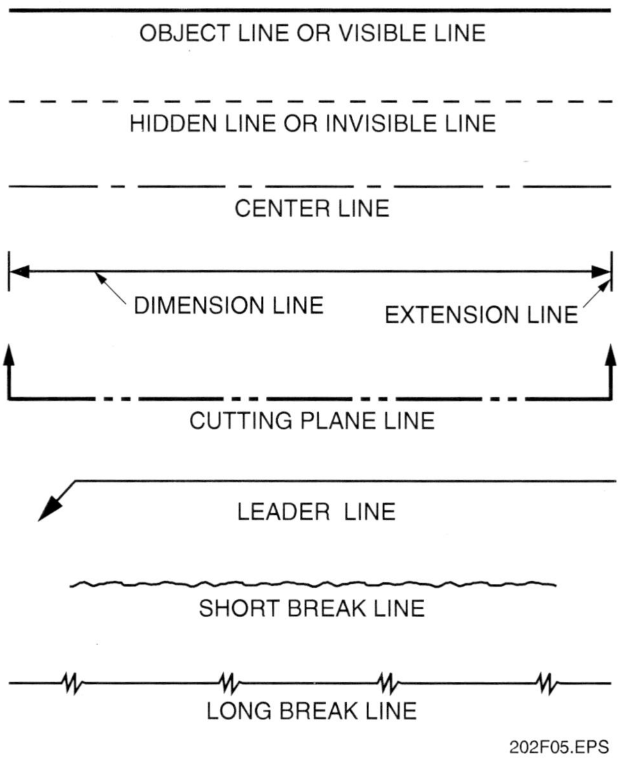

Figure 5 ◆ Common drawing lines.

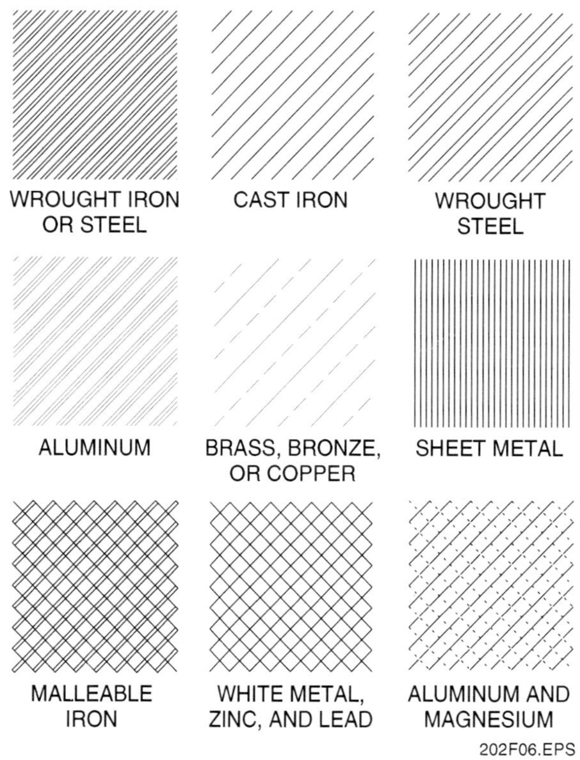

Figure 6 ◆ Material symbols.

> ***Material Symbols***
>
> Do not use material symbols shown on a drawing to select materials. They only represent the relative material. Always use the drawing notes, material list, or a bill of materials to determine the exact materials required.

3.3.0 Solid Round, Pipe, or Tubing Breaks

A solid round, pipe, or tubing break is represented by a thick, curved line that loops back on itself. The loop is filled with a material symbol to indicate an internal surface. It is used to show where a section of an object has been removed to conserve space on the drawing. *Figure 7* shows solid round, pipe, or tubing breaks.

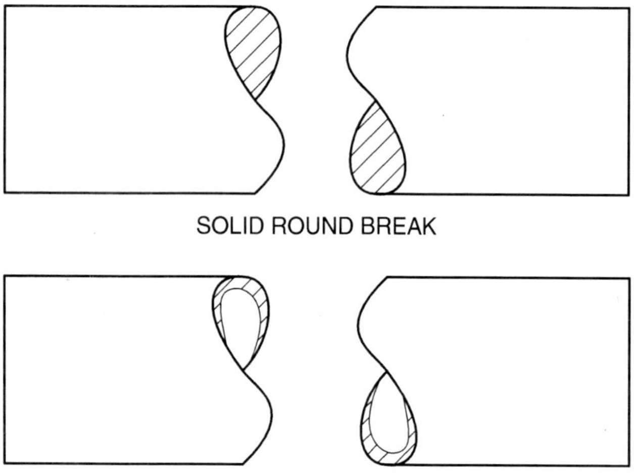

Figure 7 ◆ Solid round, pipe, or tubing break.

3.4.0 Revolved Sections

Revolved sections are similar to solid round, pipe, or tubing breaks except that the shape of the object broken is shown in the break as a section view. *Figure 8* shows revolved sections.

4.0.0 ◆ OBJECT VIEWS

Prints have to clearly show the shape of the object. In order to do this, most prints have more than one view of the object. The most commonly used views are the:

- Isometric
- Multiview
- Section

4.1.0 Isometric Views

The **isometric view** is a drawing showing a three-dimensional view of an object (*Figure 9*). The isometric (pictorial) view is an excellent way to show what an object looks like, but it is a difficult view to use when providing details such as dimensions, hole locations, and sizes or angles. For this reason, isometric views are generally not used on detail drawings.

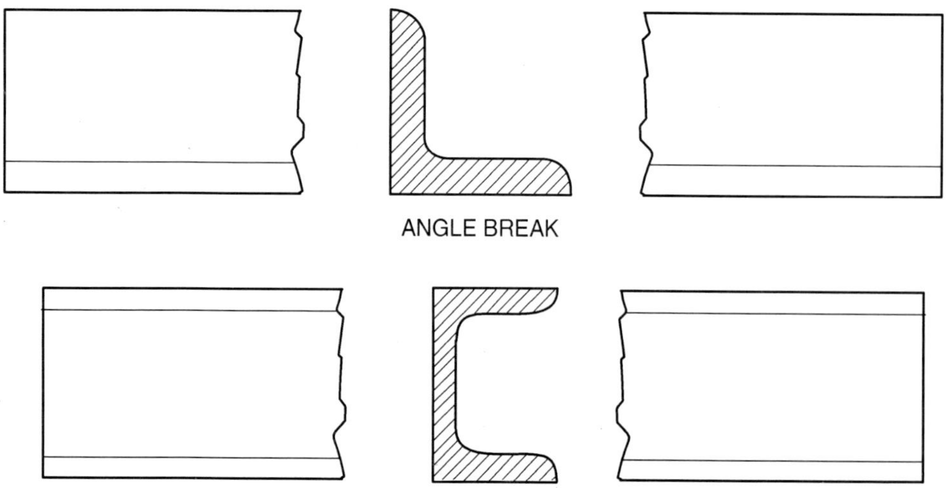

Figure 8 ◆ Revolved sections.

Hot Tip

Isometric Views

An isometric drawing of an object shows objects at 30° angles. This causes the object to appear on paper almost like it would actually appear to the eye.

4.2.0 Multiviews

Multiviews are developed by a process called **orthographic projection** and are generally used on detail drawings. Several views, each from a different direction, are developed and shown on the same drawing. The views are developed as if the object had been placed inside a transparent box. Each view is then developed by projecting points at right angles from the object onto a surface of the transparent box. If the sides of the box were folded open, a flat multiview drawing would result (*Figure 10*).

Each view in an orthographic projection has a set position in relation to the other views. In addition, they are aligned vertically and horizontally. All sides of an object can be shown, although the views used most frequently are the front, top, and right side. The front view is selected as the view that best represents the shape of the object. All other views are positioned in reference to the front view. *Figure 11* shows the relationship of possible views used with orthographic projection.

Figure 12 shows the relationship of views on a welding detail drawing.

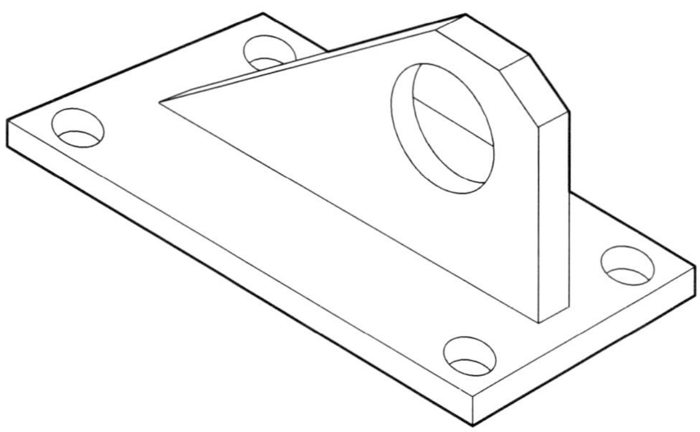

202F09.EPS

Figure 9 ◆ Isometric view.

4.3.0 Sections

A full section (or cross section) is a view of an object used to show internal details that would not normally be visible. A section line is used on one of the complete views to indicate where the part has been cut (sectioned). The arrowheads on

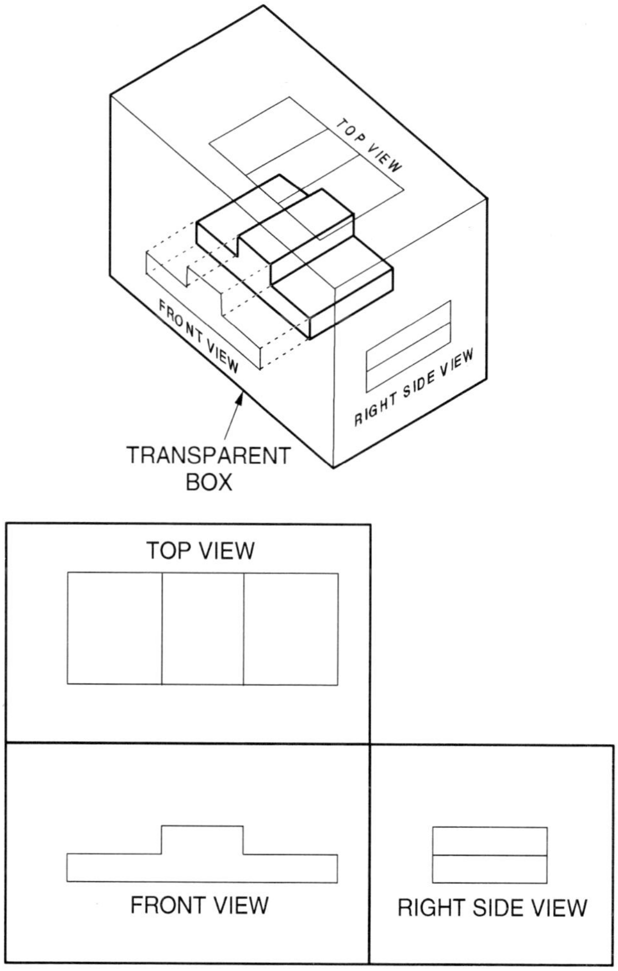

202F10.EPS

Figure 10 ◆ Orthographic projection.

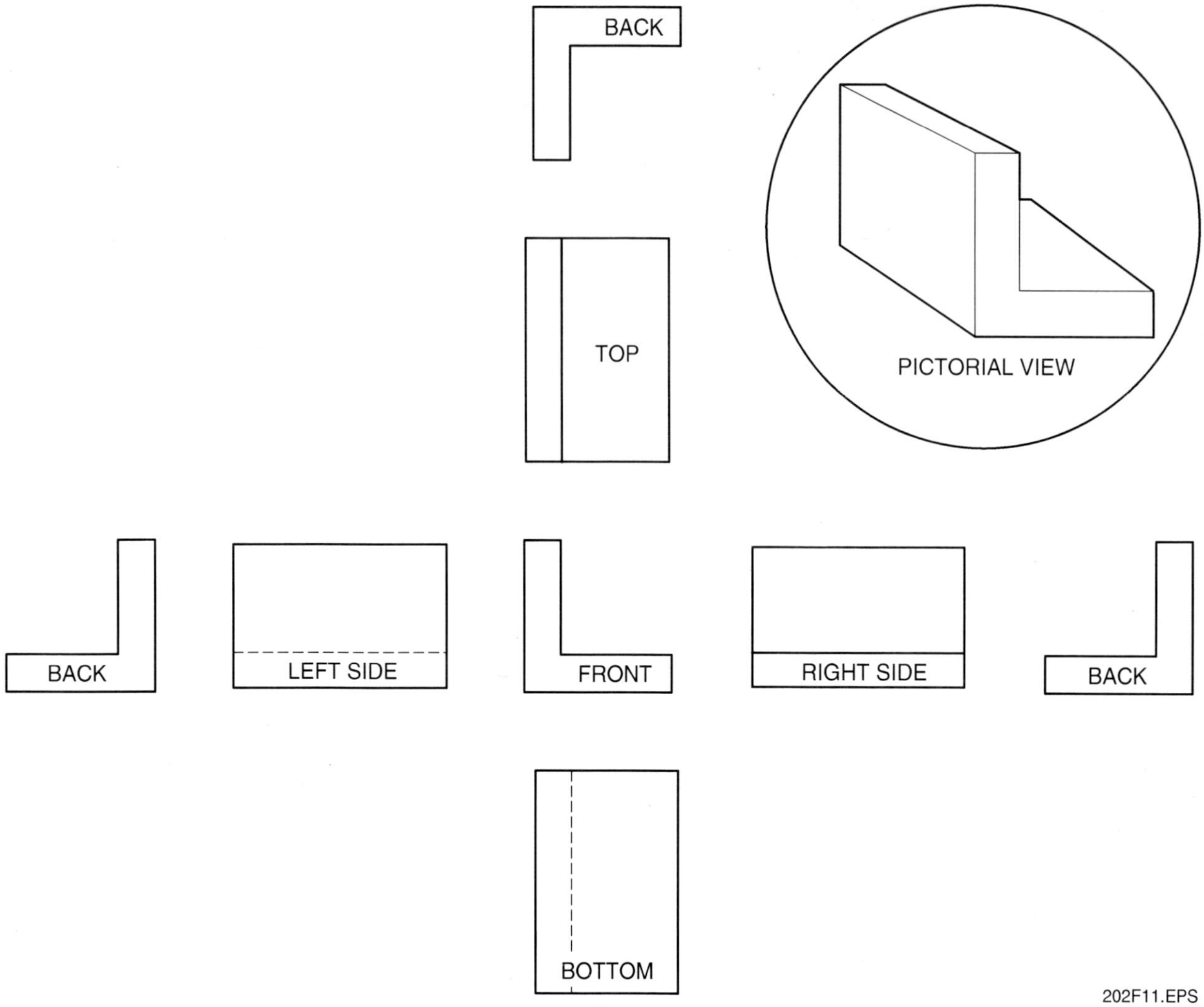

202F11.EPS

Figure 11 ◆ Relationship of possible views.

the section line point in the direction from which the sectioned object is to be viewed. Because there can be more than one section view, each is identified by placing a letter next to the arrowheads. The sectioned view is then labeled with the letters. *Figure 13* shows a section view.

A print shows a more detailed view of a part from different sections of the same part. Each of these sections is identified by pairs of letters located near the points of the cutting plane line.

5.0.0 ◆ DIMENSIONS

Dimensions identify sizes. Drawings must have dimensions if the object shown is to be fabricated. The drawing must include all dimensions in correct relationship to one another to give a complete size description of the object. The two basic methods for dimensioning a part are conventional dimensioning and baseline dimensioning.

In the conventional method of dimensioning, shown in *Figure 14*, the overall dimensions of length, width, and height are given. Dimensions A and B are not given because they are easily found by adding and/or subtracting the given dimensions. If they were given, dimensions A and B would be identified in parentheses as reference dimensions. A reference dimension, like the (1¾") dimension on the drawing, is obtained by adding or subtracting other dimensions found on the drawing. If the overall sizes of a dimensioned part are not given, then those dimensions that comprise the total must be added to find the overall size. Reference dimensions expedite the reading of drawings.

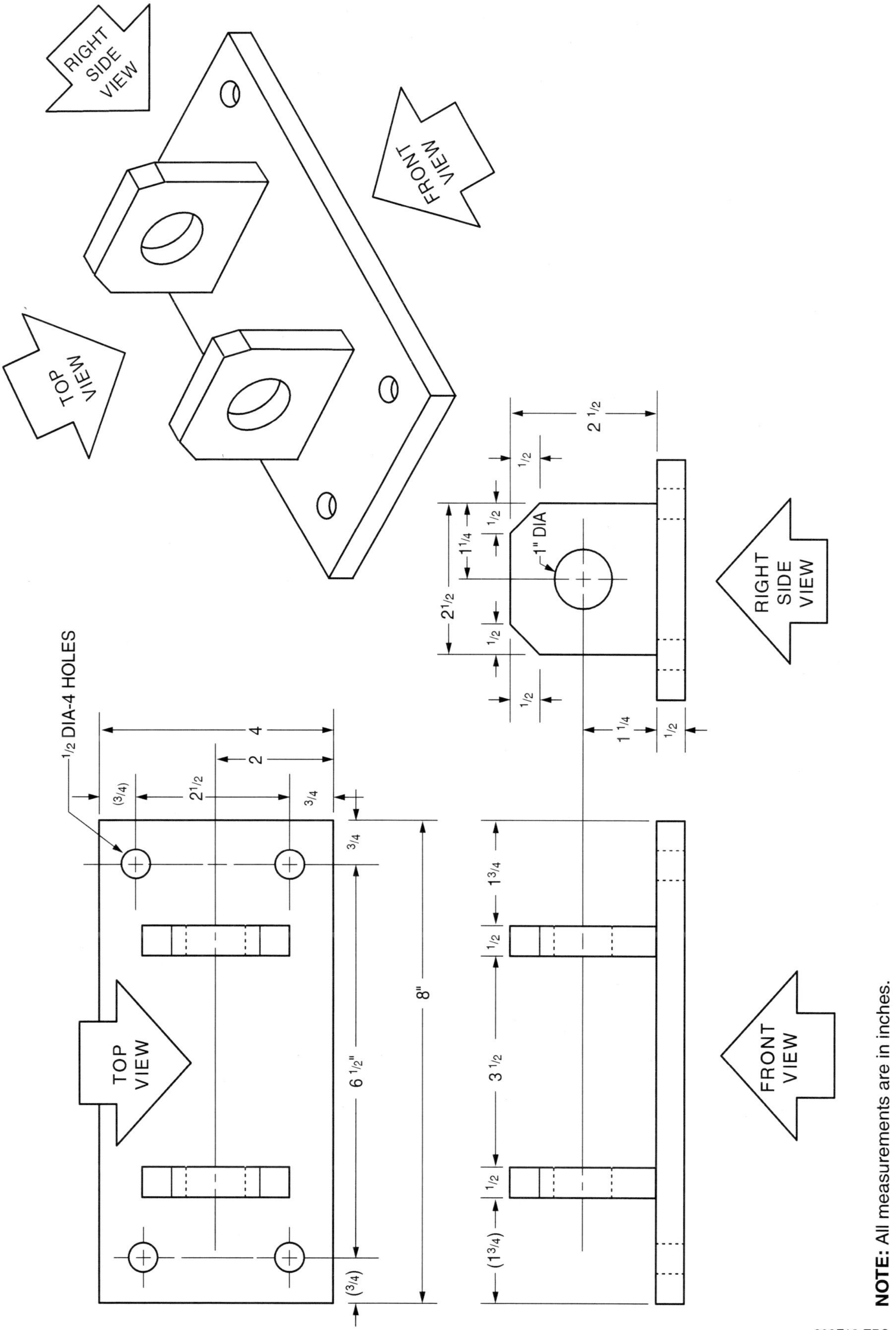

202F12.EPS

Figure 12 ◆ Relationship of views.

Dimension Calculation

THINK ABOUT IT

What is the reference dimension of a section whose overall part dimension is 4⅞"? The part consists of three sections. Two of the sections are 1⅞" and 1⅛", respectively.

Baseline dimensioning (*Figure 15*) differs from conventional dimensioning in that all dimensions originate from baselines. The dimensions are usually expressed as three-place decimals for accuracy. The baselines are usually machined surfaces that are indicated by the symbol V. Baselines may also originate from centerlines or from a critical edge. The advantage of baseline dimensioning is that it eliminates cumulative error that often occurs with conventional dimensioning. It doesn't require calculations in finding a dimension. Most welding drawings that require hand assembly and welding do not use three-place decimal dimensioning because of the difficulty in maintaining tight **tolerances**.

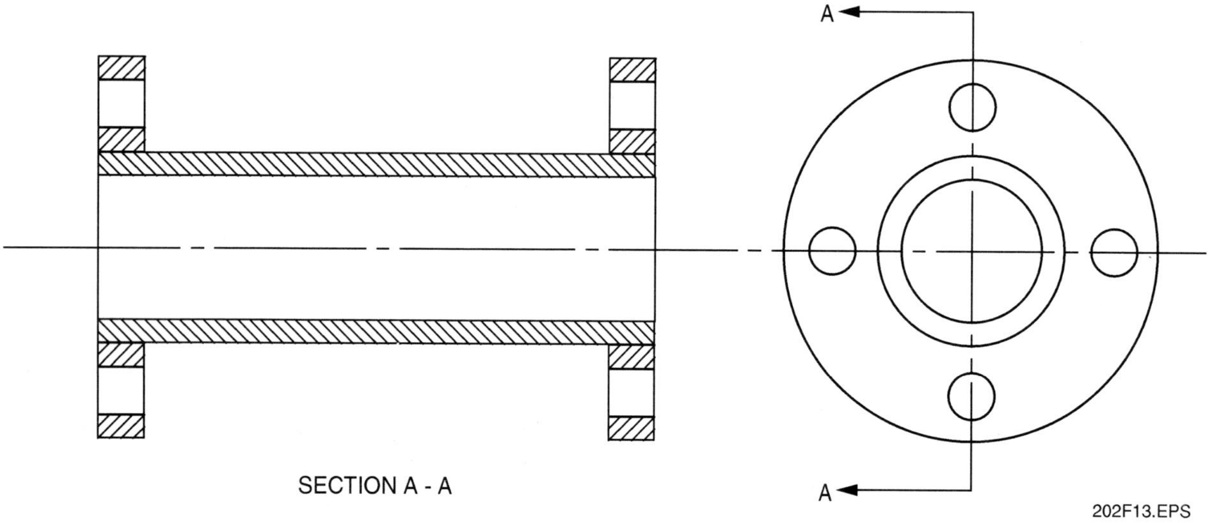

Figure 13 ◆ Section view.

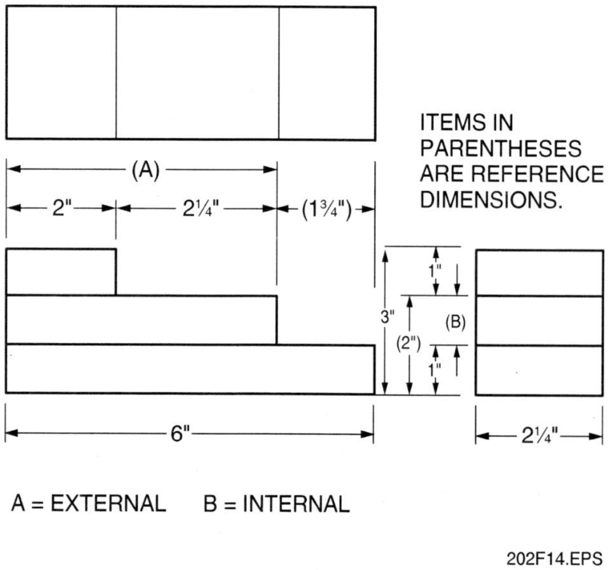

Figure 14 ◆ Conventional dimensioning.

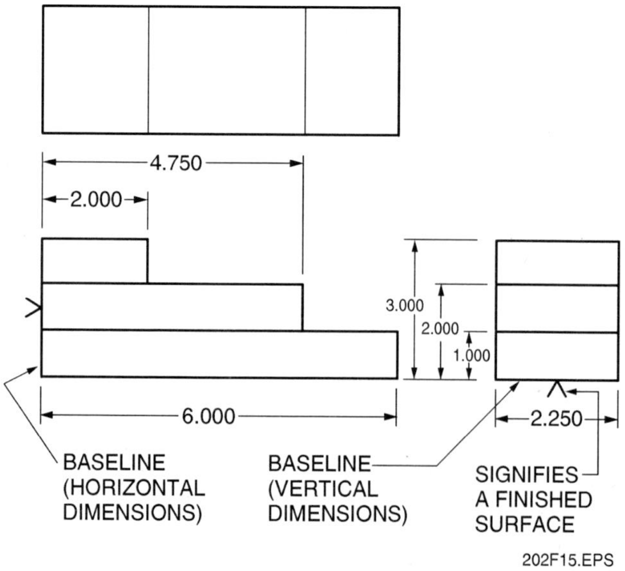

Figure 15 ◆ Baseline dimensioning.

5.1.0 Scale

Most drawings are made in **scale**, which means the drawing can be larger, the same size, or smaller than the actual object, as long as the proportions are the same. Most objects shown by welding detail drawings are too large to fit full size on a standard drawing, so they are drawn smaller than the actual object size. The scale used on the drawing is given in the title block in the form of an equation. Many different scales can be used. Some examples are:

- ¼" = 1"
- ⅛" = 1"
- ¼" = 1'

The figure on the left of the equal sign represents the measure for the drawing, while the figure on the right of the equal sign represents the measure for the object. For example, if the scale ¼" = 1" is used, an actual surface 3" long would be represented on the drawing by a line ¾" long (¼" for each actual inch). If the scale ¼" = 1' is used, an actual surface 4 feet long would be represented on the drawing by a line 1" long. *Figure 16* represents how scale is used to reduce the size of a drawing.

5.2.0 Size and Location Dimensions

The most common dimensions are size and location. Size dimensions provide the measurements for the overall size of the object. Location dimensions provide measurements to another part, hole, break, or other component of the object. *Figure 17* shows size and location dimensions.

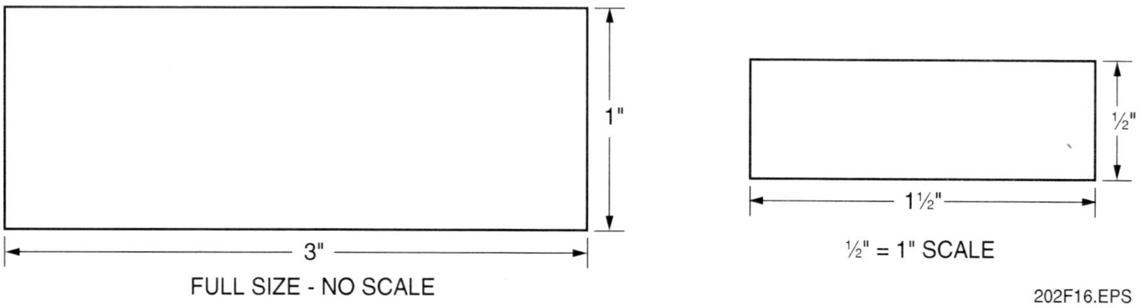

Figure 16 ◆ Drawing scale.

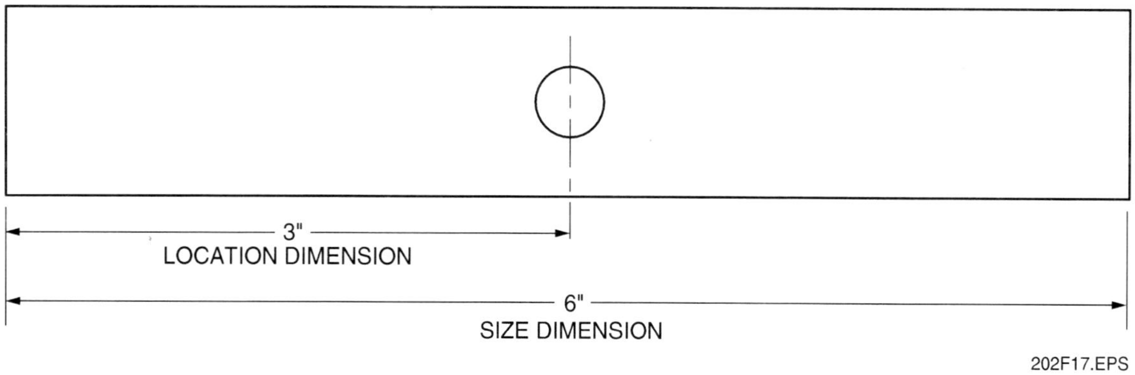

Figure 17 ◆ Size and location dimensions.

Hot Tip

Drawing Measurements

Use only dimensions provided on the drawing. If a dimension needed is not provided, add or subtract the dimensions furnished to obtain the dimension needed. Avoid measuring directly off a drawing (called scaling a drawing). Measuring directly off a drawing almost always results in errors. If a measurement cannot be determined by adding or subtracting given dimensions, check with your supervisor.

Structural Steel and Pipe Sizes

You should be aware that structural steel and pipe sizes, if given on drawings, are usually nominal (approximate) sizes only. Structural steel is classified as hot rolled or cold rolled. Hot-rolled varieties are somewhat oversized and have rounded corners and edges. The sizes of cold-rolled varieties are more precise and the corners and edges are square. Size specifications for some structural shapes including S-beams (formerly called I-beams), H-beams, and wide flange (WF) beams are nominal.

Pipe is sometimes sized through the application of a schedule number as part of a specification. The 11 schedules commonly used are: 5, 10, 20, 30, 40, 50, 60, 80, 100, 120, and 160. The schedule numbers represent the strength and refer to wall thickness. Pipe schedule tables provide the nominal wall thickness, weight, and outside diameter for each size and type of pipe.

5.3.0 Hole Dimensions

Drilled hole dimensions are shown with a leader line and a note (*Figure 18*). The note gives the hole size and the number of holes to be drilled (if more than one). If the hole is to be drilled only part way through the object (**blind hole**), then the depth will also be given. If the depth is not given, then the hole continues completely through the object (thru hole).

Holes may also be **counterbored** or countersunk. A hole that is counterbored has a square shoulder, used to recess socket head fasteners; a hole that is countersunk is tapered from the surface to the drilled hole size, used for flat head fasteners. If required, the dimensions for countersinking or counterboring are added to the note. *Figure 19* shows the dimensioning of counterbores and countersinks.

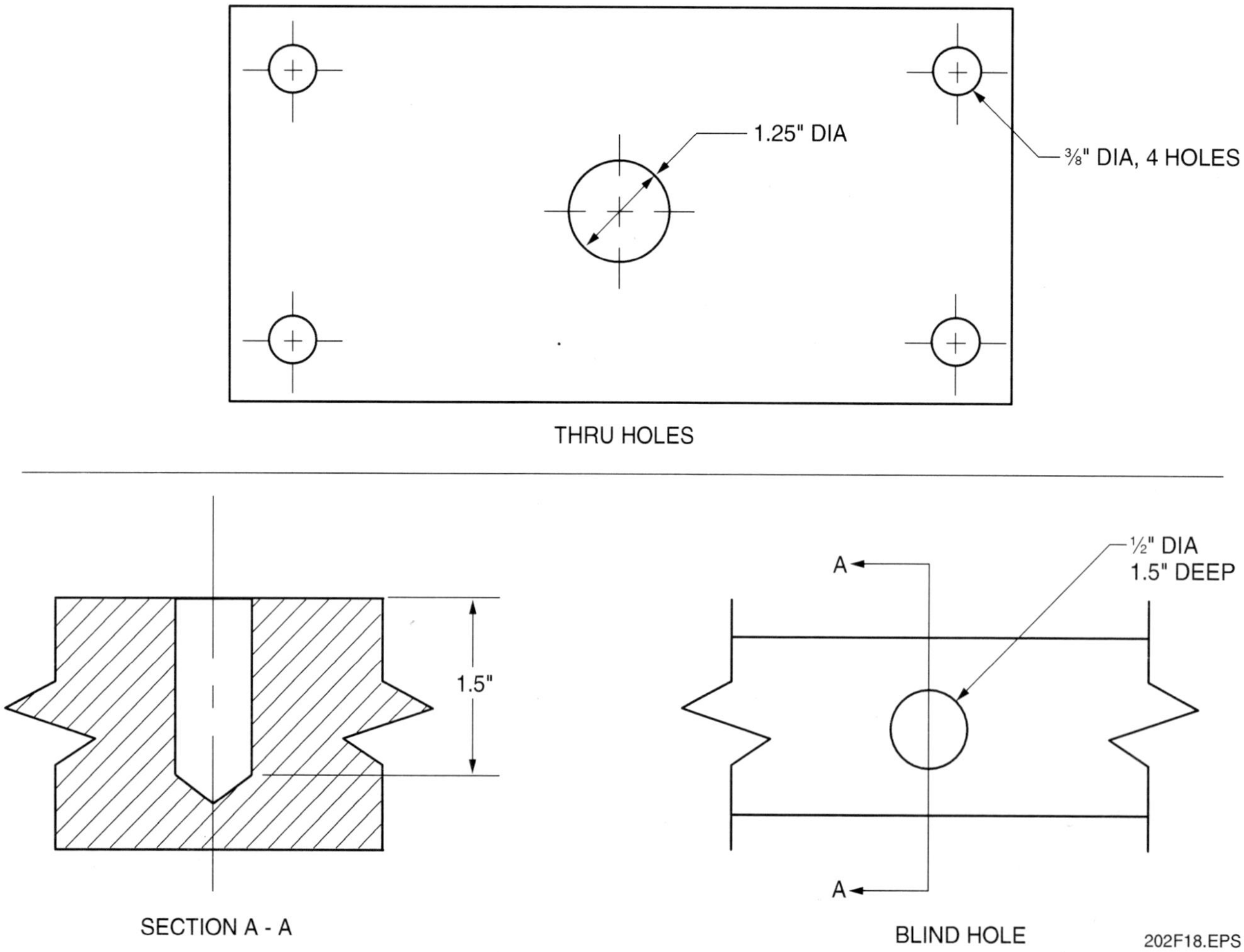

Figure 18 ◆ Dimensioning holes.

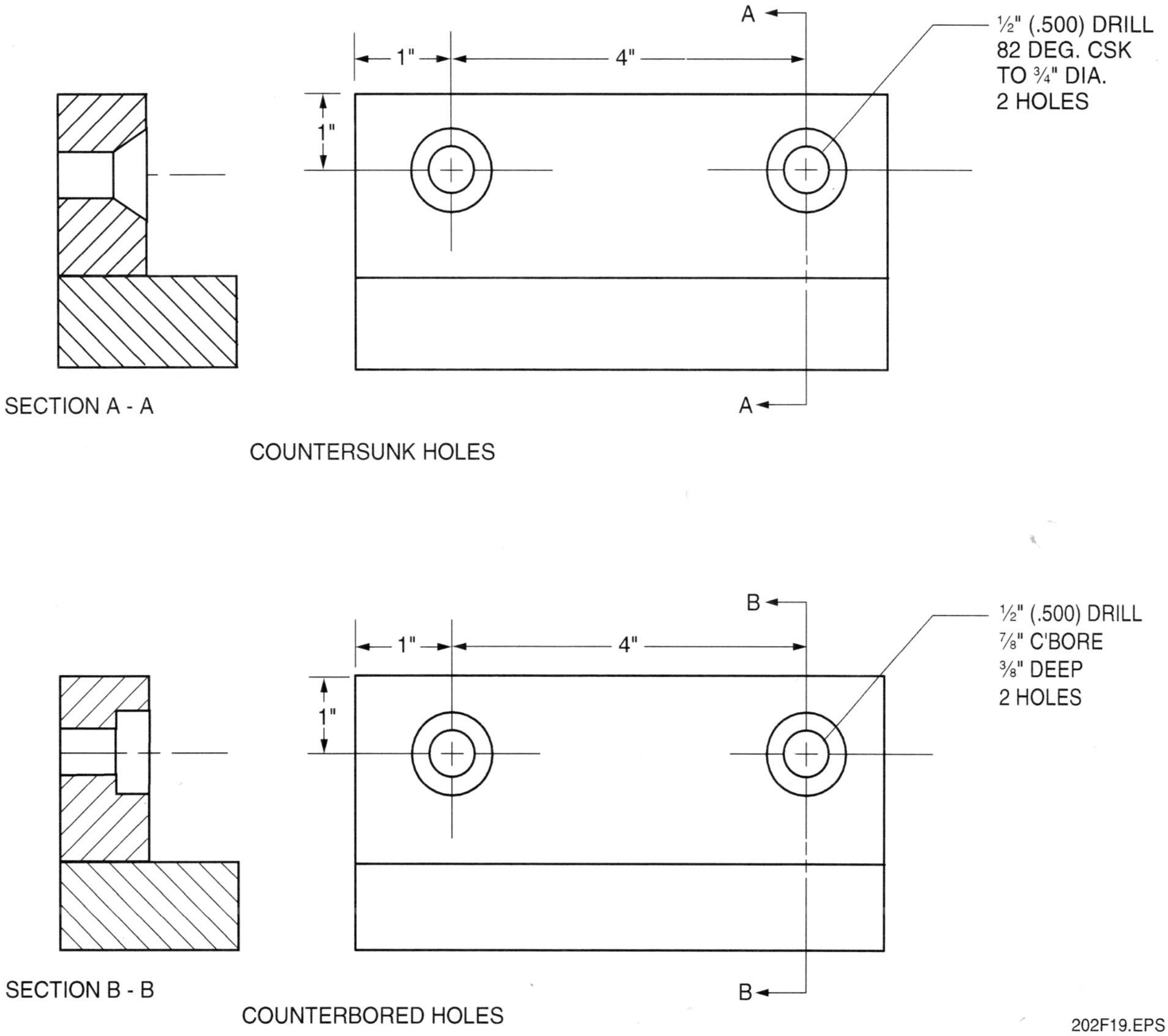

Figure 19 ◈ Dimensioning counterbores and countersinks.

5.4.0 Angle, Chamfer, and Bevel Dimensions

Angles, **chamfers**, and bevels are common on welding detail drawings because they are used with weld joint preparation. They can be dimensioned with a leader line and a note or they can be dimensioned using extension and dimension lines. The size can be given as angular, linear, or a combination of angular and linear dimensions. *Figure 20* shows dimensioning angles, chamfers, and bevels.

5.5.0 Radius and Arc Dimensions

When the ends, sides, or corners of an object are to be rounded, a radius or arc is shown and dimensioned (*Figure 21*). Typically a combination of center lines, radius, extension, and dimension lines are used. To obtain the arc required, the radius line may start outside the physical shape of the object.

5.6.0 Tolerances

Tolerance is an important part of dimensioning because it identifies the variance in size that is allowable on a part. Depending on its intended use, a part can have a large or small tolerance. If a part is fabricated outside the allowable tolerance, it will have to be reworked or scrapped at an additional cost. On the other hand, if a part with a large tolerance is fabricated to very fine tolerances, it will generally add to the cost since parts with large tolerances typically can be fabricated faster than parts with smaller tolerances. For these reasons, it is important to identify the tolerances on a drawing and then work accordingly.

Tolerances can be given following the dimension, as a note, or in a table as part of the title block. They are generally given as a plus (+) or minus (–) quantity. For example, if the dimension 10" was followed by the tolerance, + or –⅛", the object

Cut Tolerance

What is the minimum and maximum allowable size of an 18" cut that has to be made with a tolerance of ±¼"?

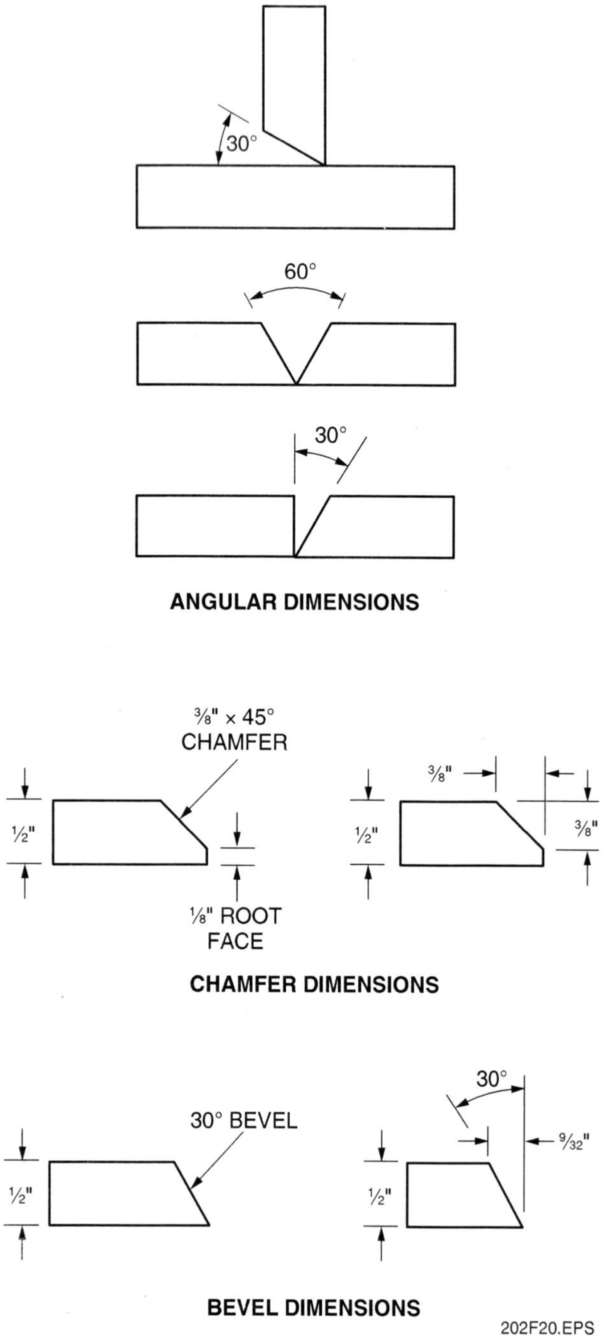

Figure 20 ◆ Dimensioning angles, chamfers, and bevels.

Manual Welding or Cutting Tolerances

Drawing tolerances for most manual welding or cutting operations are usually not less than 1⁄16". This is because manual cutting torches are usually inaccurate. Heat warpage and movement, as the result of manual welding, are also very difficult to control.

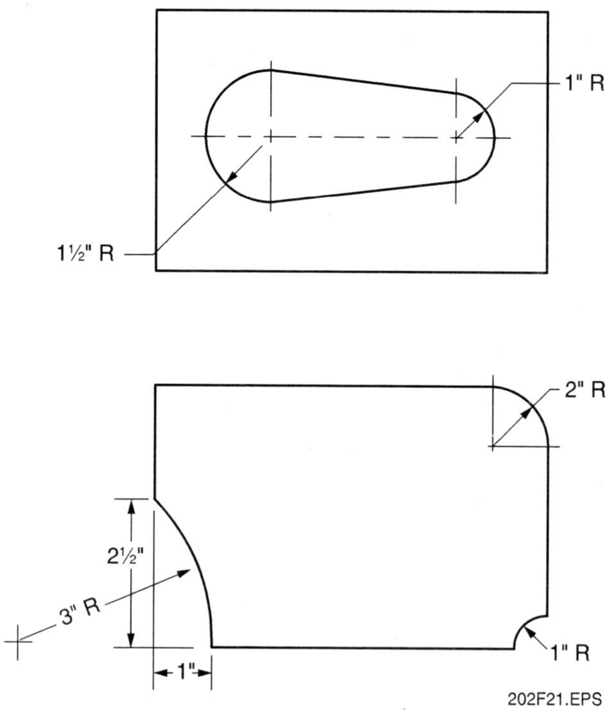

Figure 21 ◆ Dimensioning radius and arcs.

could be 9⅞" through 10⅛". The plus and minus tolerances for a size can be different. For example, a dimension of 12", +⅛" and –0, indicates that the object can be from 12" through 12⅛". *Figure 22* shows examples of dimensioning tolerances.

6.0.0 ◆ NOTES AND BILL OF MATERIALS

Additional information that can be found on many prints includes notes and a bill of materials.

6.1.0 Notes

A note is an explanatory or critical comment concerning the details of construction. Two types of notes are found on drawings: general notes and local notes. A general note applies to the drawing in its entirety and is placed in an area away from the views so that it can be noticed. On large jobs, general notes are placed on a separate drawing. On smaller jobs they are placed on detail drawings and refer to the entire drawing. General notes are grouped together and numbered. Items typically found in these general notes include:

- References to specifications, procedures, or other notes
- Identification of work responsibility
- Surface finish requirements (smoothness, type of finish)
- Special handling requirements
- Material requirements such as specification, grade, and class
- Special assembly requirements

A local note applies to a particular part on an object. It is placed near one of the views representing the part. Local notes are used for items such as:

- Thread fastener type and size
- Location of special items (part to be supplied by others)
- Reference to detail on another drawing
- **Elevations**
- Special treatments or finishes
- Drilled hole size

Figure 23 shows general and local notes.

A specification is a note that provides a detailed description of requirements for the type and size of parts/material required. For instance, it may specify the type and size of electrode and/or the welding process to be used. In *Figure 24*, the letter A refers to specification A. In this case, specification A in the tail of the arrow indicates that No. 30, ⅛" copper rod is to be used for the weld. Specifications are usually located near the views to which they refer. If there are quite a few specifications, they are attached on a separate sheet of paper and referenced to the drawing.

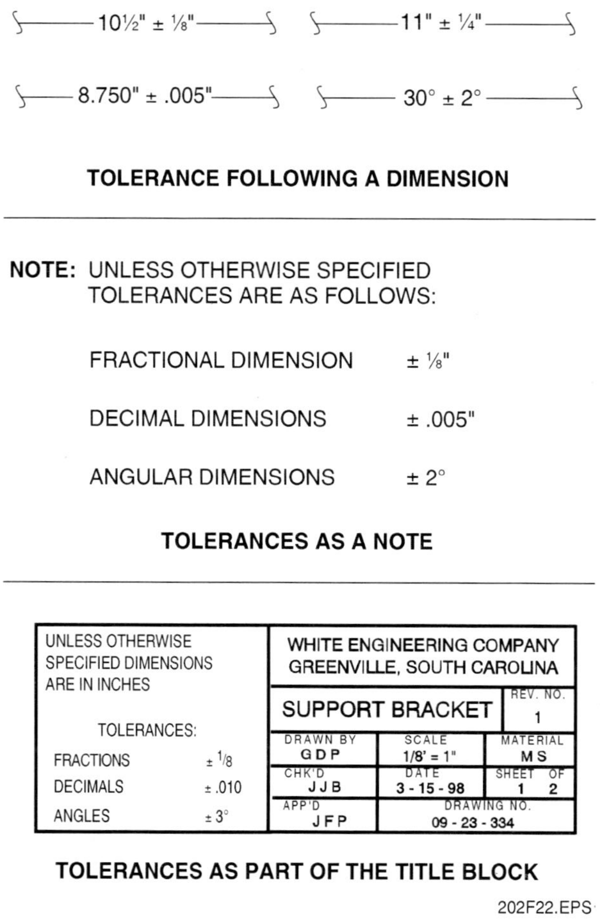

Figure 22 ◆ Dimensioning tolerances.

6.2.0 Bill of Materials

A bill of materials, also called a parts list or schedule of parts, is a list of all the parts required to fabricate the object. It generally identifies each part with a number or letter (sometimes called a **mark**) and includes a description of the part, the quantity needed for each assembly, and the part number. Some may also include size and dimensions for the part. This is often the case when the parts being described are structural shapes. Not all drawings have a bill of materials. *Figure 25* shows a drawing with a bill of materials.

GENERAL NOTES - STEEL

1. Design, fabrication, & erection of structural steel shall be in accordance with the *Manual of Steel Construction* of the American Institute of Steel Construction, Eighth Edition, unless otherwise modified on the drawings or in the specifications.

2. Material shall meet the requirements of the following specifications unless otherwise noted:

STRUCTURAL STEEL	- *ASTM A - 36 - 87*
HIGH STRENGTH BOLTS	- *ASTM A325 - 86, ⅞ INCH DIA*
WELD STEEL	- *AWS D 1.1 CLASS E70*

3. Shop and field paint shall be as called for in the specifications.

4. Steel to be encased in concrete shall not be painted.

5. Hot dipped galvanized surfaces that have been damaged by welding, cutting, burning, shearing, or other damage incurred during transit or erection shall be repaired with Z. R. C. Cold Galvanizing Compound as manufactured or approved equal repair shall be in strict accordance with the manufacturer's recommendations.

6. All shop connections shall be welded or made with high strength bolts unless noted specifically.

7. Erectors shall provide all temporary shoring & bracing needed for stability until structure is completed.

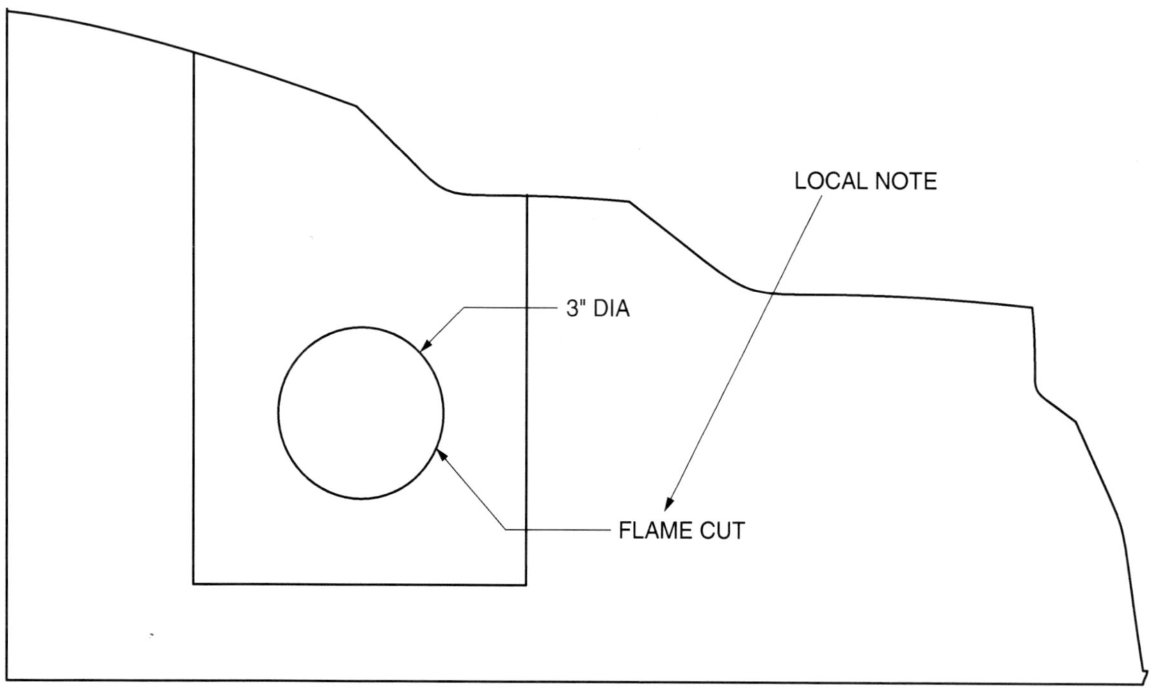

Figure 23 ◆ General notes and local notes.

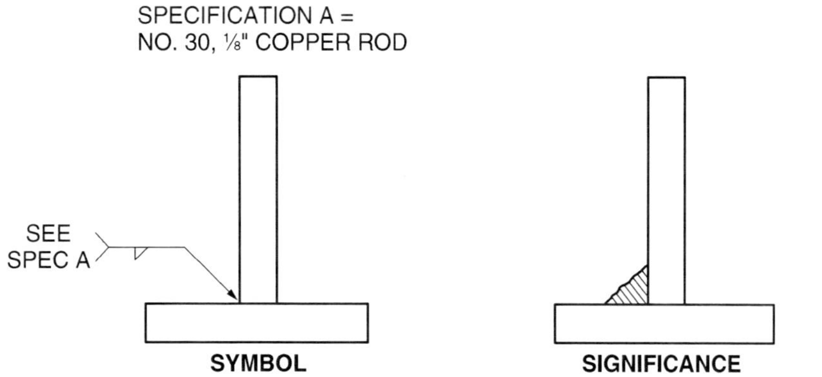

202F24.EPS

Figure 24 ◆ Application of a specification.

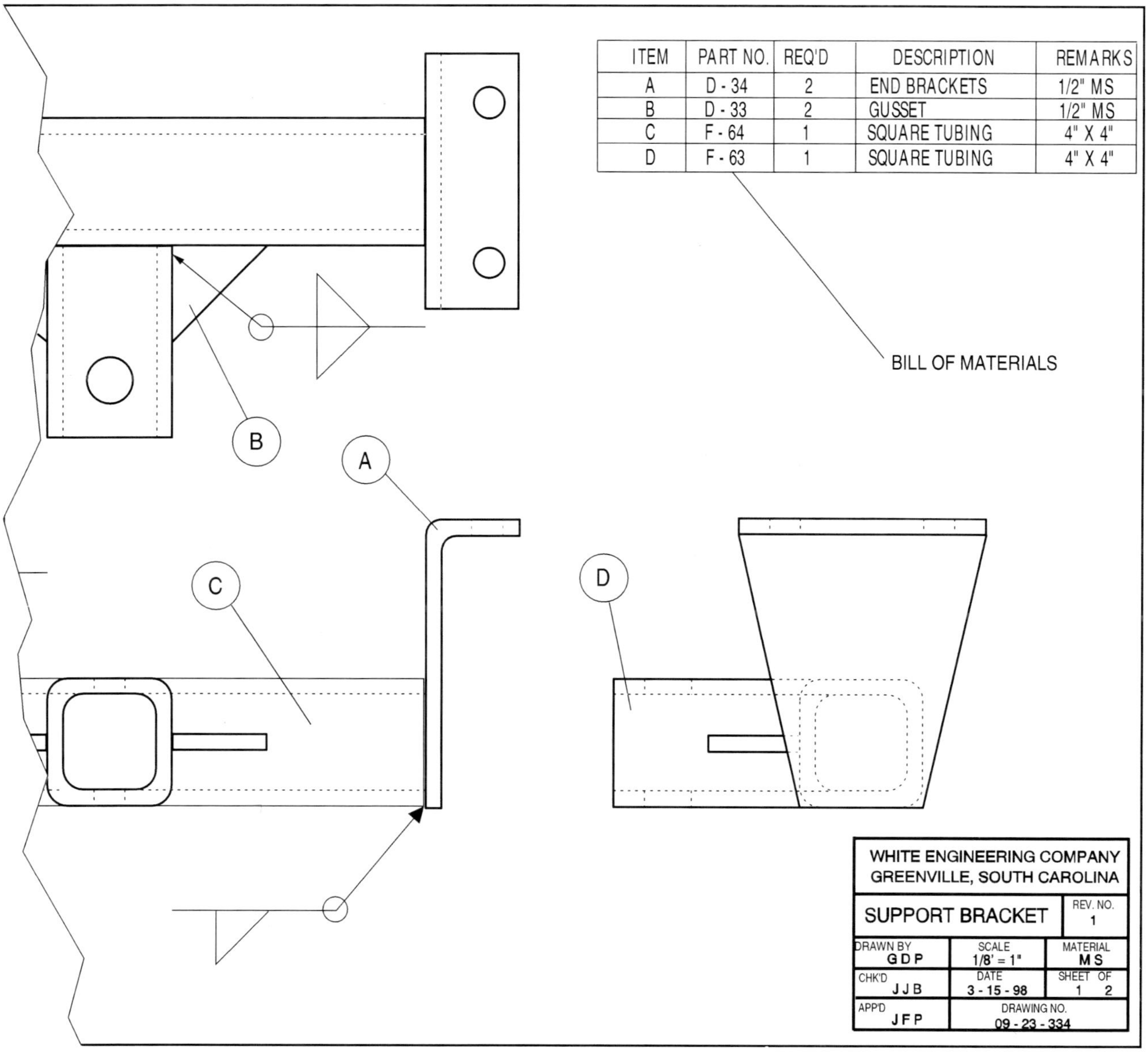

202F25.EPS

Figure 25 ◆ Bill of materials.

Summary

Welding detail drawings are used by engineers, designers, and draftsmen to precisely convey information to the welders in the field. To correctly interpret this information, welders must be able to accurately read welding detail drawings.

Reading and interpreting welding detail drawings requires an understanding of lines, fills, views, and dimensioning. It is also important to be able to read and interpret drawing notes and the bill of materials.

Review Questions

1. When an object to be fabricated is too complicated for one drawing, the complete and assembled object may be shown on a(n) _____ drawing, plus one or more detail drawings.
   a. general arrangement
   b. mechanical
   c. structural
   d. assembly

2. The most common drawing used by welders is the _____.
   a. isometric view
   b. assembly drawing
   c. section view
   d. welding detail drawing

3. Welding symbols are used to show how the pieces are to be _____.
   a. joined
   b. aligned
   c. cut
   d. machined

4. On a typical drawing, a thick solid line used to show all visible edges of a part is called a(n) _____ line.
   a. object
   b. center
   c. leader
   d. hidden

5. A medium broken line that is made up of evenly spaced dashes and used to show edges that are not actually visible is called a(n) _____ line.
   a. object
   b. center
   c. leader
   d. hidden

6. A series of thin lines that are either solid or solid and broken, and arranged in a pattern on a drawing to indicate various types of materials, is called a material _____.
   a. cutaway
   b. symbol
   c. cutting plane
   d. extension and dimension

7. Multiview drawings are developed by a process called _____ and are generally used on detail drawings.
   a. dimensional representation
   b. reflective representation
   c. orthographic projection
   d. transparent projection

8. The views developed for a multiview drawing show the object as if it had been placed inside a _____.
   a. circle
   b. square
   c. rectangle
   d. transparent box

9. The view that best represents the shape of an object is the _____ view.
   a. top
   b. back
   c. front
   d. side (left or right)

10. A section line is used on one of the complete views (of an object or part) to indicate where the part has been _____.
   a. cut
   b. welded
   c. centered
   d. rotated

11. Drawings used for fabricating an object must have _____.
   a. pictorial views
   b. section views
   c. dimensions
   d. scales

12. Drawings that can be larger, the same size, or smaller than the actual object (as long as the proportions are the same) are _____.
   a. made in scale
   b. made in draft
   c. modeled
   d. distorted

13. In a drawing scale equation, the figure on the left of the equal sign represents _____.
   a. the measure for the object
   b. the measure for the assembly
   c. the measure for the drawing
   d. the drawing sequence number

14. The tolerances used for the dimensions of a drawing can be given (shown) following the dimension, as a note, or in the _____.
   a. title block
   b. scale factor
   c. drawing title
   d. object drawing itself

15. All of the parts required to fabricate an object will normally be listed in the _____ of a drawing.
   a. bill of materials
   b. general notes
   c. local notes
   d. title notes

Refer to the following figure to answer Questions 16 through 30.

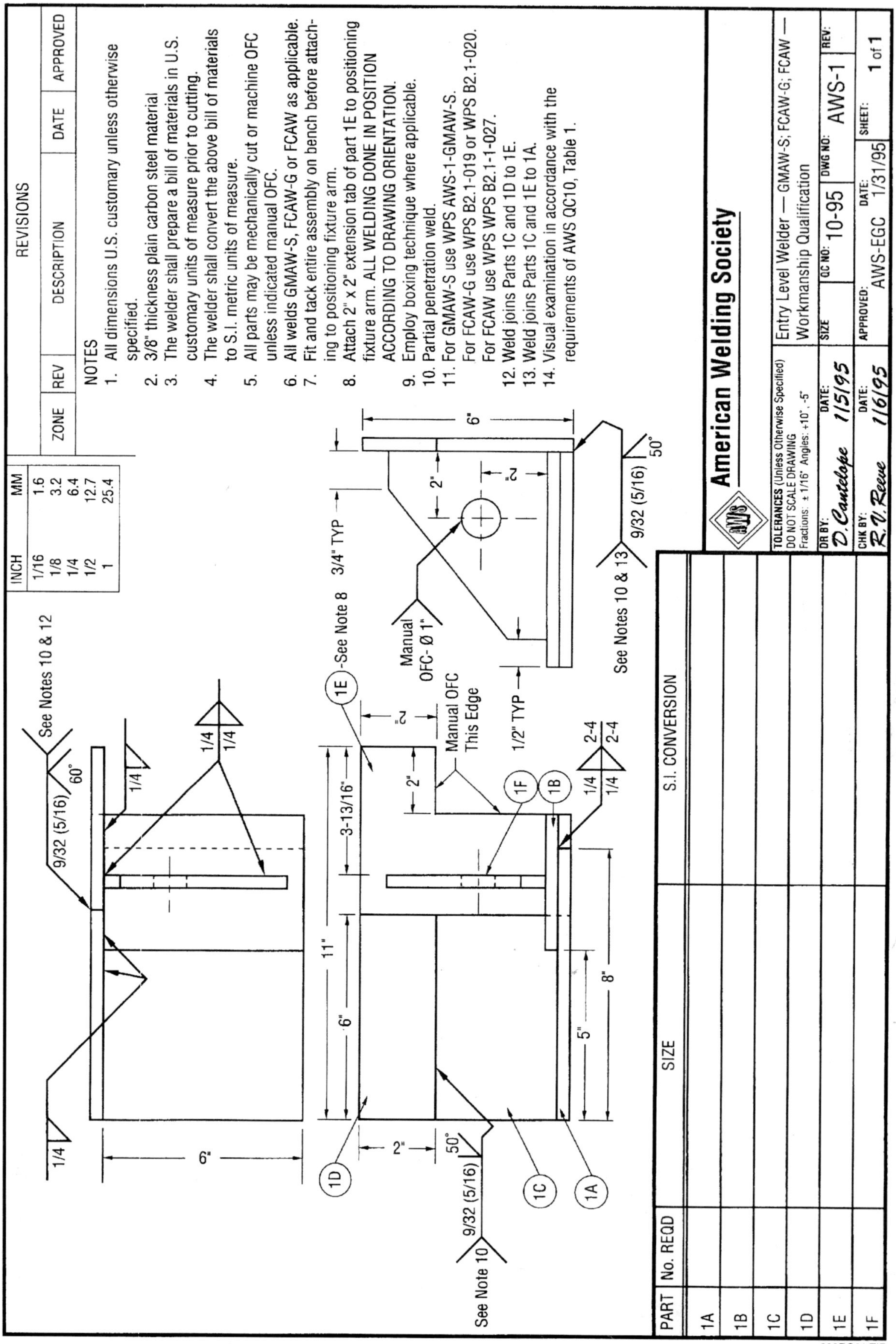

202RQ01.TIF

16. How many pieces make up this weldment?

____

17. What is the overall length and height of this weldment?

____

18. What type of weld is used to join parts 1F to 1B?

____

19. What type of weld is used to join part 1D to part 1E?

____

20. What is the angle for the bevel groove weld joining 1D and 1C?

____

21. What is the offset for the radius of the hole in part 1F?

____

22. What is the drawing number for this drawing?

____

23. What is the dimensional tolerance for all parts in this drawing?

____

24. What is the length of the fillet weld required to join parts 1A and 1B together?

____

25. What is the plate thickness used for this weldment?

____

26. On what two parts is a 50° bevel needed?

____

27. What is the height of part 1F?

____

28. What does the number (5/16) mean in this print?

____

29. On what date was this drawing approved?

____

30. What process is used to make the hole in part 1F?

____

GLOSSARY

Trade Terms Introduced in This Module

Blind hole: A drilled hole that does not go all the way through the object.

Chamfer: A cut made from the face to the side of a piece of material, usually at a 45° angle.

Counterbore: To enlarge a portion of a drilled hole so that a socket head fastener will seat flush or below the surface.

Elevation: A height measurement above sea level or other identified point; also, the vertical sides of an object in an orthographic view.

Isometric view: A drawing depicting a three-dimensional view of an object.

Mark: A system of letters and numbers used with structural steel to identify the type of member and where it is placed.

Orthographic projection: A method of developing multiple flat views of an object by projecting lines out from the object at right angles from the face.

Scale: A method of drawing an object larger or smaller than its actual size while keeping the proportions the same.

Tolerance: The allowance over or under the specified dimension, value, or other parameter.

Additional Resources

This module is intended to be a thorough resource for task training. The following reference works are suggested for further study. These are optional materials for continued education rather than for task training.

Printreading For Welders. T.E. Proctor, J.F. Grosse. Homewood, IL: American Technical Publishers.

Welding Printreading. John R. Walker and W. Richard Polanin. Tinley Park, IL: The Goodheart-Willcox Company, Inc.

Figure Credit

Figure 202RQ01 courtesy of American Welding Society

AWS Education Department, 1995, *Guide for the Training and Qualification of Welding Personnel—Entry Level Welders, AWS EG2.0-95*, American Welding Society, Figure 2.

CONTREN™ LEARNING SERIES — USER UPDATES

The NCCER makes every effort to keep these textbooks up-to-date and free of technical errors. We appreciate your help in this process. If you have an idea for improving this textbook, or if you find an error, a typographical mistake, or an inaccuracy in NCCER's Contren™ textbooks, please write us, using this form or a photocopy. Be sure to include the exact module number, page number, a detailed description, and the correction, if applicable. Your input will be brought to the attention of the Technical Review Committee. Thank you for your assistance.

Instructors – If you found that additional materials were necessary in order to teach this module effectively, please let us know so that we may include them in the Equipment/Materials list in the Instructor's Guide.

Write: Curriculum Revision and Development Department
National Center for Construction Education and Research
P.O. Box 141104, Gainesville, FL 32614-1104

Fax: 352-334-0932

E-mail: curriculum@nccer.org

Craft ____________________ Module Name ____________________

Copyright Date __________ Module Number __________ Page Number(s) __________

Description

____

____

____

(Optional) Correction

____

____

(Optional) Your Name and Address

____

____

Module 29203-03

SMAW – Stainless Steel Groove and Pipe Welds

COURSE MAP

This course map shows all of the modules in the second level of the Welding curriculum. The suggested training order begins at the bottom and proceeds up. Skill levels increase as you advance on the course map. The local Training Program Sponsor may adjust the training order.

WELDING LEVEL TWO

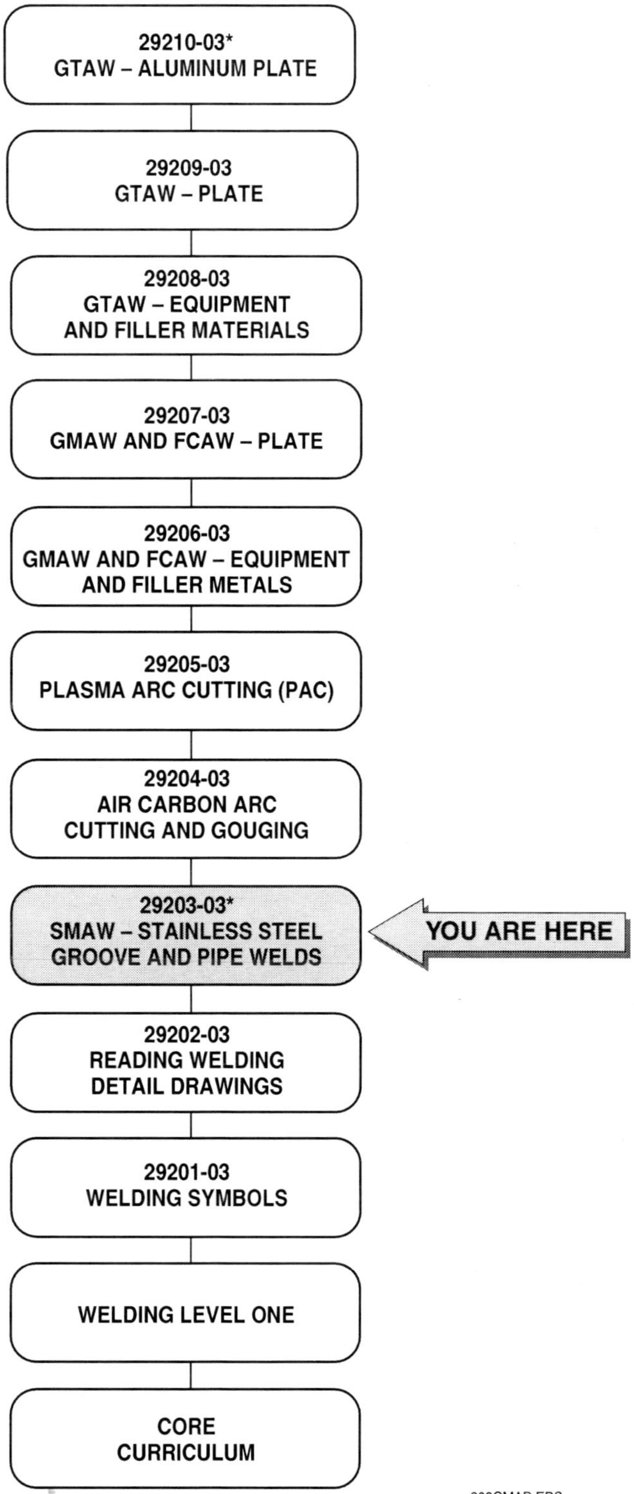

203CMAP.EPS

Please note that Modules 29203-03 and 29210-03 are electives for *Welding Level Two* but are requirements for *AWS Entry Level Welder.

MODULE 29203-03 CONTENTS

Figures

Tables

MODULE 29203-03

SMAW – Stainless Steel Groove and Pipe Welds

Objectives

When you have completed this module, you will be able to do the following:

1. Identify and explain stainless steel metallurgy.
2. Identify and explain the selection of electrodes for welding stainless steel.
3. Identify and explain welding variations for stainless steel.
4. Prepare arc welding equipment for stainless steel welds.
5. Explain stainless steel open-root V-groove welds.
6. Perform shielded metal arc welding (SMAW) on stainless steel open-root V-groove joints in the following positions:
   - Flat (1G) position
   - Horizontal (2G) position
   - Vertical (3G) position
   - Overhead (4G) position
7. Explain stainless steel open-root V-groove pipe welds.
8. Perform shielded metal arc welding (SMAW) on stainless steel open-root V-groove pipe welds in the following positions:
   - Flat (1G-ROTATED) position
   - Horizontal (2G) position
   - Multiple (5G) position
   - Inclined multiple (6G) position

Prerequisites

Before you begin this module, it is recommended that you successfully complete the following: Core Curriculum; Welding Level One; Welding Level Two, Modules 29201-03 and 29202-03.

Required Trainee Materials

1. Pencil and paper
2. Appropriate personal protective equipment

1.0.0 ◆ INTRODUCTION

Stainless steel types vary widely in chemical composition and physical characteristics. Because of this, they require different welding techniques, filler rod compositions, and heat treatments. This module explains stainless steel metallurgy and electrodes. It also explains how to perform shielded metal arc welding (SMAW) of stainless steel using V-groove joints, with and without backing, and open root V-groove pipe. For the purposes of this training module, the dimensions are representative of codes and may not be specific to one exact code. Always refer to the applicable code or site-specific specifications.

2.0.0 ◆ STAINLESS STEEL METALLURGY

Stainless steels are iron (ferrous) based alloys with a chromium content of at least 11%. Other alloying agents, including nickel and carbon, may be present in varying quantities depending upon the specific type of stainless steel. The principal physical characteristic of all stainless steels is their resistance to corrosion and heat, which varies with the category of stainless steel. Some of the stainless steels also have good low- and high-temperature mechanical properties.

2.1.0 Stainless Steel Categories

Stainless steel is categorized based upon its grain structure. The type of grain structure is determined by the specific alloy content of the stainless steel and its heat treatment during manufacture. The alloy and heat treatment also affects its mechanical and magnetic properties. Stainless steel is categorized into three basic grades depending on their crystalline grain structures:

- Austenitic
- Martensitic
- Ferritic

2.1.1 Austenitic Stainless Steels

Austenitic stainless steels are the largest of the stainless steel groups and the group most often encountered by welders in the field. They include the American Iron and Steel Institute (AISI) 200 and 300 series stainless steels, which all contain significant amounts of both chromium and nickel. The chromium content of austenitic stainless steels ranges from 16% to 26%. Nickel content ranges from 3.5% to 23%, depending upon the type. The 200 series substitutes manganese and nitrogen for part of the nickel content.

Austenitic stainless steels are nonmagnetic to very weakly magnetic. They remain stable in the austenite phase over a wide range of temperatures and do not transform to martensite when cooled. They are not hardenable by heat treatment but do harden somewhat when cold worked. They have high resistance to oxidation and offer maximum resistance to corrosion. They are extremely tough and ductile as welded and do not usually require **annealing** unless they are to be used under severely corrosive conditions. Generally, austenitic stainless steels are easily welded.

2.1.2 Martensitic Stainless Steels

Martensitic stainless steels comprise two groups: plain martensitic and chromium-martensitic. In the chromium-martensitic steels, the austenite-to-martensite grain structure is formed by slow cooling. The plain martensitic stainless steels contain only 4% to 6% chromium and no nickel. These are not considered true stainless steels although their corrosion resistance is much greater than the mild carbon steels, even at elevated temperatures. Examples of plain martensitic stainless steels include AISI types 501 and 502. The chromium-martensitic stainless steels contain from 11.5% to 18% chromium, about 1% manganese, and in

Specialty Stainless Steels

Besides the three basic types, other types of specialty stainless steels are available. These include *duplex, precipitation hardening, superaustenitic,* and *superferritic* stainless steels. Duplex stainless steels are supplied with approximately equal amounts of ferrite and austenite and contain about 24% chromium and 5% nickel. They are the most recently developed stainless steels and offer higher yield strength and greater stress corrosion cracking resistance than the 300 series steels. These steels may be welded using SMAW, gas tungsten arc welding (GTAW), gas metal arc welding (GMAW), submerged arc welding (SAW), plasma arc welding (PAW), and flux cored arc welding (FCAW) techniques.

Precipitation hardening stainless steels contain alloying additions, such as aluminum, that allow them to be hardened by a solution and aging heat treatment. Precipitation hardening types of stainless steels are classified in subgroups as martensitic, semiaustenitic, and austenitic precipitation hardening stainless steels. The weldable versions of these steels can be accomplished by GTAW, GMAW, and PAW techniques.

Superferritic stainless steels differ from ordinary ferritic stainless steels by their lower carbon and nitrogen content and higher molybdenum and/or chromium content. Their chemical composition provides much higher corrosion resistance in oxidizing and chloride environments and outstanding resistance to corrosion in sea water. Superferritic stainless steels are welded with only GTAW and GMAW techniques.

Superaustenitic stainless steels have compositions that fall between 300 series stainless steels and nickel-alloy steels. Like superferritic stainless steels, they are highly resistant to corrosion and stress corrosion cracking caused by solutions containing chloride. Many of these alloys contain deliberate amounts of copper and/or nitrogen to increase the strength and corrosion resistance. These stainless steel types are used extensively in process industry applications in place of common austenitic stainless steels that have failed due to pitting, crevice corrosion, or chloride-induced corrosion cracking. These stainless steels can be welded using SMAW, GTAW, GMAW, and SAW techniques.

some cases 0% to 2.5% nickel. Examples of chromium-martensitic stainless steels are AISI types 403, 410, 414, 416, 420, 422, 431, and 440.

All martensitic stainless steels are magnetic and hardenable by heat treatment. They are highly heat resistant and retain many of their properties up to 1,100°F. They are more corrosion-resistant than alloy steels but are not as corrosion-resistant as the other stainless steel types. Martensitic stainless steels are air-hardening and must be cooled slowly after welding to prevent cracking. They may also require heat treatment after welding.

2.1.3 Ferritic Stainless Steels

Ferritic stainless steels are straight chromium stainless steels containing 11.5 to 27% chromium, about 1% manganese, and little or no nickel. Carbon content is 0.20% or less. The ferrite grain structure is stable at all temperatures and does not transform to austenite. Examples of ferritic stainless steels include AISI types 405, 409, 430, 430F, 442, 444, and 446.

The ferritic stainless steels have a low coefficient of expansion and good resistance to corrosion. They are magnetic and cannot be hardened by heat treatment, but can be hardened somewhat by cold working. They are used for pressure vessel linings, automobile mufflers, automobile and building trim, and low-cost equipment that must be corrosion resistant. They have high resistance to nitric acid and high temperature scaling. Welds in ferritic stainless steels are typically the most difficult of all the stainless steel types. Types 430F and 430FSe are generally considered to be unweldable or very difficult to weld.

2.2.0 Carbide Precipitation and Hot Cracking

When austenitic stainless steel containing more than 0.02% carbon is heated to between 800°F and 1,600°F, carbide precipitation occurs in the heat-affected zone (HAZ). The carbon in excess of 0.02% moves to the grain boundaries where it combines with chromium, forming chromium carbide. Since chromium prevents corrosion in stainless steel, removing chromium by forming chromium carbide reduces the corrosion resistance in the stainless steel. The phenomenon is known as *sensitization* and makes austenitic stainless steel subject to intergranular corrosion when exposed to a corrosive agent.

Carbide precipitation increases with carbon content. No other property of the stainless steel is affected other than the corrosion resistance. The corrosion is often visible on the surface in the heat-affected zone after exposure to a corrosive. *Figure 1* shows the effect of carbide precipitation inside a pipe after exposure to a nitric acid solution.

The three methods to control carbide precipitation are:

- Using stabilized base metal and filler metal
- Using low carbon base metal and filler metal (L grades)
- Post-weld heat treatment

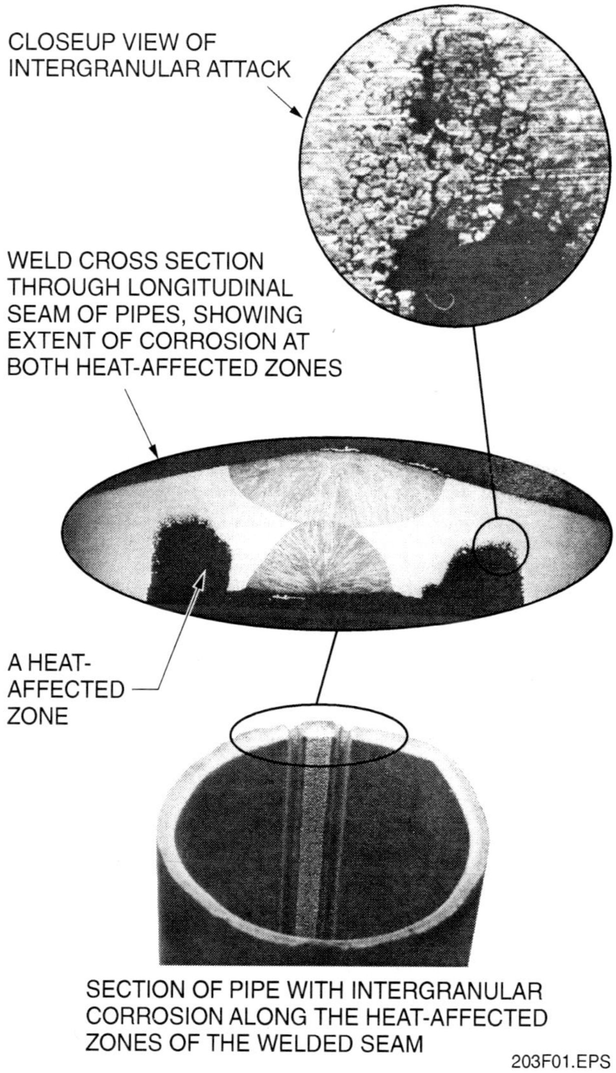

Figure 1 ◆ Effect of carbide precipitation.

2.2.1 Stabilized Base Metal and Filler Metal

Stabilized stainless steels contain columbium or titanium to inhibit carbide precipitation. Columbium and titanium are both used to stabilize base metals but only columbium is used in electrodes because columbium transfers across the arc more readily than titanium does. Both columbium and titanium have a greater attraction for carbon than chromium does. When a stabilized metal is heated to between 800°F and 1,400°F, the columbium or titanium combines with the carbon. This leaves the chromium in solution to inhibit corrosion. Stabilized stainless steels can be used at temperatures up to 1,400°F as welded without heat treatment and without affecting their corrosion resistance. AISI 321 and 347 are examples of stabilized stainless steels.

2.2.2 Low-Carbon Base Metal and Filler Metal

Chromium carbide is formed by the carbon in excess of 0.02%. Therefore, if the carbon content of the stainless steel base metal is kept closer to 0.02%, less carbide precipitation will occur. For the filler metal, keeping the carbon content below 0.03% will decrease carbide precipitation. Low-carbon stainless steels have an L in their designation to signify low carbon. The designation ELC, for extra low carbon, is sometimes also used. ANSI 304L and 316L are examples of low-carbon stainless steels.

2.2.3 Post-Weld Heat Treatment (PWHT)

The chromium carbide formed in the heat-affected zone can be returned to solution by post heating the weldment (for example to 1,950°F) and rapidly cooling it. Although this method can remedy the problem of carbide precipitation, it is typically not used because of the size of the weldment and/or the problems that can occur when the weldment is heated to such a high temperature. Some of the problems that are difficult to control are sagging and loss of dimension control. Other problems, such as excessive scaling, can be controlled with an inert gas atmosphere.

2.2.4 Hot Cracking

Hot cracking of austenitic stainless steel is caused when metallic compounds of sulfur and phosphorus, which can be present in the weld or HAZ, penetrate grain boundaries. When the weld cools, cracks will appear as shrinkage stresses occur. Hot cracking is usually prevented by deliberately introducing a small amount of ferrite in the austenitic matrix from the base and filler material. The ferrite provides ferrite-austenite grain boundaries that control sulfur and phosphorous compounds to prevent hot cracking.

2.3.0 AISI Stainless Steel Table

Table 1 summarizes some of the different AISI-type stainless steels by category.

3.0.0 ◆ STAINLESS STEEL ELECTRODES

SMAW electrodes for welding stainless steel are basically the same as SMAW electrodes for welding carbon steel. Most have a solid wire core of filler metal with an external flux coating. The wire core transfers the welding current from the electrode holder to the work. The arc at the end of the electrode melts the tip of the electrode core, the flux coating at the electrode tip, and the base metal under the arc. The molten electrode metal fuses with the base metal. The electrode metal forms the majority of the weld. As the flux coating melts and vaporizes, it:

- Generates a cloud of carbon dioxide gas that shields the molten weld by pushing away the harmful oxygen and hydrogen present in the atmosphere
- Acts as a fluxing agent to clean and deoxidize the molten metal
- Stabilizes the arc and reduces spatter
- Forms a slag covering over the weld to further protect it from the atmosphere and slow its cooling rate

Figure 2 shows stainless steel SMAW electrode action.

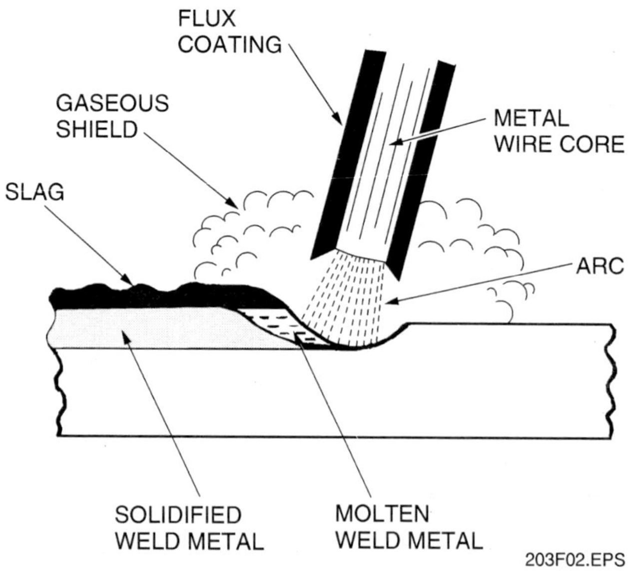

Figure 2 ◆ Stainless steel SMAW electrode action.

Table 1 Certain Basic AISI-Type Stainless Steels by Category

Austenitic			
Type	**Carbon**	**Chromium**	**Nickel**
201	0.15	16/18	3.5/5.5
202	0.15	17/19	4/6
301	0.15	16/18	6/8
302	0.15	17/19	8/10
303	0.15	17/19	8/10
304	0.08	18/20	8/12
304L	0.03	18/20	8/12
308	0.08	19/21	10/12
309	0.20	22/24	12/15
310	0.25	24/26	19/22
316	0.08	16/18	10/14
316L	0.03	16/18	10/14
321	0.08	17/19	9/12
347*	0.08	17/19	9/13
348*	0.08	17/19	9/13

* Stabilized with columbium and/or titanium

Martensitic			
Type	**Carbon**	**Chromium**	**Nickel**
403	0.15	11.5/13	—
410	0.15	11.5/13.5	—
414	0.15	11.5/13.5	1.25/2.5
416	0.15	12/14	—
420	0.15+	12/14	—
422	0.20/0.25	11/13	0.5/1.0
431	0.20	15/17	1.25/2.5
440A	0.60/0.75	16/18	—
440B	0.75/0.95	16/18	—
440C	0.95/1.20	16/18	—

Ferritic			
Type	**Carbon**	**Chromium**	**Nickel**
405	0.08	11.5/14.5	—
409	0.08	10.5/11.75	—
430	0.12	14/18	—
442	0.20	18/23	—
444	0.25	17.5/19.5	1.0
446	0.20	23/27	—

Some flux coatings also have alloying elements in them. The alloying elements add alloys to change the chemical composition and strength of the deposited weld metal. Alloys are normally added to electrodes so the deposited weld metal will match the base metal chemical composition and strength.

Some stainless electrodes such as E430-16 have a carbon steel core. The chromium, nickel, and other alloys required for the stainless steel deposited weld metal are in the flux coating. As the arc melts the carbon steel core and flux coating, the deposited weld metal becomes stainless steel.

Coated tubular electrodes are specialty stainless electrodes. These electrodes have a thin carbon steel tube filled with powdered metal containing chromium, nickel, and any other alloys. The metal tube has an external flux coating. As the arc melts the tube, the steel and powdered metals flow together forming stainless steel. These electrodes are most often used for welding specialty stainless alloys, dissimilar steels, manganese, or other wear-resistant alloys. These electrodes can operate at lower amperage ranges than solid wire electrodes to reduce the possibility of carbon precipitation and to allow welding of thin materials. At higher amperage, their disposition rates are higher than solid wire electrodes. Most of these tubular electrodes can be used in all positions.

3.1.0 AWS Filler Metal Specification System

The American Welding Society (AWS) writes specifications for welding consumables such as the electrodes, filler rods, and fluxes used with all the various welding processes. There is a series of different specifications identified as *AWS A5.XX*. These specifications are used in all major welding codes and accepted in all industries except where other approvals are specified, such as the American Bureau of Shipping (ABS) and the Coast Guard, Federal Highway Administration, Military Specifications (MIL), and Lloyds of London.

Hot Tip

Carbon Steel Cored Stainless Steel Electrodes

Do not use carbon steel cored stainless steel electrodes if the flux coating is chipped (sections missing). Without the alloys in the flux, the deposited weld metal will be mild steel.

The American Society of Mechanical Engineers (ASME) Boiler Code electrode specifications closely follow the AWS specifications.

The purpose of the specifications is to set standards that all manufacturers must follow when manufacturing welding consumables. This ensures consistency for the user, regardless of which company manufactured the product. The specifications set standards for the following:

- Classification system, identification, and marking
- Chemical composition of the deposited weld metal
- Mechanical properties of the deposited weld metal

The specification that applies to covered (SMAW) stainless steel electrodes is *AWS A5.4, Specification for Stainless Steel Electrodes for Shielded Metal Arc Welding*.

NOTE

The specification number is generally followed by a hyphen and a two-digit number, such as *A5.1-91* or *A5.5-96*. The hyphen and number indicate the year that the classification was last revised. These documents are on a five-year revision cycle. Always use the latest version.

3.2.0 Classification System

The classification system used to identify *AWS A5.4* stainless steel electrodes is based on the numbering system developed by the AISI, which uses a three- or four-digit number to identify the type of stainless steel alloy. The prefix E is added to indicate an electrode. The AISI digits are followed by a one- to four-letter alloy modification designation (if required) and by a two-digit usability designation number that is separated by a hyphen or space. This two-digit number is based on the flux coating composition, welding position, and type of welding current with which the electrode is usable. The classification system number is printed on the flux coating of each electrode near the bare end. *Figure 3* shows a typical stainless steel electrode marking, and *Figure 4* shows the classification marking system.

203F03.EPS

Figure 3 ◆ Typical stainless steel electrode marking.

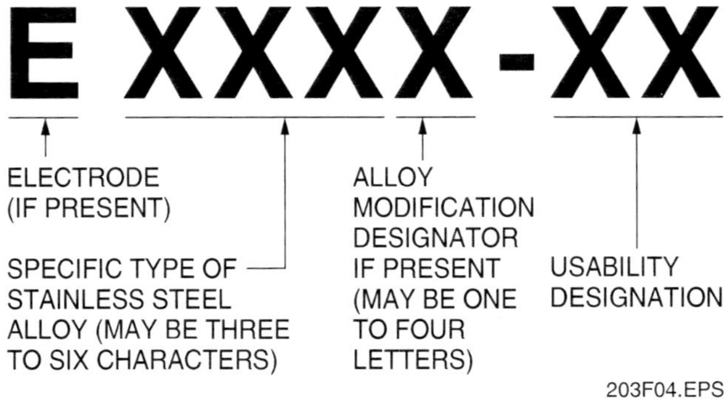

Figure 4 ◆ Stainless steel electrode classification marking system.

Five types of usability designations exist for stainless steel SMAW electrodes: -15, -16, -17, -25, and -26.

3.2.1 Usability Designation -15

The -15 designation indicates that the coating is usually a lime (lime-fluorspar or basic lime) coating. A lime coating gives the purest weld metal, which is also usually the lowest in nitrogen, oxygen, and other inclusions. The lime coating is also most resistant to moisture so it tends to produce less **porosity**. They are used for the most demanding applications. Electrodes up to 5⁄32" in diameter have good all-position weld characteristics with a fast-freeze characteristic. One disadvantage of the lime coating is that it tends to leave a secondary film on the weld when the weld flux is chipped. This secondary film must be removed with a stainless steel bristled wire brush. Suffix -15 coated electrodes are used with direct current electrode positive (DCEP) current and produce a convex bead that is ideal for root passes on heavy plate.

3.2.2 Usability Designation -16

These electrodes usually have a rutile (titania) coating that also contains some limestone. They are also called basic-rutile coatings. This is the most popular coating type and has excellent arc characteristics. The slag can be easily and completely removed without leaving a secondary film. Electrodes up to 5⁄32" in diameter can be used in all positions. However, vertical and overhead welds are somewhat more difficult than those performed with -15 electrodes because of slower freeze characteristics. The electrodes can be used with either AC or DCEP welding current, but preferably DCEP. Beads are smooth and flat, and penetration is complete and free of porosity. Ductility and strength can be better than with lime-coated electrodes.

3.2.3 Usability Designation -17

These electrodes usually use a modified -16 coating that is a titania-silica combination. The electrodes produce more of a spray arc that results in a finer rippled weld bead. They also have a slower freezing slag that has better handling qualities during drag applications. Like -16 electrodes, -17 electrodes can be used with AC or DCEP welding current, but preferably with DCEP current. They are usually used in flat or horizontal positions. They can be used in all positions with greater difficulty. If used for vertical welds, wider weave beads are required in uphill travel than is necessary for -16 electrodes. Electrodes 3⁄16" or larger are not recommended for overhead or vertical welding.

3.2.4 Usability Designation -25

These electrodes are similar in coatings and usability to -15 electrodes. The major difference is that the core wire can be very different. It may be carbon steel with the additional required alloys present in the coating. This makes the diameter of the electrode much larger than a corresponding -15 type, and may permit higher welding current. This type of electrode does not have wide commercial usage.

3.2.5 Usability Designation -26

These electrodes are similar in coatings and usability to -17 and some -16 electrodes. Like -25 electrodes, the core wire can be different, such as plain carbon steel with the required alloys contained in the coating. This makes the electrode larger in diameter than a -16 or -17 type.

Special -16 and -17 Electrodes

Some manufacturers now market specially coated solid or tubular electrodes specifically designed for vertical or overhead welding and for 5G and 6G pipe positions.

Hot Tip: Dissimilar Core Wire

It is not uncommon for stainless steel electrode core wire to be different from the base metal, with the additional required alloys contained in the coating. Some -15 and -16 electrodes, such as those used for 400 series stainless steel, may have carbon steel as core wire. Because of this, the practice of using core wire from a stainless steel electrode for GTAW rod is not recommended unless the manufacturer approves of such use.

3.2.6 Alloy Designation

Some classification numbers have additional letters following the classification digits that indicate a modification to the basic class alloy. These additional letters and their meanings include:

- *L* – Low carbon (0.03% or less). Less than 0.04% carbon is deposited in the weld. Commonly used to weld chemical plant equipment.
- *H* – High carbon (0.04% to 0.08%). Used for creep-resistant welds in 304H stainless steel.
- *HC* – Very high carbon (0.08% to .15%). Used to limit magnetic permeability to 2.0 and where corrosion is not a major factor.
- *Cb* – Columbium (0.8%). Added to electrode alloys to stabilize the chromium and inhibit carbide precipitation.
- *Mo* – Molybdenum (2.2% to 2.5%). Added to electrode alloys used to weld stainless steel containing molybdenum.
- *LR* – Low residuals. Lower carbon, silicon, phosphorus, and sulfur content and very controlled columbium and manganese content.

3.3.0 Manufacturers' Classification

All electrodes manufactured in accordance with AWS specifications must have the AWS classification number printed on them and on the container in which they are shipped. In addition, electrode manufacturers generally print their own unique classification name or number on the electrode and/or on the shipping container. Manufacturers may make more than one electrode for an AWS classification or electrodes that cover more than one AWS classification or usability designation. For example, ESAB® (Elektriska Svetsnings Aktiebolaget) manufactures two different SMAW electrodes that cover AWS classification E308 and E308L -15, -16, and -17. They are called Arcaloy® 308L and 308L Plus. Arcaloy® 308L is an AWS E308/308L -15 and -16 low-carbon electrode used to weld chemical plant equipment of type 304L stainless steel, where carbon in the weld must be under 0.04%. It is also used to weld stainless steel types 321 and 347. Arcaloy® 308L Plus is an AWS E308/208L -17 electrode that contains even less carbon than the Arcaloy® 308L, but contains more silicon. It is used to weld stainless steel types 304, 304L, 308, and 308L when bead appearance is important.

NOTE

Most electrode manufacturers publish paperback welding guides that include information on their electrodes, recommended amperage settings, and uses. These booklets can be obtained through your local distributor free or at a nominal charge.

When there is more than one electrode in an AWS classification and you are unsure which electrode to use, check with your supervisor.

Hot Tip: Electrode Sizes Versus Position

Stainless steel electrodes up to $\frac{5}{32}$" can usually be run in any position. Stainless electrodes $\frac{3}{16}$" and above should only be used in the flat and horizontal fillet weld positions. Always check the manufacturer's literature for positional usage information.

3.4.0 Storing Stainless Steel Electrodes

The flux coating on stainless electrodes is low hydrogen (low in moisture, which produces hydrogen during welding). For this reason, stainless steel electrodes must be kept dry. The dry mineral flux coatings on stainless electrodes start to absorb moisture as soon as they are exposed to the atmosphere. Moisture in the flux will cause porosity and other defects in stainless welds.

To keep the electrodes dry, stainless steel electrodes are shipped in hermetically sealed containers. Immediately after opening the sealed containers, the electrodes must be stored in special ovens at temperatures of 200°F to 250°F (93°C to 120°C), or as recommended by the manufacturer (*Figure 5*). When stainless steel electrodes are removed from the oven, they can only be exposed to the atmosphere for about four hours. Electrodes exposed for more than four hours may require redrying at higher temperatures. The exact exposure times and procedures for handling and drying filler metal at your site are typically written out in a quality standard. These procedures must be followed. If you are unsure what the requirements are at your site, check with your supervisor.

NOTE

Site quality standards are very strict. Be sure you understand them before handling electrodes. The information provided in this trainee module is of a general nature. Your site's quality standards must be followed.

203F05.EPS

Figure 5 ◆ Electric drying oven.

CAUTION

Never handle hot electrodes from an oven with bare hands or dirty gloves. Use clean gloves only. When removed from an oven for use, electrodes should be stored in a pouch or rod holder to avoid contamination.

3.5.0 Electrode Selection Considerations

Because there are so many types of electrodes available, a welder must be able to distinguish between them and select the correct electrode for the job. There are two basic considerations when selecting stainless steel electrodes: welding procedure specification (WPS) and base metal composition.

3.5.1 WPS Requirement

If there is a welding procedure specification (WPS) or site quality standard, the electrode type will be specified. It will be given as an AWS classification and/or manufacturer's standard and will sometimes include a manufacturer's name. If there is a WPS, the filler metal specified must be used.

CAUTION

Filler metal specifications in the WPS or site quality standards must be followed to maintain weld quality.

3.5.2 Base Metal Composition

As a general rule when welding stainless steel, stainless steel electrodes are selected to match the chemical composition of the base metal. Because there are so many stainless base metals and electrode types, it is important to refer to a table or chart. The manufacturers of stainless steel electrodes provide tables or charts showing which type of electrode (filler metal) works best with which types of stainless steel. *Table 2* is an example of a typical chart used to identify the correct filler metal for the base metal to be welded. Unless specified, electrodes with type -15 (EXXX-15), -16, -17, -25, and -26 usability designations can be used. *Table 3* lists the intended uses of various classes of stainless steel electrodes as identified in *ANSI/AWS 5.4*.

CAUTION

The recommendations shown in *Tables 2* and *3* may be inappropriate for some applications. Always refer to the WPS for specific electrode requirements.

Table 2 Typical Manufacturer's Chart for Welding Stainless Steels

Austenitic				
Base Metal	**Carbon**	**Chromium**	**Nickel**	**Filler Metal**
201	0.15	16/18	3.5/5.5	308
202	0.15	17/19	4/6	308
301	0.15	16/18	6/8	308
302	0.15	17/19	8/10	308
303	0.15	17/19	8/10	308[1]
304	0.08	18/20	8/12	308
304L	0.03	18/20	8/12	308L
308	0.08	19/21	10/12	308
309	0.20	22/24	12/15	309
310	0.25	24/26	19/22	310
316	0.08	16/18	10/14	316
316L	0.03	16/18	10/14	316L
321	0.08	17/19	9/12	347[2]
347[3]	0.08	17/19	9/13	347[2]
348[3]	0.08	17/19	9/13	347[2]

[1] EXXX-15 Flux coating only
[2] Stabilized with columbium
[3] Stabilized with columbium and/or titanium

Martensitic				
Base Metal	**Carbon**	**Chromium**	**Nickel**	**Filler Metal**
403	0.15	11.5/13	–	410
410	0.15	11.5/13.5	–	410
414	0.15	11.5/13.5	1.25/2.5	410
416	0.15	12/14	–	410
420	0.15+	12/14	–	410
431	0.20	15/17	1.25/2.5	430

Ferritic				
Base Metal	**Carbon**	**Chromium**	**Nickel**	**Filler Metal**
405	0.08	11.5/14.5	–	410
430	0.12	14/18	–	430
442	0.20	18/23	–	309/310
446	0.20	23/27	–	309/310

Table 3 Stainless Steel SMAW Electrode Classes and Uses (1 of 5)

AWS Classification	Intended Uses
E209-XX	These electrodes are most often used to weld AISI Type 209 (UNS S20910) base metals. The electrode is a nitrogen-strengthened austenitic stainless steel alloy with high strength and good toughness over a wide range of temperatures. Nitrogen alloying reduces the tendency for intergranular carbide precipitation in the weld area by inhibiting carbon diffusion. Nitrogen alloying coupled with the molybdenum content provides superior resistance to pitting and crevice corrosion in chloride solutions. These electrodes are also used to join dissimilar metals such as carbon steel to stainless steel and for direct overlay on carbon steel for corrosion applications.
E219-XX	These electrodes are most often used to weld AISI Type 219 (UNS S21900) base metals and have similar composition, characteristics, and uses as E209 electrodes.
E240-XX	These electrodes are most often used to weld AISI Type 240 and 241 base metals. They have similar composition, characteristics, and uses as E209 electrodes. Additionally, they exhibit an improvement in resistance to wear in particle-to-metal and metal-to-metal (galling) applications compared to the more conventional austenitic stainless steels like Type 304. Weldments in alloys AISI 240 and AISI 241, when compared to Type 304, exhibit improved resistance to transgranular stress corrosion cracking in hot chloride solutions.
E307-XX	These electrodes are used primarily for moderate strength welds with good crack resistance between dissimilar steels such as austenitic manganese steel and carbon steel forgings or castings.
E308-XX	These electrodes are most often used to weld base metal of similar composition such as AISI Types 301, 302, 304, and 305.
E308H-XX	These electrodes are the same as E308 except that the allowable carbon content has been increased to the highest portion of the range for E308. The high carbon content provides higher tensile and creep strengths at elevated temperatures. These electrodes are used for welding Type 304H base metal.
E308L-XX	These electrodes are the same as E308, except for restricted carbon content. The low carbon content of welds made by these electrodes reduces the possibility of intergranular carbide precipitation, which increases the resistance to intergranular corrosion without the use of stabilizers such as columbium (niobium) or titanium. The low carbon welds, however, are not as strong at elevated temperature as the columbium-stabilized alloys or 304H.
E308Mo-XX	These electrodes are the same as E308 except for the addition of molybdenum. The electrodes are recommended for welding ASTM CF8M stainless steel castings because they match the base metal with regard to chromium, nickel, and molybdenum. They can be used for welding wrought materials such as Type 316 stainless when increased ferrite is desired that cannot be attained with E316 electrodes.
E308MoL-XX	These electrodes are recommended for welding ASTM CF3M stainless steel castings, as they match the base metal with regard to chromium, nickel, and molybdenum. The electrodes may also be used for welding wrought materials such as Type 316L stainless when increased ferrite is desired that cannot be attained with E316L electrodes.
E309-XX	These electrodes are commonly used for welding similar alloys in wrought or cast form. They are used for welding dissimilar metals, such as joining Type 304 to carbon steel, welding the clad side of Type 304 clad steels, and applying stainless steel sheet linings to carbon steel shells. Occasionally, they are used to weld Type 304 and similar base metals where severe corrosion conditions exist requiring higher alloy weld metal.
E309L-XX	The weld metal is the same as E309 electrodes except for a restricted carbon content. The low carbon content of these weld deposits reduces the possibility of intergranular carbide precipitation, which increases the resistance to intergranular corrosion without the use of stabilizers such as columbium (niobium) and titanium. The low carbon welds are not as strong at elevated temperature as the columbium-stabilized alloys or higher carbon content Type 309 welds.

Table 3 Stainless Steel SMAW Electrode Classes and Uses (2 of 5)

AWS Classification	Intended Uses
E309Cb-XX	The weld metal is the same as Type 309, except for the addition of columbium (niobium) and a reduction in the carbon limit. The columbium (niobium) provides resistance to carbide precipitation, which increases intergranular corrosion resistance and provides higher strength in elevated temperature service. These electrodes are also used for welding Type 347 clad steels or for the overlay of carbon steel.
E309Mo-XX	The weld metal is the same as E309 electrodes except for the addition of molybdenum and a small reduction in the carbon content. These electrodes are used for welding Type 316 clad steels or for the overlay of carbon steels.
E309MoL-XX	The weld metal is the same as E309Mo electrodes except for the restricted carbon content. The lower carbon content of the weld metal reduces the possibility of intergranular corrosion.
E310-XX	These electrodes are most often used to weld base metals of similar composition.
E310H-XX	The weld metal is the same as E310 electrodes, except that the allowable carbon content has been greatly increased. These electrodes are used primarily for welding or repairing high alloy heat and corrosion resistant castings designated as Type HK by the Alloy Castings Institute. The welds have high strength at temperatures over 1,700°F. The welds are not recommended for high sulfur atmospheres or where severe thermal shock is present. Long time exposure to temperatures in the approximate range of 1,400°F to 1,600°F may induce formation of sigma and secondary carbides that can result in reduced corrosion resistance, reduced ductility, or both.
E310Cb-XX	The weld metal is the same as E310 electrodes except for the addition of columbium (niobium) and a reduction in carbon content. These electrodes are used for the welding of heat resisting castings, Type 347 clad steels, or the overlay of carbon steels.
E310Mo-XX	The weld metal is the same as E310 electrodes except for the addition of molybdenum and a reduction in carbon content. These electrodes are used for the welding of heat resisting castings, Type 316 clad steels, or for the overlay of carbon steels.
E312-XX	These electrodes were originally designed to weld cast alloys of similar composition. They have been found to be valuable in welding dissimilar metals, especially if one of them is a stainless steel high in nickel. This electrode gives a two-phase weld deposit with substantial amounts of ferrite in an austenitic matrix. Even with considerable dilution by austenite-forming elements, such as nickel, the microstructure remains two-phase and highly resistant to weld metal cracks and fissures. Applications should be limited to service temperature below 800°F (420°C) to avoid formation of secondary brittle phases.
E316-XX	These electrodes are used for welding Type 316 and similar alloys. They have been used successfully in certain applications involving special base metals for high-temperature service. The presence of molybdenum in the weld metal provides creep resistance at elevated temperatures. Rapid corrosion of Type 316 weld metal may occur when the following three factors co-exist: (1) The presence of a continuous or semi-continuous network of ferrite in the weld metal microstructure (2) A composition balance of the weld metal giving a chromium-to-molybdenum ratio of less than 8.2 to 1 (3) Immersion of the weld metal in a corrosive medium Attempts to classify the media in which accelerated corrosion will take place by attack on the ferrite phase have not been entirely successful. Strong oxidizing and mildly reducing environments have been present where a number of corrosion failures were investigated and documented. The literature should be consulted for the latest recommendations.

Table 3 Stainless Steel SMAW Electrode Classes and Uses (3 of 5)

AWS Classification	Intended Uses
E316H-XX	These electrodes are the same as E316 except that the allowable carbon content has been increased to the higher portion of the E316 range. Higher carbon content provides higher tensile and creep strengths at elevated temperatures. These electrodes are used for welding 316H base metal.
E316L-XX	These electrodes are the same as E316 except for restricted carbon content. The lower carbon content of the weld metal reduces the possibility of intergranular carbide precipitation, which increases the resistance to intergranular corrosion without the use of stabilizers such as columbium (niobium) or titanium. These electrodes are used principally for welding low carbon, molybdenum-bearing austenitic alloys. Tests have shown that the low carbon content in the weld metal gives adequate protection against intergranular corrosion in most cases. This low carbon electrode is not as strong at elevated temperatures as Type E316H.
E317-XX	The alloy content of weld metal deposited by these electrodes is somewhat higher than that of E316 electrodes, particularly in molybdenum. These electrodes are usually used for welding alloys of similar composition. They are also used in severely corrosive environments (such as those containing halogens) where crevice and pitting corrosion are of concern.
E317L-XX	The weld metal is the same as for E317 electrodes except for restricted carbon content. The lower carbon content of weld metal reduces the possibility of intergranular carbide precipitation, which increases the resistance to intergranular corrosion without the use of stabilizers such as columbium (niobium) or titanium. The low carbon welds are not as strong at elevated temperatures as the columbium (niobium)-stabilized alloys or the standard Type 317 weld metal with higher carbon content.
E318-XX	The weld metal is the same as for E316 electrodes except for the addition of columbium (niobium). Columbium (niobium) provides resistance to intergranular carbide precipitation and increases resistance to intergranular corrosion. These electrodes are used primarily for welding base metals of similar composition.
E320-XX	These electrodes have columbium (niobium) added to improve resistance to intergranular corrosion. The electrodes are primarily used to weld base metals of similar composition for applications where resistance to severe corrosion is required for a wide range of chemicals including sulfuric and sulfurous acids and their salts. These electrodes can be used to weld both castings and wrought alloys of similar composition without postweld beat treatment. A modification of this grade without columbium (niobium) is available for repairing castings that do not contain columbium. With this modified composition, solution annealing is required after welding.
E320LR-XX	The LR (low residuals) welds of E320LR electrodes are the same as E320 welds except that elements C, Si, P, and S are specified at lower maximum levels, and Cb (Nb) and Mn are controlled within narrower ranges. These changes reduce the weld metal fissuring (while maintaining the corrosion resistance) frequently encountered in fully austenitic stainless steel weld metals. Consequently, welding practices typically used to deposit ferrite-containing austenitic stainless steel weld metals can be used. Type E320LR weld metal has a lower minimum tensile strength than E320 weld metal.
E330-XX	These electrodes are commonly used where heat- and scale-resisting properties above 1,800°F are required. However, high-sulfur environments may adversely affect performance at elevated temperature. Repairs of defects in alloy castings and the welding of castings and wrought alloys of similar composition are the most common applications.
E330H-XX	The weld metal is the same as E330 electrodes except for higher carbon content. These electrodes are used primarily for the welding and repairing of high alloy heat and corrosion resistant castings that are designated HT by the Alloy Castings Institute. The weld metal can be used in temperatures up to 2,100°F in oxidizing atmospheres and at 2,000°F in reducing atmospheres. However, high-sulfur environments may adversely affect performance at elevated temperatures.

Table 3 Stainless Steel SMAW Electrode Classes and Uses (4 of 5)

AWS Classification	Intended Uses
E347-XX	These electrodes have columbium (niobium) or columbium (niobium) and tantalum added as a stabilizer. Either of these additions reduces the possibility of intergranular chromium carbide precipitation, which increases resistance to intergranular corrosion. These electrodes are usually used for welding chromium-nickel alloys that are stabilized with either columbium (niobium) or titanium. Electrodes depositing titanium as a stabilizing element are not commercially available because titanium is not readily transferred across the arc in shielded metal arc welding. Although columbium (niobium) is the stabilizing element usually specified in Type 347 alloys, it should be recognized that tantalum also is present. Tantalum and columbium (niobium) are almost equally effective in stabilizing carbon and in providing high-temperature strength. This specification recognizes the usual commercial practice of reporting columbium (niobium) as the sum of columbium (niobium) plus tantalum. If dilution by the base metal produces a low ferrite or fully austenitic weld metal deposit, crack sensitivity of the weld may increase substantially. Some applications, especially those involving high-temperature service, are adversely affected if the ferrite content is too high. Consequently, high ferrite content should not be specified for custom electrodes unless tests prove it necessary.
E349-XX	These electrodes are used for welding steels of similar composition such as AISI Type 651 or 652. The combination of columbium (niobium), molybdenum, and tungsten with chromium and nickel in the weld metal gives good high-temperature rupture strength. The weld metal also has an appreciable content of ferrite that increases the crack resistance of the weld metal.
E383-XX	These electrodes are used to weld base metal of a similar composition to itself and to other grades of stainless steel. Type E383 weld metal is recommended for sulfuric and phosphoric acid environments. The elements C, Si, P, and S are specified at low levels to minimize weld metal hot cracking and fissuring (while maintaining the corrosion resistance) frequently encountered in fully austenitic stainless steel weld metals.
E385-XX	These electrodes are used primarily for welding Type 904L materials for the handling of sulfuric acid and many chloride solutions. E385 electrodes are used to join Type 317L material where improved corrosion resistance in specific media is needed. E385 electrodes are also used for joining Type 904L base metal to other grades of stainless. The elements C, Si, P, and S are specified at low levels to minimize weld metal hot cracking and fissuring (while maintaining corrosion resistance) frequently encountered in fully austenitic weld metals.
E410-XX	This 12 Cr alloy electrode weld metal is an air-hardening steel. Preheat and postheat treatments are required to achieve welds of adequate ductility for many purposes. The most common application of these electrodes is for welding alloys of similar compositions. They are also used for surfacing of carbon steels to resist corrosion, erosion, or abrasion.
E410NiMo-XX	These electrodes are used for welding ASTM CA6NM castings or similar materials, as well as light gage Type 410, 410S, and 405 base metals. The weld metal for these electrodes is modified to contain less chromium and more nickel than the weld metal of E410 electrodes. This is to eliminate ferrite in the microstructure because ferrite has a bad effect on the mechanical properties of the alloy. Final postweld heat treatment should not exceed 1,150°. Higher temperatures may result in re-hardening due to untempered martensite in the microstructure after cooling to room temperature.
E430-XX	The weld metal of these electrodes contains between 15 and 18 Cr (wt.%). The composition is balanced by providing sufficient chromium to give adequate corrosion resistance for the usual applications and yet retain sufficient ductility in the heat-treated condition to meet the mechanical requirements of the specification. (Excessive chromium will result in lowered ductility.) Welding with an E430 electrode usually requires preheat and postheat. Optimum mechanical properties and corrosion resistance are obtained only when the weldment is heat-treated following the welding operation.
E630-XX	These electrodes are primarily designed for welding ASTM A564, Type 630, and some other precipitation-hardening stainless steels. The weld metal is modified to prevent the formation of ferrite networks in the martensite microstructure that could have a bad effect on mechanical properties. Dependent on the application and weld size, the weld may be used in an as-welded, precipitation-hardened, or solution-treated and precipitation-hardened condition.

Table 3 Stainless Steel SMAW Electrode Classes and Uses (5 of 5)

AWS Classification	Intended Uses
E16-8-2-XX	These electrodes are used primarily for welding stainless steel, such as Types 16-8-2, 316, and 347, for high-pressure, high-temperature piping systems. The weld metal usually has a Ferrite Number no higher than 5 FN. The weld metal also has good hot ductility properties, which offer relative freedom from weld or crater cracking even under high-restraint conditions. The weld is usable in either the as-welded or solution-treated condition. These electrodes depend on a very carefully balanced chemical composition to develop their fullest properties. Corrosion tests indicate that Type 16-8-2 weld metal may have less corrosion resistance than Type 316 base metal depending on the corrosive media. Where the weldment is exposed to severe corrodents, surface layers should be deposited with a more corrosion-resistant weld metal.
E2209-XX	These electrodes are used primarily to weld duplex stainless steels that contain approximately 22% chromium. The weld metal of these electrodes has a duplex microstructure consisting of an austenite-ferrite matrix. E2209 electrodes combine increased tensile strength with improved resistance to pitting, corrosive attack, and stress corrosion cracking.
E2553-XX	These electrodes are used primarily to weld duplex stainless steels that contain approximately 25% chromium. The weld metal of these electrodes has a duplex microstructure consisting of an austenite-ferrite matrix. E2553 electrodes combine increased tensile strength with improved resistance to pitting, corrosive attack, and stress corrosion cracking.

4.0.0 ◆ BASIC WELDING VARIATIONS FOR STAINLESS STEEL

Welding stainless steel is slightly more difficult than welding carbon steel. When compared to carbon steel, stainless steel has a lower melting point, conducts heat more slowly, expands more when heated, and has a higher resistance to electrical current flow. These characteristics are similar for all stainless steels but vary greatly with the metallurgical structure. Those in the same metallurgical group (austenitic, martensitic, or ferritic) have similar welding characteristics.

> **NOTE**
> Information in this trainee module is provided as a general guideline only. Refer to your WPS or site quality standards for specific requirements for a weld.

4.1.0 Welding Austenitic Stainless Steels

Austenitic stainless steel is generally easier to weld than the other types of stainless steels. It normally does not require preheating or postheating, but the **interpass temperature** must be controlled to prevent carbide precipitation.

4.2.0 Welding Martensitic Stainless Steels

Martensitic stainless steels air harden in proportion to the chromium and carbon content, which makes the weld and weld zone very hard and brittle unless it is heat treated.

Martensitic heating requirements are based on carbon content. Approximate heating requirements are as follows:

- *0% to 0.1% carbon* – No preheat or postheat is required.
- *0.1% to 0.2% carbon* – Preheat to 500°F and cool slowly.
- *0.2% to 0.5% carbon* – Preheat to 500°F and then anneal after welding (anneal to 1,500° F to 1,600°F, furnace cool to 1,100°F, and then air cool).

If preheating and postheating cannot be used, weld martensitic stainless steels with austenitic type (stainless or nickel-alloy) filler metal such as E308 or E309 (if allowed by the procedure). The heat-affected zone will still be hard and brittle, but the filler metal will be ductile and able to absorb some stresses to protect the hard, heat-affected zones.

4.3.0 Welding Ferritic Stainless Steels

Ferritic stainless steel has the poorest weldability of the stainless types because of loss of toughness due to severe grain growth. In addition, welding certain ferritic stainless steels (430, 430F, 430FSe, 434, 442, and 446, for example) causes embrittlement and cracking because of martensite formation. When heat treatment is possible, preheating ferritic stainless steel followed by annealing it is recommended.

The welding of ferritic stainless steel is usually accomplished with austenitic E308, E309, E310, or E312 electrodes. The filler metal is ductile and will resist cracking better than a less ductile filler metal. The only ferritic stainless steel electrode for SMAW that is specifically listed in AWS is E430.

4.4.0 Welding Carbon Steels to Stainless Steels

When welding carbon steels to stainless steels, a high-alloy electrode, such as E309, E312, or nickel-alloy electrodes, should be used to reduce the effect that mixing carbon steel with the stainless steel will have on the deposited weld metal. E309 or E312 are also sometimes used for welding medium- and high-carbon steels when preheating and postheating cannot be performed. E309 or E312 provide a strong, ductile weld deposit. For welding austenitic manganese steel or carbon steel to manganese steel, E308 electrodes are recommended.

5.0.0 ◆ WELDING EQUIPMENT SETUP

Before welding can take place, safety procedures and practices must be reviewed, the area has to be made ready, the welding equipment must be set up, and the metal to be welded must be prepared. The following sections explain how to set up arc welding equipment for welding.

5.1.0 Safety Practices

The following is a summary of safety procedures and practices that must be observed while cutting or welding. Keep in mind that this is just a summary. Complete safety coverage is provided in the Level One module, *Welding Safety*. If you have not completed that module, do so before continuing. Above all, be sure to wear appropriate protective clothing and equipment when welding or cutting.

5.1.1 Protective Clothing and Equipment

- Always use safety goggles with a full face shield or a helmet. The goggles, face shield, or helmet lens must have the proper light-reducing tint for the type of welding or cutting to be performed. Never directly or indirectly view an electric arc without using a properly tinted lens (*Figure 6*).
- Wear proper protective leather and/or flame-retardant clothing along with welding gloves that will protect you from flying sparks and molten metal as well as heat.
- Wear 8" or taller high-top safety shoes or boots. Make sure that the tongue and lace area of the footwear will be covered by a pant leg. If the tongue and lace area is exposed or the footwear must be protected from burn marks, wear leather spats under the pants or chaps and over the front top of the footwear.
- Wear a solid material (non-mesh) hat with a bill pointing to the rear or, if much overhead cutting or welding is required, a full leather hood with a welding face plate and the correct tinted lens. If a hard hat is required, use a hard hat that allows the attachment of rear deflector material and a face shield.

Hot Tip

Stainless Steel Welding Techniques

Welding stainless steel with SMAW is similar to welding carbon steel with low-hydrogen electrodes. The arc should be short and the coating should not touch the weld puddle. Note that certain electrodes from some manufacturers are designed to be dragged on the base metal in downhill and horizontal travel directions. Make sure to fill each crater before breaking the arc to avoid crater cracks. Use as low a current as possible to achieve sufficient penetration while reducing heat input to the work to reduce distortion and carbide precipitation. Also use stringer beads at higher speeds rather than wide beads at slower speeds. If weave beads must be used, limit the weave to 2½ times the electrode diameter. Use chill bars near the weld, both at the top and under the weld. Rapid cooling of austenitic stainless steel is beneficial and can help prevent distortion. Copper chill bars should be nickel plated to prevent contamination of the weld. Using skip welding and back-step welding will also help to minimize distortion. Make sure that interpass temperatures do not exceed 350°F to reduce sensitization of the stainless steel.

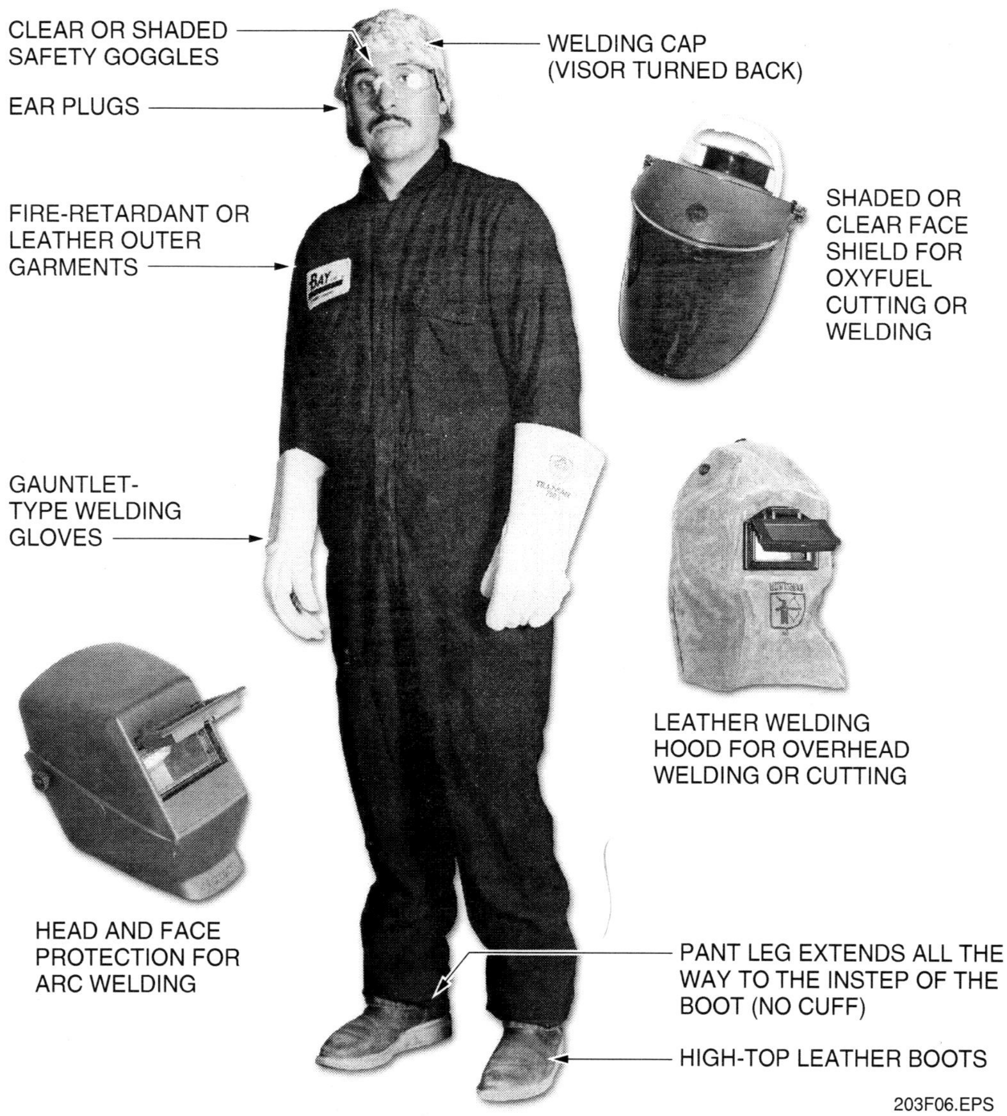

Figure 6 ◆ Personal protective equipment.

- If a full leather hood is not worn, wear a face shield and snug-fitting welding goggles over safety glasses for gas welding or cutting. Either the face shield or the lenses of the welding goggles must be an approved shade 5 or 6 filter. For electric arc welding, wear safety goggles and a welding hood with the correct tinted lens (shade 9 to 14).
- If a full leather hood is not worn, wear earmuffs or at least earplugs to protect your ear canals from sparks.

5.1.2 Fire/Explosion Prevention

- Never carry matches or gas-filled lighters in your pockets. Sparks can cause the matches to ignite or the lighter to explode, causing serious injury.
- Never perform any type of heating, cutting, or welding until a hot-work permit is obtained and an approved fire watch is established. Most work-site fires caused by these types of operations are started by cutting torches.
- Never use oxygen to blow off clothing. The oxygen can remain trapped in the fabric for a time. If a spark hits the clothing during this time, the clothing can burn rapidly and violently out of control.
- Make sure that any flammable material in the work area is moved or shielded by a fire-resistant covering. Approved fire extinguishers must be available before attempting any heating, welding, or cutting operations.
- Never release a large amount of oxygen or use oxygen as compressed air. Its presence around flammable materials or sparks can cause rapid and uncontrolled combustion. Keep oxygen away from oil, grease, and other petroleum products.

- Never release a large amount of fuel gas, especially acetylene. Methane and propane tend to concentrate in and along low areas and can ignite at a considerable distance from the release point. Acetylene is lighter than air but is even more dangerous. When mixed with air or oxygen, it will explode at much lower concentrations than any other fuel.
- To prevent fires, maintain a neat and clean work area and make sure that any metal scrap or slag is cold before disposal.
- Before cutting or welding containers such as tanks or barrels, check to see if they have contained any explosive, hazardous, or flammable materials including petroleum products, citrus products, or chemicals that decompose into toxic fumes when heated. As a standard practice, always clean and then fill any tanks or barrels with water or purge them with a flow of inert gas to displace any oxygen.

5.1.3 Work Area Ventilation

- Make sure confined space procedures are followed before conducting any welding or cutting in the confined space.
- Never use oxygen in confined spaces for ventilation purposes.
- Always perform cutting or welding operations in a well-ventilated area. Cutting or welding operations involving zinc or cadmium materials or coatings result in toxic fumes. For long-term cutting or welding of such materials, always wear an approved full-face, supplied-air respirator (SAR) that uses breathing air supplied externally of the work area. For occasional, very short-term exposure, a high efficiency particulate arresting (HEPA)-rated or metal-fume filter may be used on a standard respirator.
- Make sure confined spaces are ventilated properly for cutting or welding purposes.

5.2.0 Preparing the Welding Area

To practice welding, a welding table, bench, or stand is needed. The welding surface must be steel, and provisions must be made for placing welding coupons out of position. *Figure 7* shows a typical welding station.

To set up the area for welding, follow these steps:

Step 1 Ensure that the area is properly ventilated. Use doors, windows, and fans.

Step 2 Check the area for fire hazards. Remove any flammable materials before proceeding.

Step 3 Verify the location of the nearest fire extinguisher. Do not proceed unless the extinguisher is charged and you know how to use it.

Step 4 Position a welding table near the welding machine.

Step 5 Set up flash shields around the welding area.

> **NOTE**
> The following sections explain how to prepare welding coupons with backing (V-groove with backing) and without backing (open-root V-groove). All welding objectives in this module will be for open-root V-groove welds. Depending upon site conditions, V-groove with backing welds can be substituted with a wider root opening (¼"). The techniques for both types of welds are identical except for the open-root V-groove root pass.

5.3.0 Preparing the Practice Weld Coupons

If possible, the weld coupons should be ⅜" thick stainless steel. If this size is not readily available, ¼" to ¾" thick steel can be used for practice welds. Open root welds on stainless steel plate generally have a bevel angle of 30°.

203F07.EPS

Figure 7 ◆ Welding station.

NOTE
Under some codes for performance qualification, the base metal is not an essential variable. In these cases, carbon steel can be used in the place of more expensive stainless steel. For practice purposes, use carbon steel if stainless steel is not available.

Clean a carbon plate prior to welding by using a wire brush or grinder to remove all heavy mill scale or corrosion. If stainless steel is being used, clean with a stainless steel brush to prevent contamination of the base metal with carbon steel.

As shown in *Figure 8,* cut the metal into 3" × 5" rectangles with one (or both) of the 5" lengths beveled at 30°. Grind a 0" to ⅛" root face on the bevel. The 5" length is used to conserve material. If backing strips will be used, cut the backing strips 1" × 6" (or 8") long. The backing strips should be cut from the same type of base metal as the coupons. The backing strips can be thinner material than the coupons.

NOTE
The welding codes allow the root face and opening on open root welds to be 0" to ⅛". Adjust the root face and root opening as needed when you start your welding practices.

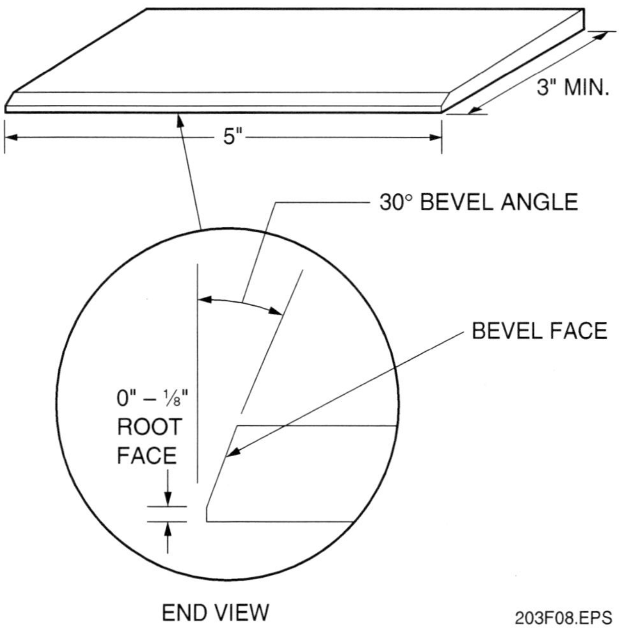

Figure 8 ◆ Metal cut for open-root V-groove weld coupons.

Follow these steps to prepare each V-groove weld coupon:

Step 1 Check the bevel face. There should be no dross and a 0" to ⅛" root face, and the bevel angle should be 30° or 37½°. An example of this is shown in *Figure 9.*

Step 2 Center the beveled strips with a 0" to ⅛" root opening and tack weld them in place. Place the tack welds on the back side of the joint. Use three to four ¼" tack welds.

Steel, especially stainless steel, for practice welding is expensive and difficult to obtain. Every effort should be made to conserve the material that is available. Reuse weld coupons until all surfaces have been welded on by cutting the weld coupon apart and reusing the pieces. Use material that cannot be cut into weld coupons to practice running beads.

5.4.0 Electrodes

Obtain a small quantity of the electrodes to be used. For the welding exercises in this module, 3/32" and ⅛" electrodes compatible with the base metal being

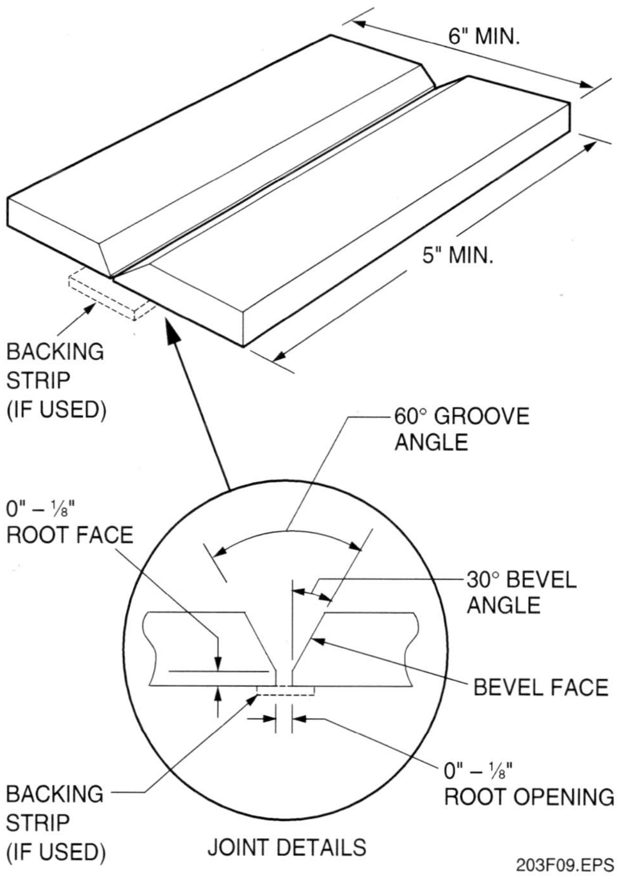

Figure 9 ◆ Open-root V-groove weld coupon.

welded must be used. Keep these general rules in mind when running stainless steel electrodes:

- Do not whip stainless steel electrodes; keep the arc in the weld puddle at all times.
- Maintain a short arc because stainless steel is very susceptible to atmospheric contamination. A longer arc will allow the atmosphere to contaminate the weld, causing porosity.
- Fill craters before breaking the arc to prevent crater cracks.
- Clean the slag thoroughly from the finish of a bead before starting another electrode.
- Remove all slag completely between passes.
- Remove only a small number of electrodes from the oven at a time because stainless steel electrodes pick up moisture quickly.
- Restart by striking the arc in the leading edge of the crater, move back into the crater to tie in, and then proceed as usual. This technique welds over the arc strike, making a smoother restart.
- Stringer beads should be no more than three times the electrode diameter.
- When running stringer beads, use a slight side-to-side motion with a slight pause at the toe to tie-in and flatten the bead.
- When running weave beads, use a slight pause at the toe to tie-in and flatten the bead.
- Do not use stainless steel electrodes with chipped or missing flux.
- Set the amperage in the lower portion of the suggested range for better puddle control when welding vertically.

Stainless steel electrodes are very expensive, so be especially careful with them. Obtain only the electrodes to be used for a particular welding exercise at one time. Have a pouch or rod holder to store the electrodes in to prevent them from becoming damaged. Never store electrodes loose on a table. They may end up on the floor, where they are a tripping hazard and can become damaged. Discard the used electrode stubs into a suitable container.

WARNING!

Do not throw electrode stubs on the floor. They roll easily, and someone could step on one, slip, and fall.

5.5.0 Preparing the Welding Machine

Identify a welding machine to use. *Figure 10* shows typical welding machines. Proceed with the following welding set-up steps:

Step 1 Verify that the welding machine can be used for DC welding.

Step 2 Check to be sure that the welding machine is properly grounded through the current receptacle.

Step 3 Check the area for proper ventilation.

Step 4 Verify the location of the current disconnect.

Step 5 Set the polarity to DCEP.

Step 6 Connect the clamp of the workpiece lead to the workpiece.

Step 7 Set the amperage for the electrode type and size to be used, as shown in *Table 4*.

NOTE

Amperage recommendations vary by manufacturer, position, current type, and electrode brand. For specific recommendations, refer to the manufacturer's recommendations for the electrode being used.

LASH

To ensure good welding results, remember LASH:

*Length of arc (**L**)* – The distance between the electrode and the base metal (usually one times the electrode diameter).

*Angle (**A**)* – Two angles are critical:

- *Travel angle* – The longitudinal angle of the electrode in relation to the axis of the weld joint
- *Work angle* – The traverse angle of the electrode in relation to the axis of the weld joint

*Speed (**S**)* – Travel speed is measured in inches per minute (IPM). The width of the weld will determine if the travel speed is correct.

*Heat (**H**)* – Controlled by the amperage setting and dependent upon the electrode diameter, base metal type, base metal thickness, and the welding position.

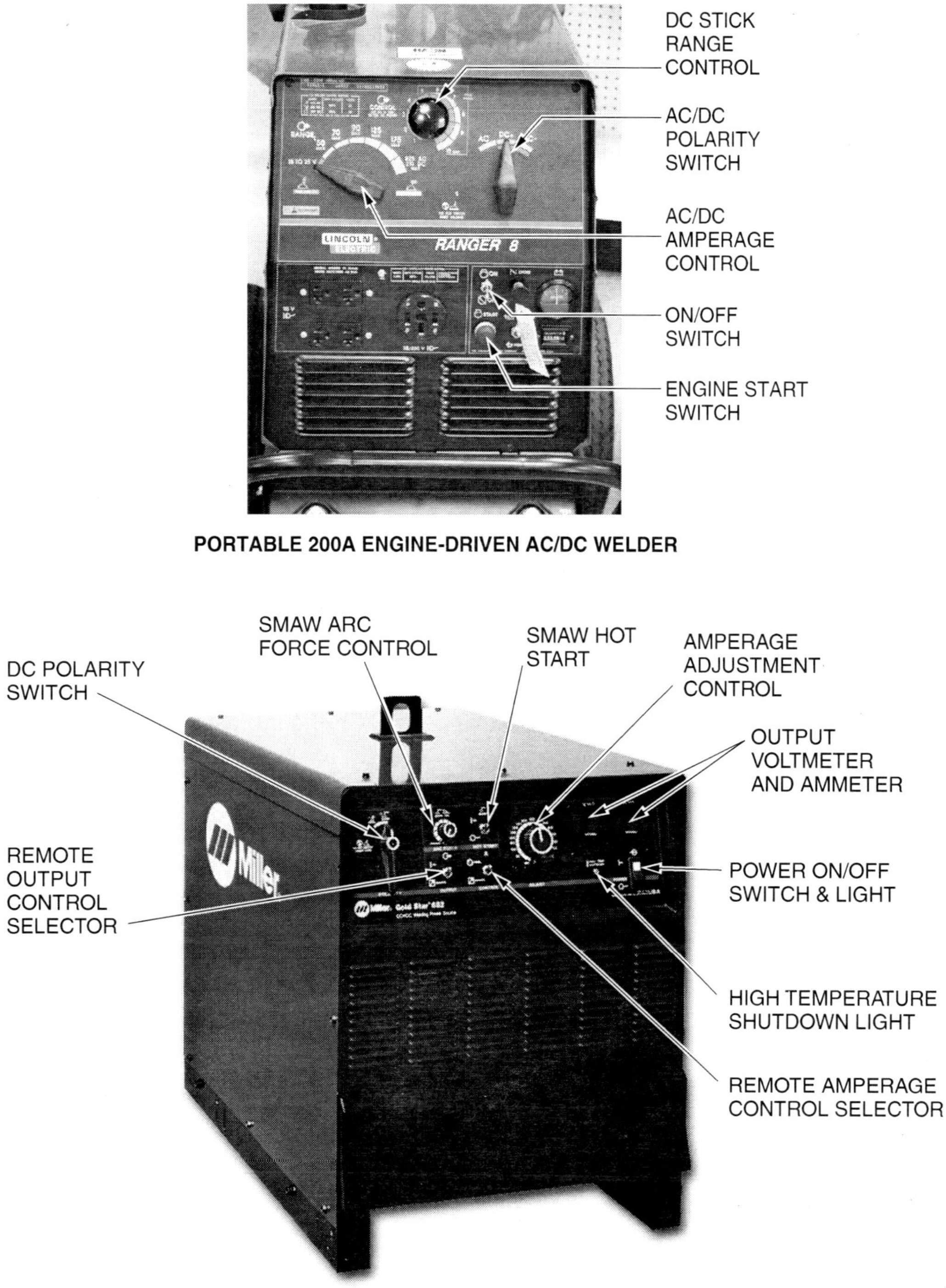

PORTABLE 200A ENGINE-DRIVEN AC/DC WELDER

SHOP TYPE 600A DC WELDER

203F10.EPS

Figure 10 ◆ Typical DC and AC/DC welding machines.

Table 4 Electrode Types and Amperage Settings

Electrode	300 Series			400 Series		
Size	**F**	**V**	**OH/HOR**	**F**	**V**	**OH/HOR**
3/32	40–70	35–50	40–60	50–75	40–55	45–65
1/8	65–100	55–70	60–85	80–115	65–80	75–100
5/32	100–145	85–100	95–120	15–160	95–110	105–130

Step 8 Check to be sure the electrode holder is not grounded.

Step 9 Turn on the welding machine.

6.0.0 ◆ OPEN-ROOT V-GROOVE WELDS

The open-root V-groove weld is a common groove weld normally made on plate and pipe. Practicing the open-root V-groove weld on plate will prepare the welder to make the more difficult pipe welds covered later.

The performance qualification requirements for this module for visual and destructive testing are based on the *ASME Boiler and Pressure Vessel Code, Section IX, Welder Certification Test*.

6.1.0 Root Pass

The most difficult part of making an open-root V-groove weld is the root pass. The root pass is made from the V-groove side of the joint and must have complete penetration, but not an excessive amount of burn-through root reinforcement. The penetration is controlled with a technique called running a keyhole. A keyhole, as shown in *Figure 11*, is a hole made when the root faces of two plates are melted away by the arc.

The molten metal flows to the back side of the keyhole, forming the weld. The keyhole should be about ⅛" to $\frac{3}{16}$" in diameter. If the keyhole is larger, there will be excess burn-through and slag inclusions. If the keyhole closes up, there will be a lack of penetration. Root reinforcement should be flush to ⅛". On thin materials, a vertical weld is usually performed downhill with a drag travel angle. On thick materials, the weld is usually performed uphill with a push travel angle.

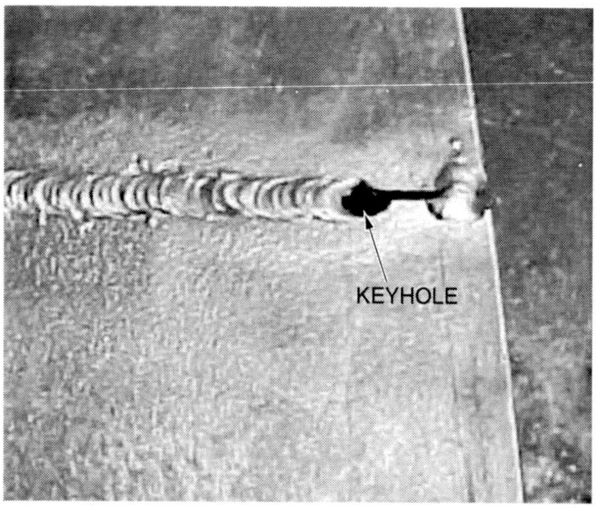

203F11.EPS

Figure 11 ◆ Root pass with an oversized keyhole.

The keyhole is controlled by oscillating the electrode tip slightly from side to side. Use care not to increase the arc length when moving the electrode. If the keyhole starts to grow, increase the side-to-side motion. If the keyhole starts to close up, decrease the side-to-side motion. Penetration and keyhole size can also be altered by changing the travel angle from push to drag to decrease penetration or vice versa to increase penetration. Other factors that affect the keyhole size are the amperage setting, root face size, and root opening. All these items can be adjusted within the parameters allowed by the welding codes or your site quality procedures.

After the root pass is run, it should be cleaned and inspected. Inspect the face of the weld for excess buildup or undercut, as shown in *Figure 12*, which could trap slag when the filler pass is run. Remove excess buildup or undercut with a hand grinder by grinding the face of the root pass with the edge of a grinding disk. Use care not to grind through the root pass or widen the groove.

> **NOTE**
> The use of grinders on stainless is discouraged. Some WPS may not allow grinding of welds. If used, the grinding wheel must not have been used on any other material.

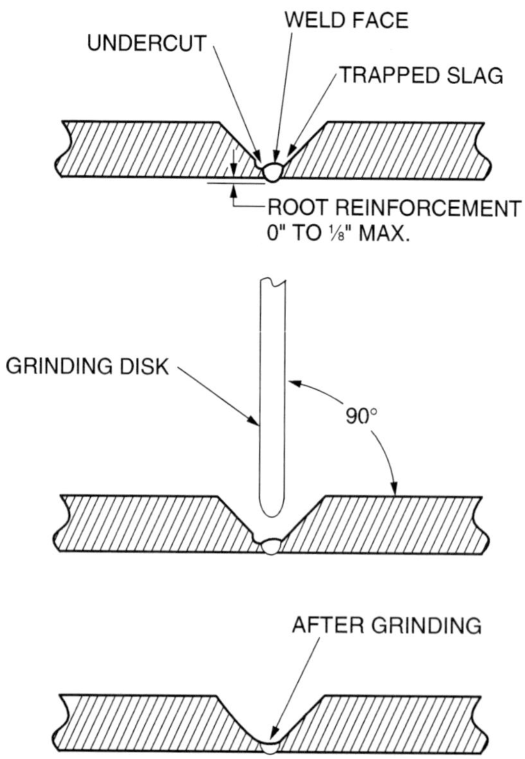

203F12.EPS

Figure 12 ◆ Grinding a root pass.

Hot Tip

Back Welds

On flat plates only, single V-groove welds can be finished with a back weld. Back welds are performed on the opposite side of a V-groove weld after the V-groove weld passes are completed. This method is somewhat simpler and faster because a melt-through root pass is not required. The edge of a grinding disk is used to cut through the root face to the root weld from the opposite side of the V-groove weld. Then the back weld pass is made.

6.2.0 Groove Weld Positions

As illustrated in *Figure 13*, groove welds can be made in all positions. The weld position for plate is determined by the axis of the weld and the orientation of the workpiece. Groove weld positions for plate are flat, or 1G (groove); horizontal, or 2G; vertical, or 3G; and overhead, or 4G.

In the 1G and 2G positions, the weld axis can be inclined up to 15°. Any weld axis inclination for the other positions varies with the rotational position of the weld face as specified in AWS standards.

6.3.0 Acceptable and Unacceptable Groove Weld Profiles

Groove welds should be made with slight reinforcement and a gradual transition to the base metal at each toe. As shown in *Figure 14*, groove welds must not have excessive reinforcement, underfill, excessive undercut, or overlap. If a groove weld has any of these defects, it must be repaired.

7.0.0 ◆ PRACTICING OPEN-ROOT V-GROOVE WELDS

This section of the module explains how to perform the following open-root V-groove welds:

- Flat (1G) welds
- Horizontal (2G) welds
- Vertical (3G) welds
- Overhead (4G) welds

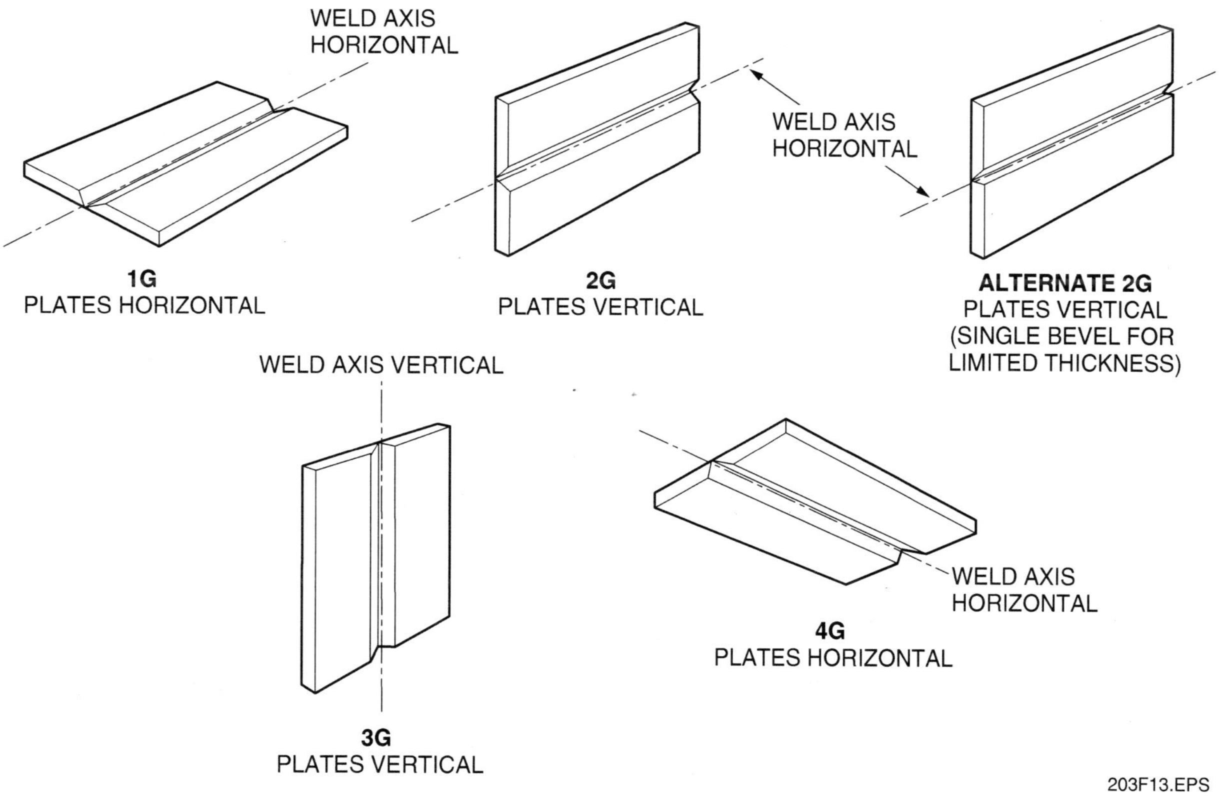

Figure 13 ◆ Groove weld positions for plate.

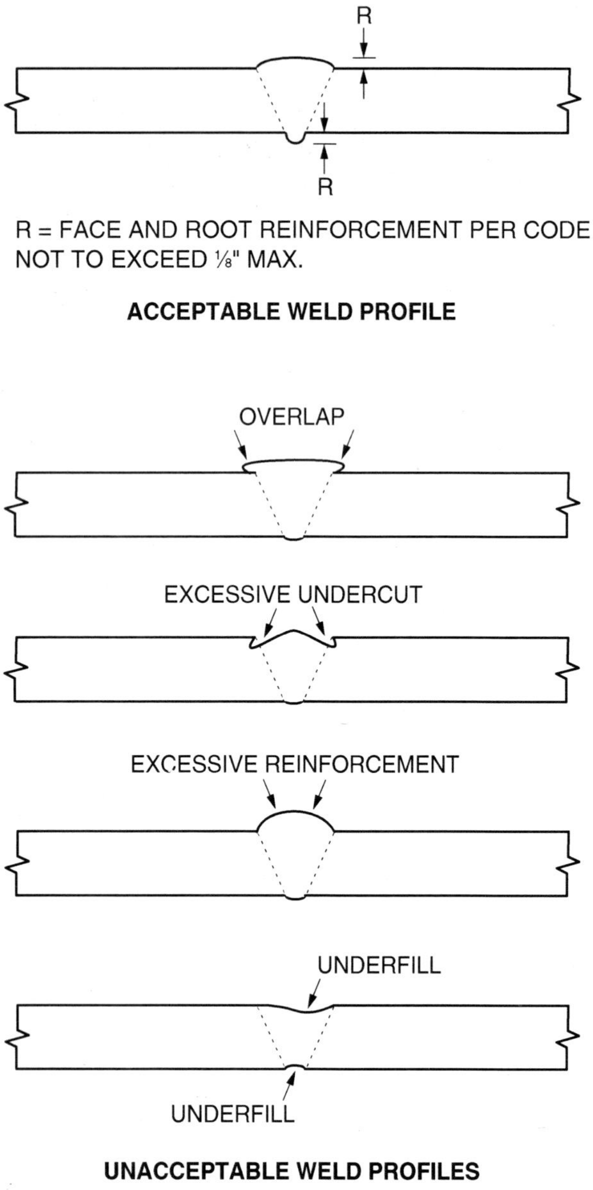

203F14.EPS

Figure 14 ◆ Acceptable and unacceptable groove weld profiles.

> 
>
>
> ***Follow Site Procedures***
>
> Refer to your site's welding procedure specifications (WPS) for specific requirements on groove welds. The information provided here is a general guideline only. Thesite WPS or quality specifications must be followed for all welds. Check with your supervisor if you are unsure of the specifications for your application.

7.1.0 Practicing Flat (1G) Position Open-Root V-Groove Welds

Practice flat open-root V-groove welds, as depicted in *Figure 15*, using stainless steel electrodes. Use stringer beads and keep the electrode angle 90° to the plate surface (0° work angle) with a 10° to 15° drag angle. When using stringer beads, use a 10° to 15° drag angle, but adjust the electrode work angle to tie in and fill the bead. Pay particular attention at the termination of the weld to fill the crater. If the practice coupon gets too hot between passes, you may have to cool it in water.

> **WARNING!**
>
> Use pliers to handle the hot practice coupons. Wear gloves when placing the practice coupon in water. Steam will rise off the coupon and can burn or scald unprotected hands.

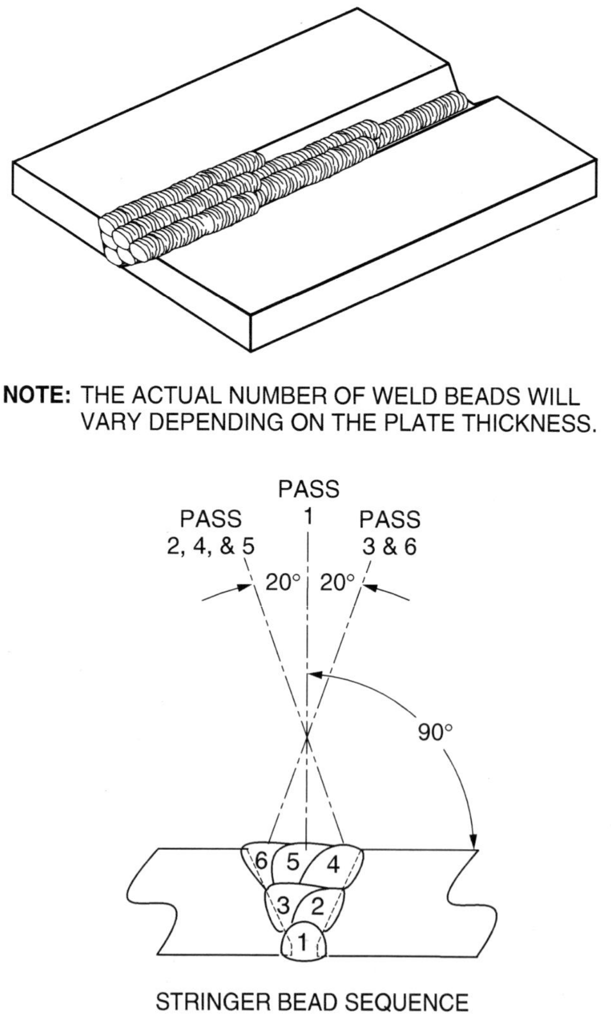

203F15.EPS

Figure 15 ◆ Multipass 1G weld sequences and work angles.

> **NOTE**
> Cooling with water is only done on practice coupons. Never cool test coupons or on-the-job welds with water. Cooling with water can cause weld cracks and affect the mechanical properties of the base metal.

Follow these steps to practice open-root V-groove welds in the flat position:

Step 1 Tack-weld the practice coupon together as explained earlier.

Step 2 Clamp or tack-weld the weld coupon in a flat position above the welding table surface.

Step 3 Run the root pass using 3/32" or 1/8" stainless steel electrodes.

Step 4 Clean the root pass. Grind if required.

Step 5 Run the remaining passes at the appropriate work angles using 3/32" or 1/8" stainless steel electrodes. Overlap each pass. Clean the weld after each pass.

7.2.0 Practicing Horizontal (2G) Position Open-Root V-Groove Welds

A horizontal open-root V-groove weld is shown in *Figure 16*. Practice horizontal open-root V-groove welds using stainless steel electrodes and stringer beads. For the electrode angles, use a 10° to 15° drag angle and adjust the work angle as required. Pay particular attention at the termination of the weld to fill the crater.

Follow these steps to practice welding open-root V-groove welds in the horizontal position:

Step 1 Tack-weld the practice coupon together as explained earlier. Use the standard or alternate horizontal weld coupon as directed by your instructor.

Step 2 Clamp or tack-weld the flat plate weld coupon into the horizontal position.

Step 3 Run the root pass using 3/32" or 1/8" electrodes.

Step 4 Clean the weld.

Step 5 Run the remaining passes at the appropriate work angles using 3/32" or 1/8" electrodes and stringer beads. Overlap each pass. Clean the weld after each pass.

NOTE: THE ACTUAL NUMBER OF WELD BEADS WILL VARY DEPENDING ON THE PLATE THICKNESS.

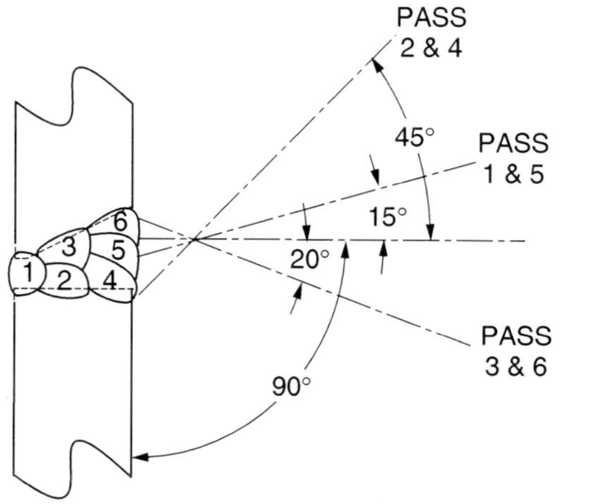

ALTERNATE JOINT REPRESENTATION

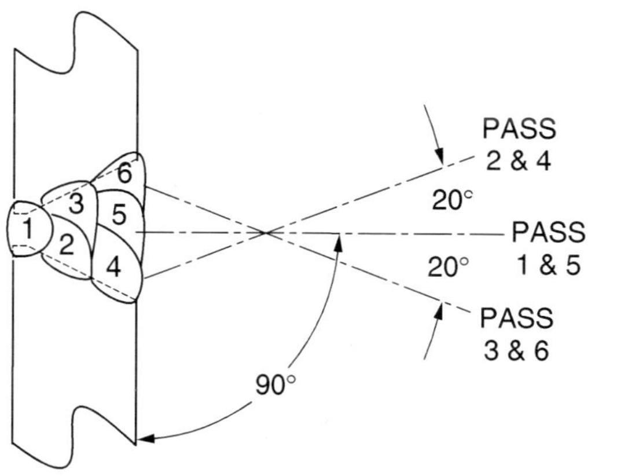

STANDARD JOINT REPRESENTATION

203F16.EPS

Figure 16 ◆ Multipass 2G weld sequences and work angles.

7.3.0 Practicing Vertical (3G) Position Open-Root V-Groove Welds

Figure 17 is an illustration of a vertical open-root V-groove weld. Practice vertical open-root V-groove welds using 3⁄32" or 1⁄8" stainless steel electrodes and stringer beads. Set the amperage in the lower part of the range for better puddle control. Use an upward travel direction with a 0°to 10° push angle, but adjust the electrode work angle to tie in the weld on one side or the other as needed. To control the molten weld puddle for the filler passes, first establish a shelf of weld metal and then use a small *A* motion. Maintain a short arc and use care not to remove the arc from the weld puddle. Pay particular attention at the termination of the weld to fill the crater.

NOTE: THE ACTUAL NUMBER OF WELD BEADS WILL VARY WITH THE METAL THICKNESS.

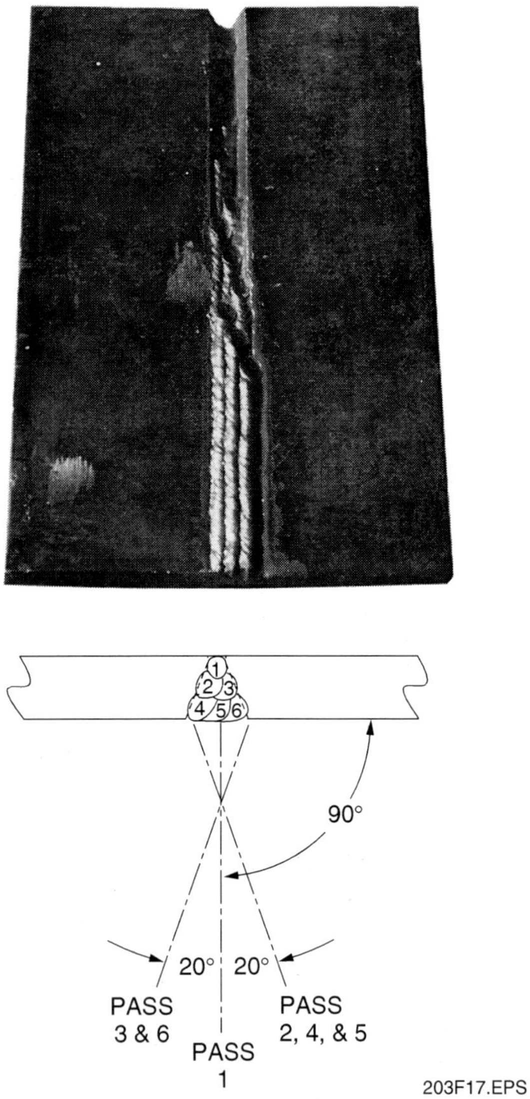

Figure 17 ◆ Multipass 3G weld sequences and work angles.

Follow these steps to practice open-root V-groove welds in the vertical position:

Step 1 Tack-weld the practice coupon together as explained earlier.

Step 2 Clamp or tack-weld the weld coupon into the vertical position.

Step 3 Run the root pass using 3⁄32" or 1⁄8" electrodes.

Step 4 Clean the weld.

Step 5 Run the remaining passes at the appropriate work angles using 3⁄32" or 1⁄8" electrodes. Overlap each pass. Clean the weld between each pass.

7.4.0 Practicing Overhead (4G) Position Open-Root V-Groove Welds

An overhead open-root V-groove weld is shown in *Figure 18*. Practice running overhead open-root V-groove welds using 3⁄32" and 1⁄8" stainless steel electrodes and stringer beads. Use a 10° to 15° drag angle, but adjust the electrode work angle to tie in the weld as needed. Pay particular attention at the termination of the weld to fill the crater.

Follow these steps to practice welding open-root V-groove welds in the overhead position:

Step 1 Tack-weld the practice coupon together as explained earlier.

Step 2 Clamp or tack-weld the weld coupon into the overhead position.

Step 3 Run the root pass using 3⁄32" or 1⁄8" electrodes.

Step 4 Clean the weld.

Step 5 Run the remaining passes at the appropriate work angles using 3⁄32" or 1⁄8" electrodes. Overlap each pass. Clean the weld after each pass.

8.0.0 ◆ STAINLESS STEEL PIPE WELDING

This section of the module will explain stainless steel pipe welding.

8.1.0 Preparing Practice Pipe Weld Coupons

Pipe weld coupons should be cut from 3" to 12" diameter Schedule 40 or Schedule 80 stainless or carbon steel pipe. Each welded joint will require two coupons of the same size and schedule pipe.

Figure 19 shows typical American Society of Mechanical Engineers (ASME), American Petroleum Institute (API), and AWS pipe bevel specifications for bevel angles, root faces, and root openings.

> **Hot Tip**
>
> ***Conserve Materials***
>
> Pipe for practice welding is expensive and difficult to obtain. Every effort should be made to conserve the material that is available. Reuse weld coupons until all surfaces have been welded on by cutting the coupon apart and reusing the pieces. Check with the instructor for the appropriate size coupon.

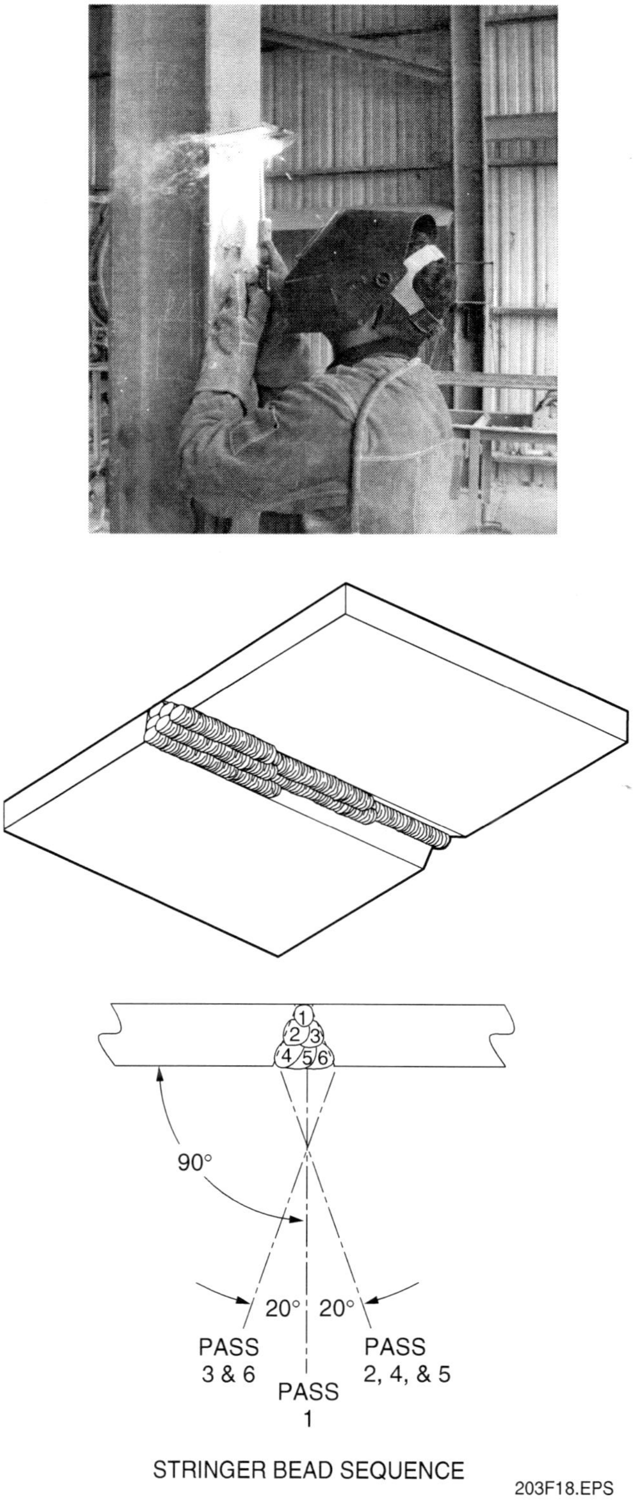

Figure 18 ◆ Multipass 4G weld sequences and work angles.

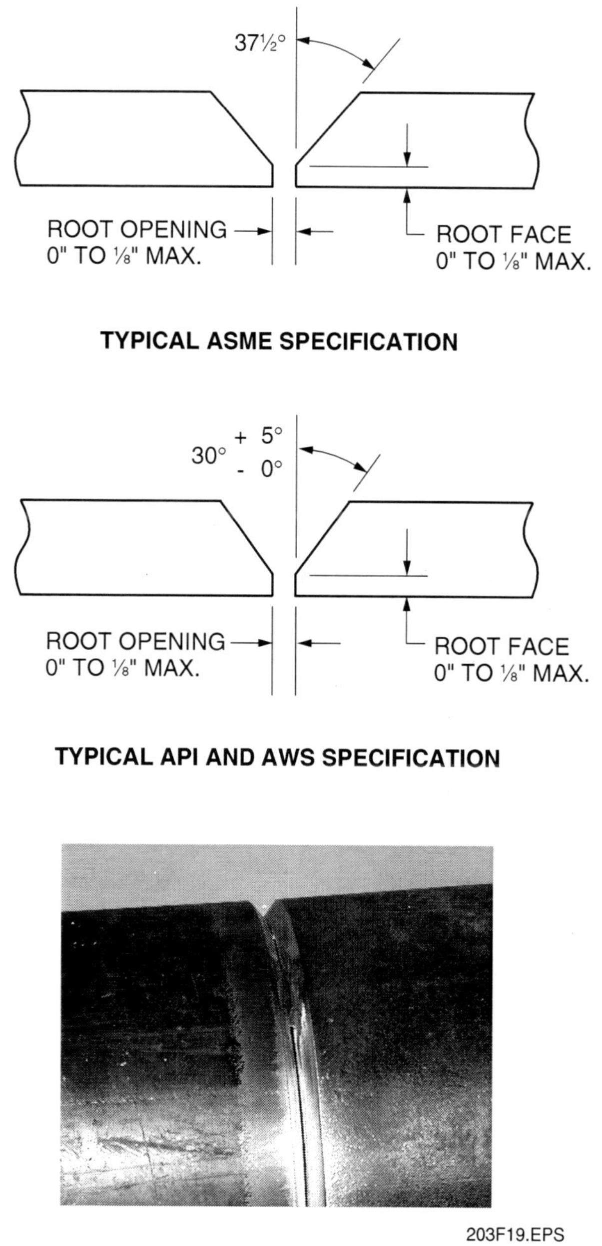

Figure 19 ◆ ASME, API, and AWS pipe bevel specifications.

To prepare weld coupons for open-root V-groove weld joints, follow these steps:

Step 1 Clean heavy rust or mill scale from the inside and outside of the carbon steel pipe with a grinder or wire brush. Clean for a distance of ½" from the weld joint. If stainless steel is used, clean with a stainless steel brush.

Step 2 Bevel the end of the pipe to 30° or 37½° by any acceptable beveling method, such as mechanical cutting, flame cutting, or grinding.

Step 3 Cut off a section of beveled pipe end (1½" minimum).

NOTE
For 1G-ROTATED welding, the coupons may have to be cut longer to fit the rollers, if used.

Follow these steps to prepare the open-root V-groove weld pipe coupon:

Step 1 Check the bevel. There should be no dross and the bevel angle should be 30° or 37½°.

Step 2 Grind or file a 0" to ⅛" root face on the bevel as specified by your instructor.

NOTE
Welding codes allow the root face on open root welds to be from 0" to ⅛" wide and the root opening to be from 0" to ⅛". Select and adjust the root face and opening as needed when you start the welding practices.

Hot Tip — *Welding Larger Diameter Pipe*

The practice coupons on larger diameter pipe must be greater than 1½" in length. The coupons on larger size pipe need more metal to help absorb the increased heat needed to make these welds.

Step 3 Align the two pipe sections so that the inside diameter (ID) surfaces are even all around. Small-diameter pipe can be aligned by clamping both pieces to a piece of angle iron. Larger diameter pipe can be aligned with the aid of a pipe alignment jig or by holding a straightedge across the joint, parallel to the pipe axis. The straightedge must be used all around the pipe in case one or both sections are distorted.

Step 4 Gap the root opening with pieces of filler wire, metal shims, or pieces of bare welding electrode of the correct diameter.

Step 5 When the root opening is correct and the pipe ends are aligned, make the first of four tack welds of no greater than 1".

NOTE
Longer tacks or more tacks may be required on heavy wall or large diameter pipe.

Step 6 After the first tack weld, check the root opening opposite the first tack weld. Adjust the gap if necessary and make the second tack weld on the opposite side from the first tack.

Step 7 Check the root opening again and weld the third tack midway between the first two tacks.

Step 8 Weld the fourth tack opposite the third tack and midway between the first and second tacks. There should now be four tack welds evenly spaced (90°) around the pipe coupon. This is illustrated by *Figure 20*.

Hot Tip — *Beveling*

The end of the joint must meet squarely with the mating pipe. To accomplish that, the sharp, inside edge (root face) of the bevel must be ground flat and square to the length of pipe. In order for the root gap to be uniform, final shaping should be done by mechanical means.

Step 9 Clean and feather the tack welds with the edge of a grinding wheel. Feathering the ends of the tack welds with a grinder will help to fuse the tack welds into the root pass.

9.0.0 ◆ OPEN-ROOT V-GROOVE PIPE WELDS

The open-root V-groove joint is the most common groove weld normally made on pipe.

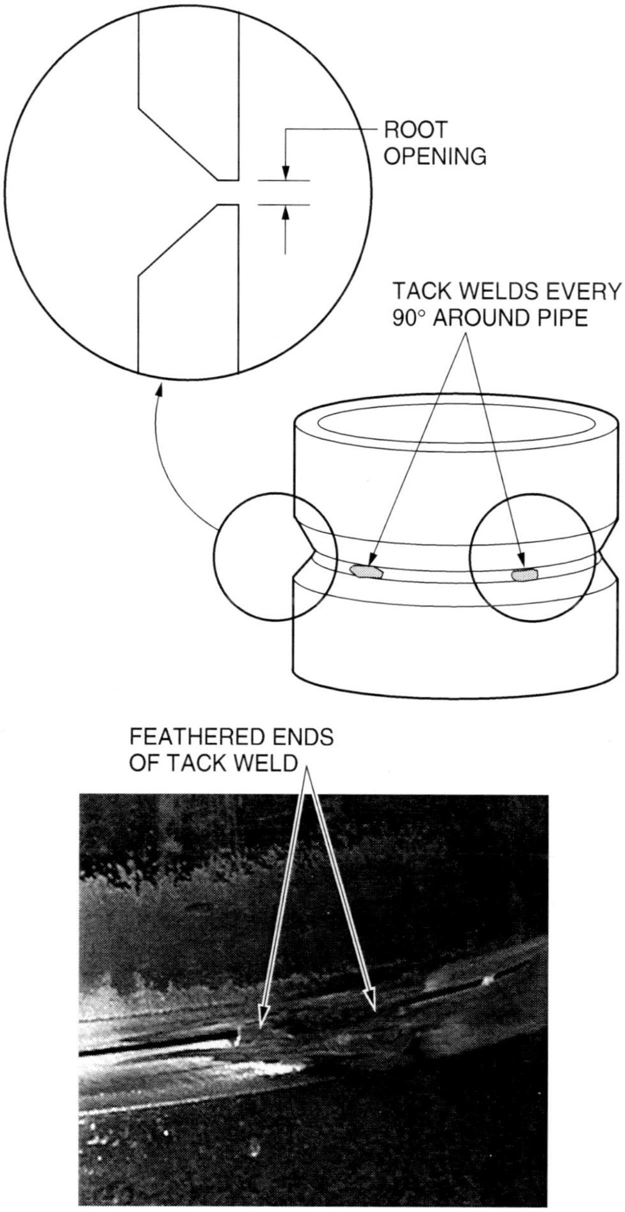

Figure 20 ◆ Tacked open-root V-groove weld coupon.

9.1.0 Root Pass

The most difficult part of making an open-root V-groove weld is the root pass. The root pass must have complete penetration, but not an excessive amount of burn-through or root reinforcement. The extent of penetration is significantly important. If penetration is too little, it will create a weak joint. However, if there is too much penetration and excessive root reinforcement is created, the inside diameter of the pipe will be reduced and the flow will be restricted. The penetration is controlled with a technique called running a keyhole. A keyhole-shaped hole is formed when the root faces on the two pipe edges are melted away by the arc (*Figure 21*). The molten metal flows to the back side of the keyhole, forming the weld. The keyhole should be about 1/8" to 3/16" in diameter. If the keyhole is larger, there will be excess burn-through with slag inclusions. If the keyhole closes up, a lack of penetration will result. Root reinforcement should be flush to 1/8".

Hot Tip

Arc Strikes

When welding pipe, do not make arc strikes outside the weld groove on the surface of the pipe. An arc strike can cause a hardened spot that can crack as the pipe expands and contracts. An arc strike on the pipe surface is considered a defect, and will require repair or rework.

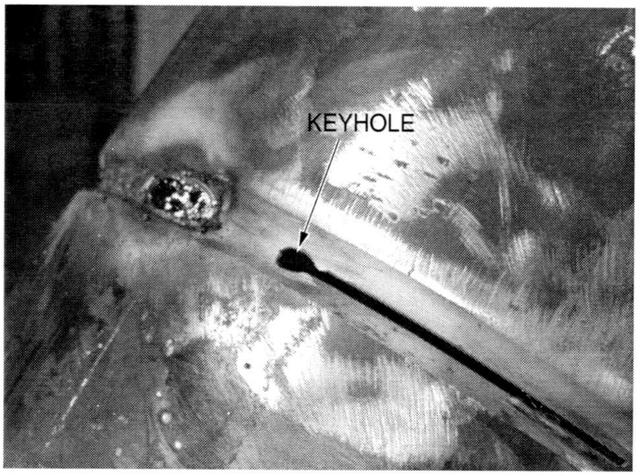

Figure 21 ◆ Root bead keyhole.

The keyhole is controlled by using a slight side-to-side oscillation. Use care not to increase the arc length when moving the electrode. If the keyhole starts to grow, increase the side-to-side motion. If the keyhole starts to close up, decrease the side-to-side motion. Other factors that affect the keyhole size are the amperage setting, root face size, and root opening. All these items can be adjusted within the parameters allowed by the welding codes or your site quality procedures.

> **NOTE**
> Some WPS do not allow grinding of stainless steel welds.

After the root pass is run, it should be cleaned and inspected. Check the face of the weld for excess buildup or undercut that could trap slag when the filler pass is run.

Remove excess buildup or undercut with a hand grinder by grinding the face of the root pass with the edge of a grinding disk. Most of the slag along the sides of the weld bead is removed by this grinding action, thus making it easier to add the remaining passes. Use care not to grind through the root pass or widen the groove. *Figure 22* shows

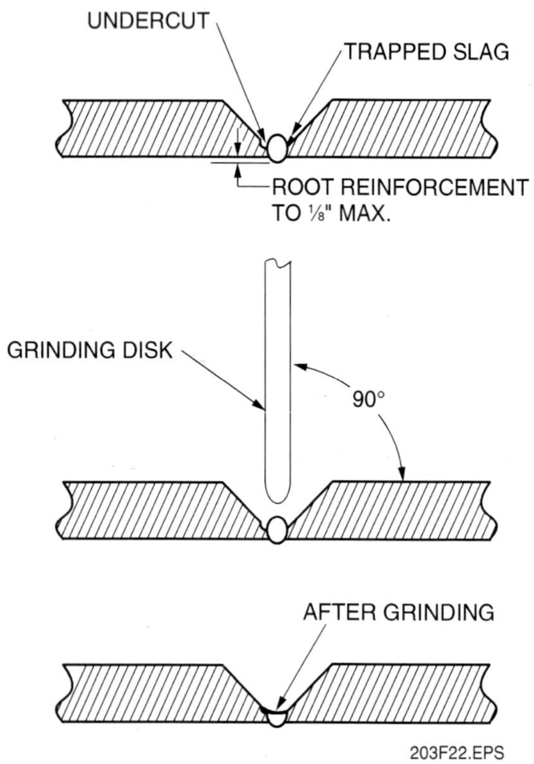

Figure 22 ◆ Grinding a root pass.

the effect of grinding a root pass. A narrow grinding wheel (not a cutoff wheel) is effective for this purpose.

It should be noted that root passes are not always ground for pipe that is to be used in low- and medium-pressure systems. Grinding each root pass takes a long time and oxide contaminants from the grinding wheel can be embedded in the weld. In these cases, the root pass slag is removed only by chipping and brushing with a stainless steel brush.

9.2.0 Pipe Groove Weld Test Positions

Groove welds may be made in all positions on pipe. The weld position is determined by the axis of the pipe. Four standard weld test positions, shown in *Figure 23*, are used with pipe:

- *Flat (1G-ROTATED) welding position* – The pipe axis is horizontal and the pipe is slowly rotated while being welded on the top; weld beads are flat.
- *Horizontal (2G) welding position* – The pipe axis is vertical and the pipe rotation is fixed; weld beads are horizontal.
- *Multiple (5G) welding position* – The pipe axis is horizontal and the pipe rotation is fixed; weld beads are flat, vertical, and overhead.
- *Inclined multiple (6G) welding position* – The pipe axis is inclined, nominally at 45° ±5°, and pipe rotation is fixed; weld beads are horizontal, vertical, and overhead.

The pipe axis positions for 1G-ROTATED, 2G, and 5G can vary ±15° from the basic position. The pipe axis position for 6G can vary ±5° from the nominal 45° angle. See *Table 5.*

9.2.1 Tack Welds and Restarts on 5G and 6G Coupons

When 5G or 6G welds are to be destructively tested, the test specimens must be cut from the four regions illustrated in *Figure 24*, midway between the 12-o'clock and 3-o'clock, 3-o'clock and 6-o'clock, 6-o'clock and 9-o'clock, and 9-o'clock and 12-o'clock positions. (The 12-o'clock position is the top of the coupon groove and 6-o'clock is the bottom.)

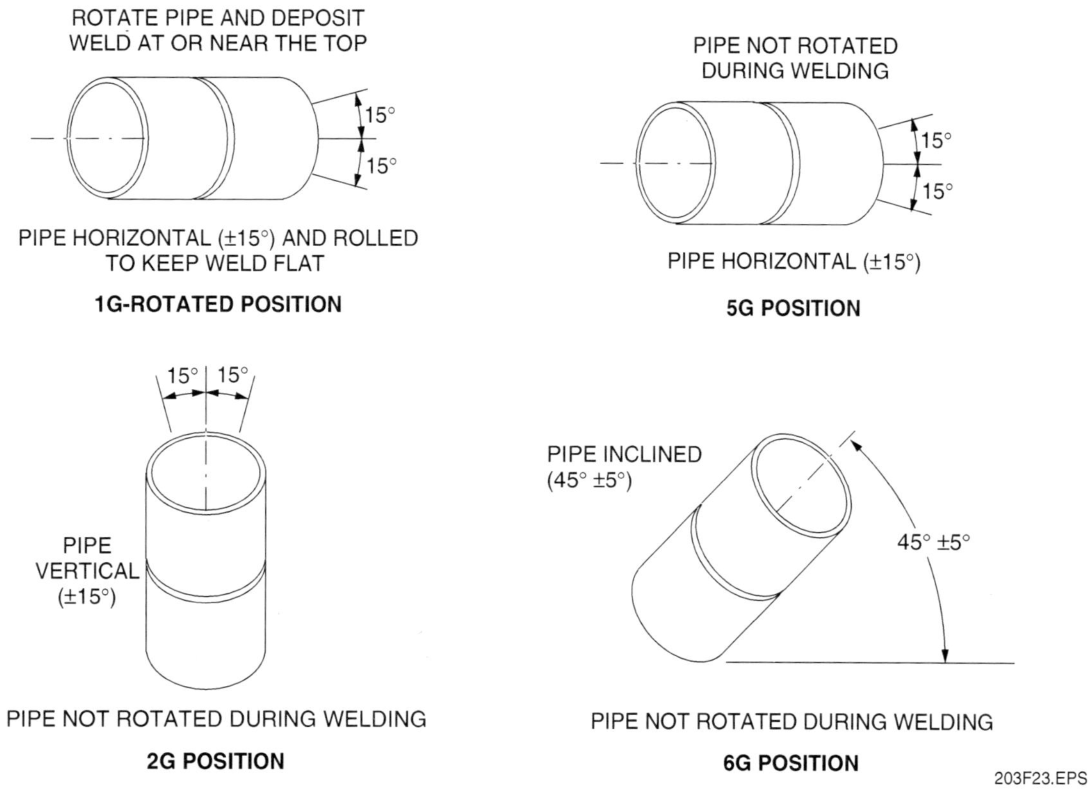

Figure 23 ◆ Four basic pipe groove weld test positions.

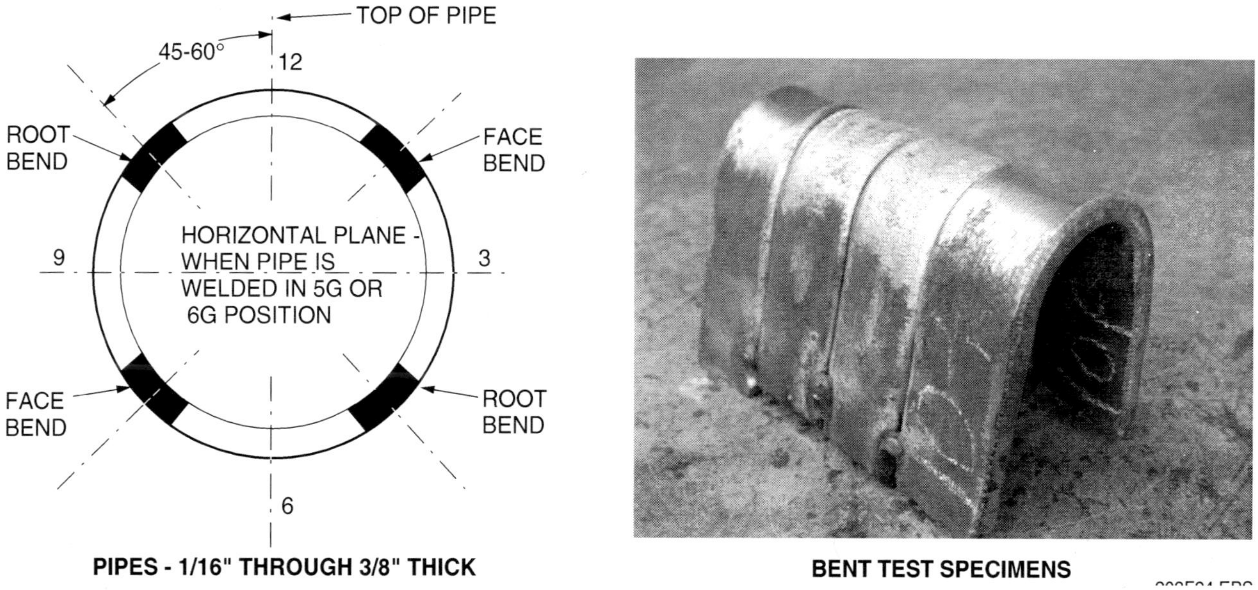

Figure 24 ◆ Test specimen regions of 5G or 6G position pipe.

Table 5 Pipe Groove Weld Test Positions

Weld Test Position	Pipe Axis	Weld Beads	Pipe Position
1G-ROTATED	Horizontal	Flat	Slowly Rotated
2G	Vertical	Horizontal	Fixed
5G	Horizontal	Flat, Vertical, Overhead	Fixed
6G	Inclined 45°	Horizontal, Vertical, Overhead	Fixed

9.3.0 Acceptable and Unacceptable Pipe Weld Profiles

Pipe groove welds should be made with slight reinforcement (not exceeding ⅛") and a gradual transition to the base metal at each toe. The root pass should have complete penetration. The root reinforcement on the inside of the pipe ranges from being flush to a maximum of ⅛". Pipe groove welds must not have excess reinforcement or underfill at the face or root, excessive undercut, or overlap. Excessively large cover passes will reduce the pipe's strength. They cause the stresses in the pipe to be concentrated along the sides of the weld and will not permit the pipe to expand and contract in a uniform manner along its length. If a weld has any of these defects, as shown in *Figure 25,* it must be repaired.

10.0.0 ◆ PRACTICING OPEN-ROOT V-GROOVE WELDS

This section of the module explains how to perform the following open-root V-groove welds:

- Flat (1G-ROTATED) position welds
- Horizontal (2G) position welds
- Multiple (5G) position welds
- Inclined multiple (6G) position welds

Hot Tip *6GR Test Position*

Welding pipe in the 6G position is a common welding test. A ring may be added to test for restricted accessibility (6GR).

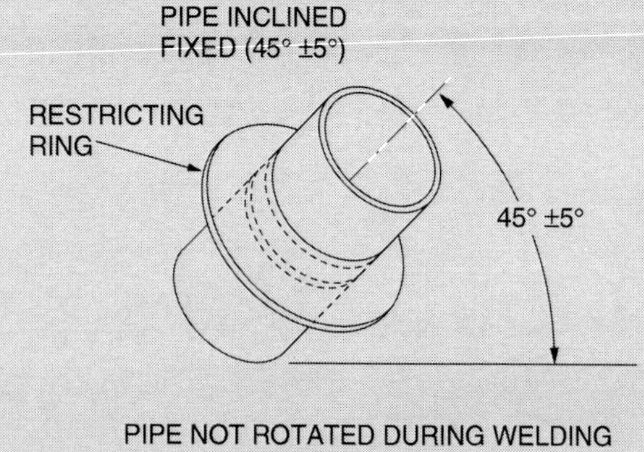

6GR POSITION

203SA01.EPS

10.1.0 Practicing Flat (1G-ROTATED) Position Open-Root V-Groove Pipe Welds

Practice the 1G-ROTATED position open-root V-groove pipe welds using $\frac{3}{32}$" and ⅛" stainless steel electrodes.

Keep the electrode angle 90° (0° work angle) to the pipe axis with a 10° to 15° drag angle for the root pass. Use a side-to-side oscillation to run a keyhole for the root pass, paying particular attention to tie into the tack welds and restarts. Clean the face of the root pass with a file, grinder, or stainless steel brush.

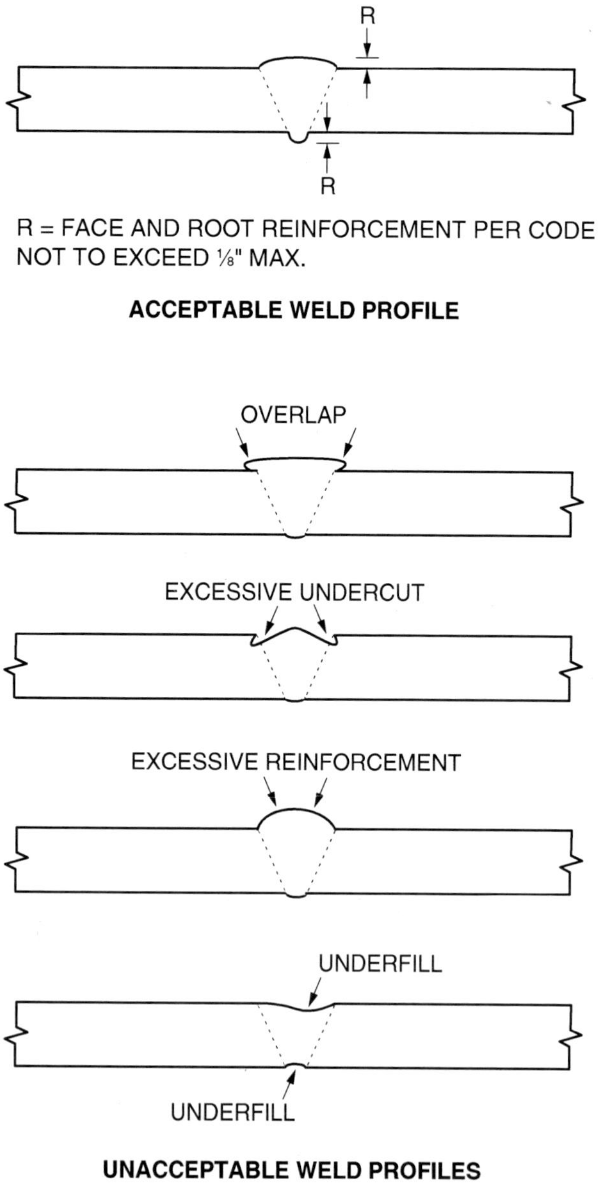

203F25.EPS

Figure 25 ◆ Acceptable and unacceptable pipe groove weld profiles.

Hot Tip: Filler Pass

Follow these guidelines when running the filler pass:

- Fill the crater to preclude cracking.
- Clean the crater of slag before restarting the arc; failure to do so will result in inclusions.
- Ensure high-strength, high-pressure pipe cleanliness by slightly grinding or filing the crater.
- The location of the start and stop spots for each weld must be staggered.
- The weld groove should be filled level with spot beads prior to the cover pass.

When running the filler and cover passes, use stringer beads with a slight side-to-side motion to ensure tie-in at the toes of the weld bead. Take particular care at the termination of the weld to fill the crater. Run all passes at or near the top of the pipe as the pipe is rotated.

Follow these steps to practice open-root V-groove pipe welds in the 1G-ROTATED position:

Step 1 Tack-weld the practice pipe weld coupon together as explained earlier.

Step 2 Position the pipe weld coupon horizontally on two sets of rollers at a comfortable welding height. *Figure 26* shows roller supports commonly found in pipe welding shops.

Step 3 Make sure the workpiece clamp is attached directly onto the pipe coupon. This will prevent the welding current from passing through the roller bearings or arcing between the rollers and the pipe coupon.

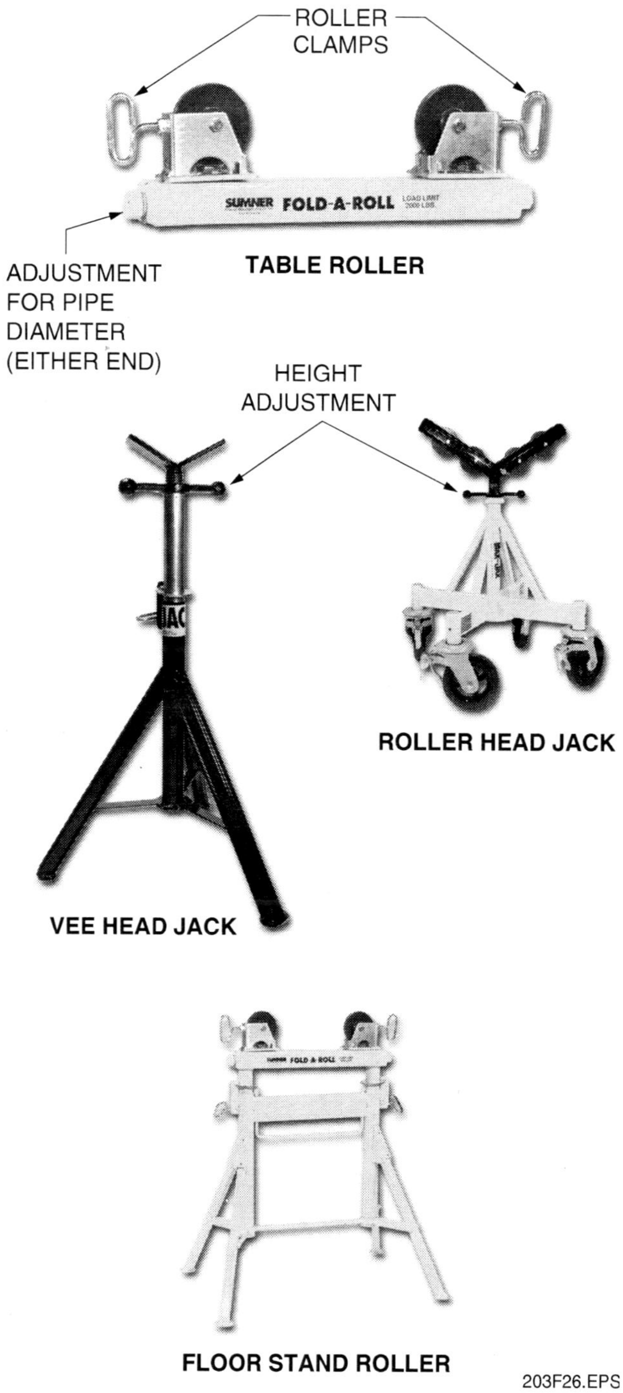

Figure 26 ◆ Pipe roller supports.

CAUTION

Failure to attach the workpiece clamp to the pipe coupon can result in variations in welding current, damage to the roller bearings, and arcing on the rollers and pipe coupon.

Hot Tip: Homemade Jack Stands

OSHA prohibits the use of homemade jack stands. Only stands that are manufactured to industry standards are acceptable.

Step 4 Run the root pass using 3⁄32" stainless steel electrodes and a side-to-side oscillation. Position a tack weld at the 11 o'clock position, start the weld bead on the tack weld, and advance toward the 12 o'clock position.

Step 5 Roll the pipe as necessary to keep the weld in the flat position. Chip the restart between beads.

Step 6 Chip, grind, file, and/or brush the root pass to remove trapped slag.

Step 7 Use the same rolling procedure to make the remaining passes. Use 3⁄32" or 1⁄8" stainless steel electrodes and stringer beads. Pay particular attention at the tie-ins to prevent slag inclusions or excess buildup. Overlap the passes and clean the weld after each pass. *Figure 27* shows the bead sequence and work angles.

Hot Tip

Powered Pipe Roller

A pipe roller that is operated by an electric motor activated by a foot switch is shown here. These devices are primarily used for large, heavy pipe sections.

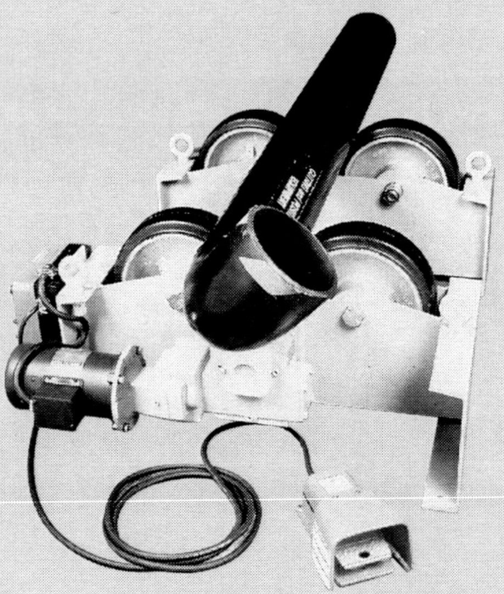

203SA02.EPS

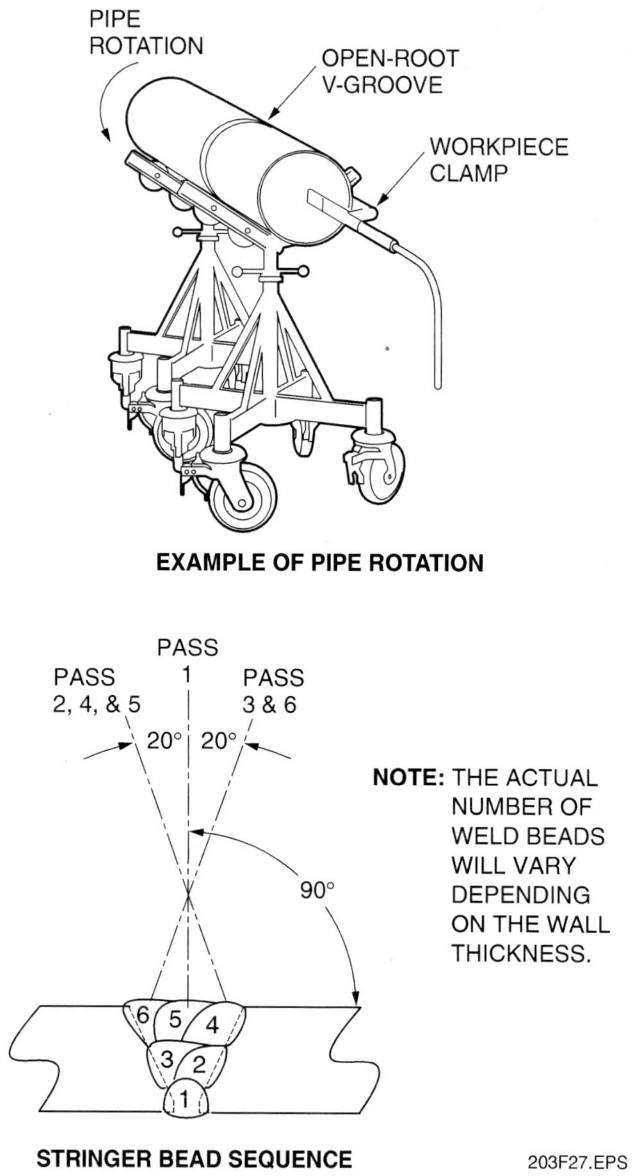

Figure 27 ◆ Multiple pass 1G-ROTATED weld sequences and work angles.

Think About It

Cover Pass

What effect will excessively large cover passes have on the strength of a pipe?

10.2.0 Practicing Horizontal (2G) Position Open-Root V-Groove Pipe Welds

Practice 2G (horizontal) position open-root V-groove pipe welds using 3/32" and 1/8" stainless steel electrodes.

The electrode should be at a 10° to 15° drag angle and at 90° (0° work angle) to the surface of the pipe for the root pass. To prevent the weld puddle from sagging, the electrode work angle can be dropped slightly, but no more than 10°. Use a side-to-side oscillation to run a keyhole for the root pass, paying particular attention to tie into the tack welds and restarts.

When running the fill/cover passes, use stringer beads and a slight side-to-side motion to ensure tie-in at the toes of the weld bead. Take particular care at the termination of the weld to fill the crater. *Figure 28* shows the 2G pipe weld coupon position and bead sequence.

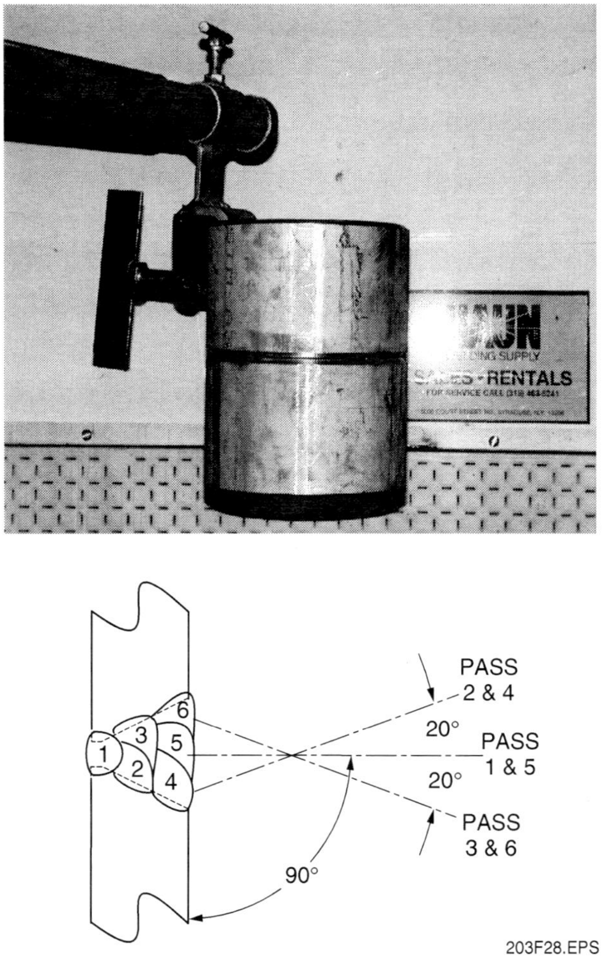

Figure 28 ◆ Multiple pass 2G weld sequences and work angles.

Follow these steps to practice open-root V-groove pipe welds in the 2G position:

Step 1 Tack-weld the practice pipe weld coupon together as explained earlier.

Step 2 Clamp or tack-weld the pipe coupon with the axis of the pipe vertical.

Step 3 Run the root pass using 3/32" stainless steel electrodes and a side-to-side oscillation. Start the root pass on a tack weld. Clean and chip the crater between restarts. Continue this procedure until the root pass is completed.

Step 4 Chip, grind, file, and/or brush the root pass to remove trapped slag.

Step 5 Run the remaining passes at the appropriate work angles using either 3/32" or 1/8" stainless steel electrodes. Overlap the passes and clean the weld after each pass.

10.3.0 Practicing Multiple (5G) Position Open-Root V-Groove Pipe Welds

Practice the multiple (5G) position open-root V-groove pipe welds using 3/32" and 1/8" stainless steel electrodes.

Weld uphill by starting at the bottom of the pipe and welding toward the top. Use a side-to-side oscillation to run a keyhole for the root pass, paying particular attention to tie into the tack welds and restarts.

When running the fill and cover passes, use a slight side-to-side motion to ensure tie-in at the toes of the weld bead. Pay particular attention at the termination of the weld to fill the crater. *Figure 29* shows the 5G pipe weld coupon and bead sequence.

Follow these steps to practice open-root V-groove pipe welds in the 5G position:

Step 1 Tack-weld the practice pipe weld coupon together as explained earlier.

Step 2 Clamp or tack-weld the pipe weld coupon in position with the pipe axis horizontal.

Step 3 Run the root pass using 3/32" electrodes. Start the root pass at the bottom of the pipe and weld uphill. When welding at the top and bottom of the coupon (flat and overhead positions), the electrode should usually point toward the pipe axis. When welding up the coupon sides (vertical welding), a slight electrode push angle (maximum of 10°) can be used to control the weld puddle.

Step 4 Chip and grind, file, or brush the root pass to remove trapped slag.

Step 5 Run the remaining passes at the appropriate work angles using either $\frac{3}{32}$" or $\frac{1}{8}$" stainless steel electrodes. Overlap the passes and clean the weld after each pass.

10.4.0 Practicing Inclined Multiple (6G) Position Open-Root V-Groove Pipe Weld

Practice the inclined (45°) multiple (6G) position open-root V-groove pipe welds using $\frac{3}{32}$" and $\frac{1}{8}$" stainless steel electrodes.

Weld uphill by starting at the bottom of the pipe and welding toward the top. Use a side-to-side oscillation to run a keyhole for the root pass paying particular attention to tie into the tack welds and restarts.

When running the fill and cover passes, use stringer beads and a slight side-to-side motion to ensure tie-in at the toes of the weld bead. Pay particular attention at the termination of the weld to fill the crater. *Figure 30* shows the 6G pipe weld coupon and bead sequence.

Follow these steps to practice open-root V-groove pipe welds in the 6G position:

> **NOTE**
> If required for training purposes, a restricting ring may be added to the 6G position coupon to form a 6GR position coupon.

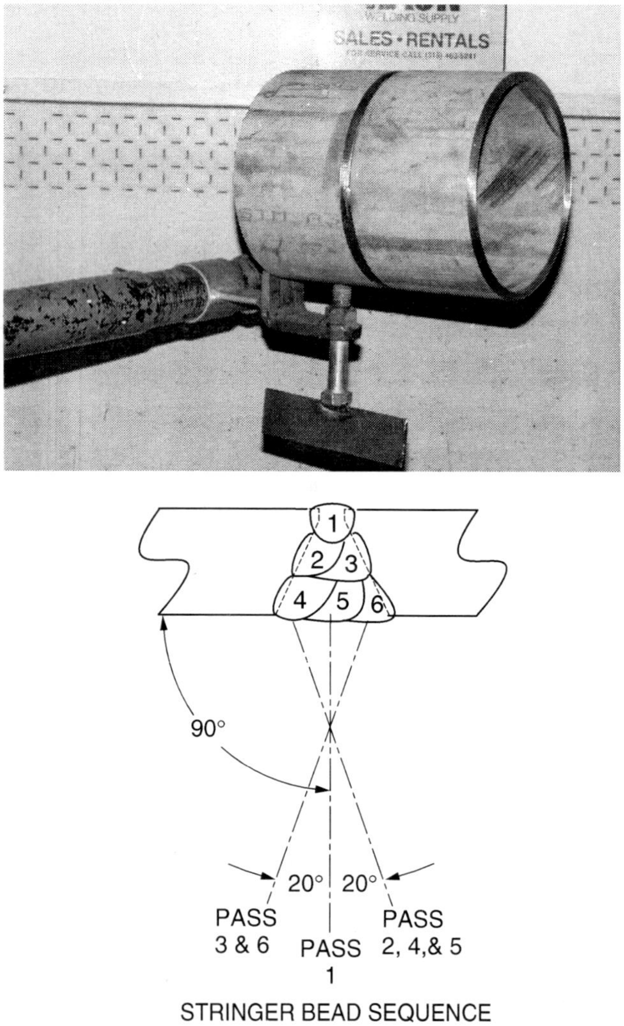

203F29.EPS

Figure 29 ◆ Multiple pass 5G bend sequences and work angles.

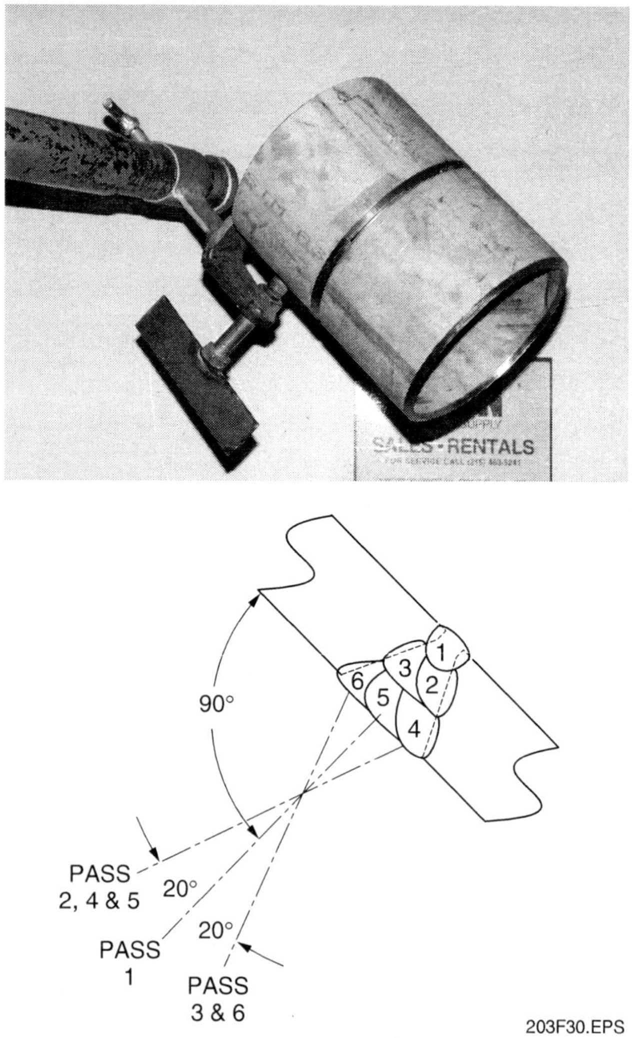

203F30.EPS

Figure 30 ◆ Multiple pass 6G bead sequences and work angles.

Step 1 Tack-weld the practice pipe weld coupon together as explained earlier.

Step 2 Clamp or tack-weld the pipe weld coupon in position with the pipe axis inclined 45° to the horizontal plane.

Step 3 Run the root pass using 3⁄32" stainless steel electrodes. Start the root pass at the bottom of the pipe and weld uphill. When welding at the top and bottom of the coupon (flat and overhead welds), the electrode should point toward the pipe axis. When welding up the coupon sides (vertical welds), a slight electrode push angle (maximum of 10°) can be used to control the weld puddle.

Step 4 Chip and grind, file, or brush the root pass to remove trapped slag.

Step 5 Run the remaining passes at the appropriate work angles using 3⁄32" or 1⁄8" stainless steel electrodes. Overlap the passes and clean the weld after each pass.

Summary

Being able to make open-root V-groove welds on plate using stainless steel electrodes is an essential skill a welder must have to perform stainless steel welding jobs and to progress on to more difficult open-root V-groove stainless steel pipe welding procedures. Welders must understand the differences between stainless and carbon steel and must be able to identify and select proper electrodes. The arc welding equipment must be properly set up to produce acceptable welds. There are several stainless steel joints, and you must be able to evaluate acceptable and unacceptable welds after they are completed. Practice these welds until you can consistently produce acceptable welds.

Review Questions

1. The principal physical characteristic of all stainless steels is their resistance to corrosion and heat, which varies with the _____ of stainless steel.

a. size
b. shape
c. length
d. category

2. The specific alloy content of stainless steel and its _____ treatment during manufacture determine the stainless steel's type of grain structure.

a. air
b. heat
c. water
d. steam

3. The American Iron and Steel Institute's (AISI) 200 and 300 series of stainless steels are included in the _____ group of stainless steels.

a. martensitic
b. crystalline
c. austenitic
d. ferritic

4. The plain martensitic group of stainless steels contains only _____ % chromium and no nickel.

a. 2 to 3
b. 4 to 6
c. 7 to 9
d. 11 to 13

5. The chromium content of ferritic stainless steels ranges from _____.

a. 11.5% to 27%
b. 16.5% to 26%
c. 28% to 30%
d. 31% to 36%

6. When austenitic stainless steel containing more than 0.02% carbon is heated to between 800 and 1,400°F, _____ occurs in the heat affected zone (HAZ).

a. rust
b. warping
c. hot cracking
d. carbide precipitation

7. When a stabilized metal is heated to between 800 and 1,400°F, the columbium or titanium combines with the _____ in the metal.
   a. iron
   b. zinc
   c. carbon
   d. nickel

8. When welding stainless steel, the majority of the weld is formed by _____.
   a. the base metal
   b. an equal mix of electrode metal and base metal
   c. the flux coating of the electrode
   d. the electrode metal

9. Electrodes that have a thin carbon steel tube filled with powdered metal containing chromium, nickel, and other alloys, along with an external flux material are called _____ electrodes.
   a. fast-fill
   b. fast-freeze
   c. coated tubular
   d. low-hydrogen

10. The American Welding Society (AWS) specification that applies to covered (SMAW) stainless steel electrodes is _____.
    a. *A5.1*
    b. *A5.2*
    c. *A5.3*
    d. *A5.4*

11. Usability designations for stainless steel SMAW electrodes include _____.
    a. -11 and -12
    b. -12 and -13
    c. -13 and -14
    d. -15 and -16

12. When the basic class alloy inside a stainless steel electrode has been modified with 0.04% to 0.08% carbon, the markings on the side of electrode will include the additional letter(s) _____ in the middle of the overall marking.
    a. L
    b. H
    c. HC
    d. Cb

13. Of the different stainless steel types, the _____ stainless steels are generally easiest to weld.
    a. ferritic
    b. austenitic
    c. crystalline
    d. martensitic

14. When stainless steel base metal is being used for welding coupons, use _____ to clean the coupons.
    a. oil
    b. a grinder
    c. chemicals
    d. a stainless steel wire brush

15. Practicing open-root V-groove joint welds on stainless steel plate will prepare the welder to make the more difficult _____ welds.
    a. flat
    b. pipe
    c. vertical
    d. horizontal

16. The most difficult part of making an open-root V-groove weld is the _____.
    a. bead
    b. overlap
    c. keyhole
    d. root pass

17. Root pass penetration is controlled with a technique called _____.
   a. running a keyhole
   b. root inclusion
   c. burn-through
   d. whipping the root

18. When performing flat open-root V-groove welds with stringer beads, use a _____ angle.
   a. 10° to 15° push
   b. 10° to 15° drag
   c. 40° to 50° push
   d. 40° to 50° drag

19. When performing vertical open-root V-groove welds, use a(n) _____.
   a. downward travel direction with a 10° to 15° drag angle
   b. downward travel direction with a 40° to 50° push angle
   c. upward travel direction with a 0° to 10° push angle
   d. upward travel direction with a 10° to 15° drag angle

20. Welding codes allow the root face on open-root welds to be from _____.
   a. 0" to ⅛"
   b. 1⁄16" to 3⁄16"
   c. ⅛" to 3⁄16"
   d. 1⁄32" to 5⁄32"

21. When the axis of the pipe being welded is horizontal, and the pipe is slowly rotated while being welded on the top, the weld position is called _____.
   a. 1G-ROTATED
   b. 2G
   c. 5G
   d. 6G

22. When the axis of the pipe being welded is horizontal, and the pipe is fixed, the weld position is called _____.
   a. 1G-ROTATED
   b. 2G
   c. 5G
   d. 6G

23. Pipe positions can vary ±15 degrees from the basic position for the _____ pipe positions.
   a. 1G-ROTATED, 2G, and 6G
   b. 1G-ROTATED, 2G, and 5G
   c. 5G and 6G
   d. 1G-ROTATED and 6G

24. When performing an open-root V-groove pipe root-pass weld in the 1G-ROTATED position, keep the electrode angle _____ to the pipe axis.
   a. 10° to 15°
   b. 45°
   c. 55°
   d. 90°

25. When performing open-root V-groove pipe welds in the 5G position, weld _____.
   a. toward the bottom (6 o'clock)
   b. uphill by starting at the top of the pipe
   c. uphill by starting at the bottom of the pipe
   d. downhill by starting at the top of the pipe

GLOSSARY

Trade Terms Introduced in This Module

Annealing: A heat treatment process that reduces the brittleness of the work.

Interpass temperature: The temperature of the base metal in the general area of the weld before the next pass is started. The definition of the area varies by the applicable welding code.

Porosity: Cavities caused by gas entrapment during weld solidification.

APPENDIX

Performance Accreditation Tasks

The Performance Accreditation Tasks (PATCs) correspond to and support the learning objectives in *AWS EG2.0-95, Guide for the Training and Qualification of Welding Personnel: Entry-Level Welder*.

PATCs provide specific acceptable criteria for performance and help to ensure a true competency-based welding program for students.

The following tasks are designed to evaluate your ability to run open-root V-groove welds on plate in all positions and on pipe in the 6G position using stainless steel electrodes and SMAW equipment. Perform each task when you are instructed to do so by your instructor. As you complete each task, take it to your instructor for evaluation. Do not proceed to the next task until told to do so by your instructor.

For AWS 6G certifications, refer to *AWS EG4.0* for bend test requirements.

OPEN-ROOT V-GROOVE WELDS IN THE FLAT POSITION

Note: It is preferable to use stainless steel base metal. However, substituting carbon steel is permitted by the codes.

Using stainless steel electrodes, make an open-root V-groove weld on plate in the flat (1G) position, as shown.

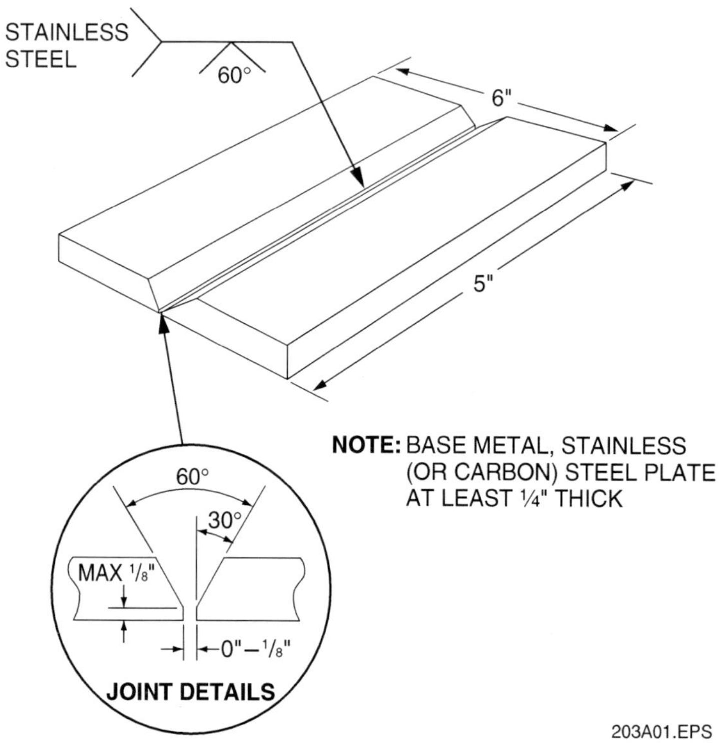

203A01.EPS

Criteria for Acceptance

- Uniform rippled appearance on the bead face ____________
- Craters and restarts filled to the full cross section of the weld ____________
- Uniform weld size ±1/16" ____________
- Acceptable weld profile in accordance with the *ASME Boiler and Pressure Vessel Code* ____________
- Smooth transition with complete fusion at the toes of the weld ____________
- Complete uniform root reinforcement at least flush with the base metal to a maximum buildup of 3/32" ____________
- No porosity ____________
- No overlap ____________
- No excessive undercut ____________
- No inclusions ____________
- No cracks ____________
- No pinholes (fisheyes) ____________

OPEN-ROOT V-GROOVE WELDS IN THE HORIZONTAL POSITION

Using stainless steel electrodes, make an open-root V-groove weld on plate in the horizontal (2G) position, as shown.

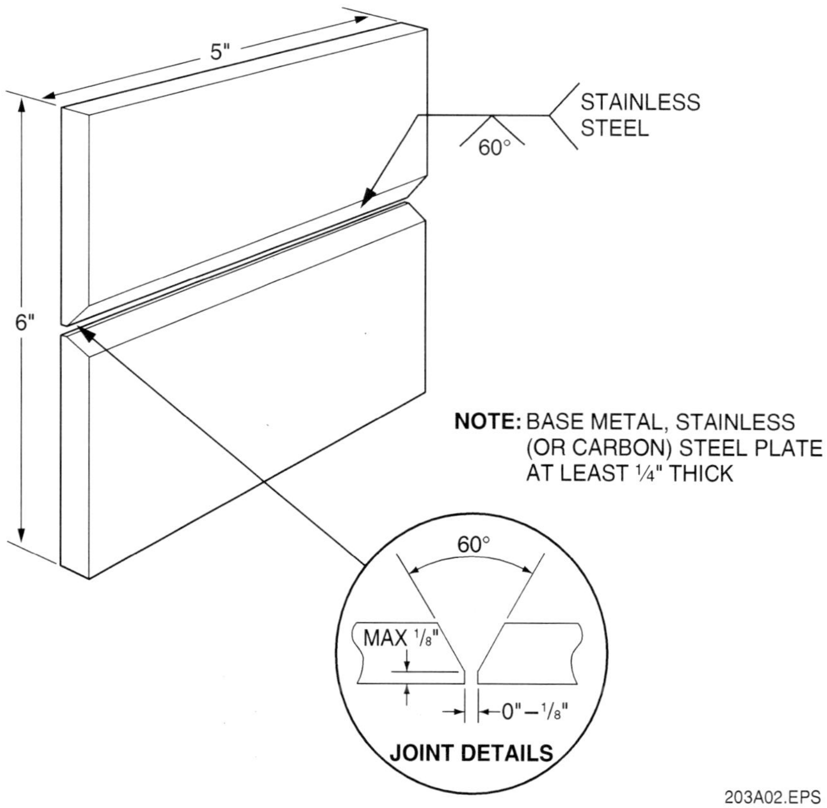

203A02.EPS

Criteria for Acceptance

- Uniform rippled appearance on the bead face __________
- Craters and restarts filled to the full cross section of the weld __________
- Uniform weld size ±1⁄16" __________
- Acceptable weld profile in accordance with the *ASME Boiler and Pressure Vessel Code* __________
- Smooth transition with complete fusion at the toes of the weld __________
- Complete uniform root reinforcement at least flush with the base metal to a maximum buildup of 3⁄32" __________
- No porosity __________
- No overlap __________
- No excessive undercut __________
- No inclusions __________
- No cracks __________
- No pinholes (fisheyes) __________

OPEN-ROOT V-GROOVE WELDS IN THE VERTICAL POSITION

Using stainless steel electrodes, make an open-root V-groove weld on plate in the vertical (3G) position, as shown.

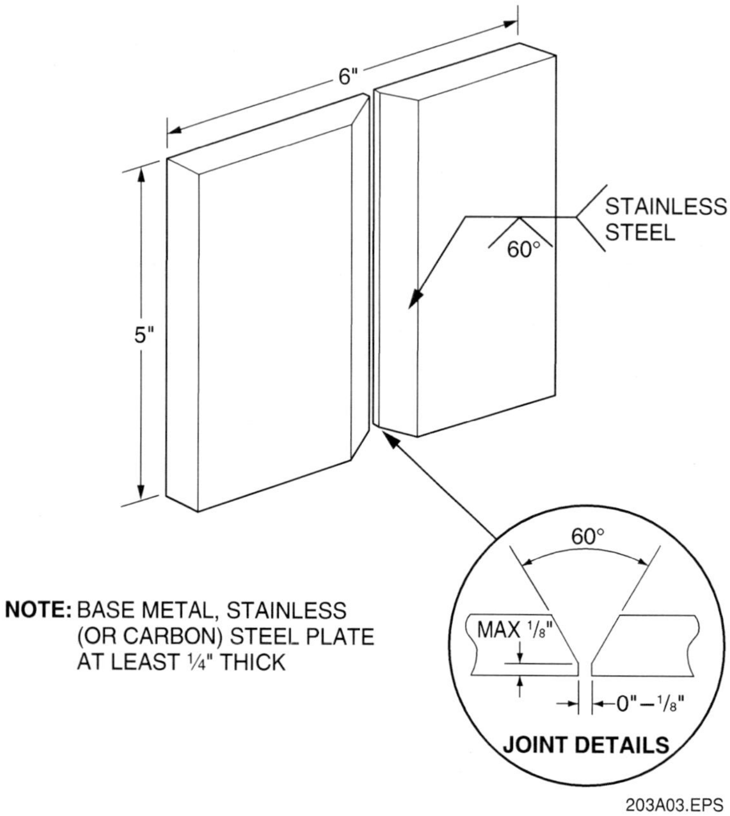

Criteria for Acceptance

- Uniform rippled appearance on the bead face ____________
- Craters and restarts filled to the full cross section of the weld ____________
- Uniform weld size ±1/16" ____________
- Acceptable weld profile in accordance with the *ASME Boiler and Pressure Vessel Code* ____________
- Smooth transition with complete fusion at the toes of the weld ____________
- Complete uniform root reinforcement at least flush with the base metal to a maximum buildup of 3/32" ____________
- No porosity ____________
- No overlap ____________
- No excessive undercut ____________
- No inclusions ____________
- No cracks ____________
- No pinholes (fisheyes) ____________

OPEN-ROOT V-GROOVE WELDS IN THE OVERHEAD POSITION

Using stainless steel electrodes, make an open-root V-groove weld on plate in the overhead (4G) position, as shown.

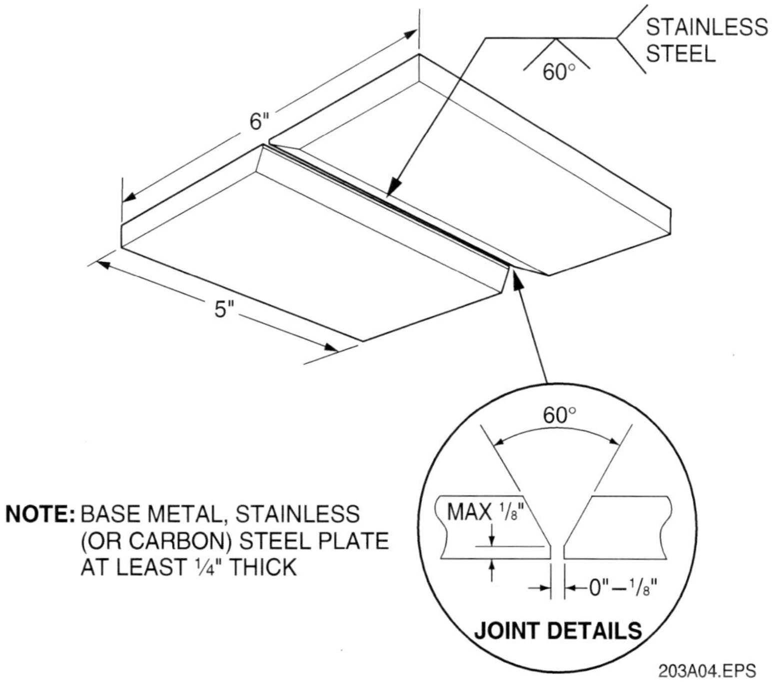

203A04.EPS

Criteria for Acceptance

- Uniform rippled appearance on the bead face __________
- Craters and restarts filled to the full cross section of the weld __________
- Uniform weld size ±1/16" __________
- Acceptable weld profile in accordance with the *ASME Boiler and Pressure Vessel Code* __________
- Smooth transition with complete fusion at the toes of the weld __________
- Complete uniform root reinforcement at least flush with the base metal to a maximum buildup of 3/32" __________
- No porosity __________
- No overlap __________
- No excessive undercut __________
- No inclusions __________
- No cracks __________
- No pinholes (fisheyes) __________

OPEN-ROOT V-GROOVE PIPE WELDS IN THE INCLINED MULTIPLE POSITION

Using stainless steel electrodes, make an open-root V-groove weld on pipe in the inclined (6G) multiple position, as shown. If required, a restricting ring may be added to the 6G position to form a 6GR position coupon.

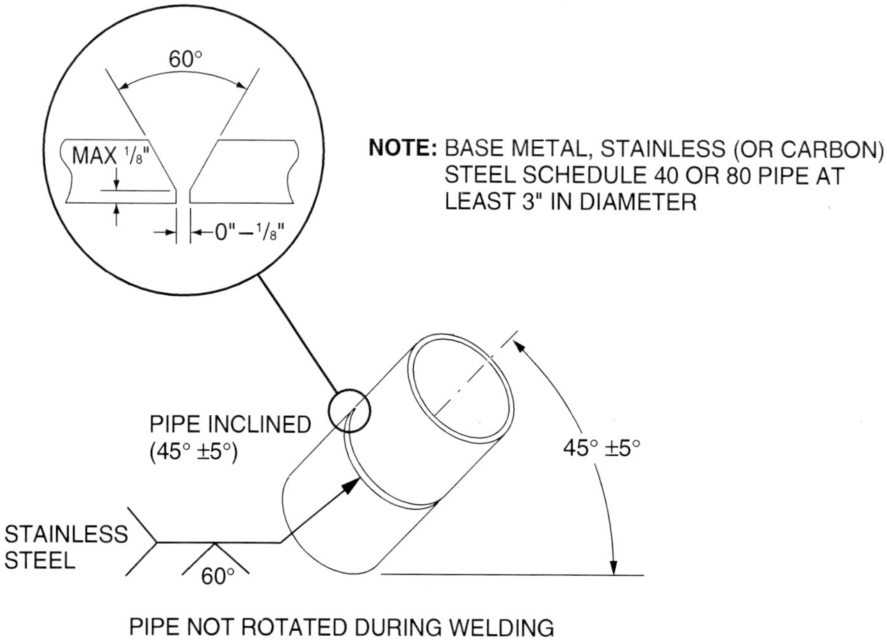

6G POSITION

203A05.EPS

Criteria for Acceptance

- Uniform rippled appearance on the bead face __________
- Craters and restarts filled to the full cross section of the weld __________
- Uniform weld size ±1⁄16" __________
- Acceptable weld profile in accordance with the *ASME Boiler and Pressure Vessel Code* __________
- Smooth transition with complete fusion at the toes of the weld __________
- Complete uniform root reinforcement at least flush with the base metal to a maximum buildup of 3⁄32" __________
- No porosity __________
- No overlap __________
- No excessive undercut __________
- No inclusions __________
- No cracks __________
- No pinholes (fisheyes) __________

Additional Resources

This module is intended to be a thorough resource for task training. The following reference works are suggested for further study. These are optional materials for continued education rather than for task training.

High Quality Stainless Steel Wires & Electrodes for All Welding Processes, Latest Edition. Hanover, PA: Alloy Rods Corporation.

Modern Welding, Latest Edition. A.D. Althouse, C.H. Turnquist, W.A. Bowditch, K.E. Bowditch. Tinley Park, IL: The Goodheart-Willcox Company, Inc.

The Procedure Handbook of Arc Welding, Latest Edition. Cleveland, OH: The Lincoln Electric Company.

Welding Pressure Pipe Lines and Piping Systems, Latest Edition. Cleveland, OH: The Lincoln Electric Company.

Welding Skills, Latest Edition. R.T. Miller. Homewood, IL: American Technical Publishers, Inc.

Welding Technology, Latest Edition. Gower A. Kennedy. Indianapolis, IN: The Bobbs-Merrill Company, Inc.

Figure Credits

Figure 1 courtesy of American Welding Society

Oates, W.R., and A. Saitta, eds., 1998, *Materials and Applications—Part 2*, Volume 4 of Welding Handbook, Miami: American Welding Society, Figures 5.14 and 5.15.

Gerald Shannon	203F03, 203F07, 203F10 (upper), 203F19, 203F20, 203F28, 203F29, 203F30
Lincoln Electric Co.	203F05
Charles Rogers	203F06
Miller Electric Mfg. Company	203F10 (lower)
Rex Ball	203F11
John Murray	203F16, 203F17
John Yochum	203F18, 203F21, 203F24 (photo)
Sumner Manufacturing Company, Inc.	203F26
Koike Aronson, Inc.	203SA02

CONTREN™ LEARNING SERIES — USER UPDATES

The NCCER makes every effort to keep these textbooks up-to-date and free of technical errors. We appreciate your help in this process. If you have an idea for improving this textbook, or if you find an error, a typographical mistake, or an inaccuracy in NCCER's Contren™ textbooks, please write us, using this form or a photocopy. Be sure to include the exact module number, page number, a detailed description, and the correction, if applicable. Your input will be brought to the attention of the Technical Review Committee. Thank you for your assistance.

Instructors – If you found that additional materials were necessary in order to teach this module effectively, please let us know so that we may include them in the Equipment/Materials list in the Instructor's Guide.

Write: Curriculum Revision and Development Department
National Center for Construction Education and Research
P.O. Box 141104, Gainesville, FL 32614-1104

Fax: 352-334-0932

E-mail: curriculum@nccer.org

Craft ______________________ Module Name ______________________

Copyright Date ____________ Module Number ____________ Page Number(s) ____________

Description

____

____

____

(Optional) Correction

____

____

(Optional) Your Name and Address

____

____

Module 29204-03

Air Carbon Arc Cutting and Gouging

COURSE MAP

This course map shows all of the modules in the second level of the Welding curriculum. The suggested training order begins at the bottom and proceeds up. Skill levels increase as you advance on the course map. The local Training Program Sponsor may adjust the training order.

WELDING LEVEL TWO

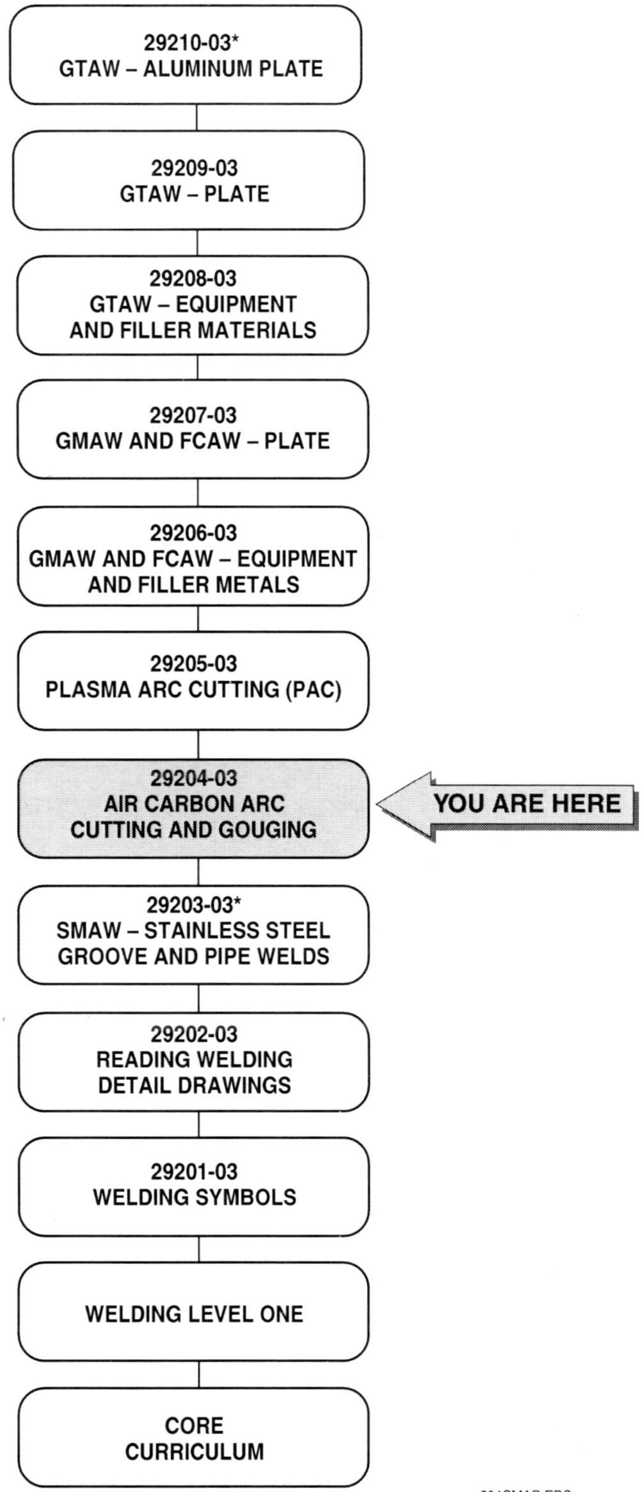

204CMAP.EPS

Please note that Modules 29203-03 and 29210-03 are electives for *Welding Level Two* but are requirements for *AWS Entry Level Welder.

MODULE 29204-03 CONTENTS

Figures

Tables

MODULE 29204-03

Air Carbon Arc Cutting and Gouging

Objectives

When you have completed this module, you will be able to do the following:

1. Identify and explain the air carbon arc cutting (CAC-A) process and equipment.
2. Select and install CAC-A electrodes.
3. Prepare the work area and CAC-A equipment for safe operation.
4. Use CAC-A equipment for washing and gouging activities.
5. Perform storage and housekeeping activities for CAC-A equipment.
6. Make minor repairs to CAC-A equipment.

Prerequisites

Before you begin this module, it is recommended that you successfully complete the following: Core Curriculum; Welding Level One; Welding Level Two, Modules 29201-03 through 29203-03.

Required Trainee Materials

1. Pencil and paper
2. Appropriate personal protective equipment

1.0.0 ◆ INTRODUCTION

Air carbon arc cutting (CAC-A) is an arc cutting and gouging process. The base metal is melted with an electric arc formed between a carbon electrode and the base metal. A jet of compressed air is used to blow the molten metal away. The process can be used in all positions on any metal that will conduct electricity, such as carbon steel, stainless steel, cast iron, and copper (and their alloys). It is primarily used to gouge weld grooves, to back gouge welds, to gouge out defective welds, and to prepare cracks for welding. In addition, CAC-A is often used in **washing** and beveling operations. *Figure 1* shows a groove being cut by the CAC-A process.

This module will explain CAC-A equipment, its operation, and its use for cutting and gouging metals.

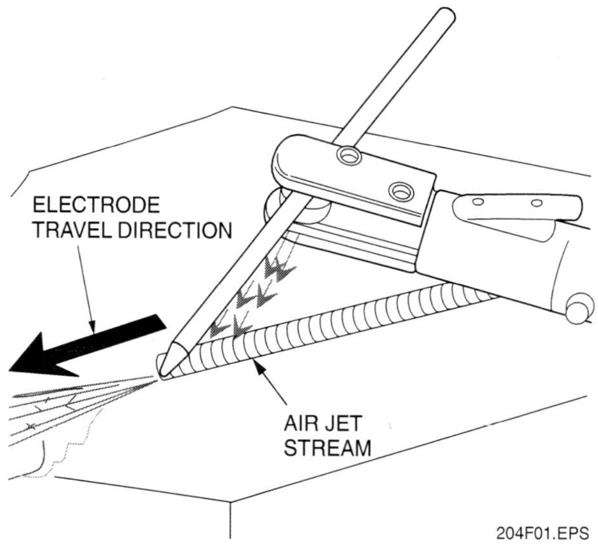

Figure 1 ◆ Cutting a groove with the CAC-A process.

Carbon Electrodes

Carbon electrode arcs produce cleaner cuts than metal electrode arc cutting methods because no foreign metals are introduced at the arc.

2.0.0 ◆ AIR CARBON ARC CUTTING PROCESS AND EQUIPMENT

This section explains the air carbon arc cutting process and the equipment used to perform air carbon arc cutting and gouging.

> **WARNING!**
> Because of the high temperature and the intensity of the arc, choose a shade of helmet lens that is darker than the normal shade you would use for welding on the same thickness of metal. A number 12 or 14 lens shade is recommended for carbon arc cutting and gouging.

2.1.0 Air Carbon Arc Cutting Process

Three common applications of air carbon arc cutting include:

- Cutting or gouging out defective metal
- Removing old or inferior welds
- Creating a groove for a welding pool

The air carbon arc cutting process uses a high-current electric arc between the base metal and the end of a **carbon-graphite electrode** to melt the base metal. The currents used with CAC-A electrodes are much higher than those required for comparably sized shielded metal arc welding (SMAW) electrodes. Air jet orifices are located in the special torch heads. When gouging or beveling, the orifices direct jets of compressed air under the CAC-A electrode, between the electrode and the base metal, to blow the molten metal pool away from the cut. When the electrode is moved along the surface of the base metal, a groove or bevel can be made.

A groove's depth and width are controlled by the following:

- Electrode size
- Arc amperage
- Electrode angle to the base metal
- Electrode advancement rate
- Electrode movement
- Regulated air pressure

2.2.0 Air Carbon Arc Cutting Equipment

Air carbon arc cutting equipment includes an air carbon arc torch and its attachments, a welding power source, and a supply of compressed air.

2.2.1 Air Carbon Arc Torch and Attachments

Typical CAC-A equipment consists of a specially designed carbon-graphite electrode gripped in a specialized torch handle (electrode holder). The torch handle is equipped with V-groove heads containing air jet holes and a combination power cable and compressed air hose. *Figure 2* shows a typical air carbon arc torch. *Figure 3* is a simplified diagram of a CAC-A setup.

The electrode is gripped between the torch heads (electrode contacts). The electrode tip should be no further than 6" from the electrode tip to the contacts to prevent overheating. The contacts are mounted in the lower and upper torch heads, which are spring-loaded to clamp onto the electrode. Depending on torch design, one head may be flat and the other V-grooved, or both may be V-grooved.

> **Hot Tip**
>
> ***CAC-A Cutting***
>
> Since CAC-A cutting is not dependent on oxidation, it can cut metals that normally cannot be cut with oxyacetylene cutting methods.

> **Hot Tip**
>
> ***Air Pressure***
>
> Air carbon arc cutting air pressure should be regulated to the lowest air pressure that cleanly blows away the molten metal, usually a range of 40 psig to 100 psig.

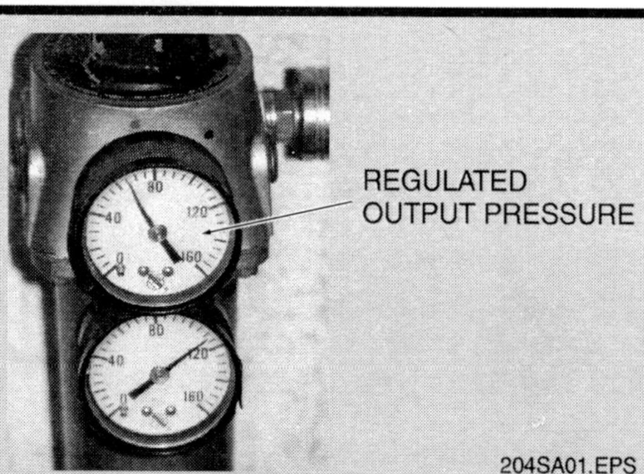

204SA01.EPS

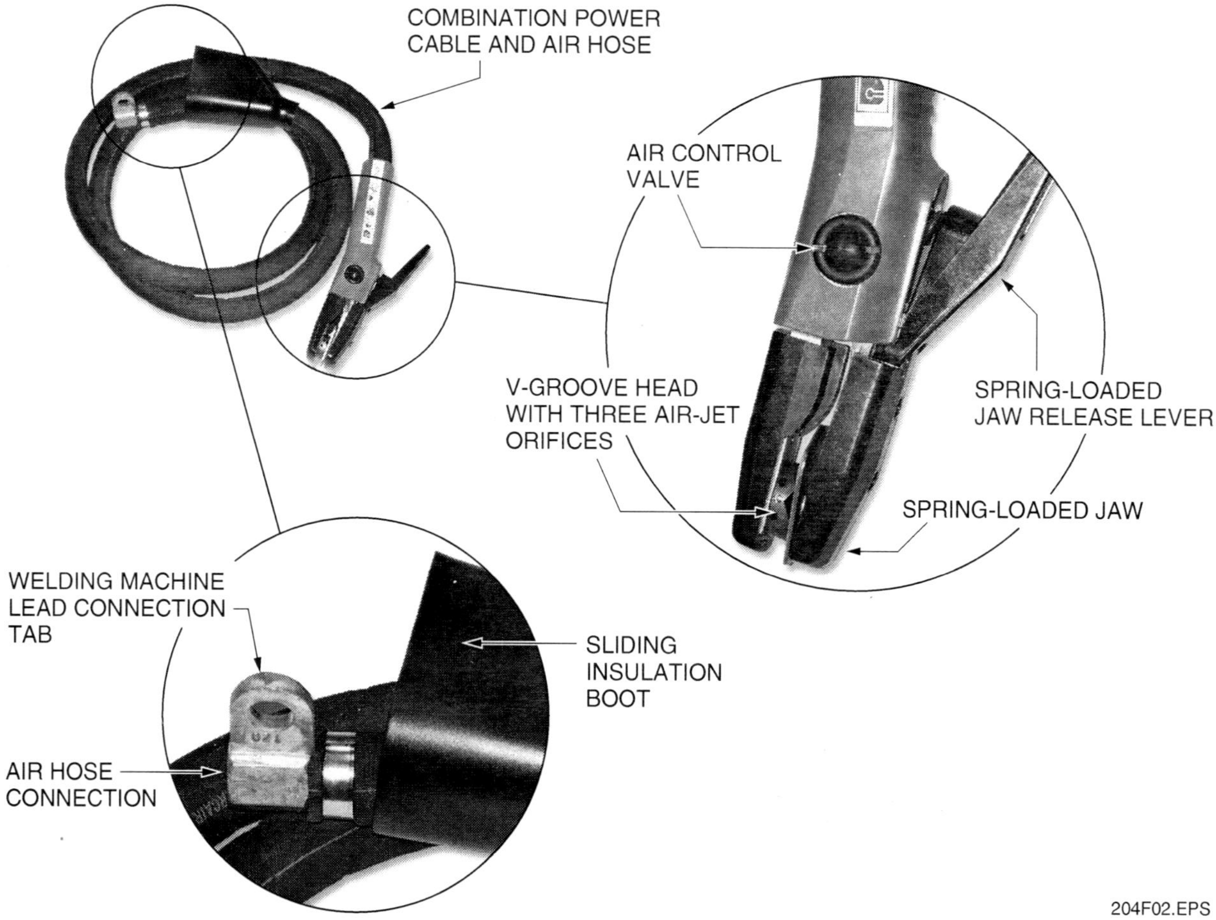

Figure 2 ◆ Typical air carbon arc torch.

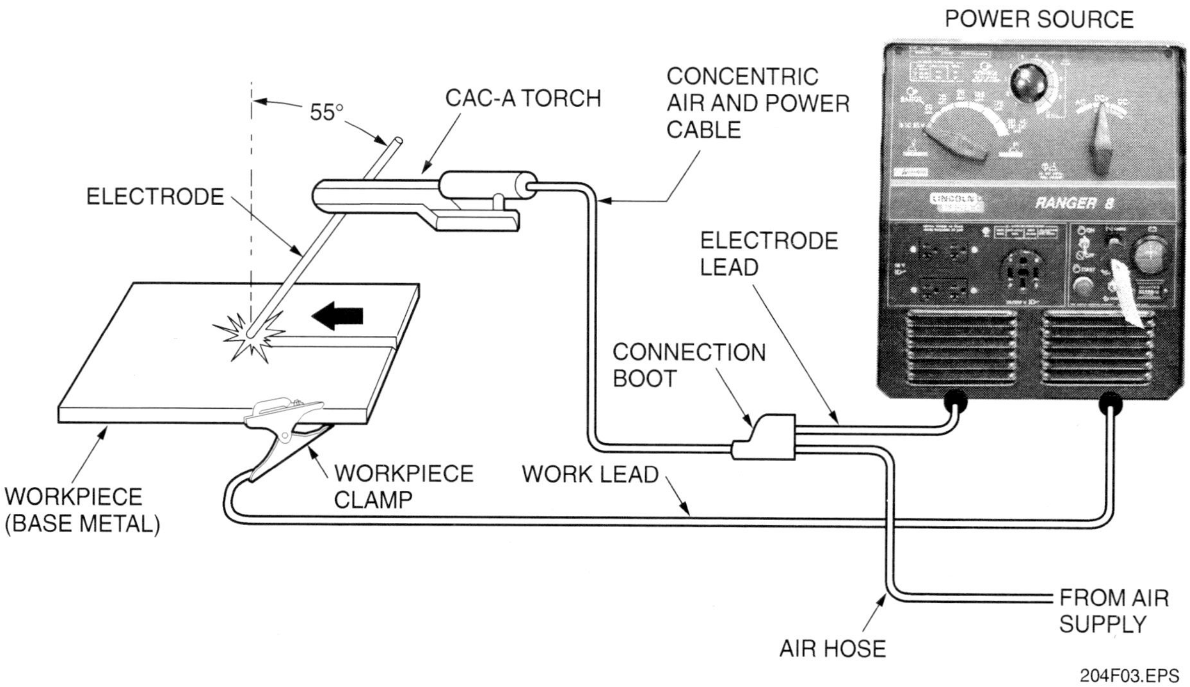

Figure 3 ◆ CAC-A torch connected to welding machine.

If both are V-grooved, one or both heads will contain air jet orifices.

The air jet orifices direct jets of air parallel to the electrode to blow the molten metal away. If required, heads may be replaced or head positions exchanged to alter orifice positions for different operations. When cutting and gouging, the jets must be located under the electrode (between the electrode and the base metal surface). For washing, the jet orifices should be located on one or both sides of the electrode. This helps prevent cutting into the base metal surface. *Figure 4* shows different head types and arrangements for a heavy-duty torch. These particular heads are available in different sizes to accommodate ⅜" to 1" electrodes.

At least one V-groove is needed to position the electrode in the center of the head. The V-grooved head provides about twice as much electrical contact surface as the flat head. A flat head is usually fixed in place, but a V-grooved head is free to rotate. This allows the electrode and grooved head to be swiveled within the torch handle for the best or most comfortable work angle. Pressing the lever on the top of the torch handle raises the upper arm and head, releasing the electrode for repositioning or replacement.

Air passages within the torch supply the head orifices with compressed air. A push-push air control spool valve (air jet control valve) extends through the torch handle. Pushing it from one side of the torch handle opens the valve and air flows to the head orifices; pushing it back from the other side closes the valve.

WARNING!
Always cut away from yourself to avoid injury from sprays of molten metal.

Figure 4 ◆ Head types and arrangements.

The connecting cable of the CAC-A torch contains both an electrical cable and a compressed air hose. For low-current applications, the torch cable consists of a cable and a hose within a common cover. For high-current applications, the torch cable is a concentric conductor that has an air conduit in the center of a hollow electric cable to provide cooling for the electric cable. Concentric

Hot Tip

Foundry Torch

The heads shown in *Figure 4* are designed for interchangeable use in a heavy-duty foundry torch, such as the one shown here. This allows the torch to be used for defect removal or padwashing, or as a general fine removal or piercing tool.

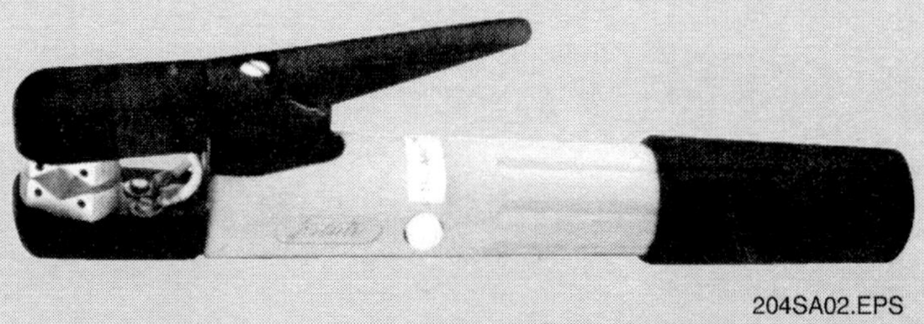

cables may include a swivel on the torch end to give the operator more freedom to maneuver the torch. Some heavy-duty and continuous-duty torches also have additional hoses or lines to carry cooling water to and from the torch, especially with currents exceeding 300A. *Figure 5* shows sections of the two cable types.

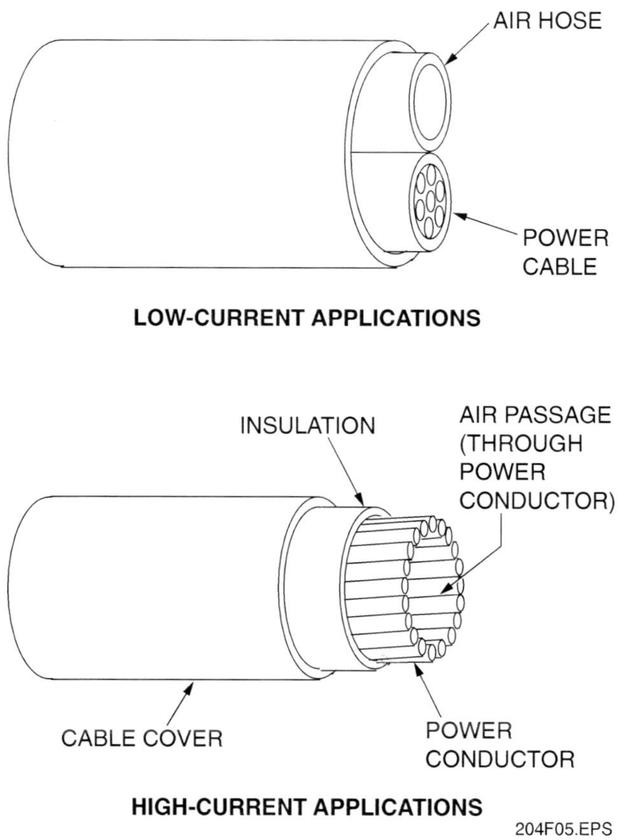

Figure 5 ◆ Torch cable configurations.

2.2.2 CAC-A Cutting Power Supply Options

Either AC or DC constant current welding machines may be used to supply power to CAC-A torches, but DC welding machines are usually preferred. The type of welding current used (AC or DC) depends upon the base metal to be cut or gouged and the electrode type used.

Light-duty gouging and cutting can be done with a 200A machine. Medium-duty gouging and cutting will require a 600A machine. Heavy-duty cutting may require from 1,200A to 2,000A.

NOTE

Avoid using constant voltage (CV) welding machines with CAC-A processes because of their voltage and duty cycle limitations.

Table 1 lists common base metals along with the recommended welding current and polarity to be used with each metal.

Table 1 Welding Current Types for Different Metals

Base Metal	Electrode Type	Current Type and Polarity
Copper alloys	AC DC	AC or DCEN* DCEP**
Iron (cast, ductile, or malleable)	AC DC	AC or DCEN DCEP (High Amps)
Nickel alloys	AC	AC or DCEN
Stainless steel	AC DC	DCEP AC
Steel	DC AC	AC DCEP
Aluminum	DC AC	DCEP AC or DCEN

* Direct current electrode negative (DCEN)
** Direct current electrode positive (DCEP)

2.2.3 CAC-A Torch Air Supply

A CAC-A torch requires regulated compressed air to blow the molten metal from the cut. Compressed air is supplied to the torch through a hose attached to a fitting on the end of the torch cable. While air pressure and volume settings can be over a wide range, they must be high enough that all molten metal will be blown from the cut, and **dross** or slag will not be left in the groove. Light-duty and medium-duty cutting and gouging typically require from 8 cfm to 35 cfm at 40 psig to 80 psig. Heavy-duty cutting can require up to 50 cfm at 80 psig to 100 psig. The regulator should be set for the lowest pressure that will reliably do the job.

Compressor Sizes

A 1hp air compressor can typically supply sufficient air pressure and volume for smaller electrodes, while larger electrodes may require an air compressor of 10hp or greater to supply sufficient air pressure and volume. The torch manufacturer's recommendations for air pressure and flow rates should be followed.

2.2.4 Safety Practices

The following is a summary of safety procedures and practices that must be observed while cutting or welding. Keep in mind that this is just a summary. Complete safety coverage is provided in the Level One module, *Welding Safety*. If you have not completed that module, do so before continuing. Always wear appropriate protective clothing and equipment when welding or cutting.

Here are the safety procedures to remember about protective clothing and equipment:

- Always use safety goggles with a full face shield or a helmet. The goggles, face shield, or helmet lens must have the proper light-reducing tint for the type of welding or cutting to be performed. Never directly or indirectly view an electric arc without using a properly tinted lens.
- Wear proper protective leather and/or flame-retardant clothing along with welding gloves that will protect you from flying sparks and molten metal, as well as heat.
- Wear 8" or taller high-top safety shoes or boots. Make sure that the tongue and lace area of the footwear will be covered by a pant leg. If the tongue and lace area is exposed or the footwear must be protected from burn marks, wear leather spats under the pants or chaps and over the front top of the footwear.
- Wear a solid material (non-mesh) hat with a bill pointing to the rear or, if much overhead cutting or welding is required, a full leather hood with a welding face plate and the correct tinted lens. If a hard hat is required, use a hard hat that allows the attachment of rear deflector material and a face shield.
- For carbon arc cutting, wear a welding hood with the correct tinted lens (shade 12 to 14).
- Wear earplugs to protect ear canals from noise and sparks.

Fire and explosions are a risk when welding and may result in personal injury or death. Always keep these procedures and warnings about fire and explosion prevention in mind while welding:

- Never carry matches or gas-filled lighters in your pockets. Sparks can cause the matches to ignite or the lighter to explode, causing serious injury.
- Never perform any type of heating, cutting, or welding until a hot-work permit is obtained and an approved fire watch is established. Most work-site fires caused by these types of operations are started by cutting torches.
- Never use oxygen to blow off clothing. The oxygen can remain trapped in the fabric for a time. If a spark hits the clothing during this time, the clothing can burn rapidly and violently out of control.
- Make sure that any flammable material in the work area is moved or shielded by a fire-resistant covering. Approved fire extinguishers must be available before attempting any heating, welding, or cutting operations.
- Always comply with any site requirement for a hot-work permit and/or a fire watch.
- Never release a large amount of oxygen or use oxygen as compressed air. Its presence around flammable materials or sparks can cause rapid and uncontrolled combustion. Keep oxygen away from oil, grease, and other petroleum products.
- Never release a large amount of fuel gas, especially acetylene. Methane and propane tend to concentrate in and along low areas and can ignite at a considerable distance from the release point. Acetylene is lighter than air but is even more dangerous. When mixed with air or oxygen, it will explode at much lower concentrations than any other fuel.
- To prevent fires, maintain a neat and clean work area and make sure that any metal scrap or slag is cold before disposal.
- Before cutting or welding containers such as tanks or barrels, check to see if they have contained any explosive, hazardous, or flammable materials including petroleum products, citrus products, or chemicals that decompose into toxic fumes when heated. As a standard practice, always clean and then fill any tanks or barrels with water, or purge them with a flow of inert gas to displace any oxygen.

Always make sure to consider work area ventilation when thinking about safety. Here are some things to keep in mind:

- Make sure confined space procedures are followed before conducting any welding or cutting in the confined space.
- Always perform cutting or welding operations in a well-ventilated area. Cutting or welding operations involving materials, coatings, or electrodes containing cadmium, mercury, lead, zinc, chromium, and beryllium result in toxic fumes. For long-term cutting or welding of such materials, always wear an approved full face, supplied-air respirator (SAR) that uses breathing air supplied externally of the work area. For occasional, very short-term exposure, a high-efficiency particulate arresting (HEPA)-rated or metal-fume filter may be used on a standard respirator.

- Make sure confined spaces are ventilated properly for cutting or welding purposes.
- Never use oxygen in confined spaces for ventilation purposes.

3.0.0 ◆ SELECTION AND INSTALLATION OF CAC-A ELECTRODES

Before selecting and installing CAC-A electrodes, you should have a basic understanding of the construction and applications of various CAC-A electrodes.

3.1.0 CAC-A Electrodes

Electrodes used for air carbon arc cutting and gouging are composed of a blend of carbon and graphite with a binder material, which is baked to produce a uniform structure. They may be plain or coated with copper (copper clad). Electrodes designed for AC current are copper-coated and contain a rare-earth arc-stabilizing ingredient. Coated and uncoated electrodes are available in various types, styles, and sizes. *Figure 6* shows the two types of coated electrodes that are most commonly used.

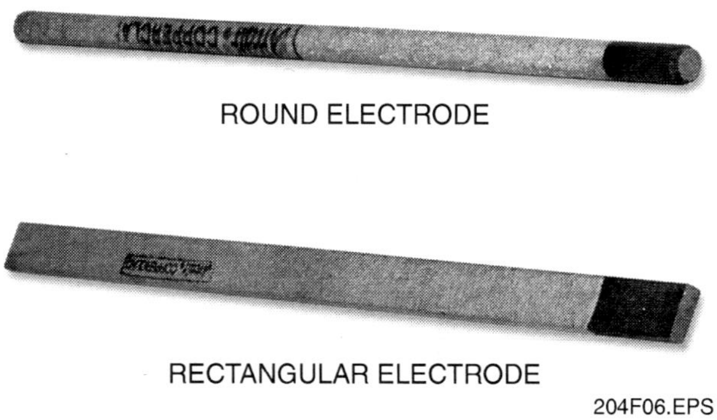

Figure 6 ◆ Copper-coated electrodes.

3.1.1 Electrode Types

CAC-A electrodes are of three basic types:

- *Plain* – Plain carbon electrodes for direct current use do not have copper coating. They run at hotter surface temperatures, **oxidize** along their length, and lose diameter (thickness) between the torch and the arc tip. They are also less expensive, carry less current, and burn more easily.
- *Copper-coated for direct current* – Copper-coated (copper clad) electrodes for DC use are the most common. These electrodes are superior for most applications because they can carry more current, can operate at cooler temperatures, and do not lose diameter through oxidation. However, they cannot be used where copper contamination could be a problem. They are also more costly.
- *Copper-coated for alternating current* – Copper-coated AC electrodes differ from copper-coated direct current electrodes because they contain rare earth materials that help stabilize the arc; copper-coated DC electrodes do not. This type of electrode is sometimes used for direct current electrode negative (DCEN) when cutting cast iron.

Commonly Available Carbon Electrodes

The carbon electrodes that are usually readily available from general welding suppliers are the copper-coated round and rectangular electrodes used with DCEP current. Other types and styles, including jointed, plain, semiround, or copper-coated AC versions, are not commonly used and must be special-ordered from suppliers.

3.1.2 Electrode Styles

Electrodes are manufactured in several styles, each with a range of sizes:

- *Round electrodes* – Round electrodes are the most common type used for manual cutting or gouging. They are manufactured with an uncoated tip that tapers as they begin to be consumed.
- *Round-jointed electrodes* – Round-jointed electrodes are designed for use with continuous-feed automatic and semi-automatic torches. Each electrode has a slightly conical tenon machined on one end and a matching socket machined into the opposite end. These ends allow the electrodes to be joined end-to-end to allow continuous cutting or gouging without interruptions for electrode changes. This system also eliminates stub waste.
- *Special shapes* – Special-shape electrodes are designed to make special-shaped grooves or cuts. Special shapes include flat (rectangular) and half-round. The flat electrodes are designed for close tolerance metal removal. Their uses include removing weld buildups,

shaping dies, removing welded lugs, and beveling edges. Half-round electrodes are designed for cutting wide grooves with torches designed for smaller electrodes. *Figure 7* shows the cross-section shapes of special electrodes.

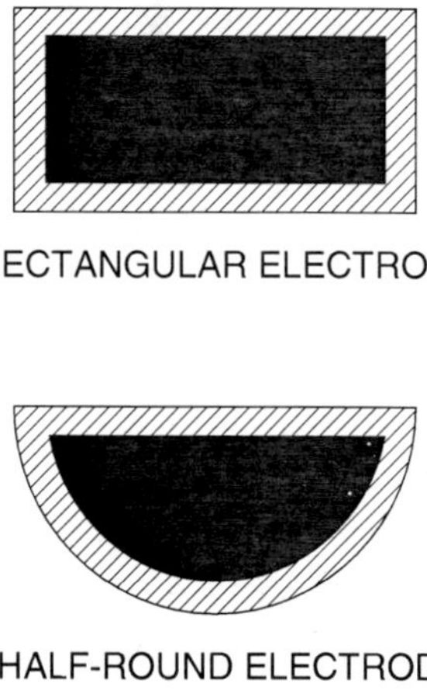

Figure 7 ◆ Cross-section shapes of special electrodes.

3.1.3 Electrode Sizes and Amperages

Round electrode sizes are specified by diameter; flat electrodes are specified by the widths of their widest side.

Plain and copper-coated AC electrodes range in size from 5⁄32" to 5⁄8". They are 12" long.

DC copper-coated electrodes range in size from 1⁄8" to 1". The smaller-sized round and flat electrodes are 12" long. Some of the larger-sized jointed electrodes are 17" long. AC copper-coated electrodes range in size from 3⁄16" to 1⁄2".

Electrode current ratings increase with size. Plain electrodes and AC copper-coated electrodes operate at somewhat lower amperages than the same size DC copper-coated electrodes. AC copper-coated electrodes also operate at substantially lower maximum amperages than the same size DC copper-coated electrodes.

Table 2 shows some typical current ratings for plain DC and copper-covered AC electrodes. These ratings will vary with different manufacturers. For DC copper-covered electrodes, add 10% to the maximum amperage values shown in the table.

3.2.0 Electrode Selection

When performing CAC-A cutting or gouging, the minimum groove width is controlled by the size of the electrode. Groove depth and contour are controlled by electrode angle, travel speed, electrode side-to-side movement, current setting, and air pressure.

With copper-coated electrodes up to 1⁄2" in size on carbon steel, the minimum groove width will run about 1⁄8" wider than the electrode size (diameter).

Table 3 lists copper-coated electrode sizes and travel speeds for various groove sizes in carbon steel.

DCEP Electrode Consumption

When making a gouge that equals the electrode diameter, you will get approximately 8" of groove for every inch of electrode consumed. However, this will vary widely depending on electrode angle, speed, current, and the material being cut.

Table 2 Electrode Current Ratings

ELECTRODE TYPE	ELECTRODE SIZE	MINIMUM CURRENT	MAXIMUM CURRENT
PLAIN DC OR COPPER-CLAD AC	1⁄8"	30 AMP	60 AMP
	5⁄32"	90 AMP	150 AMP
	3⁄16"	150 AMP	200 AMP
	1⁄4"	200 AMP	400 AMP
	5⁄16"	250 AMP	450 AMP
(FOR COPPER-CLAD DC, ADD 10% TO MAXIMUM CURRENT VALUE)	3⁄8"	350 AMP	600 AMP
	1⁄2"	600 AMP	1000 AMP
	5⁄8"	800 AMP	1200 AMP
	3⁄4"	1200 AMP	1600 AMP
	1"	1800 AMP	2200 AMP

204T02.EPS

Table 3 Groove Size, Electrode Size, Amperes, and Travel Speed

GROOVE WIDTH (IN, MM)	GROOVE DEPTH (IN, MM)	ELECTRODE SIZE (DIA.) (IN, MM)	DIRECT CURRENT (AMPERES)	TRAVEL SPEED (IN / MIN)
1/4, 6.4	1/16, 1.6	3/16, 4.8	200	82
5/16, 7.9	1/8, 3.2	1/4, 6.4	300	51
5/16, 7.9	1/4, 6.4	1/4, 6.4	320	29
5/16, 7.9	3/8, 9.5	1/4, 6.4	300	15
3/8, 9.5	1/8, 3.2	5/16, 7.9	320	65
3/8, 9.5	1/4, 6.4	5/16, 7.9	420	31
3/8, 9.5	1/2, 12.7	5/16, 7.9	540	27
7/16, 11.1	1/8, 3.2	3/8, 9.5	560	82
7/16, 11.1	1/4, 6.4	3/8, 9.5	560	75
7/16, 11.1	1/2, 12.7	3/8, 9.5	560	15
7/16, 11.1	11/16, 17.5	3/8, 9.5	560	12
9/16, 14.3	1/4, 6.4	1/2, 12.7	1200	22
9/16, 14.3	3/8, 9.5	1/2, 12.7	1200	21
9/16, 14.3	1/2, 12.7	1/2, 12.7	1200	18
9/16, 14.3	3/4, 19.1	1/2, 12.7	1200	12
13/16, 20.6	1/4, 6.4	5/8, 15.9	1300	29
13/16, 20.6	1/2, 12.7	5/8, 15.9	1300	14
13/16, 20.6	3/4, 19.1	5/8, 15.9	1300	11

204T03.EPS

To select an electrode, follow these steps:

Step 1 Identify the base metal type or alloy.

WARNING!

Never cut or burn cadmium material without proper ventilation and respiratory protection. The process produces poisonous gases that can cause severe personal injury.

CAUTION

Most manufacturers do not recommend using inverters or electronically controlled welding machines for carbon arc cutting. Peak currents are produced that are not usually filtered by the machine, which can damage the internal circuitry.

Step 2 Select a welding machine with the recommended current type (AC or DC), adequate ampere rating, and duty cycle for the base metal type and cut or gouge size.

Step 3 Select an electrode type compatible with the weld current type.

Step 4 Select the electrode size to produce the desired groove width.

3.3.0 Electrode Installation

An air carbon arc electrode is usually gripped about 6" from the arc tip, as shown in *Figure 8*. For aluminum, a 3" stickout is recommended. As the electrode is consumed, it is periodically advanced in the torch handle to maintain the proper projection.

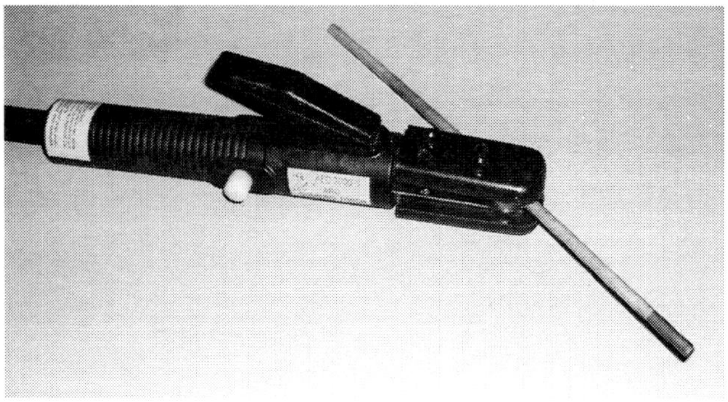

204F08.EPS

Figure 8 ◆ Properly gripped electrode.

4.0.0 ◆ PREPARING WORK AREA AND CAC-A EQUIPMENT

When preparing to perform air carbon arc cutting, the work area must be prepared, along with the CAC-A equipment. The CAC-A equipment should also be tested before being used.

4.1.0 Preparing Work Area for CAC-A

The air carbon arc cutting equipment should be set up at a convenient distance from the welding machine. The area should be well ventilated and clear of all combustible material. The work to be cut or gouged should be placed at a comfortable working position.

Air carbon arc cutting and gouging will blow molten metal considerable distances. It is important that everything combustible be removed from the range of the blown molten metal. If necessary, flame-resistant shields or curtains must be erected to protect any workers or equipment within the range of the molten metal. It is also important to station a fire watch in the area. Make sure a fully charged fire extinguisher is at hand.

Air carbon arc cutting and gouging generates loud noise and heavy smoke and fumes.

> **WARNING!**
>
> Ear protection is essential to protect ears from the noise. Other personal protective equipment must be worn to protect the operator from hot metal and slag.

If the site ventilation is not adequate to keep the smoke and fumes away from the operator, an approved respirator must be worn.

4.2.0 Setting Up CAC-A Equipment

Air carbon arc equipment consists of the arc torch and its attached cable. The torch cable must be connected to a welding machine and a compressed air source. Some torch cables contain a separate air hose and electrical cable within a common cover. Others use a special concentric cable that routes the compressed air through the center of a hollow electrical cable. With the concentric system, a unique combination fitting is used to connect the torch cable to welding power and compressed air, as shown in *Figure 9*. It consists of a special female connector assembly and a hose nut with a pierced tab extending from one side. (Heavy-duty, high-current foundry torches may have two tabs on the nut.) The special fitting is electrically connected to the hollow electrical conductor in the torch cable. A standard air hose threaded end screws into the connector and a welding machine's electrode cable lug bolts to the lug on the side of the fitting. A protective rubber boot on the torch cable covers the connections.

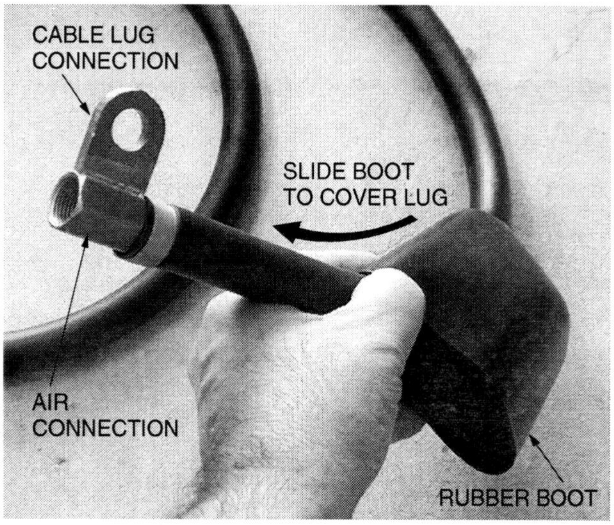

204F09.EPS

Figure 9 ◆ CAC-A torch combination power and air fitting.

4.3.0 Test Operating CAC-A Equipment

Air carbon arc equipment is simple to operate. To operate the equipment, follow these steps:

Step 1 Set the compressed air regulator to the desired pressure and open all valves supplying the CAC-A torch. Operate the air valve to ensure proper flow.

Step 2 Adjust the welding machine for the desired polarity and amperage and turn it on.

Step 3 Put on a cutting helmet. Position the helmet on top of the head, ready to be lowered over the face when needed.

> **WARNING!**
> Use proper hearing protection and appropriate personal protective equipment.

> THINK ABOUT IT
>
> *Ear Protection*
>
> Why is ear protection required when using CAC-A equipment?

Step 4 Put on heat-insulating gauntlet-type welding gloves.

Step 5 Attach the welding machine workpiece clamp to the workpiece.

> **CAUTION**
> Be sure to position the ground clamp where the welding current will not pass through seals, bearings, or other contacting surfaces that could be damaged from arcing.

Step 6 Open the air jet spool valve. Position the torch at a 55° push angle with the air jets below the electrode and between the workpiece and the electrode.

Step 7 Strike an arc by touching the electrode to the workpiece. Do not draw the electrode back after the arc is established. Maintain a short arc and begin washing or gouging as described in the next section.

5.0.0 ◆ WASHING AND GOUGING WITH CAC-A EQUIPMENT

Before using CAC-A equipment for washing or gouging activities, always evaluate the work to be done and plan the best way to achieve the desired results.

5.1.0 CAC-A Planning

Before using CAC-A equipment, always be sure you understand how to position the CAC-A electrode to be most effective. It is also important to inspect the surfaces needing to be cut or gouged. Planning the work saves time, effort, and materials.

5.1.1 Controlling Groove Depth and Contour

Groove depth and contour are controlled by electrode travel angle, tip side movement, and travel speed. For a shallow gouge, the electrode is advanced at a rapid rate. To increase the groove depth, decrease the advance speed, or increase the current setting. To increase the width of the groove, move the electrode tip from side to side or use a larger electrode.

5.1.2 Inspecting CAC-A Cut Surfaces

Inspect the CAC-A cut surface for the following conditions:

- Smooth edges that are relatively free of notches
- No dross
- No carbon deposits

Notches left in a base metal have a tendency to trap slag during welding. Dross or carbon deposits left on a surface to be welded will combine with the base metal to form hard, brittle zones and porosity in the weld zone.

> **CAUTION**
> Check your site's quality standards for the base metal being cut. Cleaning is always necessary after CAC operations to ensure that all carbon and/or dross left by the electrode is removed prior to welding.

5.2.0 CAC-A Washing and Gouging Activities

CAC-A equipment is normally used for washing and gouging activities.

> Hot Tip
>
> *Vertical Washing and Gouging*
>
> Positioning the workpiece in a vertical position simplifies metal ejection by using gravity in addition to the air jets. Make sure that adequate shoes and shoe top coverings are worn if vertical washing and gouging positions are used.

5.2.1 CAC-A Washing

Washing is the process of cutting objects or projections from the surface of a base metal, such as bolt or rivet heads, old welds, and other unwanted materials, to leave a flush surface. Typically, the electrode is used at a push angle of about 55° to the surface that contains the projection. The air orifices must blow air between the electrode and the base metal. The washing is performed by weaving the electrode from side to side across the projection. On some torches with special washing heads, the air jet orifices are located on one side of the electrode. This arrangement reduces the tendency to gouge, but limits the torch to cutting from only one direction.

> **NOTE**
> Special pad washing heads are available.

To perform washing with a CAC-A torch, follow these steps:

> **WARNING!**
> Wear eye and ear protection and protective clothing during this procedure.

Step 1 Select the proper electrode and set the correct polarity and amperage at the welding machine.

Step 2 Turn on the welding machine and start the air flowing through the jet air orifices.

Step 3 Holding the electrode at about a 55° push angle, strike an arc at the edge of the projection to be removed. As the arc melts the metal, swing it gradually across the projection. Continue the process until the projection is removed flush with the surface.

5.2.2 CAC-A Gouging

Gouging is a process of cutting a groove into a surface, usually to prepare it for welding. With CAC-A gouging, the electrode size determines the minimum groove width. Groove depth and bottom contour are determined by the electrode travel angle and speed. Special-shaped electrodes can be used to cut special-shaped grooves. The slower the advance speed, the deeper the groove. Groove width can be increased by weaving the electrode tip slightly from side to side. *Figure 10* shows the result of CAC-A gouging at a 55° push angle, a 0° work angle, and with proper airflow and pressure.

Hot Tip

Beveling

Edge beveling a workpiece for V-groove welds can be accomplished using CAC-A electrodes. One of two methods can be employed. On thin plate, a round electrode can be positioned as shown here and the tip moved back and forth slightly at the bevel angle to accomplish the bevel. For thicker plate, the electrode can be positioned at a 90° work angle to the bevel edge with a push angle equal to the bevel angle. The electrode is then moved along the edge to be beveled in a sawing motion to cut the bevel. The use of a rectangular electrode instead of a round electrode helps to keep the bevel surfaces flat.

204SA03.EPS

204F10.EPS

Figure 10 ◆ CAC-A gouging.

To gouge with a CAC-A torch, follow these steps:

WARNING!
Wear eye and ear protection and protective clothing during this procedure.

Step 1 Select the proper size and shape electrode and set the correct polarity and amperage at the welding machine.

Step 2 Turn on the welding machine and start the air flowing through the air jet orifices.

Step 3 Holding the electrode at about a 55° push angle, strike an arc where the groove is to begin. As the arc melts the metal, advance it steadily along the base metal. Keep the electrode at a 0° work angle with respect to the groove direction.

6.0.0 ◆ PROPER CAC-A EQUIPMENT STORAGE AND HOUSEKEEPING

Proper equipment storage and housekeeping are essential for work efficiency and safety. Good housekeeping requires the following steps:

Step 1 Disconnect the air carbon arc torch from the welding machine cable and air hose and store the torch in its assigned location.

Step 2 Store the air hose, welding cables, and any tools used in their assigned locations. *Figure 11* shows proper and improper hose/cable storage.

Step 3 Clean off the welding table (if used) and sweep up the slag and debris that were blown around the area.

6.1.0 Repair of CAC-A Equipment

Following proper equipment storage and housekeeping procedures usually provides for trouble-free cutting and minimal repairs. However, cables and air hoses may become damaged and require

Work Angle

This illustration shows the effects of an improper work angle. Dross from the gouging operation has blown to one side instead of off the surface of the workpiece.

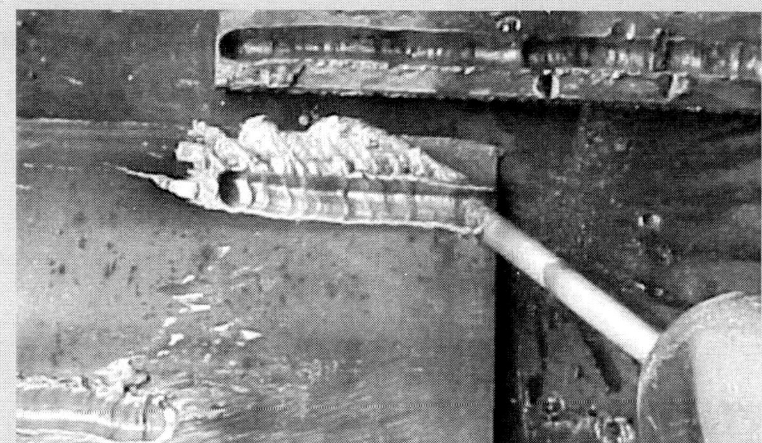

204SA04.EPS

PROPER AIR HOSE STORAGE
(CAC-A TORCH REMOVED)

IMPROPER AIR HOSE STORAGE
(CAC-A TORCH NOT REMOVED)

204F11.EPS

Figure 11 ◆ Hose/cable storage.

replacement, especially if they have been run over by forklifts or other mobile equipment, dragged across hot metal, or burned by molten-metal cutting sparks.

Before replacing a cable, always make sure the electrical disconnect is in the OFF position. Remove the cable from the machine using the correct size combination wrench. If an electrode holder becomes cracked or worn, it may be possible to replace only that part.

Summary

Air carbon arc cutting is very useful for washing surfaces flush or for gouging out cracks to prepare them for welding repairs. However, because the torch air jets are continuously blasting molten metal from the arc, good site fire-prevention practices and use of appropriate personal protective equipment are essential. Ear protection is especially important to prevent noise damage from the air blast and to prevent sparks or slag from entering the ears.

Review Questions

1. The CAC-A process uses a high-current electric arc between the base metal and the end of a(n) _____ to melt the base metal.

a. carbon-graphite electrode
b. high-frequency pilot arc
c. plasma jet
d. idle arc

2. During the CAC-A process, jets of compressed air are used to blow the _____ away from the cut.

a. carbon dioxide
b. molten metal
c. nitrogen
d. carbon arc

3. When cutting and gouging with CAC-A equipment in the flat position, the jet orifices of the torch head(s) must be located _____ the electrode.

a. beside
b. above
c. under
d. near

4. The electrode tip should be no farther than _____ from the electrode tip to the contacts to prevent overheating.
   a. 1"
   b. 2"
   c. 4"
   d. 6"

5. Light-duty gouging or cutting is typically done with a _____ welding machine.
   a. 10A
   b. 200A
   c. 600A
   d. 1,200A

6. Heavy-duty cutting can require a torch air supply up to _____ cfm at 80 psig to 100 psig.
   a. 20
   b. 30
   c. 40
   d. 50

7. Compressed air is supplied to the torch through a hose attached to a fitting on the _____.
   a. welding machine
   b. end of the torch cable
   c. air jet orifices
   d. work cable

8. Electrodes designed for AC current contain a rare-earth _____ ingredient.
   a. magnet
   b. undercurrent
   c. amplifier
   d. arc-stabilizing

9. The most common style of CAC-A electrode used for manual cutting or gouging is the _____ electrode.
   a. half-round
   b. flat
   c. round
   d. round-jointed

10. When performing CAC-A cutting or gouging, the minimum groove width is controlled by the _____ being used.
   a. size of the torch
   b. size of the electrode
   c. speed of travel
   d. angle of the electrode holder

11. Because air carbon cutting and gouging activities tend to blow molten metal considerable distances, it is important to station a(n) _____ in the CAC-A work area.
   a. observer
   b. slag trap
   c. oxygen monitor
   d. fire watch

12. To wash a projection from a metal surface using CAC-A equipment, the electrode is typically used at a push angle of about _____ to the base metal surface.
   a. 30°
   b. 45°
   c. 55°
   d. 90°

13. Torches with special washing heads in which the air jet orifices are located on one side of the electrode are limited to cutting from _____ direction(s).
   a. one
   b. two
   c. three
   d. four

14. In CAC-A gouging, groove depth is determined by the _____.
   a. electrode travel angle and speed
   b. electrode size
   c. electrode weave
   d. electrode length

15. In CAC-A gouging, bottom contour is determined by the _____.
   a. electrode travel angle and speed
   b. electrode size
   c. electrode weave
   d. electrode length

GLOSSARY

Trade Terms Introduced in This Module

Carbon-graphite electrode: An electrode composed of a mixture of soft amorphous carbon and hard graphite carbon that may be coated with copper.

Dross: Any solidified, oxidized, metallic material adhering to a workpiece adjacent to a cut surface after a thermal cutting operation.

Oxidize: To combine with oxygen such as in burning (rapid oxidation) or rusting (slow oxidation).

Washing: A term used to describe the process of thermally cutting objects or projections from the surface of a base metal.

APPENDIX

Performance Accreditation Tasks

The Performance Accreditation Tasks (PATCs) correspond to and support the learning objectives in *AWS EG2.0-95, Guide for the Training and Qualification of Welding Personnel: Entry-Level Welder.*

PATCs provide specific acceptable criteria for performance and help to ensure a true competency-based welding program for students.

The following tasks are designed to test your competency with an air carbon arc cutting torch. Do not perform a cutting task until directed to do so by your instructor.

AIR CARBON ARC WASHING AND GOUGING

PERFORM AIR CARBON ARC WASHING

Using any of the materials identified below, perform air carbon arc washing to remove the portion identified by the instructor. Materials that can be used for this task include:

- Steel backing strip on a butt weld
- Excess buildup on the face of a weld
- Rivets or bolts in a plate
- Blocks, angles, clips, eyes, D-rings, or items welded to a plate

Criteria for Acceptance

- Material removed flush with the base metal surface __________
- No notching in the surface of the base metal __________

PERFORM AIR CARBON ARC GOUGING

Using mild steel plate ½" thick or thicker, gouge a U-groove at least 8" long as shown in the figure.

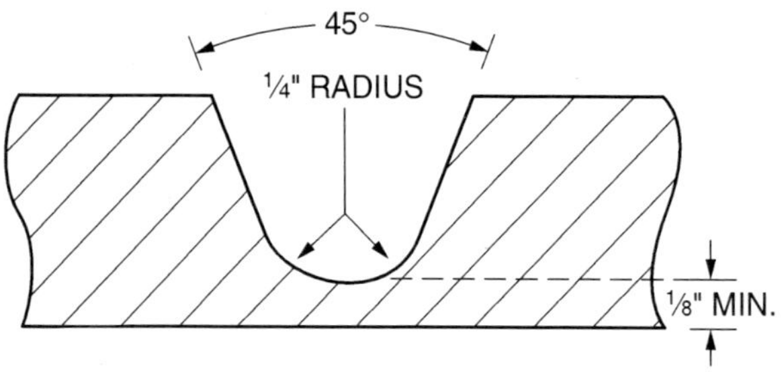

204A01.EPS

Criteria for Acceptance

- Groove width and depth are uniform to ±⅛" __________
- Root face minimum is ⅛" __________
- Groove bottom radius minimum is ¼" __________
- Included angle of groove walls is 45°, ±10° __________
- Groove walls are smooth and uniform __________
- No slag within the groove __________

Additional Resources

This module is intended to be a thorough resource for task training. The following reference works are suggested for further study. These are optional materials for continued education rather than for task training.

Modern Welding. W.A. Bowditch, K.E. Bowditch, A.D. Althouse, and C.H. Turnquist. Tinley Park, IL: The Goodheart-Willcox Company, Inc.

Modern Welding Technology. Howard B. Cary. Englewood Cliffs, NJ: Prentice-Hall, Inc.

Welding Skills Workbook. J.W. Giachino and W.R. Weeks. Homewood, IL: American Technical Publishers, Inc.

Welding Technology. Gower A. Kennedy. Indianapolis, IN: The Bobbs-Merrill Company, Inc.

ANSI C5.3, Latest Edition. Recommended Practices for Air Carbon Arc Gouging and Cutting.

Figure Credits

Gerald Shannon	204SA01, 204F02, 204F03, 204F06
Thermadyne Industries	204F04, 204SA02
Chuck Rogers	204F08, 204F09
Rex Ball	204F10, 204SA03, 204SA04, 204F11

CONTREN™ LEARNING SERIES — USER UPDATES

The NCCER makes every effort to keep these textbooks up-to-date and free of technical errors. We appreciate your help in this process. If you have an idea for improving this textbook, or if you find an error, a typographical mistake, or an inaccuracy in NCCER's Contren™ textbooks, please write us, using this form or a photocopy. Be sure to include the exact module number, page number, a detailed description, and the correction, if applicable. Your input will be brought to the attention of the Technical Review Committee. Thank you for your assistance.

Instructors – If you found that additional materials were necessary in order to teach this module effectively, please let us know so that we may include them in the Equipment/Materials list in the Instructor's Guide.

Write: Curriculum Revision and Development Department
National Center for Construction Education and Research
P.O. Box 141104, Gainesville, FL 32614-1104

Fax: 352-334-0932

E-mail: curriculum@nccer.org

Craft ____________________ Module Name ____________________

Copyright Date ____________ Module Number ____________ Page Number(s) ____________

Description

____

____

____

(Optional) Correction

____

____

(Optional) Your Name and Address

____

____

Module 29205-03

Plasma Arc Cutting (PAC)

COURSE MAP

This course map shows all of the modules in the second level of the Welding curriculum. The suggested training order begins at the bottom and proceeds up. Skill levels increase as you advance on the course map. The local Training Program Sponsor may adjust the training order.

WELDING LEVEL TWO

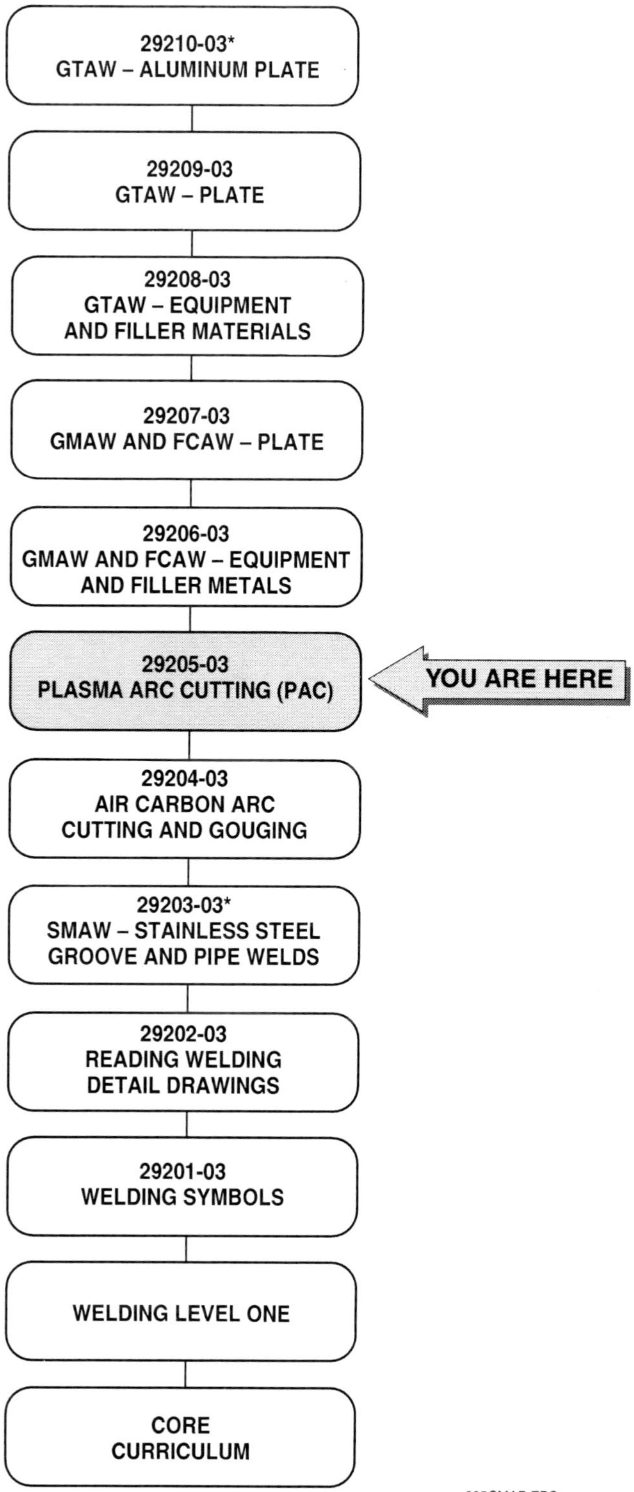

205CMAP.EPS

Please note that Modules 29203-03 and 29210-03 are electives for *Welding Level Two* but are requirements for *AWS Entry Level Welder.

MODULE 29205-03 CONTENTS

Figures

Table

MODULE 29205-03

Plasma Arc Cutting (PAC)

Objectives

When you have completed this module, you will be able to do the following:

1. Identify and understand plasma arc cutting processes.
2. Identify plasma arc cutting equipment.
3. Prepare and set up plasma arc cutting equipment.
4. Use plasma arc cutting equipment to make various types of cuts.
5. Properly store equipment and clean the work area after use.

Prerequisites

Before you begin this module, it is recommended that you successfully complete the following: Core Curriculum; Welding Level One; Welding Level Two, Modules 29201-03 through 29204-03.

Required Trainee Materials

1. Pencil and paper
2. Appropriate personal protective equipment

1.0.0 ◆ INTRODUCTION

Plasma arc cutting (PAC) uses a jet of **plasma** to pierce, cut, and gouge metal. The plasma is created by superheating gas in an electric arc. The temperature of the plasma jet is hot enough to melt any metal. The process can be set up for manual or mechanized operation (*Figure 1*) and in most cases is faster and more efficient than any other cutting method. PAC offers the following advantages:

- Cuts both ferrous and nonferrous metals
- Produces minimal dross
- Creates a very minimal heat-affected zone

MANUAL PLASMA ARC CUTTING

MECHANIZED PLASMA ARC CUTTING

205F01.EPS

Figure 1 ◆ Plasma arc cutting (PAC).

- Causes very little or no distortion
- Possesses high cutting speeds
- Produces a very narrow kerf

The plasma arc cutting process requires an electrical power supply and a cutting gas, which can include compressed air. Depending on the equipment, a separate shielding or cooling gas may also be required. PAC torches convert the cutting gas into plasma by passing the gas through an electric arc inside the torch. The torch has a relatively small opening (orifice), which constricts the arc and high-pressure gas flow. This results in the plasma stream exiting the torch as a supersonic jet hotter than any flame. Depending on the current flow, the plasma cutting jet may reach temperatures of 30,000°F or higher. When the jet contacts metal, it transfers its tremendous heat, causing the metal to instantly melt. The high-velocity jet blasts the molten metal away, forming a hole, groove, or gouge.

Figure 2 shows plasma arc cutting.

1.1.0 Safety Practices

The following is a summary of safety procedures and practices that must be observed while cutting or welding. Keep in mind that this is just a summary. Complete safety coverage is provided in the Level One module, *Welding Safety*. If you have not completed that module, do so before continuing. Above all, be sure to wear appropriate protective clothing and equipment when welding or cutting.

1.1.1 Protective Clothing and Equipment

- Always use safety goggles with a full face shield or a helmet. The goggles, face shield, or helmet lens must have the proper light-reducing tint for the type of welding or cutting to be performed. Never directly or indirectly view an electric arc without using a properly tinted lens.

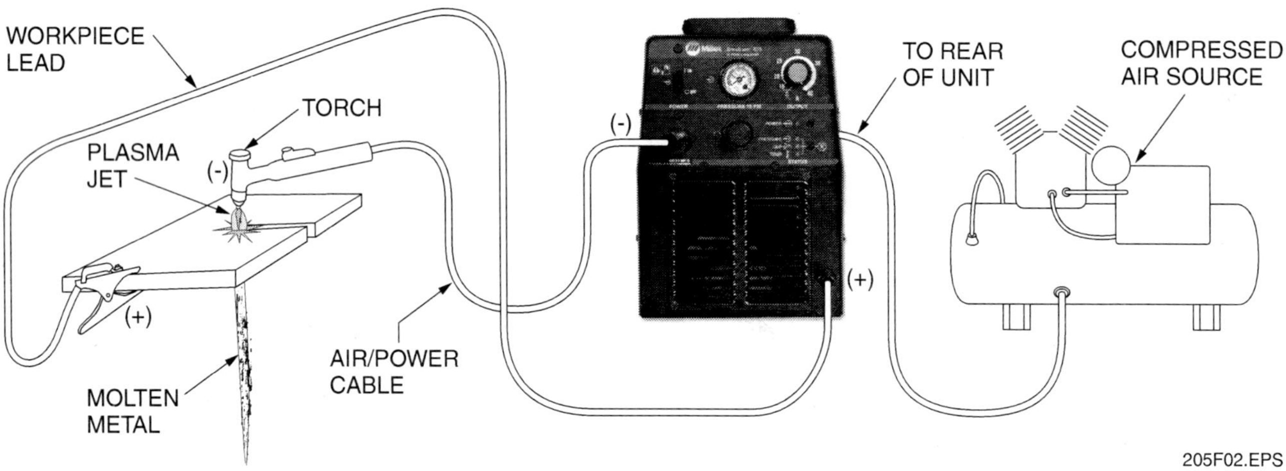

Figure 2 ◆ Typical plasma arc cutting setup diagram.

PAC Safety

A plasma arc cutting process that uses air or nitrogen and is set up correctly is generally safer than oxyfuel processes. There is no chance of flashbacks and other dangers caused by flammable gases, like those found in oxyfuel processes. However, the voltages used, the extremely high temperatures of the cutting jet, and the intense ultraviolet radiation of the arc are dangerous.

In addition to the safety practices given in the following section, the practices listed below should be observed:

- Make sure that the PAC power source is turned off before disassembling a plasma torch to replace any consumable components. Open-circuit DC voltages as high as 400 volts may be present between the torch components and the workpiece to which the workpiece lead is connected.
- The extremely high temperatures of the plasma cutting jet can quickly cut through gloves and flesh, resulting in severe injuries. Never place your hands under the workpiece near the cutting jet.

- Wear proper protective leather and/or flame retardant clothing along with welding gloves that will protect you from flying sparks and molten metal, as well as heat.
- Wear 8" or taller high-top safety shoes or boots. Make sure that the tongue and lace area of the footwear will be covered by a pant leg. If the tongue and lace area is exposed or the footwear must be protected from burn marks, wear leather spats under the pants or chaps and over the front top of the footwear.
- Wear a solid material (non-mesh) hat with a bill pointing to the rear or, if much overhead cutting or welding is required, a full leather hood with a welding face plate and the correct tinted lens. If a hard hat is required, use a hard hat that allows the attachment of rear deflector material and a face shield.
- If a full leather hood is not worn, wear a face shield and snug-fitting welding goggles over safety glasses for gas welding or cutting. Either the face shield or the lenses of the welding goggles must be an approved shade five or six filter. For electric arc welding or cutting, wear safety goggles or a welding hood with the correct tinted lens (shade 8 to 14 for PAC).
- If a full leather hood is not worn, wear earplugs to protect your ear canals from sparks.

1.1.2 Fire/Explosion Prevention

- Never carry matches or gas-filled lighters in your pockets. Sparks can cause the matches to ignite or the lighter to explode causing serious injury.
- Never perform any type of heating, cutting, or welding until a hot-work permit is obtained and an approved fire watch is established. Most work-site fires caused by these types of operations are started by cutting torches.
- Never use oxygen to blow off clothing. The oxygen can remain trapped in the fabric for a time. If a spark hits the clothing during this time, the clothing can burn rapidly and violently out of control
- Make sure that any flammable material in the work area is moved or shielded by a fire-resistant covering. Approved fire extinguishers must be available before attempting any heating, welding, or cutting operations.
- Always comply with any site requirements for a hot-work permit and/or a fire watch.
- Never release a large amount of oxygen or use oxygen as compressed air. Its presence around flammable materials or sparks can cause rapid and uncontrolled combustion. Keep oxygen away from oil, grease, and other petroleum products.
- Never release a large amount of fuel gas, especially acetylene. Methane and propane tend to concentrate in and along low areas and can ignite at a considerable distance from the release point. Acetylene is lighter than air but is even more dangerous than methane. When mixed with air or oxygen, it will explode at much lower concentrations than any other fuel.
- To prevent fires, maintain a neat and clean work area and make sure that any metal scrap or slag is cold before disposal.
- Before cutting or welding containers such as tanks or barrels, check to see if they have contained any explosive, hazardous, or flammable materials, including petroleum products, citrus products, or chemicals that decompose into toxic fumes when heated. As a standard practice, always clean and then fill any tanks or barrels with water or purge them with a flow of inert gas to displace any oxygen.

1.1.3 Work Area Ventilation

- Make sure confined space procedures are followed before conducting any welding or cutting in the confined space.
- Always perform cutting or welding operations in a well-ventilated area. Cutting or welding operations involving materials, coatings, or electrodes containing cadmium, mercury, lead, zinc, chromium, and beryllium result in toxic fumes. For long-term cutting or welding of such materials, always wear an approved full face, supplied-air respirator (SAR) that uses breathing air supplied outside of the work area. For occasional, very short-term exposure, a high-efficiency particulate arresting (HEPA)-rated or metal-fume filter may be used on a standard respirator.
- Make sure confined spaces are ventilated properly for cutting or welding purposes.
- Never use oxygen in confined spaces for ventilation purposes.

2.0.0 ◆ PLASMA ARC CUTTING (PAC) PROCESS

There are two types of plasma arc cutting processes: transferred arc and nontransferred arc. The transferred arc process is the most common and is used to cut materials that are electrically conductive, such as metals. The nontransferred arc process is less common and is used to cut materials that are not electrically conductive, such as ceramics and concrete.

2.1.0 Transferred Arc Process

In the transferred arc process, the workpiece is part of the electrical circuit as in other arc processes. The arc is established between the electrode and the work via an arc transfer process. A number of different arc initiation methods are used for the transfer process. These include contact starting, discharge starting to the workpiece, and discharge starting using a pilot arc. Of these methods, discharge starting using a pilot arc is the most common. In this method, a low-voltage and low-current, high-frequency (HF) arc is established between the electrode and the torch nozzle inside the torch. This small arc between the torch electrode and torch nozzle is called a *pilot arc*. The pilot arc is used to ionize the air jet and results in a plasma. When the torch is near the workpiece, a higher-voltage, current-cutting arc is established (transferred) between the torch electrode and the work via the plasma. The higher-voltage and current-cutting arc maintains the plasma jet and the low-current, high-frequency pilot arc is shut off. For the transferred arc process to work, one lead of the power supply (the positive lead) must be connected to the metal to be cut. *Figure 3* shows the transferred arc process.

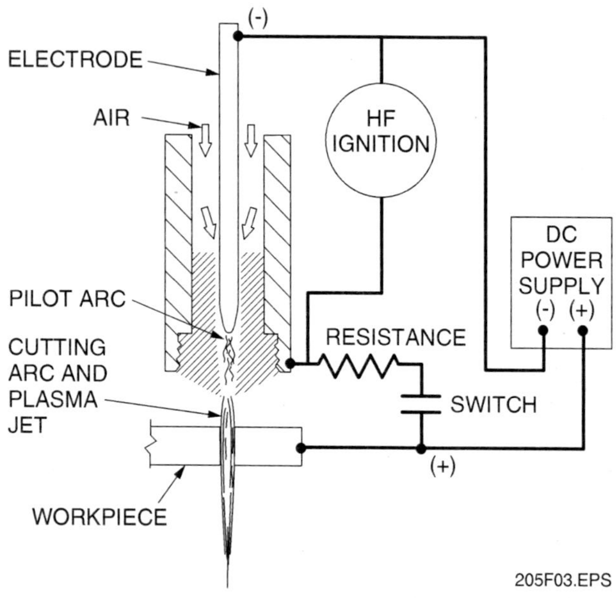

Figure 3 ◆ Schematic of transferred arc process.

2.2.0 Nontransferred Arc Process

In the nontransferred arc process, the entire arc is established within the torch, between the torch electrode and the torch nozzle. The material being cut is not electrically connected to the arc circuit. The nontransferred arc process is not normally found in facilities that cut only metal.

Figure 4 shows the nontransferred arc process.

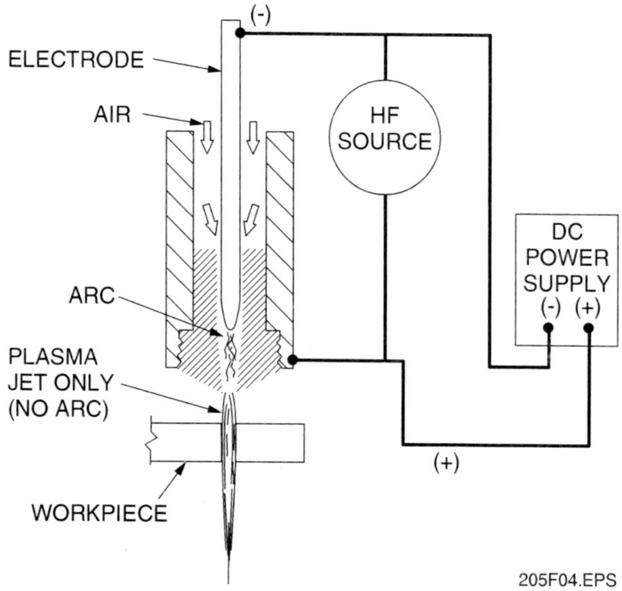

Figure 4 ◆ Schematic of nontransferred arc process.

3.0.0 ◆ PLASMA ARC CUTTING EQUIPMENT

Basic plasma arc cutting equipment includes the following items:

- Power source control unit
- Plasma arc cutting torch with torch cable
- Workpiece lead and lead assembly
- Plasma, shielding and cooling gases, and gas control components

There are many different manufacturers of PAC equipment. The following sections will provide general information common to most types. For specific information on controls, operation, and parts, refer to the manufacturer's manuals for specific equipment at your site.

3.1.0 PAC Power Source-Control Units

Plasma arc cutting is performed with DCEN current. The power sources are special units designed specifically for plasma arc cutting and are available in many sizes and configurations. The smallest and

The Hazards of PAC

Ultraviolet radiation, particle matter, and noise are hazards of PAC, but these are manageable with the proper equipment. Devices such as water tables and water mufflers use water to control such hazards in mechanized plasma applications.

Cutting Ratings

Manufacturers may rate the cutting capability of their equipment in different manners. Some may rate at the listed duty cycle and others may rate at the maximum cutting (severing) capability at a much reduced duty cycle. When selecting a power supply, make sure that you understand the manufacturer's rating system.

simplest PAC units are designed for light-duty manual cutting of sheet metal and light gauge plate. They typically plug into a single-phase 115VAC or 230VAC outlet. Typical maximum cutting amperages for these light-duty units range from 14 to 30 amperes. Their **duty cycle** is usually in the 35% range. The simplest units use filtered compressed air for plasma and cooling gas. They have a high frequency generator for easy arc starting. Some of these small air units even contain their own air compressor to supply gas for cooling and plasma generation. Commercial-grade portable units are available that can operate on various single-phase or three-phase voltages and are rated for cutting material up to 1" thick at duty cycles of 50% to 60%. At maximum current and reduced duty cycles, they are capable of severing material up to 1½" thick. These units normally require an external air or gas supply for the torch. The size of the unit needed is determined by the type and thickness of the metal to be cut.

Console controls are simple. They usually include most of the following controls and indicators:

- Power ON/OFF switch
- Power ON and/or unit-ready indicator
- Air or gas ON/OFF switch or setting (for checking and adjusting torch air or gas flow, or for purging lines)
- Output current control
- Power source trouble indicator(s)

Figure 5 shows a typical commercial-grade plasma arc cutting unit.

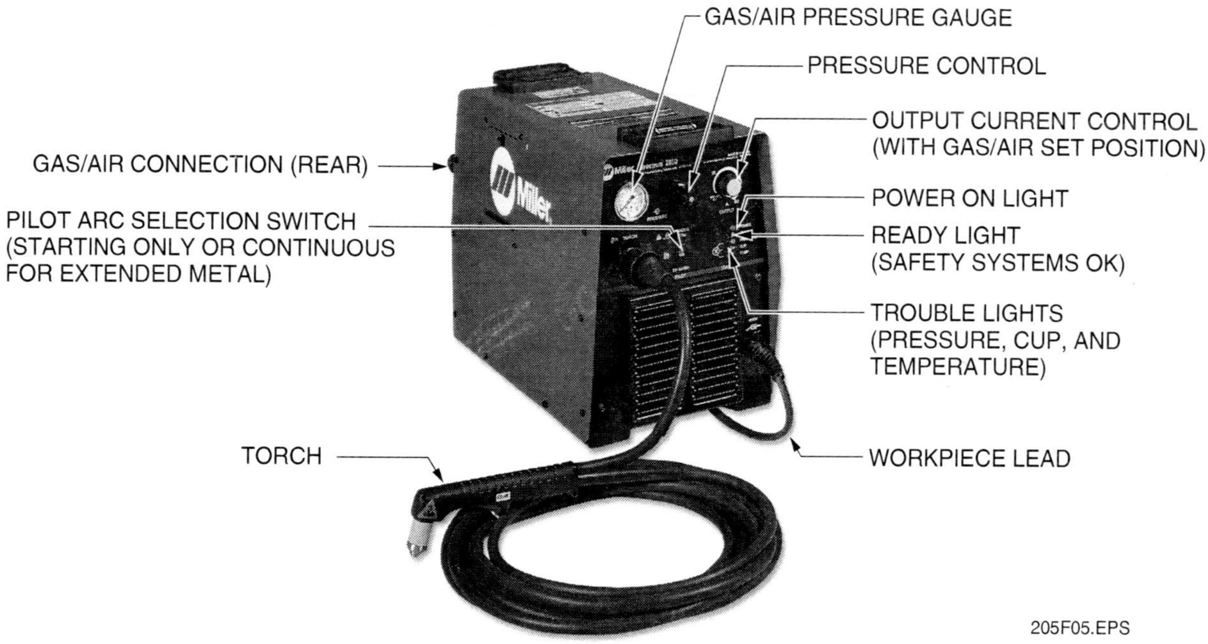

Figure 5 ◆ Typical commercial-grade plasma arc cutting unit.

Hot Tip

Control Interlocks

Interlocks may be incorporated in the controls of PAC systems. An interlock common among PAC systems monitors the air supply pressure. If the plasma torch is run without an adequate supply of air, internal arcing may damage the torch. For this reason, a gas pressure switch interlock is included in the circuit. Adequate gas pressure must be present before the torch will operate. The interlock also shuts down the torch if there is a gas supply failure during cutting. Another interlock is an overtemperature switch that shuts down the unit if the duty cycle is exceeded and the unit overheats. Some units are also equipped with a retaining cup switch that shuts down the unit if the retaining cup holding the tip, nozzle, and swirl ring in the torch is loose.

Larger industrial units are used for heavy-duty manual cutting and general-duty mechanized cutting. They usually operate on a three-phase 230/460 VAC input power and can produce DC output currents up to 750 amperes. Their duty cycle is usually 100%. They can cut stainless steel and aluminum up to 2½" thick and carbon steel up to 3" thick. Some units are water cooled, often containing their own closed-loop torch cooling system.

The larger units are often dual-flow mechanized units that operate with several types of gases. One gas is used for cutting and the other is used as a shield around the cutting area to improve the quality of the cut. Some units may use only one of the gases for the pilot arc.

Although gouging is not the primary use for plasma arc torches, some commercial and industrial units may be used for gouging when fitted with a gouging torch or a torch with gouging components. Plasma arc gouging is sometimes referred to with the nonstandard term PAG.

Controls on the larger units typically include a power ON/OFF switch, output current control and ammeter, several gas flow controls and gauges, gas selector controls, local/remote switch, and open circuit voltmeter. *Figure 6* shows a typical industrial manual plasma arc cutting unit. Note the leather cable cover accessory used to protect the torch cable from damage due to hot or molten metal.

3.2.0 PAC Torches

The most common plasma arc cutting torch uses a pilot arc and transferred arc system. In many torches, a high-frequency pilot arc is started between the electrode and the nozzle. A small amount of gas is injected into the arc chamber where it is heated to a plasma and escapes through the nozzle as a fine jet. However, the torch current circuit is designed so that the full voltage (**potential**) and current is not available in the torch between the electrode and nozzle. The base metal to be cut is connected to the control unit power source by the workpiece clamp. This gives the workpiece a higher potential difference than the nozzle. When the torch is near the workpiece, the

205F06.EPS

Figure 6 ◆ Typical industrial manual plasma arc cutting unit.

Hot Tip

Auto Voltage Power Sources

Some newer PAC units sense the input voltage and frequency when they are connected to a power source. The units continuously and automatically reconfigure themselves to operate properly with the connected power. This eliminates improper initial power connections and provides continuous adjustment to changing input voltage so that the PAC arc current remains stable during cutting operations.

higher-current arc extends (transfers) to the workpiece because of its higher potential. The control unit automatically increases the amperage and gas flow(s) to produce the longer cutting arc and shuts off the pilot arc. *Figure 7* shows the plasma transferred arc cutting process that takes place between the torch and the workpiece.

205F07.EPS

Figure 7 ◆ A transferred plasma cutting arc.

Plasma arc torches may be handheld for manual cutting or machine mounted for mechanized cutting. The heaviest duty torches are usually mechanized and water cooled. Handheld torches are usually light-duty to medium-duty and may be water cooled in the heavier duty versions. Some torches are equipped with separate passages and orifices for cooling and shielding gases. Torches have a button or lever on the torch handle to start the pilot arc. A foot pedal can also be used to start the arc. When a foot pedal is used, the switch on the control panel is set for remote operation.

The typical plasma arc cutting torch usually contains a replaceable retaining cup, copper electrode with a hafnium or thoriated tungsten insert, a gas swirl ring, and a nozzle (tip). Nozzles have various-sized holes for cutting metals of different thicknesses. The torch contains coolant passages for gas or water cooling. *Figure 8* shows a typical handheld PAC torch and its consumable components.

Mechanized PAC torches are designed to be mounted in automated carriers. Usually, they can be used in the same carriers designed for oxyfuel cutting heads. However, the oxyfuel carriers must be capable of moving at the much greater speeds of plasma arc cutting.

Some types of PAC torches (water injection torches) inject pressurized water to create a whirlpool of water around the jet, below the point where the arc narrows in the torch's arc cavity. The water swirls around the plasma jet at high velocity and compresses the jet even tighter. The narrower jet cuts with a more uniform melt rate and leaves the kerfs of stainless steel and titanium free of slag. This type of torch is normally mechanized.

As shown in *Figure 8*, several styles of shrouds, heat shields, nozzles (tips), and electrodes are usually available for a given model of PAC torch. These components vary in style and size to adapt to different metal types and thicknesses. Sometimes nozzles and parts for gouging are offered to fit some of the heavier duty torches. Electrode types and styles vary with torch manufacturers. The most common parts that are changed in a torch are the electrode and the nozzle. The electrode is changed when it becomes contaminated, and the nozzle is changed when it becomes worn (orifice washed out) or when metal of a different type or thickness is to be cut.

Contact-Start or Discharge-Start Torches

High-frequency pilot arc starting of a PAC torch can interfere with nearby electronic circuits, including computers and microprocessor controllers. To eliminate the use of a high-frequency start, some newer PAC torches use an internal contact or capacitor-discharge pilot arc starting system. One such contact-start system has a spring-loaded electrode that is in contact with the nozzle when the torch is off. When a trigger is pressed, the PAC unit senses the shorted electrode via high-current flow and opens the gas solenoid. Gas pressure then forces the electrode away from the nozzle. This action draws an arc that establishes the plasma flow. The arc instantly transfers to the workpiece because of the higher voltage difference between the electrode and the workpiece.

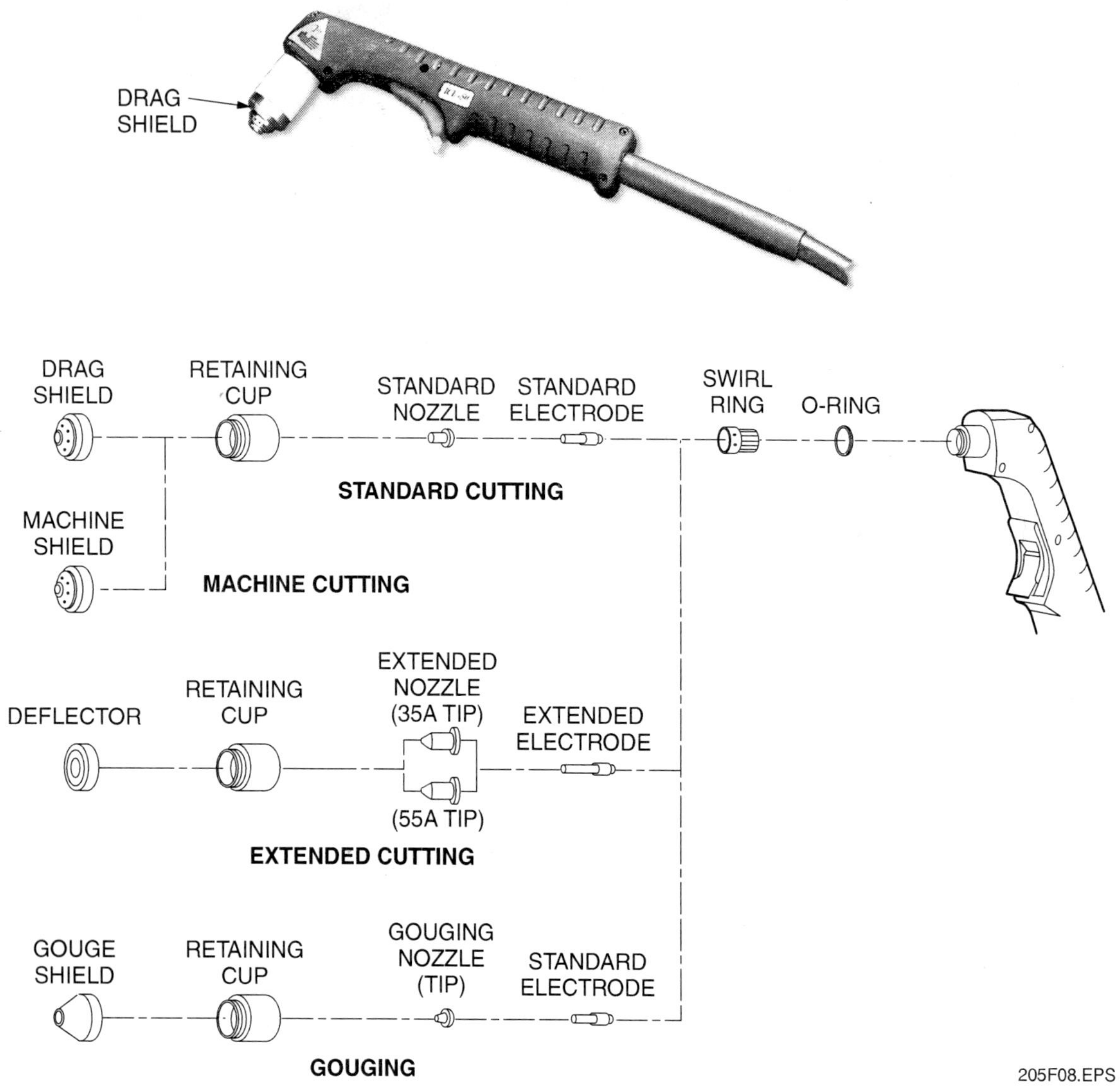

Figure 8 ◆ Typical handheld torch and consumables.

Hot Tip

Standoff Guides and Shields

When performing low-amperage cutting, it is possible to drag the torch on the workpiece using a drag shield without causing damage or sacrificing the cut. Cutting at current levels above 60 amps or extended cutting operations usually require a standoff guide, which attaches to the cup and holds the tip the required distance from the workpiece. The standoff distance (torch tip to work distance) should be kept between 1/16" and 1/4", depending on the thickness of the workpiece.

STANDOFF ROLLER GUIDE

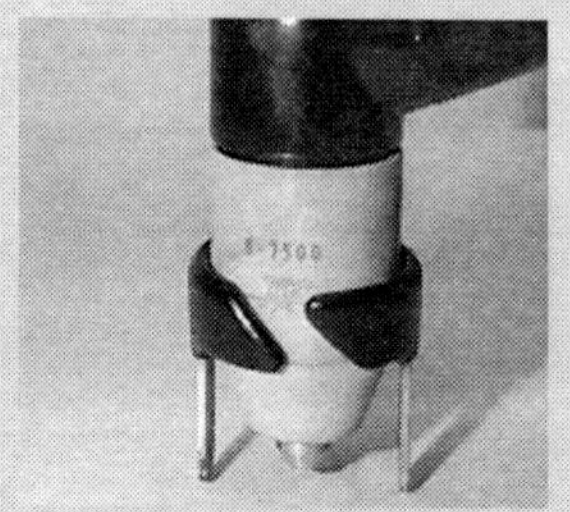

STANDOFF POST GUIDE

205SA01.EPS

Hot Tip

Replacing Consumables

The torch nozzle and electrode are considered consumables (components requiring frequent replacement due to deterioration), and both should be replaced at the same time. It's a good idea to keep a replacement torch nozzle and electrode handy. This will reduce downtime when cutting with the plasma unit.

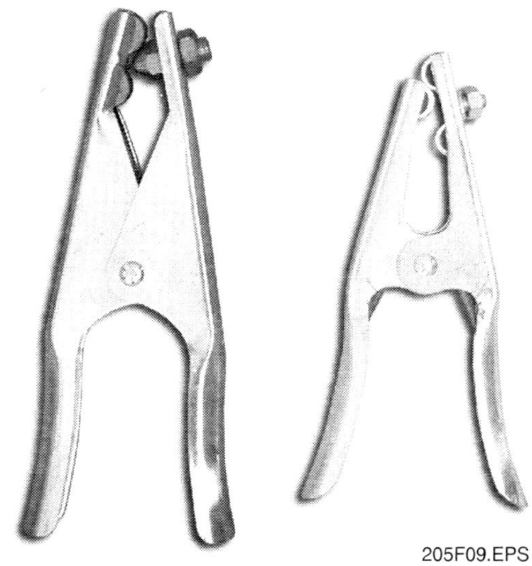

205F09.EPS

Figure 9 ◆ Typical workpiece clamps.

3.3.0 Workpiece Clamp

The workpiece clamp provides the connection between the end of the workpiece lead and the workpiece. Workpiece clamps are mechanically connected to the welding lead and come in a variety of shapes and sizes. The size of a workpiece clamp is the rated amperage that it can carry without overheating. If a workpiece clamp needs to be replaced, be sure to select one that is rated at least the same as the rated capacity of the power source it will be used on. *Figure 9* shows typical workpiece clamps.

3.3.1 Locating the Workpiece Lead

When PAC is to be used to cut on equipment containing electrical or electronic components, batteries, bearings, or seals, the workpiece clamp must be properly located to prevent damage to associated components. If the cutting current passes through any type of bearing, seal, valve, contacting surface, or electrical or electronic component, it could cause severe damage to the item from arcing and overheating. This will result in costly replacement. If cutting is near a battery, the battery must be removed to prevent hydrogen gas explosions caused by a cutting spark.

Carefully check the area to be cut and position the workpiece clamp so that the cutting current will not pass through any connecting surface. If in doubt, ask your supervisor for assistance before proceeding.

Hot Tip

Plasma Torch Cutting Guides

Many manufacturers furnish cutting guide accessories that are usable on several models of their torches. A circle cutting guide kit that can also be used as a freehand standoff roller guide is shown here.

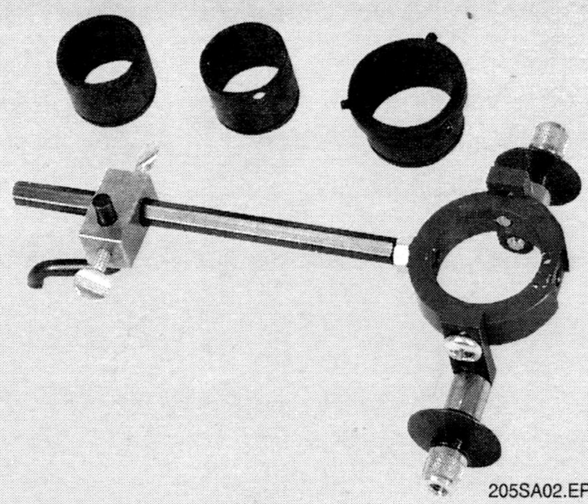

205SA02.EPS

> **WARNING!**
> To prevent damaging arcing across gapped or contacting surfaces, make sure the path from the workpiece clamp to the torch arc does not pass through these surfaces. Cutting current can severely damage bearings, seals, valves, or contacting surfaces. Position the workpiece clamp to prevent cutting current from passing through them.

> **WARNING!**
> Disconnect the grounded battery lead on any mobile equipment that is being repaired using the plasma arc cutting process. This prevents the cutting current from causing a battery explosion or battery damage.
>
> Do not cut near batteries. A spark could cause a battery to explode, showering the area with battery acid. Always remove the batteries. Workpiece leads must never be connected to pipes carrying flammable or corrosive materials. The cutting current could cause overheating or sparks, resulting in an explosion or fire.

> **CAUTION**
> The cutting current can destroy electronic or electrical equipment. Have an electrician check the equipment and, if necessary, isolate the system before cutting.

3.4.0 PAC Gases and Controls

All plasma arc cutting units require one or more types of gases. The specific type of gas required and the controls necessary to set and adjust these gases vary with the type, size, and manufacturer of the PAC equipment.

3.4.1 PAC Gases

Several different gases and gas mixtures are used with plasma arc cutting units. Gases used include air, nitrogen, oxygen, argon and argon mixtures, hydrogen, and carbon dioxide. The simplest units use clean, compressed air to cut carbon steel, stainless steel, and aluminum up to about 3⁄16" thick. Most heavy-duty air plasma units are rated to cut carbon steel and stainless steel up to about 1¼" thick and sever up to 1¾" thick. Some dual-flow PAC units require two different gases. One gas may be used for the pilot arc and for cutting, and another for shielding or cooling. *Table 1* provides general recommendations for plasma and shield gases.

Specific gas requirements for a particular PAC unit or system are specified in the manufacturer's operating instructions manual and/or information tag(s) on the unit. Refer to these specifications, and use only the gases specified for the type of equipment being used.

3.4.2 PAC Gas Controls

Plasma arc cutting gases are used to generate plasma, cool the torch, and shield the cut against corrosion. All gases must have their pressures

Table 1. Recommended Plasma/Shield Gas Combinations

Material	Air/Air	O_2/Air	N_2/CO_2	N_2/Air	H35/N_2
Carbon Steel	Most Economical Good Cut Quality Good Speed Good Gouging Good Weldability	Best Cut Quality Maximum Cut Speed Best Weldability	Some Dross Long Electrode Life	Not Recommended	Best Gouging Long Electrode Life Some Dross
Stainless Steel	Most Economical Good Speed Some Dross	Not Recommended	Good Cut Quality Good Gouging Minimal Dross Long Electrode Life	Long Electrode Life Lowest Shield Gas Cost	Best Cut Quality Best Gouging Minimal Dross Long Electrode Life Cuts Thicker Material
Aluminum	Most Economical Good Speed Some Dross	Not Recommended	Good Cut Quality Good Gouging Minimal Dross Long Electrode Life	Not Recommended	Best Cut Quality Best Gouging Minimal Dross Long Electrode Life Cuts Thicker Material

O_2 = Oxygen N_2 = Nitrogen CO_2 = Carbon Dioxide H35 = Mixture of 35% Hydrogen and 65% Oxygen

205T01.EPS

reduced before they can be used. The type of regulator used to control the gas pressure depends on the type of PAC equipment being used and the base metal being cut.

When a shop compressed air supply is being used, a heavy-duty air filter/dryer is required to eliminate oil and moisture from the pressurized air. The high-pressure shop air must be reduced externally, or by the PAC internally, for the required working pressure. Also, depending on the equipment being used, the shop air must be provided at a flow rate, generally measured in cubic feet per hour (cfh), that is sufficient for the cutting or gouging operation.

Gases other than compressed air are normally supplied in high-pressure cylinders at 1,500 psi to 2,000 psi. The high cylinder pressures are reduced by pressure regulators to a lower working pressure. A gas hose connected to the working pressure outlet of the regulator delivers the gas to the PAC console. Most heavy-duty systems will require two high-pressure cylinders (and two pressure regulators), one containing the cutting gas and the other a shielding or cooling gas. Some systems use a combination of compressed shop air and high-pressure cylinders. *Figure 10* shows an industrial dual-flow PAC with two internal gas pressure regulators.

Stop-start flow control to the torch is usually controlled by a solenoid valve that electrically opens and closes a gas valve. Solenoid valves are usually mounted inside the unit and are controlled by a pushbutton on the torch. There is one solenoid for each gas source.

Many smaller plasma arc cutting units come equipped with a built-in air compressor. These light-duty machines are portable and run on 110V power, which makes them ideal for use at remote job sites. *Figure 11* shows a portable plasma arc cutter.

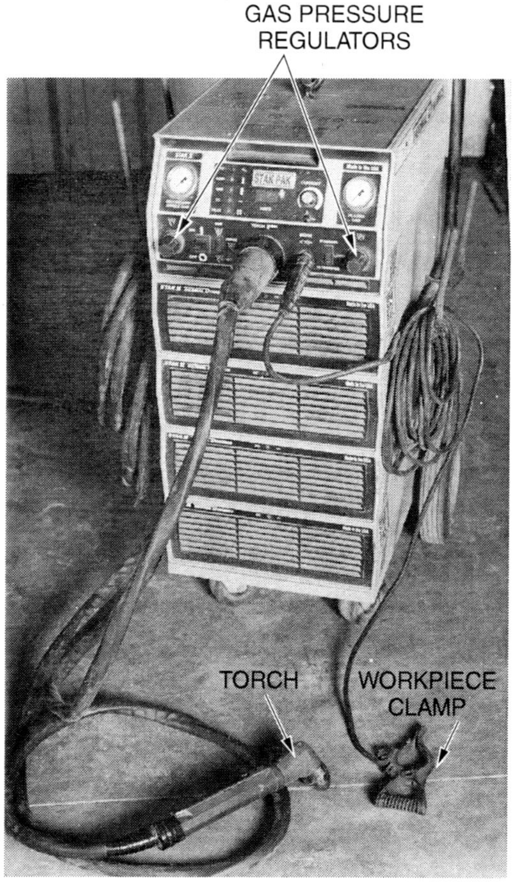

205F10.EPS

Figure 10 ◆ Industrial dual-flow PAC unit.

4.0.0 ◆ PREPARING THE WORK AREA FOR PAC

The plasma arc cutting control unit and work must be located close enough to each other for the torch cable to comfortably reach the workpiece. If possible, the workpiece should be located at a comfortable height and position for the torch

> **Hot Tip**
>
> ***Minimum Shop Airflow Rate***
>
> Always check to be sure that the shop air supply will provide the minimum flow rate required for the PAC equipment being used. If the airflow rate is too low, it will cause poor cuts and overheating of the torch. Refer to the manufacturer's specifications for the required flow rate for the equipment being used.
>
> ***Dry Air***
>
> To conserve consumables, make sure to have air driers in your compressed air system. Moisture in the compressed air can also damage the equipment.
>
> ***Inert Gas***
>
> Originally, the PAC process was developed to cut nonferrous metals using inert gases. Advances in technology led to using air and/or oxygen as the plasma gas so steel could be cut using PAC.

205F11.EPS

Figure 11 ◆ Light-duty PAC with internal air compressor.

operator. The cutting area should be well ventilated and cleared of all combustible material.

Plasma arc cutting or gouging can spray molten metal for considerable distances, sometimes 25' or more. It is important that everything combustible be removed from the range of the sprayed metal. If necessary, flame-resistant shields or curtains should be erected to protect any nearby workers or equipment from ultraviolet arc rays and metal splash. It is also important to station a fire watch in the area.

Eye and ear protection are essential when performing PAC due to noise and to prevent metal spray from entering the operator's eyes and ears. If the site ventilation is not adequate to keep the smoke and fumes away from the operator, a respirator must be used.

WARNING!

PAC produces fumes and gases that can be harmful to your health. The composition and rate of generation of fumes and gases depend on factors such as arc current, cutting speed, material being cut, and gases used. The fume and gas byproducts usually consist of ozone, oxides of the metal being cut, and oxides of nitrogen. These fumes must be removed from the work area through an exhaust system. Check all applicable local codes; some may require that exhaust be filtered before it is vented to the atmosphere.

5.0.0 ◆ SETTING UP PAC EQUIPMENT

PAC equipment is easy to set up because it is supplied as a complete system. However, there are some important considerations and preparations to complete before cutting is attempted:

- The unit must have the rated power and duty cycle to cut the intended metal type and thickness.
- The required primary power (phase, voltage, and amperage) must be available for the control unit.
- The required nozzle (tip) must be installed in the torch.
- The required gas or gases must be on-line and at the required pressures and flow rates.
- Spare consumable components must be on hand.

5.1.0 Setting Correct Cutting Amperage

The cutting amperage depends on the type of equipment being used, the type and thickness of the material being cut, and the type of gas being used (*Figure 12*). Light-gauge sheet steel may require as little as 7 amps, while 2" aluminum plate may require 250 amps. Always refer to the manufacturer's recommendations to identify the correct amperage for the equipment and job to be performed. Using the incorrect amperage will result in a poor-quality cut, and if the amperage is too high, severe damage to the torch could occur.

205F12.EPS

Figure 12 ◆ Gas and amperage settings vary according to material type and thickness.

5.2.0 Installing Gas Cylinders and Setting Gas Parameters

Several different gases may be used on the same system for cutting the same or different type and thickness of metals. In many cases, the same gas or mixture is not used for both thick and thin metals of the same type. Refer to the manufacturer's instructions or the tag on the unit for specific gas recommendations.

5.2.1 Installing a Gas Cylinder

When installing a gas cylinder on the PAC unit, follow these steps:

Step 1 After locating the correct gas type, move it to the unit and secure it in an upright position.

Step 2 Remove the protective valve cap.

WARNING!
If oxygen is being used, be sure any cloth used on or near the fittings or regulators does not have any oil or grease on it. Oil or grease compressed with oxygen will explode and can cause personal injury or death.

Hot Tip

Safety Caps

Safety caps should be kept with cylinders at all times. Immediately after removing a cylinder's regulator assembly, reinstall the safety cap over the cylinder valve. This protects the cylinder valve if the cylinder falls over. If the cylinder valve is broken, the cylinder may propel itself like a rocket. This could result in serious injury or death.

205SA03.EPS

Step 3 If dirt is visible in the connection seat, wipe it out with a clean, dry cloth.

WARNING!
Always stand to one side of the regulators when opening valves to avoid personal injury. The regulator adjustment screws can fly off. Wear protective eyewear.

Step 4 While standing to one side of the valve with safety goggles in place, momentarily crack open the valve slightly to blow any dirt or debris from the valve.

Step 5 Connect the regulator to the cylinder valve, and tighten with the proper wrench. Connect the hose from the PAC unit to the regulator.

Step 6 Open the cylinder valve very slowly at first, then all the way.

CAUTION
Do not use pliers or pipe wrenches to tighten the regulator's connections. Also be careful not to overtighten the connections because they are made of soft bronze and can be easily damaged.

6.0.0 ◆ OPERATING PAC EQUIPMENT

Plasma arc cutting equipment is simple to operate. The cutting unit's operating instructions usually provide information for properly setting air or gas pressure and/or flow rates, output current, torch standoff distance (torch tip to work distance), and torch travel speed for various metal types and thicknesses.

Hot Tip

Gas Pressure and Flow Rates

PAC unit gas pressures and/or flow rates vary with equipment design, gas type, and cut depth (metal thickness). Typical gas flow rates vary from 15 cfh for nitrogen when cutting 0.1" aluminum, carbon steel, or stainless steel to 62 cfh of argon and 31 cfh of nitrogen when cutting 3" carbon steel or 2½" aluminum or stainless steel. Refer to the manufacturer's instructions for the correct gas pressure and flow rates.

NOTE

The following steps are provided for general reference only. Refer to the manufacturer's operation manual for specific instructions on the PAC equipment being used.

To operate the equipment, follow these steps:

Step 1 Make sure the capacity of the PAC power source matches or exceeds the work to be done.

Step 2 Identify the location of the primary disconnect for the electrical outlet to be used.

Step 3 If not already wired into a disconnect box, plug the PAC power source into the outlet.

Step 4 Check that the recommended gas(es) is available. If pressurized cylinder(s) are used, check that the cylinder(s) have adequate pressure for the work to be done. Connect the compressor's air line if compressed air is to be used.

Step 5 Turn on the PAC power source.

Step 6 In accordance with the PAC manufacturer's instructions, adjust the gas regulator pressure(s) and/or flow meter(s) rates to the values recommended for the type and thickness of metal being cut.

Step 7 Set the output current (amperage) control to the value recommended in the manufacturer's instructions for the type and thickness of metal being cut.

Step 8 Attach the work lead clamp to the workpiece to be cut.

CAUTION

If cutting on machinery or other equipment, be sure to position the work lead clamp so that the cutting current will not pass through seals, bearings, or other contacting surfaces that could be damaged from heat or arcing. Also, isolate any electrical components.

Step 9 Prepare the work area as previously described, stationing a fire watch.

Step 10 Put on the appropriate personal protective equipment (hood, gloves, and ear protection).

Step 11 Hold the torch at the recommended standoff distance above the point where the cut is to begin, lower your hood, and press the torch arc start button or foot peddle. As soon as cutting begins, move the torch at a 0° travel angle with the recommended speed along the cutting line.

6.1.0 Square-Cutting Metal

Square edges are cut with PAC by holding the torch at 90° (0° work angle) to the metal surface while it is advanced smoothly with no side-to-side movement. A straightedge or metal angle can be used to guide and steady the torch end. When cutting thicker metals, the cut does taper slightly as it narrows with depth. This can result in cuts where one or both faces are not quite square with the plate face. The angle of the face is known as the cut angle. To compensate for the cut angle, the torch should be slightly slanted to produce a square edge on the desired piece. All the taper will then be on the scrap side. *Figure 13* shows how to compensate for the taper in thicker metal cuts.

Standoff Distance

Instead of manually holding the PAC torch, use a drag shield or standoff guide to establish and maintain the correct standoff distance.

Starting the Torch on a Thick Workpiece

Like oxyfuel cutting of thick workpieces, it is easier to begin a manual cut at the edge by rotating the PAC torch slightly so that the plasma cutting jet begins cutting at the top edge only. The torch is then rotated vertically so that the cutting jet slices into the remaining thickness before the torch is moved along the cutting line.

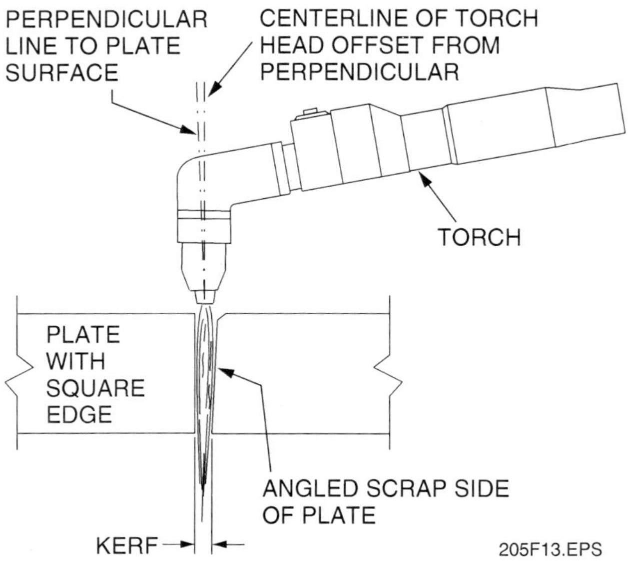

Figure 13 ◆ Compensating for taper in thicker metal cuts.

6.2.0 Bevel-Cutting Metal

Bevel cuts are made with the same technique as for square cuts, except that the torch is held at the required bevel angle. A length of angle iron laid on its open face and shimmed as necessary can be used to support the torch's side at the required angle. *Figure 14* shows how to use a metal angle to hold a bevel angle.

6.3.0 Piercing and Slot-Cutting in Metal

PAC can be used to pierce and cut slots in metal in any position. Full fire-resistant body protection, including ear protection, is recommended when piercing or cutting metal where out-of-position orientations could cause the metal splash to fall on the torch operator.

To pierce very thin metal, hold the torch directly over the point to be pierced at a 0° work angle and press the arc button. As soon as the jet passes completely through the workpiece, move the torch head in a smooth circle (or other pattern) to produce the desired diameter or shape hole. When piercing thicker materials, rotate the torch to a 10° work angle to prevent metal from splashing directly up into the torch. When the jet passes through the workpiece, rotate the torch to a 0° work angle and complete the cut.

When cutting slots, a template or straightedge can be used to precisely guide the torch tip.

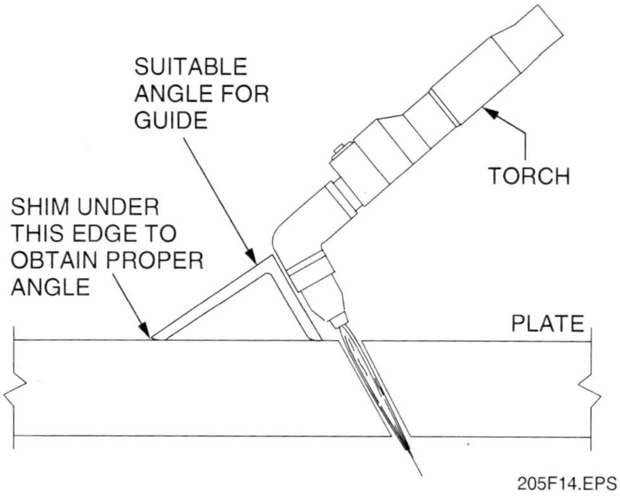

Figure 14 ◆ Using angle iron to set and hold a bevel angle.

7.0.0 ◆ PROPER EQUIPMENT STORAGE AND HOUSEKEEPING

Proper equipment storage and housekeeping are essential for work efficiency and safety. When finished using the PAC equipment, follow these steps to store the equipment and for good housekeeping:

Step 1 Turn off the power source.

Step 2 Unplug and coil the power source cable.

Step 3 Coil the torch cable.

Step 4 Close the cylinder valves and/or disconnect and coil the compressed air line.

Swirl Ring Torches

On PAC torches equipped with swirl rings, most swirl rings produce a clockwise tangential plasma swirl as viewed from the torch to the work surface. If the torch is at a 0° work angle, the clockwise swirl causes the right side to the kerf to be essentially square and the left side to be angled. For a counterclockwise swirl, the opposite is true.

> **WARNING!**
> If cylinders are to be removed, be sure to remove the pressure regulators and replace the cylinder caps before releasing and moving them.

Step 5 Return the PAC equipment to its proper storage location.

Step 6 Clean off the welding bench, if used, and sweep up the slag and debris that were blown around the area.

Figure 15 shows a properly stored PAC unit.

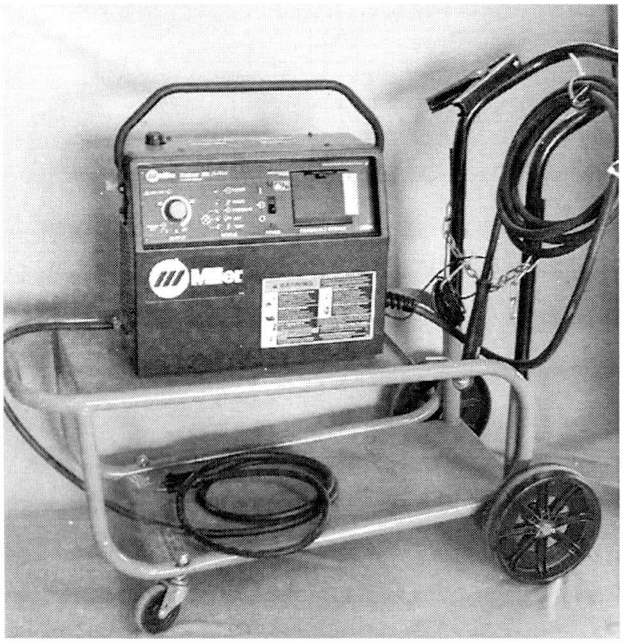

205F15.EPS

Figure 15 ◆ Properly stored PAC unit.

8.0.0 ◆ REPAIR OF PLASMA ARC CUTTING (PAC) EQUIPMENT

As with all equipment, regular maintenance should be performed according to the manufacturer's recommendations. The cutting equipment should not be altered or modified in any way, except for those changes identified in the manufacturer's literature. The plasma machine requires relatively little maintenance, but its consumables must be inspected regularly and replaced as needed. Minor external repairs usually involve these basic parts only.

8.1.0 Understanding Basic Operating Concerns

If compressed air is used with a PAC unit, shop air may be used as long as it is dry and free of oil. When properly set, the air pressure regulator and filter/dryer on the power source will maintain the correct pressure and flow rate for the system.

> **WARNING!**
> Always switch off the machine power and unplug the power cord before changing consumable parts or performing any repairs on the power source. If the power source is directly connected to a line disconnect switch, move the line disconnect switch to the OFF position.

Your ability to change and replace parts will depend on your knowledge of the procedures contained in the manufacturer's literature for the equipment in question.

Follow the manufacturer's guidelines for removal and replacement of worn or damaged parts:

- Check the shield for any signs of wear.
  - The shield should be clean and clear of metal debris.
  - Unscrew the shield and check the inside for wear. If the holes are blocked, attempt to clean the holes using a tip cleaner.
  - If the shield cannot be cleaned, replace it.
- Inspect the retaining cup for damage. Replace the retaining cup as necessary.
- Inspect the nozzle for wear or damage. If the hole in the nozzle is worn or oval-shaped, replace the nozzle.
- Inspect the electrode. Remove the electrode and check the tip for pitting. If pits exist and are more than 1⁄16" deep, replace the electrode.
- Inspect the swirl ring. It should be clean, and the holes on the sides should not be clogged. If the swirl ring is damaged, replace it.
- Inspect the O-ring. Apply a small amount of lubricant (provided in a repair kit) to the O-ring before reinstalling it. The O-ring should make a firm seal when all parts are assembled and tight.

8.2.0 Moisture

A significant concern when using PAC equipment is that moisture may enter the unit through the air line. This can cause the torch to sputter or hiss. If your PAC equipment experiences moisture problems, you must do the following:

- Drain the filter bowl.
- Replace the filter element, if the unit has one. Refer to the manufacturer's literature for cleaning and replacing this filter.

8.3.0 Cooling Filter

Some power sources contain a cooling air filter located in the back of the unit. Refer to the manufacturer's literature and clean the filter regularly as required. A cooling air filter that is neglected may cause the PAC unit to overheat and shut down.

Summary

Plasma arc cutting is very useful for cutting and piercing many types of metals. Unlike oxyfuel cutting, it can be used to cut aluminum, magnesium, copper, nickel, and stainless steel. It is usually much faster than oxyfuel cutting and produces little or no distortion or alteration zone changes. The cuts are fast and clean, with little or no dross. There is no chance of carbon inclusions as can happen with carbon arc cutting. Cut surfaces are usually ready for welding without further preparation. However, because the torch jet produces metal spray, good site fire-prevention practices and the wearing of the appropriate personal protective equipment are essential. Ear protection is important when cutting out of position to prevent sparks or molten metal from entering the ears.

Review Questions

1. Plasma arc cutting uses a jet of plasma that can reach temperatures as high as _____.
   a. 10,000°F
   b. 20,000°F
   c. 25,000°F
   d. 30,000°F

2. There are _____ types of plasma arc cutting.
   a. two
   b. three
   c. four
   d. five

3. In the transferred arc process, the pilot arc is used to _____.
   a. cut the metal
   b. ionize the air jet
   c. establish contact with the metal
   d. keep the torch heated

4. In a nontransferred arc process, the material being cut is _____ the arc circuit.
   a. adjacent to
   b. dependent on
   c. electrically connected to
   d. not electrically connected to

5. Nontransferred arc is not normally found in facilities that cut _____.
   a. non-metal materials
   b. only metal materials
   c. non-conducting materials
   d. insulating materials

6. Light-duty PAC power supplies typically require a single-phase, _____ outlet.
   a. 115VAC only
   b. 220VAC only
   c. 115VAC or 230VAC
   d. 280VAC or 440VAC

7. The heaviest duty PAC torches are usually _____.
   a. water cooled
   b. 115V systems
   c. used for cutting thin materials
   d. handheld

8. Mechanized PAC torches are designed to be _____.
   a. handheld
   b. mounted in a vise
   c. used at a slow speed
   d. mounted on automated carriers

9. The size of a workpiece clamp is the rated _____ that it can carry without overheating.
   a. voltage
   b. capacitance
   c. amperage
   d. power

10. Severe damage from arcing and overheating can be caused by cutting current that passes through a _____.
   a. workpiece
   b. cable
   c. cutting table
   d. bearing

11. Gases other than compressed air are normally supplied in high-pressure cylinders at _____.

a. 200 psi to 400 psi
b. 800 psi to 1,000 psi
c. 1,000 psi to 1,500 psi
d. 1,500 psi to 2,000 psi

12. If site ventilation is not adequate to keep fumes away from the operator, a(n) _____ must be used.

a. fan
b. respirator
c. fire watch
d. air line

13. Light-gauge sheet steel may require a cutting amperage as low as _____ amps.

a. 7
b. 10
c. 15
d. 20

14. Square edges are cut with PAC by holding the torch at _____ to the metal surface.

a. 30°
b. 45°
c. 75°
d. 90°

15. If cylinders are removed from the system, the _____ must also be removed.

a. cylinder caps
b. workpiece clamp
c. pressure regulators
d. power source

GLOSSARY

Trade Terms Introduced in This Modlule

Duty cycle: The amount of time a machine can cut without overheating within a 10-minute cycle.

Plasma: A fourth state of matter (not solid, liquid, or gas), created by heating a gas to such a high temperature that it boils electrons off the gas molecules (ionization).

Potential: The relative electrical voltage between two points of reference.

APPENDIX

Performance Accreditation Tasks

The Performance Accreditation Tasks (PATCs) correspond to and support the learning objectives in *AWS EG2.0-95, Guide for the Training and Qualification of Welding Personnel: Entry-Level Welder*.

PATCs provide specific acceptable criteria for performance and help to ensure a true competency-based welding program for students.

The following task is designed to test your competency with a plasma arc cutting torch. Do not perform this cutting task until directed to do so by your instructor.

PLASMA ARC CUTTING

PERFORM PLASMA ARC CUTTING

Using carbon steel, lay out and cut the shape shown in the figure.

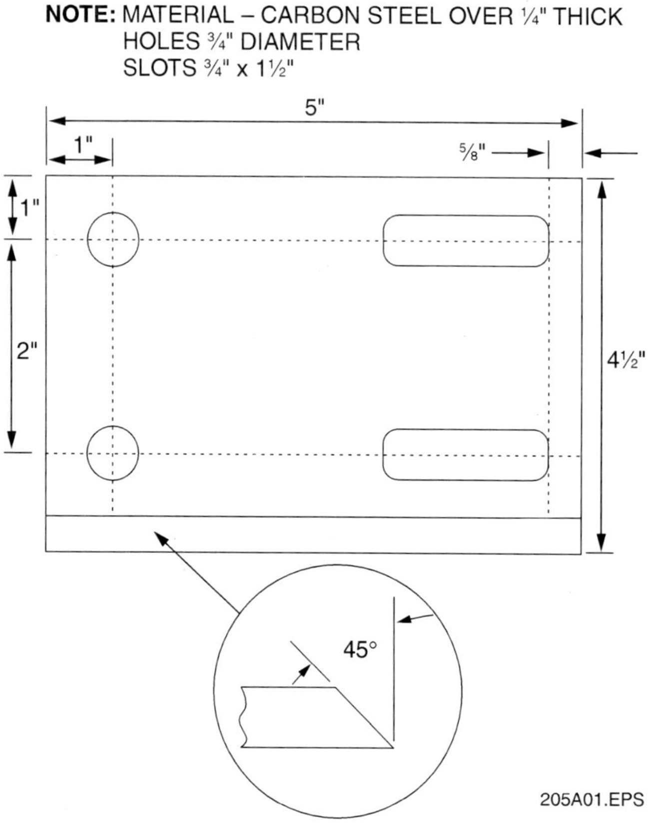

205A01.EPS

Criteria for Acceptance

- Outside dimensions ±1/16" ______
- Inside (holes and slots) dimensions ±1/16" ______
- Square ±2° ______
- Bevel ±2° ______
- Minimal amount of dross sticking to plate which can be easily removed ______
- Square kerf face with minimal notching not exceeding 1/16" deep ______

Additional Resources

This module is intended to be a thorough resource for task training. The following reference works are suggested for further study. These are optional materials for continued education rather than for task training.

Modern Welding, Latest Edition. A.D. Althouse, C.H. Turnquist, W.A. Bowditch, K.E. Bowditch. Tinley Park, IL: The Goodheart-Willcox Company, Inc.

Recommended Practices for Plasma Arc Cutting, Latest Edition. Miami, FL: American Welding Society.

Figure Credits

Miller Electric Mfg. Co.	Figures 205F01 (upper), 205F02 (photo), 205F05, 205F06, 205F08 (photo), 205SA01 (upper), 205SA02, 205F11
BUG-O Systems	Figure 205F01 (lower)
ESAB Welding and Cutting Products	Figures 205F07, 205F12 (left)
Thermadyne Industries	Figures 205SA01 (lower), 205F12 (right)
Chuck Rogers	Figures 205F09, 205F10, 205F15
Gerald Shannon	Figure 205SA03

CONTREN™ LEARNING SERIES — USER UPDATES

The NCCER makes every effort to keep these textbooks up-to-date and free of technical errors. We appreciate your help in this process. If you have an idea for improving this textbook, or if you find an error, a typographical mistake, or an inaccuracy in NCCER's Contren™ textbooks, please write us, using this form or a photocopy. Be sure to include the exact module number, page number, a detailed description, and the correction, if applicable. Your input will be brought to the attention of the Technical Review Committee. Thank you for your assistance.

Instructors – If you found that additional materials were necessary in order to teach this module effectively, please let us know so that we may include them in the Equipment/Materials list in the Instructor's Guide.

Write: Curriculum Revision and Development Department
National Center for Construction Education and Research
P.O. Box 141104, Gainesville, FL 32614-1104

Fax: 352-334-0932

E-mail: curriculum@nccer.org

Craft ______________________ Module Name ______________________

Copyright Date ______________ Module Number ______________ Page Number(s) ______________

Description

____

____

____

____

(Optional) Correction

____

____

____

(Optional) Your Name and Address

____

____

____

Module 29206-03

GMAW and FCAW – Equipment and Filler Metals

COURSE MAP

This course map shows all of the modules in the second level of the Welding curriculum. The suggested training order begins at the bottom and proceeds up. Skill levels increase as you advance on the course map. The local Training Program Sponsor may adjust the training order.

WELDING LEVEL TWO

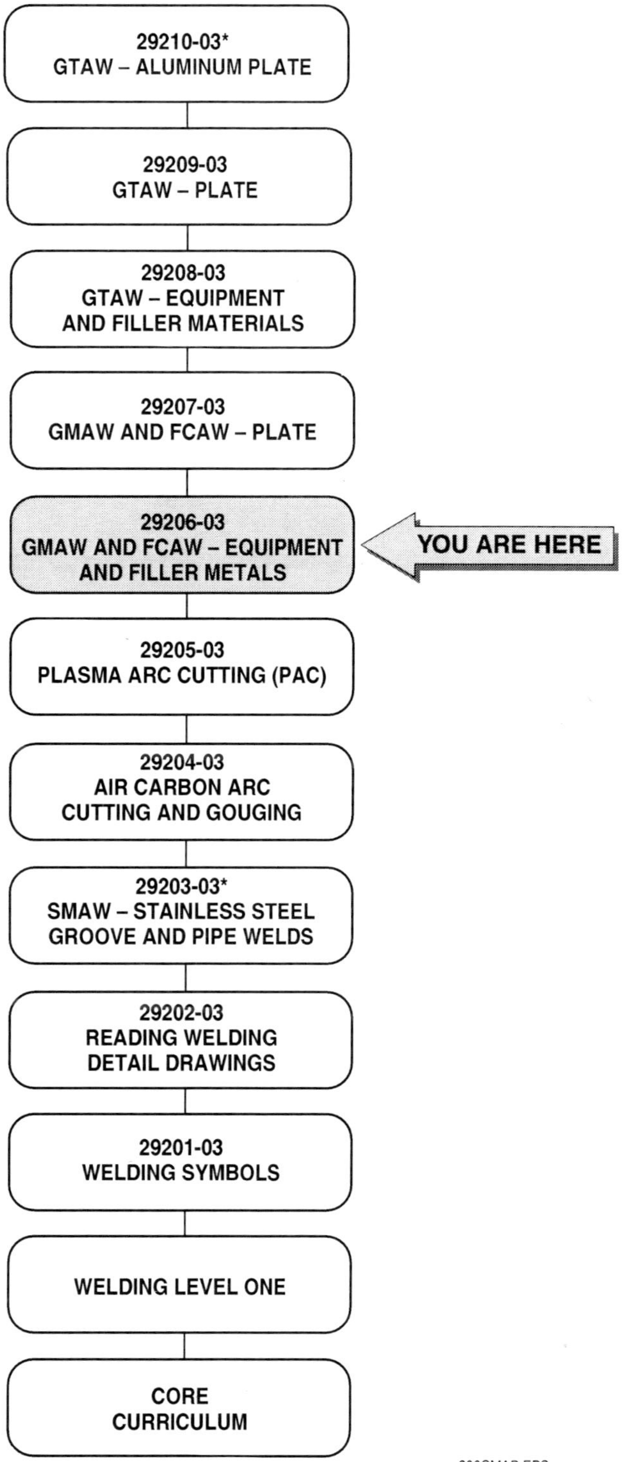

206CMAP.EPS

Please note that Modules 29203-03 and 29210-03 are electives for *Welding Level Two* but are requirements for *AWS Entry Level Welder.

MODULE 29206-03 CONTENTS

Figures

Tables (*Appendix*)

MODULE 29206-03

GMAW and FCAW – Equipment and Filler Metals

Objectives

When you have completed this module, you will be able to do the following:

1. Explain gas metal arc welding (GMAW) and flux cored arc welding (FCAW) safety.
2. Explain the characteristics of welding current and power sources.
3. Identify and explain the use of GMAW and FCAW equipment:
   - Spray transfer
   - Globular
   - Short circuiting
   - Pulse
4. Identify and explain the use of GMAW and FCAW shielding gases and filler metals.
5. Set up GMAW and FCAW equipment and identify tools for weld cleaning.

Prerequisites

Before you begin this module, it is recommended that you successfully complete the following: Core Curriculum; Welding Level One; Welding Level Two, Modules 29201-03 through 29205-03.

Required Trainee Materials

1. Pencil and paper
2. Appropriate personal protective equipment

1.0.0 ◆ INTRODUCTION

Gas metal arc welding (GMAW) uses an electric arc to melt a consumable wire electrode and fuse it with the base metal. The arc generates an intense heat of 6,000 to 10,000 degrees Fahrenheit (°F). The arc is stabilized and the molten filler and base metals are protected from oxidation by a shielding gas dispensed from a GMAW gun nozzle. The wire is automatically fed into the weld puddle as it melts. The process usually operates with direct current electrode positive (DCEP). *Figure 1* shows the basic GMAW process.

Flux cored arc welding (FCAW) uses an electric arc to melt a flux cored (tubular) wire electrode and fuse it with the base metal to form a weld. A wire feeder automatically feeds the flux cored wire as it is being consumed in the arc. The equipment used for FCAW is basically the same as that used for GMAW. The main difference between the two processes is that GMAW uses solid wire or composite core wire electrodes with a shielding gas, while FCAW uses a tubular flux-cored wire electrode with or without shielding gas.

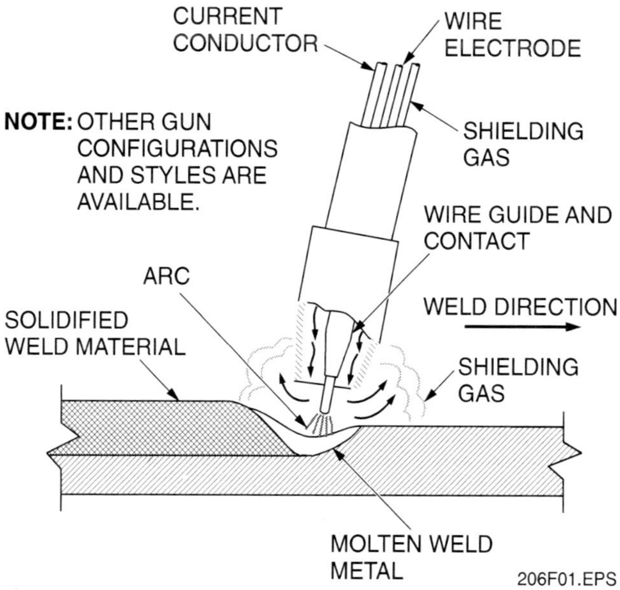

Figure 1 ◆ Basic GMAW process.

In the two basic FCAW processes, one uses an external shielding gas while the other is self-shielding. Whether a shielding gas is used is determined by the type of flux cored wire that is used. When the flux cored wire does not require an external shielding gas, the process is called self-shielding FCAW or FCAW-*S*. When an external shielding gas is required, the process is called gas-shielded FCAW or FCAW-*G*. Equipment specifically designed for self-shielding FCAW does not have provisions for shielding gas (gas connections, solenoids, and gas pre-flow and post-flow timers), and the guns do not have gas nozzles. Equipment used for gas-shielded FCAW welding is the same as that used for GMAW except that the wire feed drive assemblies have to be able to handle the core wire, which is softer and generally larger in diameter than solid wire. *Figure 2* shows the basic FCAW process.

This module provides an overview of the different items of equipment required for gas metal arc welding. Topics include safety and equipment, power sources, wire feeders, guns, equipment setup, and filler metals. Upon completion of this unit, trainees will be aware of the safety concerns associated with both GMAW and FCAW. They will be able to select and set up both GMAW and FCAW equipment safely and efficiently.

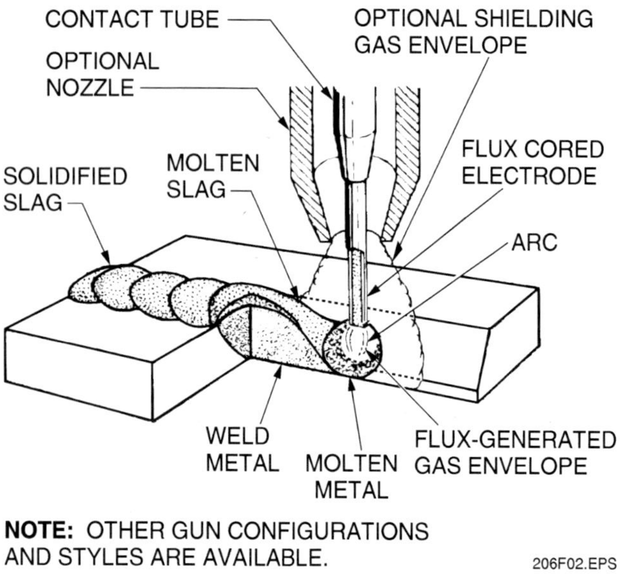

Figure 2 ◆ Basic FCAW process.

2.0.0 ◆ SAFETY PRACTICES

The following is a summary of safety procedures and practices that must be observed while cutting or welding. Keep in mind that this is just a summary. Complete safety coverage is provided in the Level One module, *Welding Safety*. If you have not completed that module, do so before continuing. Above all, be sure to wear appropriate protective clothing and equipment when welding or cutting.

2.1.0 Protective Clothing and Equipment

- Always use safety goggles with a full face shield or a helmet. The goggles, face shield, or helmet lens must have the proper light-reducing tint for the type of welding or cutting to be performed. Never directly or indirectly view an electric arc without using a properly tinted lens.
- Wear proper protective leather and/or flame retardant clothing along with welding gloves that will protect you from flying sparks and molten metal, as well as heat.
- Wear 8" or taller high-top safety shoes or boots. Make sure that the tongue and lace area of the footwear will be covered by a pant leg. If the tongue and lace area is exposed or the footwear must be protected from burn marks, wear leather spats under the pants or chaps and over the front top of the footwear.
- Wear a solid material (non-mesh) hat with a bill pointing to the rear or, if much overhead cutting or welding is required, a full leather hood with a welding face plate and the correct tinted lens. If a hard hat is required, use a hard hat that allows the attachment of rear deflector material and a face shield.

FCAW

FCAW is presently limited to the welding of ferrous metals. These include low- and medium-carbon steels, some low-alloy steels, cast irons, and some stainless steels. When critical welds are required, the gas-shielded FCAW process is often used because the combination of flux and separate shielding gas produces welds of very high quality.

- If a full leather hood is not worn, wear a face shield and snug-fitting welding goggles over safety glasses for gas welding or cutting. Either the face shield or the lenses of the welding goggles must be an approved shade 5 or 6 filter. For electric arc welding or cutting, wear safety goggles or a welding hood with the correct tinted lens (shade 5 to 14).
- If a full leather hood is not worn, wear earplugs to protect your ear canals from sparks.

2.2.0 Fire/Explosion Prevention

- Never carry matches or gas-filled lighters in your pockets. Sparks can cause the matches to ignite or the lighter to explode, causing serious injury.
- Never perform any type of heating, cutting, or welding until a hot-work permit is obtained and an approved fire watch is established. Most work-site fires caused by these types of operations are started by cutting torches.
- Never use oxygen to blow off clothing. The oxygen can remain trapped in the fabric for a time. If a spark hits the clothing during this time, the clothing can burn rapidly and violently out of control.
- Make sure that any flammable material in the work area is moved or shielded by a fire-resistant covering. Approved fire extinguishers must be available before attempting any heating, welding, or cutting operations.
- Always comply with any site requirement for a hot-work permit and/or fire watch.
- Never release a large amount of oxygen or use oxygen as compressed air. Its presence around flammable materials or sparks can cause rapid and uncontrolled combustion. Keep oxygen away from oil, grease, and other petroleum products.
- Never release a large amount of fuel gas, especially acetylene. Methane and propane tend to concentrate in and along low areas and can ignite at a considerable distance from the release point. Acetylene is lighter than air but is even more dangerous than methane. When mixed with air or oxygen, it will explode at much lower concentrations than any other fuel.
- To prevent fires, maintain a neat and clean work area and make sure that any metal scrap or **slag** is cold before disposal.
- Before cutting or welding containers such as tanks or barrels, check to see if they have contained any explosive, hazardous, or flammable materials, including petroleum products, citrus products, or chemicals that decompose into toxic fumes when heated. As a standard practice, always clean and then fill any tanks or barrels with water or purge them with a flow of inert gas to displace any oxygen.

2.3.0 Work Area Ventilation

- Make sure confined space procedures are followed before conducting any welding or cutting in the confined space.
- Never use oxygen in confined spaces for ventilation purposes.
- Always perform cutting or welding operations in a well-ventilated area. Cutting or welding operations involving zinc or cadmium materials or coatings result in toxic fumes. For long-term cutting or welding of such materials, always wear an approved full-face, supplied-air respirator (SAR) that uses breathing air supplied outside of the work area. For occasional, very short-term exposure, a high-efficiency particulate arresting (HEPA)-rated or metal-fume filter may be used on a standard respirator.
- Make sure confined spaces are ventilated properly for cutting or welding purposes.

3.0.0 ◆ CHARACTERISTICS OF WELDING CURRENT

The current produced by a welding machine has different characteristics than the current flowing through utility power lines. Welding current has low voltage and high amperage, while the power line current has high voltage and low amperage.

3.1.0 Voltage

Voltage is the measure of the electromotive force or pressure that causes current to flow in a circuit. There are two types of voltage associated with welding current: open-circuit voltage and operating voltage. Open-circuit voltage is the voltage present when the machine is on but no welding is being done. There are usually ranges of open-circuit voltages that can be selected, up to about 40 volts. Operating voltage, or arc voltage, is the voltage after the arc is struck.

This voltage is generally slightly lower than the open-circuit voltage. The arc voltage is typically 2 to 3 volts lower than the open-circuit voltage for each 100 amperes of current, but it depends on the range selected.

3.2.0 Amperage

Amperage is the electric current flow in a circuit. The unit of measure for amperage is the ampere (amp).

In welding, the current flows in a closed loop through two welding cables; the ground lead connects the power source to the base metal, and the other cable connects the power source to the wire electrode. During welding, an arc is established between the end of the wire electrode and the work. The arc generates intense heat at 6,000 to 10,000°F, melting the base metal and the wire electrode to form the weld. The amount of amperage produced by the welding machine determines the intensity of the arc and the amount of heat available to melt the work and the electrode.

4.0.0 ◆ WELDING POWER SOURCES

DC power sources are usually designed to produce either constant current (CC) or constant voltage (CV) DC. A constant current machine produces a constant current output over a wide voltage range. This is typical of a SMAW welding machine. A constant voltage machine maintains a constant voltage as the output current varies. This is typical of a GMAW or FCAW machine. The output voltages and output currents of a power source can be plotted on a graph to form a curve. These curves show how the output voltage relates to the output current as either changes. *Figure 3* shows constant current and constant voltage output curves.

4.1.0 Types of Welding Power Sources

Different types of welding machines (power sources) are available for GMAW, including:

- Transformer-rectifier power sources
- Engine-driven generator and alternator power sources
- Inverter power sources

4.1.1 Transformer-Rectifier Power Sources

The transformer-rectifier power source uses a transformer to convert the primary current to welding current and a rectifier to change the current from AC to DC. Transformer-rectifier power sources can be designed to produce either AC and DC or DC only. Transformer rectifiers that produce both AC and DC are usually lighter duty than those that produce DC only. Transformer-rectifier power sources that produce DC only are sometimes called rectifier power sources or just rectifiers. Depending on the size, transformer-rectifier power sources may require 120/240-volt single-phase power, 208-volt three-phase power, or 480-volt three-phase power.

> **WARNING!**
> Coming into contact with the primary voltage of a welding machine can cause shock or electrocution. Ensure that welding machines are properly grounded to prevent injury.

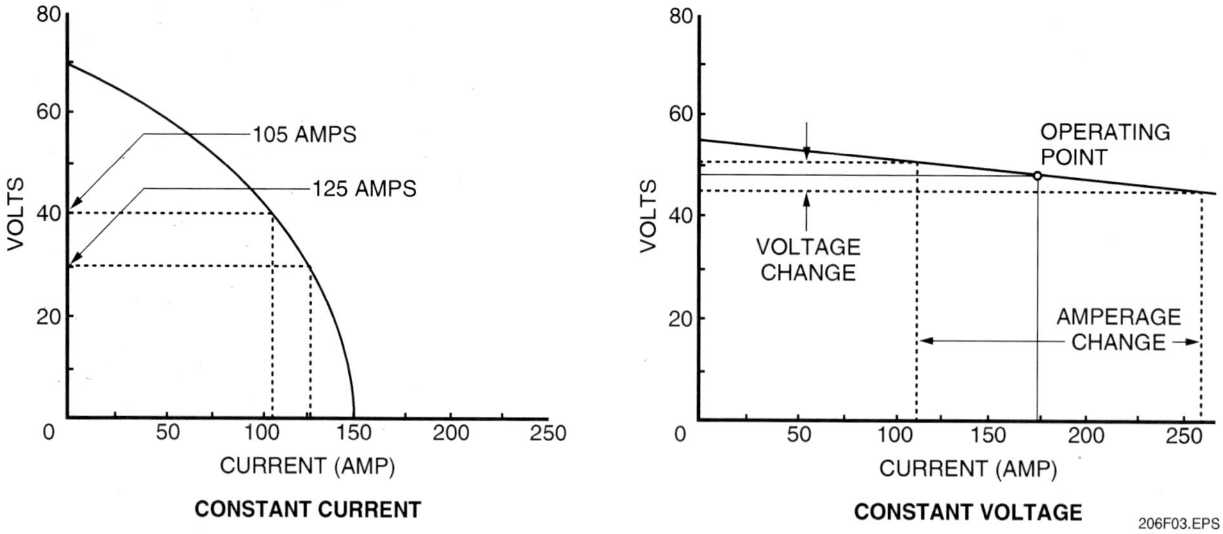

Figure 3 ◆ Constant current and constant voltage output curves.

Hot Tip

Constant Current/Constant Voltage Power Sources

Transformer rectifiers can be designed to produce constant current (CC) or constant voltage (CV) DC. Some will produce either variable or constant voltage by setting a switch. These machines are sometimes referred to as multiprocess power sources, since they can be used for any welding process, including GMAW and FCAW.

Transformer rectifiers used for GMAW and FCAW have an on/off switch and a voltage control. If the machine has a CC/CV or mode switch, it also has an amperage control for SMAW or GTAW. When the switch is set to CV for either GMAW or FCAW, the amperage control is disabled or becomes the voltage control. The welding cables (electrode cable and workpiece cable) are connected to terminals marked ELECTRODE and GROUND, or POSITIVE (+) and NEGATIVE (–). They often have selector switches to select direct current electrode negative (DCEN) or direct current electrode positive (DCEP). If there is no selector switch, the cables must be manually changed on the machine terminals to select the type of current desired.

Figure 4 shows a typical industrial transformer-rectifier CC/CV power source. These units usually have a 100% duty cycle at their rated output and a 60% duty cycle at their maximum output. They require a wire feeder and GMAW/FCAW gun or spool gun. These devices are described in a later section.

4.1.2 Engine-Driven Generator and Alternator Power Sources

Power sources can also be powered by gasoline or diesel engines. The engine can be connected to a generator or to an alternator. Engine-driven generators produce DC. Engine-driven alternators produce AC, which is fed through a rectifier to produce DC.

The size and type of engine used depend on the size of the power source. Single-cylinder engines are used to power small rectifier alternators, and multi-cylinder engines are used to power larger generators.

To produce welding current, the generator or alternator must turn at a required number of rpm (revolutions per minute). The engines that power alternators and generators have governors to control the engine speed. Most governors have a welding speed switch. The switch can be set to idle the engine when no welding is taking place. When the electrode is touched to the base metal, the governor automatically increases the speed of the engine to the required rpm for welding. If no welding takes place for about 15 seconds, the engine will automatically return to idle. The switch can also be set for the engine to run continuously at the welding speed.

Figure 5 shows an example of a diesel engine-driven power source. This particular unit can produce CC/AC or DC and CV/DC outputs up to 300 amps at a 100% duty cycle for SMAW, GTAW, GMAW, and FCAW. For GMAW and FCAW, it is available with a companion weatherproof wire

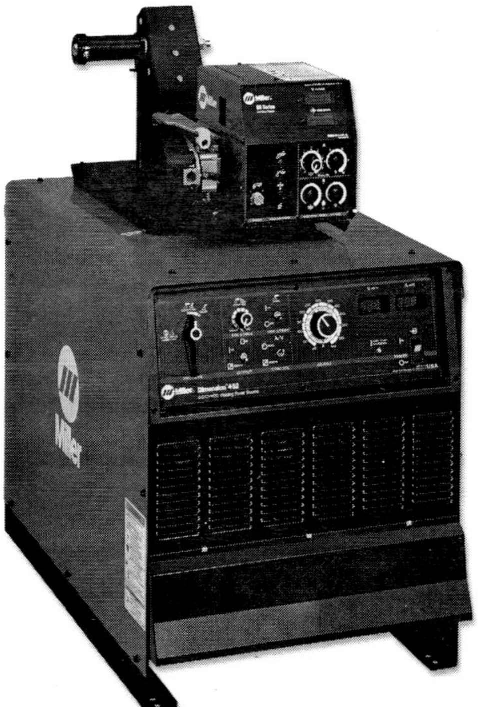

206F04.EPS

Figure 4 ◆ Industrial transformer-rectifier CC/CV power source with wire feeder.

feeder and a 24V control module for the wire feeder motor that enables an engine idle mode when no welding is being performed. The wire feeder control operates from the welding circuit voltage of the power source. When welding is started, wire feed speed is ramped up slowly while the engine reaches welding speed. The wire feeder is shown with an FCAW-only (FCAW-S) gun connected; however, it also functions with an external gas supply and gas-shielded GMAW/FCAW-G guns, as well as spool guns.

Engine-driven power sources often have an auxiliary power unit that produces 120 volts alternating current (VAC) for lighting, power tools, and other electrical equipment. When 120 VAC is required, the engine-driven generator or alternator must run continuously at the welding speed.

Engine-driven generators and alternators have engine controls and welding current controls. The engine controls vary with the type and size but normally include the following:

- Starter
- Voltage meter
- Temperature gauge
- Fuel gauge
- Hour meter

Engine-driven generators and alternators have a voltage/amperage control. There may be a polarity or mode selector switch, or you may have to manually change the welding cables at the welding current terminals to change the polarity. If the machine has a CC/CV or mode selector switch, it also has an amperage control. When the switch is set to CV or GMAW/FCAW mode, the amperage control is disabled and becomes a voltage control.

The advantage of engine-driven generators and alternators is that they are portable and can be used in the field where electricity is not available for other types of welding machines. The disadvantage is that engine-driven generators and alternators are costly to purchase, operate, and maintain.

Figure 5 ◆ Diesel engine-driven power source with wire feeder.

Hot Tip

GMAW/FCAW Feeders for Engine-Driven Power Sources

If possible, select wire feeders or spool guns that operate from 24VDC battery power and arc voltage. If AC-powered feeder or spool gun controls are used, the power source will have to operate in the continuous run mode to supply AC power to operate the feeder or spool gun.

4.1.3 Inverter Power Sources

Inverter power sources (*Figure 6*) increase the frequency of the incoming primary power. This provides a smaller, lighter power source with a faster response time and more control for pulse welding. Inverter welding machines are used where space is limited and portability is important. The controls on these welding machines vary according to size and application.

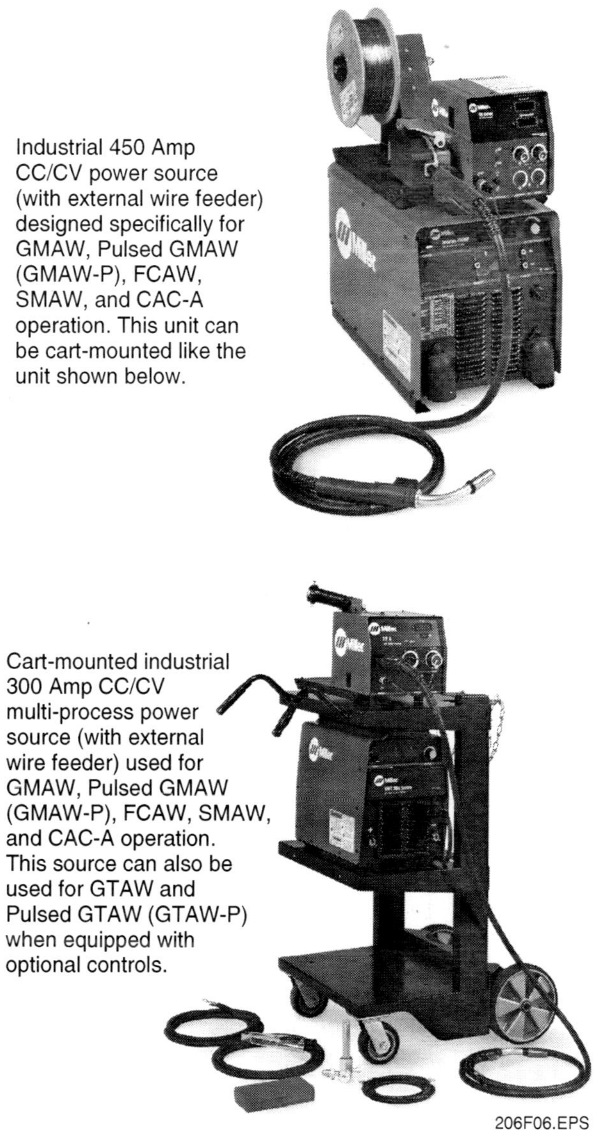

Figure 6 ◆ Inverter power sources.

Typical industrial advanced inverter power sources designed specifically for GMAW or FCAW operations usually have controls similar to those shown in *Figure 7*. These controls can include the following:

- *Power switch* – Applies power to unit.
- *Select pushbutton* – Moves a cursor up each line of 4-line display.
- *Display and change increment/decrement pushbuttons* – Contains up to 4 lines of alphanumeric text. When a line is selected, it can be changed using the increment/decrement pushbuttons.

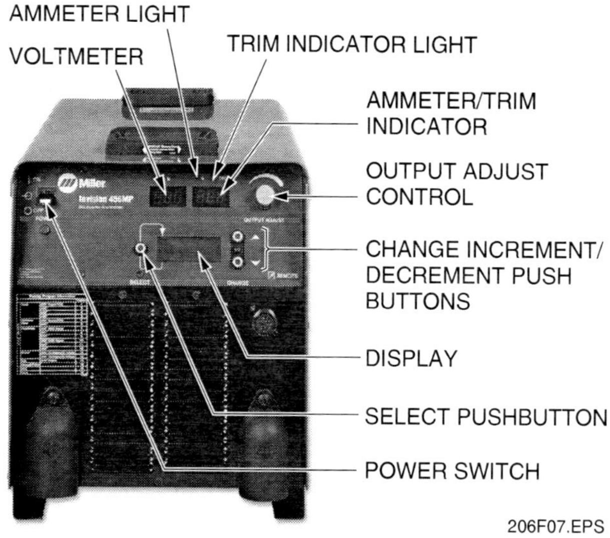

Figure 7 ◆ Typical industrial inverter power source controls.

> **NOTE**
>
> The display and the select/change increment/decrement pushbuttons are used to change the operational mode of the power source (GMAW, GMAW-P, FCAW, SMAW, Manual Pulse, etc.) and to select among a number of internal preset or programmable welding programs for various metals and shielding gases.

- *Voltmeter* – Indicates voltage during welding and voltage selected by Output Adjust Control for applicable modes or programs.
- *Ammeter/trim indicator* – Indicates amperage during welding and welding amperage or pulse trim amperage (for arc length) as selected by Output Adjust Control for applicable modes or programs.
- *Ammeter/trim indicator lights* – Indicate whether the ammeter/trim indicator is displaying welding amperage or pulse trip amperage.

A typical program sequence for one of the internal welding programs available in an advanced inverter power source is shown in *Figure 8*.

4.2.0 Power Source Ratings

The rating (size) of a welding machine is determined by the amperage output of the machine at a given duty cycle. The duty cycle of a welding

		TRIM 0-99	VOLTS 10.0-38.0	INDUCTANCE 0-99%	IPM 50-780	SECONDS
1. WELD	GMAW-P	X			X	0-100.0
	GMAW/FCAW		X	X	X	
2. CRATER	GMAW-P	X			X	0-5.00
	GMAW/FCAW		X		X	
3. MELTBACK	GMAW-P					0-0.25
	GMAW/FCAW		X			
4. & 5. POSTFLOW/PREFLOW						0-9.9
6. RUN-IN					25-780	
6. START	GMAW-P	X			X	0.00-5.00
	GMAW/FCAW		X		X	

X = Setting Available

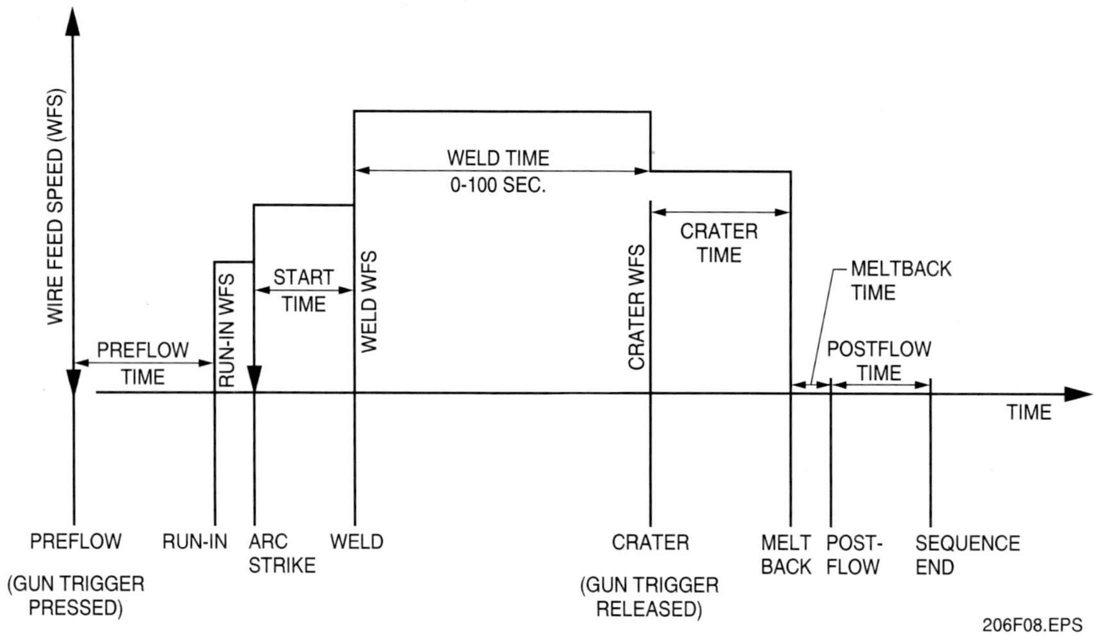

Figure 8 ◆ Typical welding program sequence.

machine is based on a 10-minute period. It is the percentage of 10 minutes during which the machine can continuously produce its rated amperage without overheating. For example, a machine with a rated output of 300 amps at 60% duty cycle can deliver 300 amps of welding current for 6 minutes out of every 10 without overheating.

The duty cycle of a welding machine will be 10%, 20%, 30%, 40%, 60%, or 100%. A welding machine with a duty cycle of 10% to 40% is considered a light- to medium-duty machine. Most industrial, heavy-duty machines for manual welding have 60 to 100% duty cycle ratings. Machines designed for automatic welding operations have 100% duty cycle ratings.

With the exception of 100% duty cycle machines, the maximum amperage that a welding machine can produce is always higher than its rated capacity. A welding machine rated 300 amps at a 60% duty cycle will generally put out a maximum of 375 to 400 amps. But, since the duty cycle is a function of its rated capacity, the duty cycle will decrease as the amperage is raised over 300 amps. Welding at 375 amps with a welding machine rated 300 amps at 60% duty cycle will lower the duty cycle to about 30%. If welding continues for more than 3 out of 10 minutes, the machine will overheat.

If the amperage is set below the rated amperage, the duty cycle increases. Setting the amper-

Hot Tip

All-in-One Welding Machines

Hand portable, light industrial GMAW/FCAW welding machines are also available. The CV/DC machine shown is rated at 130 amps with a 30% duty cycle. It is equipped with an internal wire feeder.

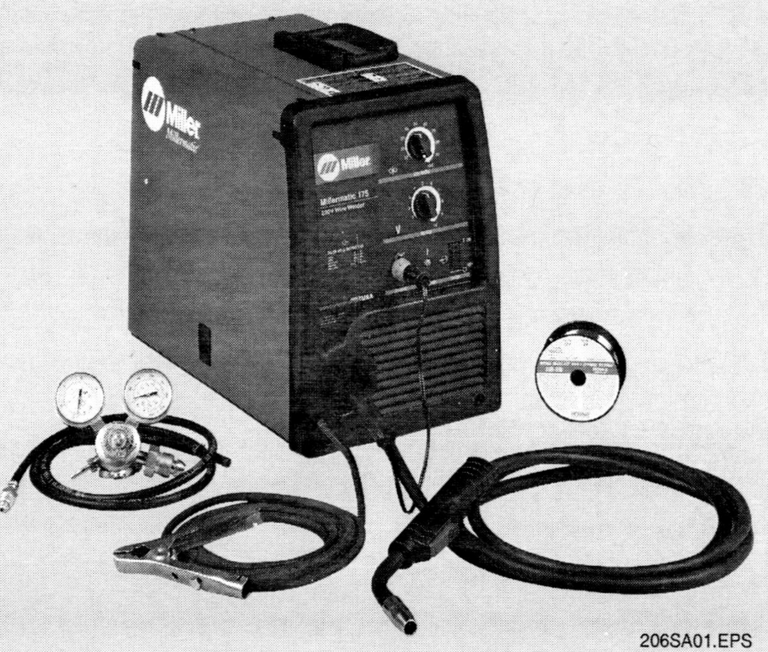

206SA01.EPS

> **NOTE**
> Most welding machines have a heat-activated circuit breaker that shuts off the machine automatically when it overheats. The machine cannot be turned back on until it has cooled.

age at 200 amps for a welding machine rated 300 amps at 60% duty cycle will increase the duty cycle to 100%. *Figure 9* shows the relationship between amperage and duty cycle.

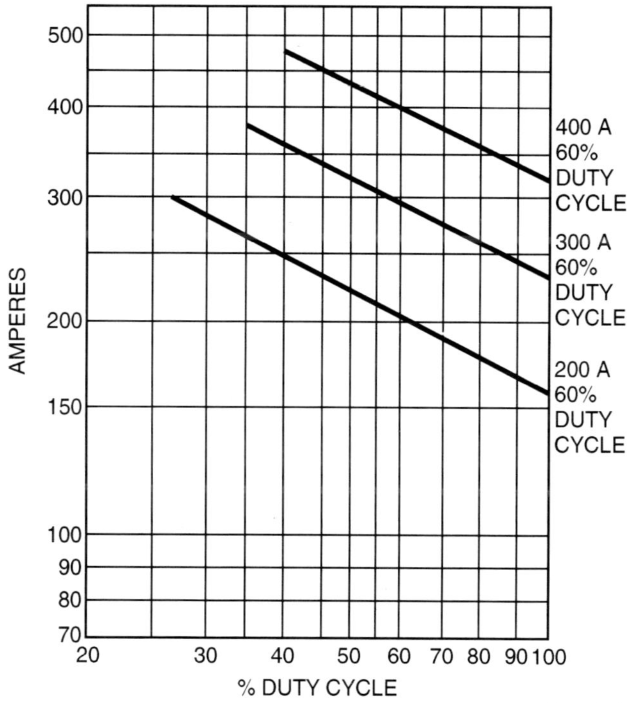

206F09.EPS

Figure 9 ◆ Relationship between amperage and duty cycle.

5.0.0 ◆ GMAW AND FCAW EQUIPMENT

Typical GMAW and FCAW systems are made of four interconnected pieces of equipment, including the following:

- DC power source (welding machine)
- Wire feeder (if required)
- Welding gun
- Shielding gas supply (if required)

In some systems, equipment functions may be integrated or combined. For example, the wire feeder can be built into the same cabinet as the power source or it may be incorporated into the welding gun. Welding guns are shown in a later section of this module.

Most systems have a separate wire feeder (push system) that pushes the wire from the feeder through the gun cable and gun. Other systems may incorporate the wire feeder into the welding gun (pull system) or do both (push-pull system). There are some systems that have synchronized wire feeders located at intervals along very long feeder conduits. Various types of wire feeders are covered in a later section. Heavy-duty, water-cooled guns may also be equipped with a closed loop water cooling system. Systems that are used for self-shielding FCAW do not require gas cylinders.

Figure 10 shows a simplified diagram of a typical GMAW/FCAW system.

5.1.0 GMAW Metal Transfer Modes

GMAW can use any of several modes (types of metal transfer) to conduct the filler metal to the molten weld pool. The principal modes follow:

- **Spray transfer**
- **Globular transfer**
- **Short-circuiting transfer**
- **Pulsed transfer**

5.1.1 Spray Transfer

In the spray transfer mode (*Figure 11*), the welding voltage is set in the higher ranges to keep the end of the wire electrode away from the work and to maintain a continuous arc. The high heat of the arc melts the end of the wire electrode, and fine droplets are carried by the arc **axially** to the molten weld puddle on the base metal. The spray transfer deposits filler metal at a high rate and produces very little spatter. However, the weld puddle is relatively large, and the mode works best in the flat or horizontal fillet positions. Also, only certain shielding gases or gas combinations can be used with spray transfer. Shielding gases are covered later in this module.

5.1.2 Globular Transfer

In the globular transfer mode, the welding voltage is set lower than it is for spray transfer. The end of the wire electrode remains a short distance from the work to maintain a continuous arc. Because of

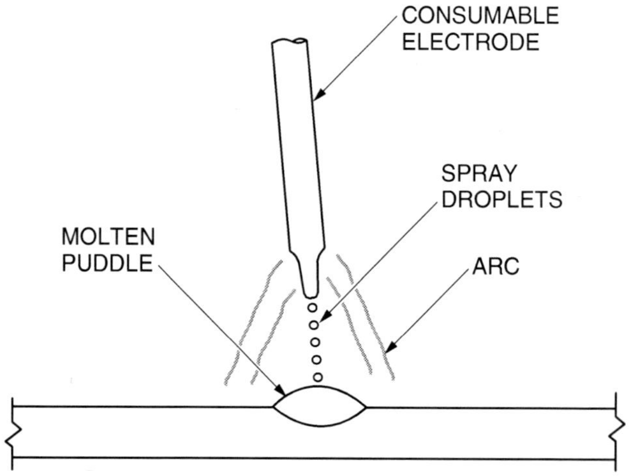

Figure 11 ◆ Spray transfer.

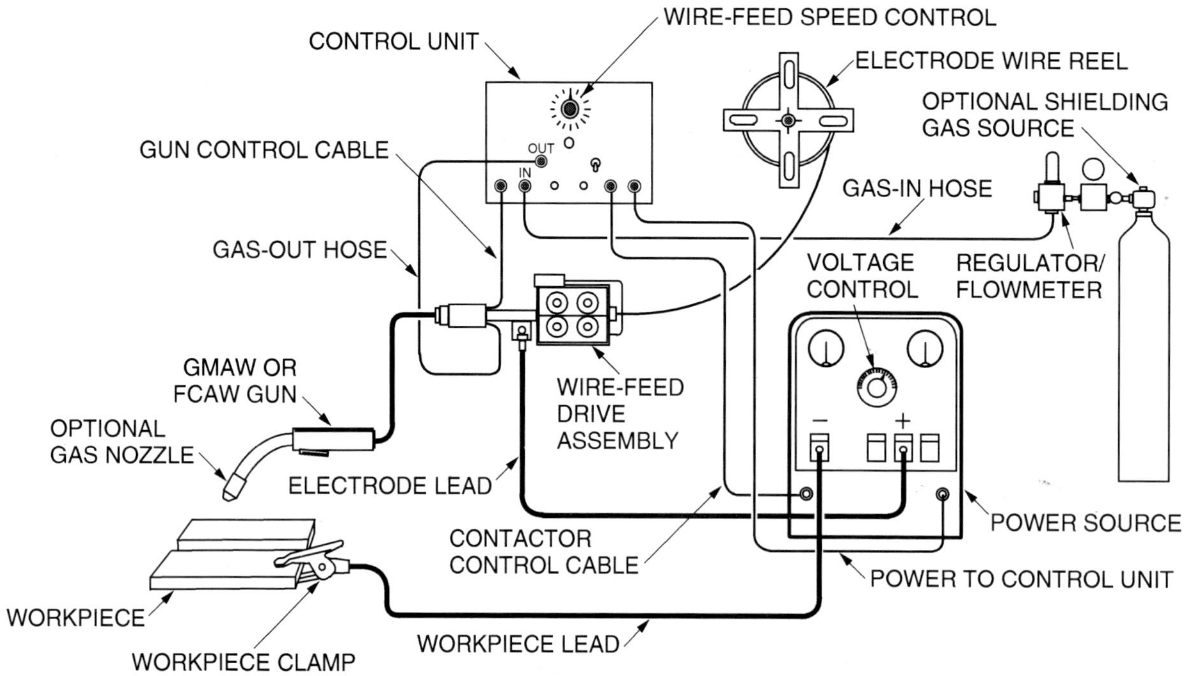

NOTE: THE POLARITY OF THE GUN AND WORKPIECE LEADS IS DETERMINED BY THE TYPE OF FILLER METAL AND APPLICATION.

206F10.EPS

Figure 10 ◆ Simplified GMAW/FCAW system diagram.

the weight, the electrode wire tip periodically melts off in a large glob, which the arc then carries to the workpiece, where it mixes with the weld puddle. Globular transfer can be used to deposit filler metal at high rates and is capable of producing high-quality welds in all positions. However, bead appearance can be rough, and spatter may be heavy in some applications. *Figure 12* shows globular transfer.

5.1.3 Short-Circuiting Transfer

Short-circuiting transfer (GMAW-S) uses the lowest voltage settings of the three modes. It is the best mode for welding light-gauge sheet metal. The wire electrode is fed against the workpiece; when contact is made, current flow heats the wire electrode until it melts enough for a droplet to separate and mix with the weld puddle. When the droplet separates, a gap is formed and an arc is initiated. The continuously feeding wire closes the gap and contacts the workpiece, shorting out the arc. The current again heats the wire until another droplet separates, and the arc is again established. This process repeats many times each second. The short-circuiting process can be used to weld thin materials and is suitable for all positions. However, it does not penetrate deeply and may produce incomplete fusion on thicker materials because of the low heat input. *Figure 13* shows short-circuiting transfer.

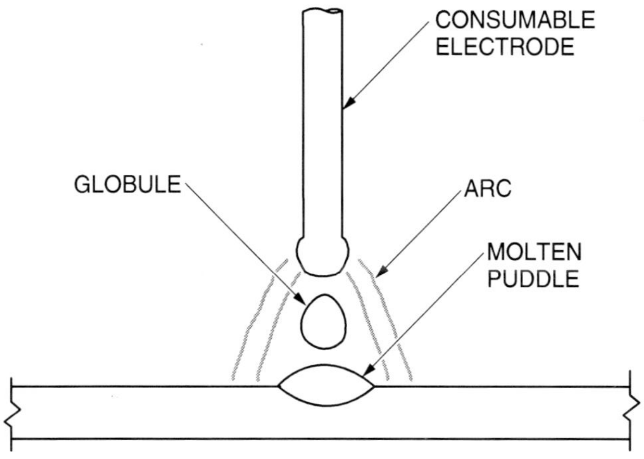

Figure 12 ◆ Globular transfer.

5.1.4 Pulsed Transfer

Pulsed transfer (GMAW-P) welding is a modified combination of spray transfer welding and short-circuit transfer arc welding. The process transfers molten metal in droplets while maintaining an arc with the base metal. The pulsed GMAW process also allows the transfer action to occur while using lower current levels. The advantages of the lower current levels are less spatter and better penetration without melting through the base metal.

In pulsed transfer welding, the welding power is electronically controlled by a special power source that pulses the output power to the consumable electrode (wire). Although different GMAW machines may have different types of special power sources, the results are the same for each. The welding current is generated in pulses of high current at specified periods of time. The frequency of these pulses varies from one power source to the next, but some power sources are capable of producing pulses over a wide range. During the period of time between each pulse, the welding current is reduced but is still maintained at a high enough level to keep the wire heated and an arc established. This lowered current level is called the background current. *Figure 14* shows output pulses of high welding current occurring over time.

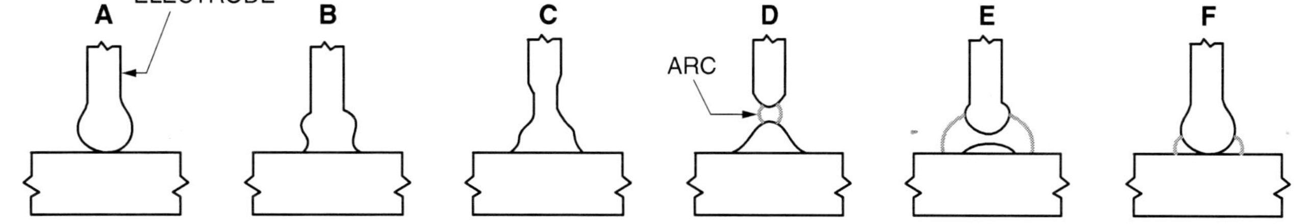

Figure 13 ◆ Short-circuiting transfer.

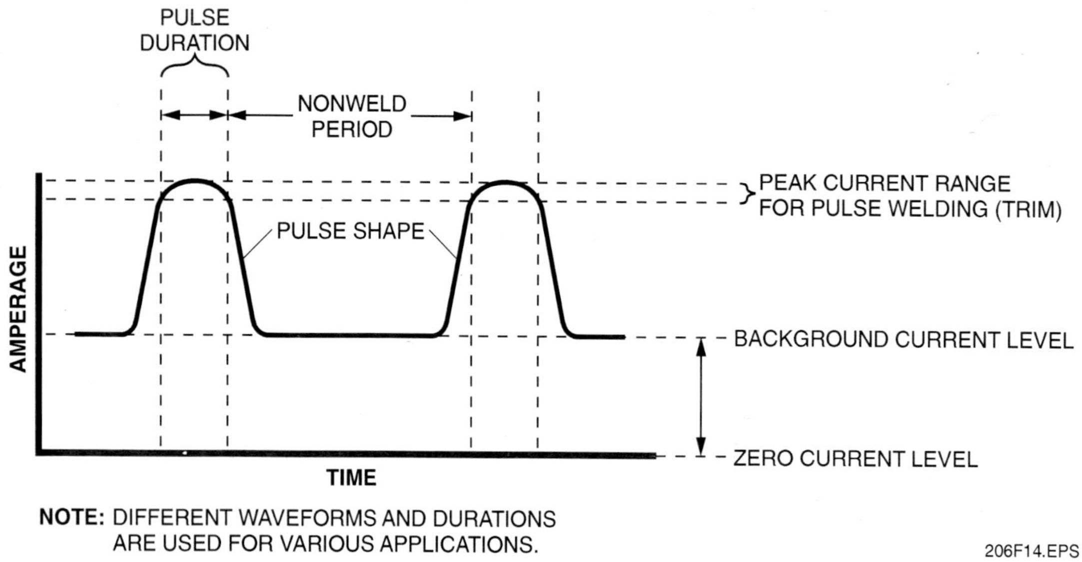

Figure 14 ◆ Output pulses of high welding current occurring over time.

After an arc is established and maintained between the consumable electrode (wire) and the base metal, the welding current to the electrode is momentarily increased to form a pulse and then returns to its lower background level. Each pulse of welding current must peak strongly enough, and have enough duration, to make a single drop of molten metal separate from the wire that is maintaining an arc with the base metal.

Figure 15 shows the pulsed welding process as associated with one of the pulse shapes. Points A through E on the diagram relate to the electrode arc illustrations, labeled A through E, above the diagram.

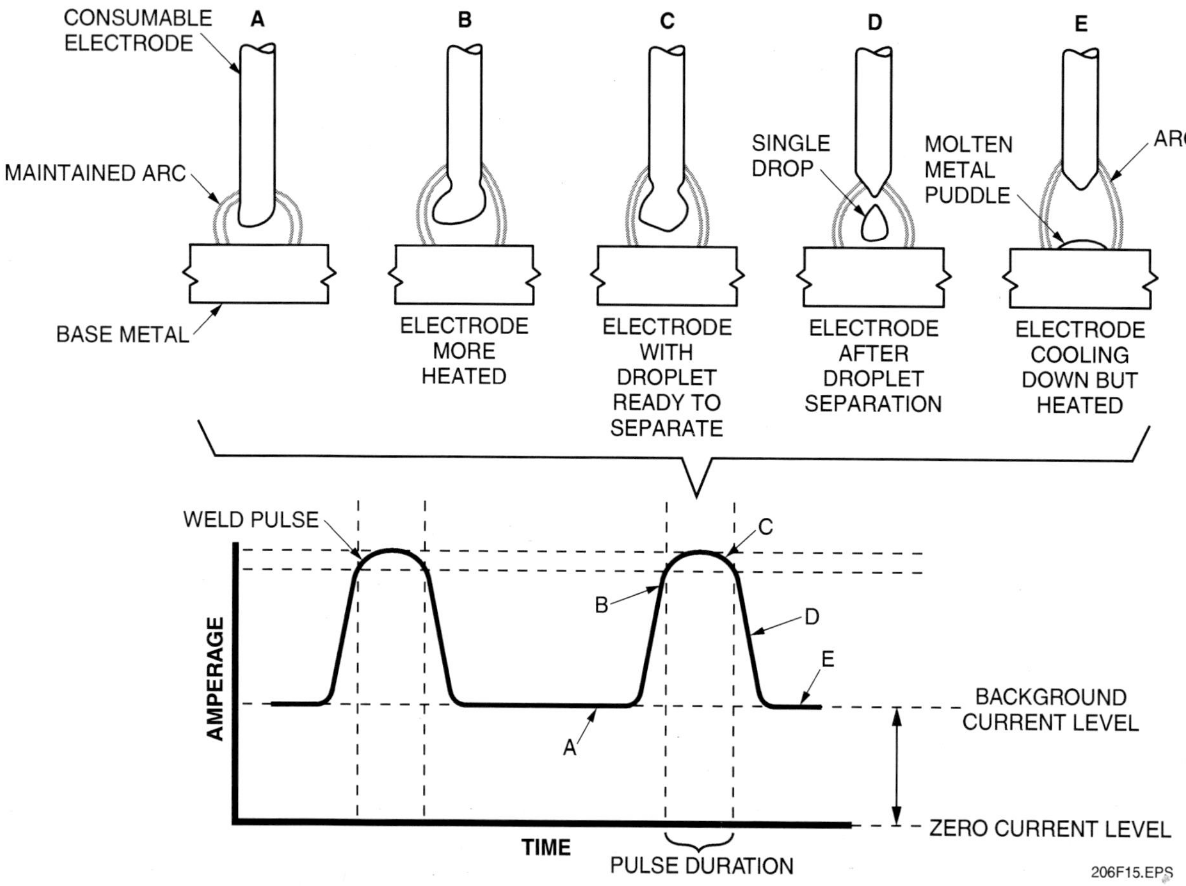

Figure 15 ◆ Typical pulsed welding process associated with one pulse shape.

Hot Ti...

frequenc...
wire spe...
charact...

When ... from th... to form... transfe... the ch... are re... a hig... the w...

5.2 ...

Cu... rous base metals. ... tric arc to melt the filler and ba... transfer fine droplets of filler metal to the weld puddle. The electromagnetic force that propels the metal spray is strong, but the process produces very little spatter. The flux inside the core contains ingredients to perform several functions:

- Ionizers to stabilize the arc
- Deoxidizers to purge the weld of gases and slag
- Additional metals and elements to enhance the quality of the weld metal
- Ingredients to generate shielding gas
- Ingredients to form a protective slag over the weld bead

...e gas shield generated by flux in the wire ...as well as that of the external gas shield pro... the arc and molten weld metal. The slag ...er produced by the flux helps to support the ...d metal, permitting the process to be used for ...king high-quality welds in all positions. With ...ger diameter flux cored wires, deposition rates ...25 pounds per hour can be achieved. *Figure 16* ...ows the FCAW process.

...5.2.1 FCAW Weld Penetration

FCAW welds made with carbon dioxide (CO_2) shielding have much deeper penetration than either SMAW or FCAW without shielding gas. For this reason, edge preparation is often not needed for double-welded butt joints on plate up to ⅜-inch thick. When making FCAW fillet welds using CO_2 shielding, a smaller size fillet weld will have as much strength as a larger fillet made with SMAW because of the deep penetration achieved with FCAW and the carbon dioxide shielding.

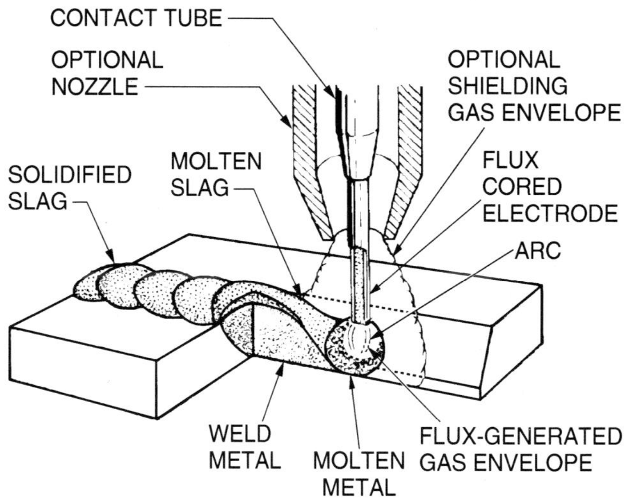

NOTE: OTHER GUN CONFIGURATIONS AND STYLES ARE AVAILABLE.

206F16.EPS

Figure 16 ◆ FCAW process.

> **Hot Tip**
>
> ***Pulse Transfer Advantages***
>
> The following are advantages of pulsed transfer welding over welding processes using continuous currents:
>
> - Rivals GTAW for weld quality.
> - Permits all-position welding.
> - Allows smooth welds to be made on thin materials without melt-through.
> - Controls thick plate weld pool more easily.
> - Reduces cost because larger diameter electrode wire can be used.
> - Can be used on base metals such as aluminum, stainless steel, copper alloy, and nickel steels as well as carbon steels.
> - Reduces weld contamination from surface oxides because less surface area is exposed during the weld.

5.2.2 FCAW Joint Design

FCAW uses the same basic joint designs as SMAW. However, when making groove welds with FCAW and CO_2 shielding gas, the square groove (square butt) joint can be used for plate thicknesses up to ⅜ of an inch. Above this thickness, beveled joints should be used.

Because FCAW electrode wire is much thinner than comparable SMAW electrodes, and FCAW penetration with carbon dioxide is superior to SMAW, the included angles of V joints can be reduced to half the angle required for SMAW. For example, a SMAW included angle of 60° can be reduced to 30° when using FCAW with CO_2 shielding. This can save approximately 50 percent on filler metal and considerable welding time when compared with using SMAW. *Figure 17* shows joint designs for FCAW with CO_2 shielding.

CAUTION

Always refer to your WPS or Site Quality Standards for specific information on joint requirements. Information in this manual is provided as a general guideline only.

5.3.0 GMAW Power Sources

As described in earlier sections, GMAW and FCAW are best performed with a constant-voltage (CV) or constant voltage/constant current (CV/CC) DC power. In constant-voltage welding machines, the open-circuit voltage and the welding voltage are nearly the same. Constant-current variable voltage DC welding machines, the type used for SMAW, in which the output voltage varies between high open-circuit voltage and lower arc voltage, are sometimes used but require specially designed voltage-sensing wire feeders. CV/CC welding machines are required for GMAW-P operation.

Power sources usually contain controls for adjusting arc voltage, slope, and inductance and have meters to monitor amperage and voltage. The simplest machines are power sources with built-in wire feeders. A simple combination welder-wire feeder may have only a wire-feed speed control and a voltage control.

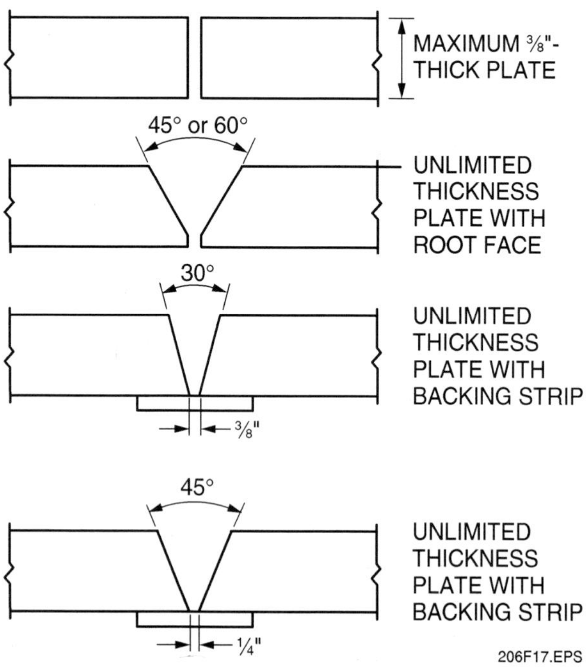

Figure 17 ◆ Joint designs for FCAW with CO_2 shielding.

5.3.1 Slope

The constant voltage power sources used for GMAW and FCAW are not truly constant in their voltage output, because the voltage always drops a little as the current (amperage) increases. This voltage-to-amperage relationship forms a slight curve when plotted as a graph. The general angle of the volt-ampere curve is known as the slope. This slope is adjustable on some power sources. A flatter (more horizontal) slope is better for spray transfer welding, while a steeper slope is better for short-circuiting welding. *Figure 18* shows flat slope and steep slope volt-ampere curves for GMAW.

5.3.2 Pinch Effect and Inductance

When using short-circuiting welding in GMAW, the current passing through the wire electrode heats and then melts the wire. This is sometimes

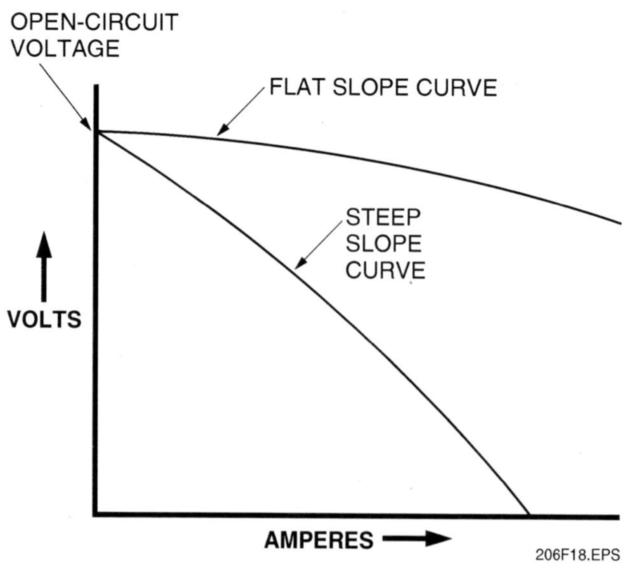

Figure 18 ◆ Flat slope and steep slope volt-ampere curves for GMAW.

called the pinch effect because the molten wire appears to be pinched off, forming a droplet of molten metal. *Figure 19* shows the pinch effect.

Low inductance produces high pinch effect, and high inductance produces low pinch effect. With low circuit inductance, the current rise is very fast, and the high pinch effect can cause the wire droplet to explode or spatter. When the circuit inductance is high, the current rises more slowly, decreasing the number of short circuits per second and extending the arc duration. The pinch effect forms the droplet more gently, which results in a more fluid puddle, smoother weld, and less spatter.

Constant voltage power sources are self-regulating in that they automatically produce current as it is required. Inductance controls the rate at which the current output rises. Some inductance is desirable in FCAW because it prevents explosive arc starts by slowing down the current rise rate. Many FCAW power sources have a control to adjust the inductance.

5.3.3 Arc Blow

When operating in GMAW spray transfer mode or in FCAW, arc blow can sometimes be a problem. Arc blow is the deflection of the arc from its normal course because of the attraction or repulsion of the arc's magnetic field with the weld current's magnetic field in the base metal. The amount and direction of arc deflection depend upon the relative position, direction, and density of the base metal's magnetic field. Arc blow can result in excess spatter and weld defects. It can be minimized by relocating the workpiece clamp or changing the weld angle of the gun.

5.4.0 Welding Cable

Cables used to carry welding current are designed for maximum strength and flexibility. The conductors inside the cable are made of fine strands of copper wire. The copper strands are covered with layers of rubber reinforced with nylon or dacron cord. *Figure 20* shows a cutaway section of welding cable.

The size of a welding cable is based on the number of copper strands it contains. Large-diameter cable has more copper strands and can carry more welding current. Typical welding cable sizes (*Appendix, Table 1*) range from #4 to #3/0 (3 aught).

When selecting welding cable size, the amperage load as well as the distance the current will travel must be considered. The longer the distance the current has to travel, the larger the cable must be to reduce voltage drop and heating caused by the electrical resistance in the welding cable. When selecting welding cable, use the rated capacity of the welding machine for the cable amperage requirement. For the distance, measure

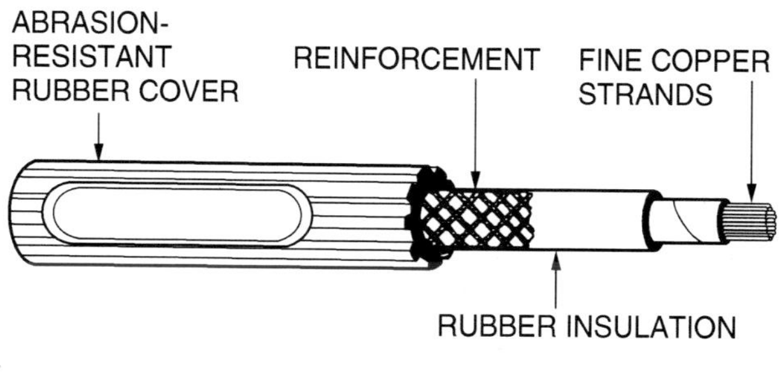

Figure 20 ◆ Cutaway section of welding cable.

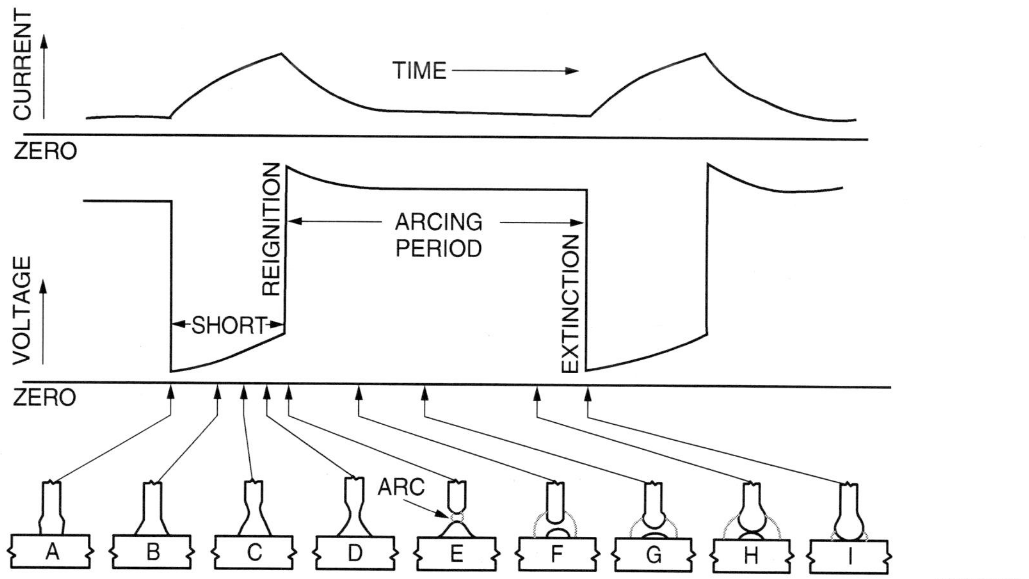

Figure 19 ◆ Pinch effect.

both the electrode and the workpiece leads, and add the two lengths together. To identify the welding cable size required, refer to a recommended welding cable size table, furnished by most welding cable manufacturers.

5.4.1 Welding Cable End Connections

For efficient use, welding cables must be equipped with the proper end connections or terminals.

> **CAUTION**
>
> If the end connection is not tightly secured to the cable, the connection will overheat and oxidize. An overheated connection will cause variations in the welding current and permanent damage to the connector and/or cable. Check connections for tightness and repair loose or overheated connections.

Lugs are used at the end of the welding cable to connect the cable to the welding machine current terminals. The lugs come in various sizes to match the welding cable size and are mechanically connected onto the welding cable.

Quick disconnects are also mechanically connected to the cable ends. They are insulated and serve as cable extensions for splicing two lengths of cable together. Quick disconnects are connected or disconnected with a half twist. When using quick disconnects, care must be taken to ensure that they are tightly connected to prevent overheating or arcing in the connector. *Figure 21* shows typical lugs and quick disconnects used as welding cable connectors.

The workpiece clamp (see *Figure 22*) provides the connection between the end of the workpiece lead and the workpiece. Workpiece clamps are mechanically connected to the welding cable and come in a variety of shapes and sizes. The size of a workpiece clamp is the rated amperage that it can carry without overheating. Workpiece clamps must be rated at least the same as the rated capacity of the power source on which they will be used.

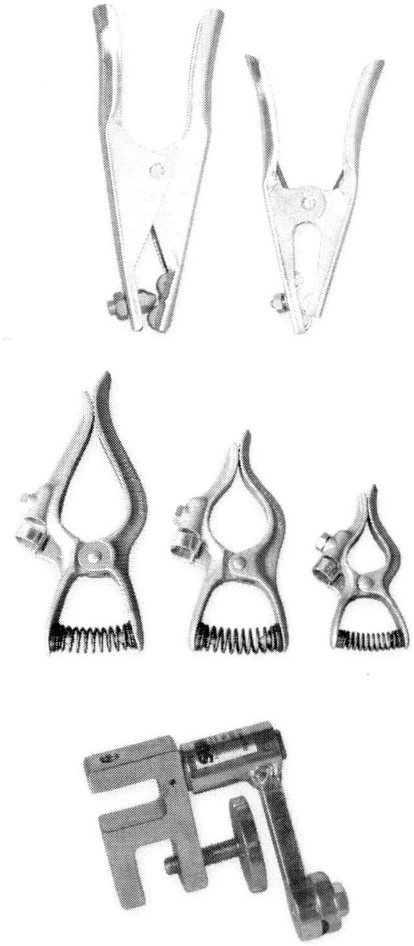

206F22.EPS

Figure 22 ◆ Workpiece clamps.

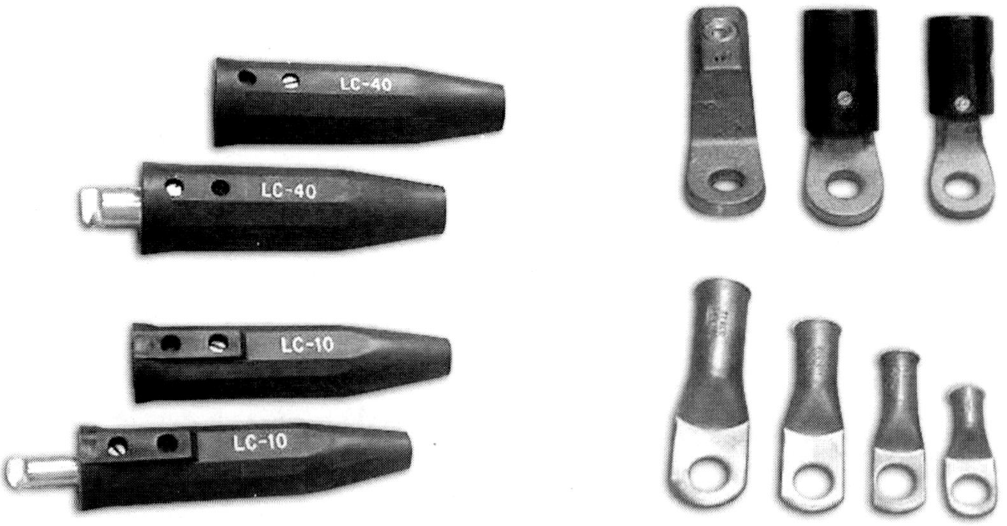

206F21.EPS

Figure 21 ◆ Lugs and quick disconnects.

5.5.0 External Wire Feeders

An external wire feeder pulls the electrode wire from a spool and pushes it through the gun cable and the gun. It consists of a wire spool holder with a drag brake, an electric motor that drives either one or two sets of opposing slotted rollers, and various controls. The wire feeder also usually contains the shielding gas connections, the gas control solenoid, and (optionally) the cooling-water flow-control solenoid. Some controls that may be located on a wire feeder include the following (not all these controls are present, and other controls may exist, on some wire feeders):

- *Spool drag* – This is a manual adjustment used to adjust the drag on the wire spool to prevent the uncontrolled unwinding of the wire.
- *Wire feed speed control* – This is a variable control used to set the wire feed speed. Since GMAW power sources are self-regulating (they produce the required current on demand), increasing or decreasing the wire feed speed will cause the welding machine to automatically increase or decrease its current output.
- *Voltage control* – This control is used to set the welding voltage.
- *Wire jog* – The manual wire jog switch is used to start and stop the wire feeder without energizing the welding current.
- *Wire meltback timer (burnback)* – This is an adjustable time delay that causes the welding current to continue after the trigger is released. This consumes the length of wire that continues to feed after the trigger is released so that the proper **stickout** is maintained to start the next weld.
- *Gas purge* – The manual gas purge switch is used to operate the gas solenoid to purge the shielding gas system prior to actual welding. It is also used to adjust the flowmeter.
- *Trigger hold* – This switch enables or disables a trigger hold function activated by the gun trigger. This function enables welding without holding the trigger.
- *Gas preflow timer* – This is an adjustable time delay that prevents the welding current contactor from closing until the shielding gas is flowing.
- *Gas postflow timer* – This is an adjustable time delay that is used to keep shielding gas flowing for a short period after the gun trigger is released.

Figure 23 shows typical basic wire feeder units, and *Figure 24* shows other types of wire feeders.

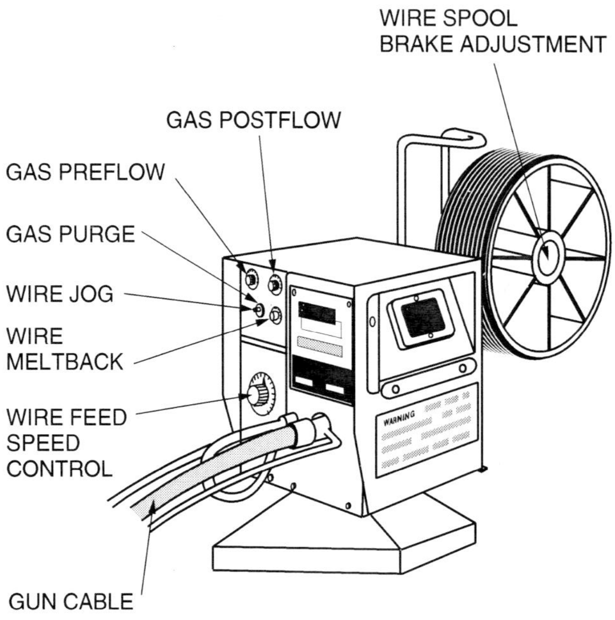

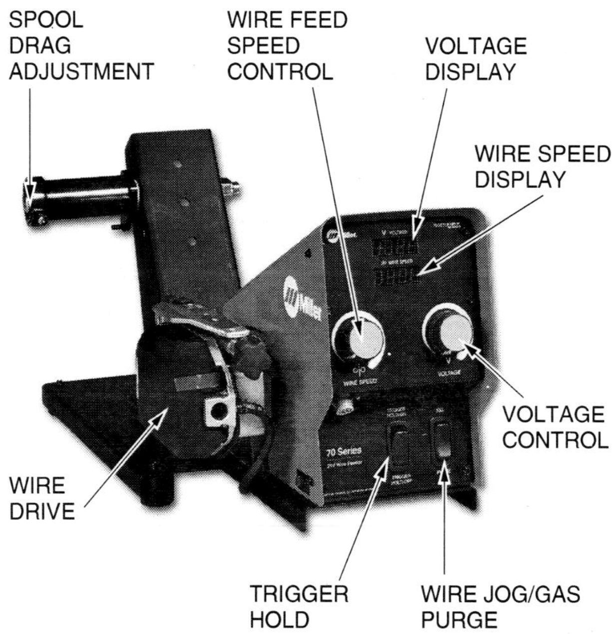

Figure 23 ◆ Basic GMAW/FCAW wire feeder units.

The wire feeder grips the wire in the grooves of opposing grooved rollers. There can be one or two sets of opposed rollers (*Figure 25*). The grooves can be V-grooved or U-grooved in cross section. V grooves are used for hard wire. One roller slot in each V-groove set may be serrated for better gripping. U-groove rollers are used for soft wire or soft-shelled cored wire. For extremely soft wire or soft-shelled cored wire, cogged U-groove rollers are used.

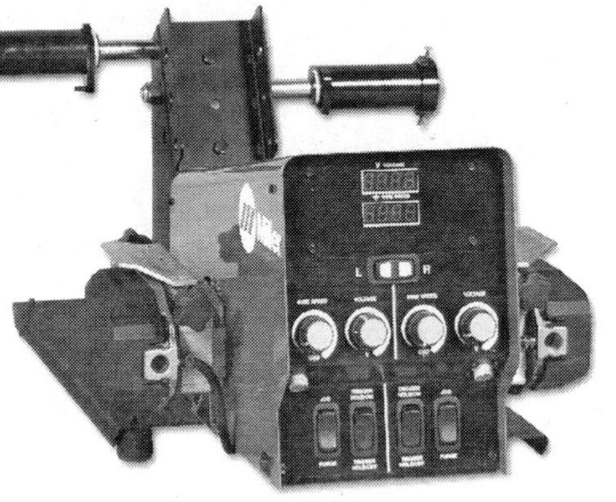

DUAL REEL BENCH GMAW/FCAW WIRE FEEDER

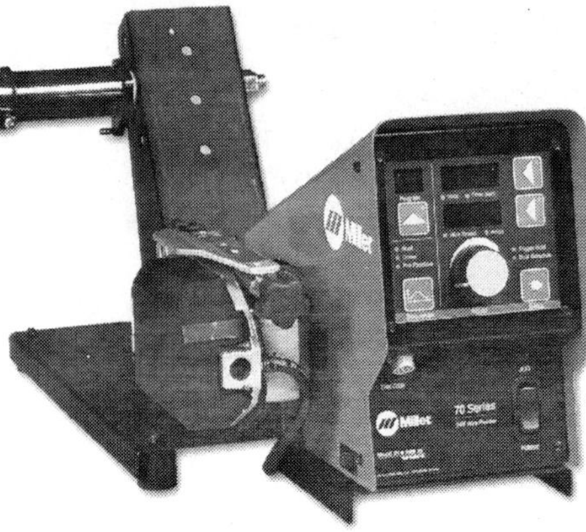

SINGLE REEL PROGRAMMABLE BENCH WIRE FEEDER FOR GMAW-P

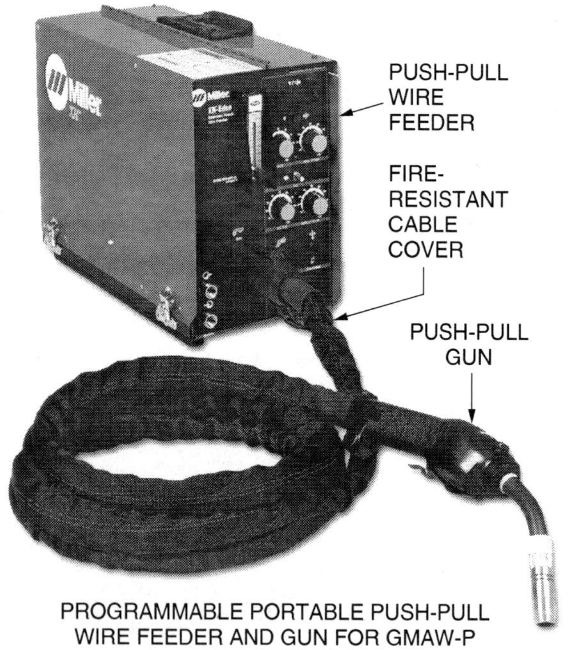

PROGRAMMABLE PORTABLE PUSH-PULL WIRE FEEDER AND GUN FOR GMAW-P OF ALUMINUM

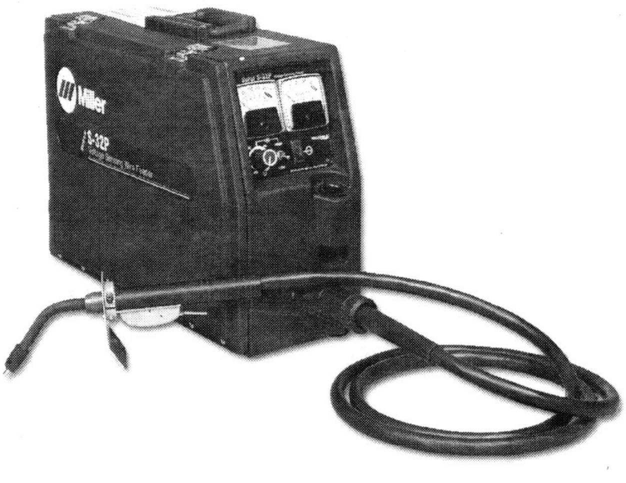

WEATHERPROOF PORTABLE GMAW/FCAW WIRE FEEDER WITH FCAW-S GUN

206F24.EPS

Figure 24 ◆ Various types of wire feeders.

TWO-ROLL DRIVE SYSTEM

FOUR-ROLL DRIVE SYSTEM

206F25.EPS

Figure 25 ◆ Wire feeder drive roll systems.

Some rollers are made in two pieces with different-size half-grooves formed on the outside edges of each half-roller. The half-rollers are assembled to fit two different ranges of wire size. *Figure 26* shows groove styles and combination rollers.

When dust or dirt is a problem, a felt wiper can be added to the wire between the spool and the inlet guide to the feed rollers. The felt wipers are treated with a silicone lubricant that cleans and lubricates the wire. Special liquid silicone cleaner/lubricants can be added to the felt wiper occasionally.

CAUTION

Use only special cleaner/lubricants made specifically for wire feed wipers. Use of other cleaners or lubricants may cause porosity in the weld.

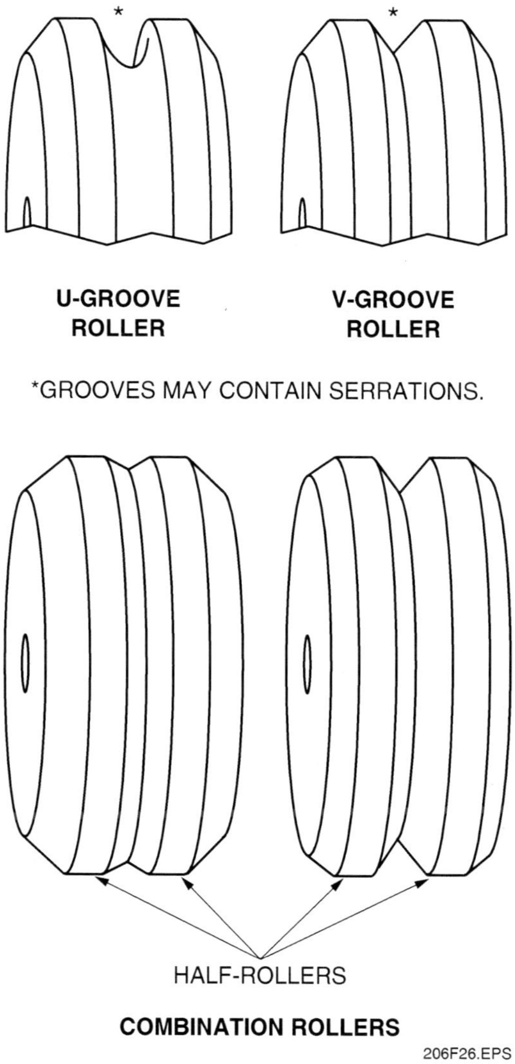

Figure 26 ◆ Groove styles and combination rollers.

Hot Tip

Wire Straighteners

A wire straightener can be added to some wire feeders. This device can be adjusted to remove the spool cast (curl) imparted to solid wire or hard-shelled cored wire to prevent wire jams and wear in the gun cable.

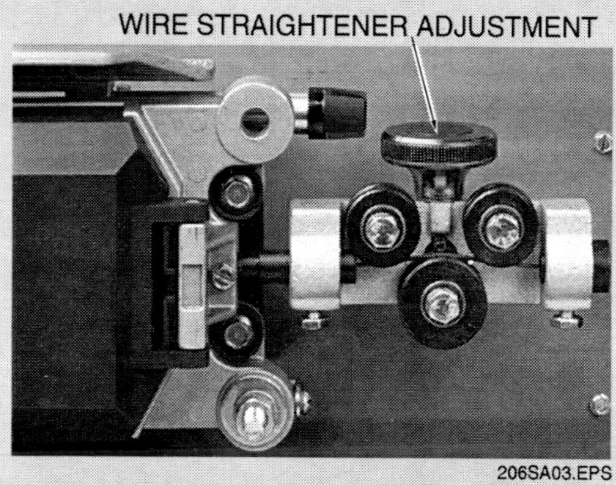

Hot Tip

Knurled Roller

In addition to the U-groove rollers, knurled V-groove rollers are used for hard-shelled cored wire. Knurled rollers have small serrations formed into the edge of the groove to help the roller grab the wire and push it out to the gun without crushing it.

5.6.0 GMAW/FCAW-G Guns

The GMAW/FCAW-G gun supports and guides the wire electrode and provides the electrical contact between it and the welding machine electrode lead. The gun also contains a remote switch to start and stop the welding current and the shielding gas flow. The gun tip contains a nozzle to direct the shielding gas around the arc. The tip of the electrode wire extends through the contact tube, located within the center of the nozzle. The contact tube is replaceable and must be matched to the electrode wire size. Because the contact tube is the welding current's electrical connection with the electrode wire, it is also subject to wear from friction with the wire and electrical erosion and must be replaced periodically. GMAW guns are usually rated for amperage at a 100% duty cycle using CO_2 and are derated from 10% to 50% for mixed gases, depending on gas mixture and welding parameters. *Figure 27* shows the construction of a typical GMAW/FCAW-G gun.

Some types of guns contain motor-driven wire pullers within the gun handle (*Figure 28*). These work in conjunction with a push wire feeder located with the power source. This system was developed for feeding soft wire, like aluminum, that tends to bend very easily.

Additionally, there are spool guns designed for aluminum wire. These are guns that contain both a wire feeder and a small spool of wire. In some applications, an auxiliary gun control unit is required for proper operation. *Figure 29* shows one style of spool gun.

Light-duty guns are gas-cooled by the shielding gas flow. Heavy-duty guns are water cooled. The cooling water flow may be controlled by a solenoid in the wire feeder or by a separate dedicated cooling

---

Hot Tip

Programmable Wire Feeders

Programmable wire feeders are normally used with compatible CC/CV power sources for GMAW-P applications. Custom welding programs can be created, or preset welding programs can be stored in the feeder or in the power source. The program can be activated by the various controls and displays on the feeder.

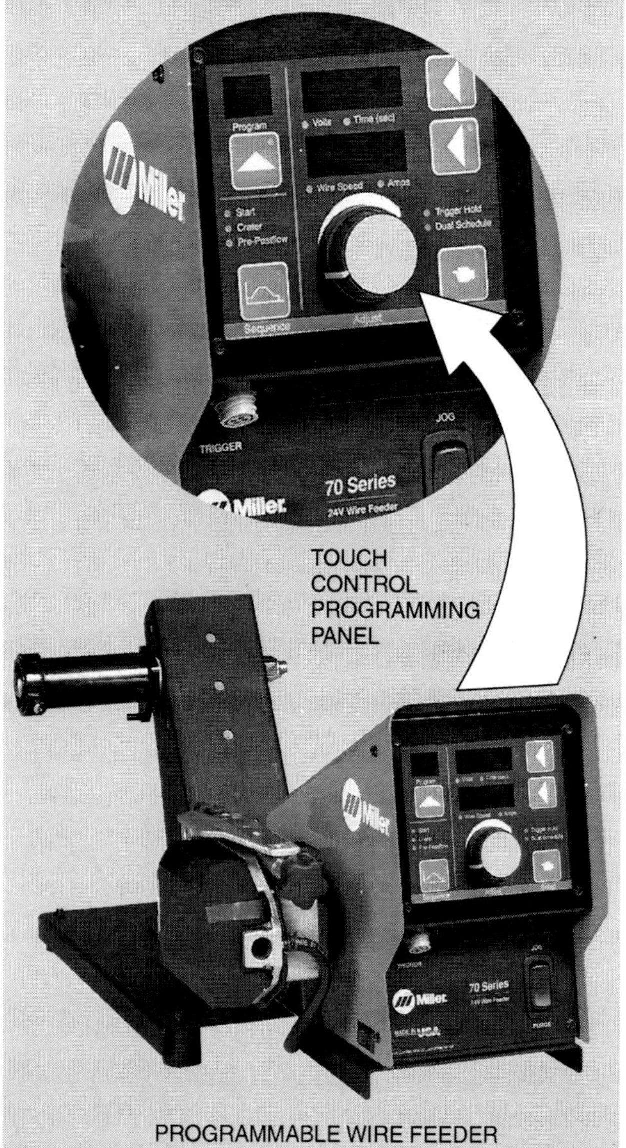

206SA04.EPS

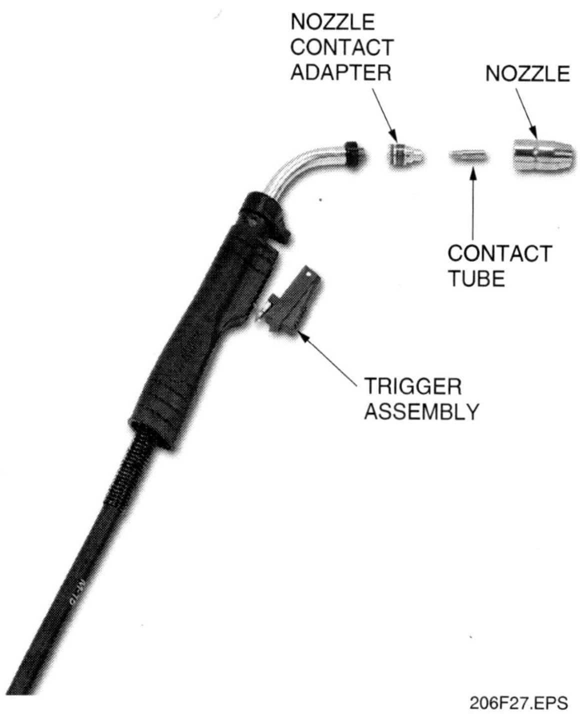

206F27.EPS

Figure 27 ◆ Construction of a typical GMAW/FCAW-G gun.

system. Cooling water is piped to and from the gun through flexible tubing integrated into the gun cable. Some cooling systems use domestic water to cool the gun. Others use a closed-loop system (see *Figure 30*) that circulates de-mineralized water or special cooling fluid that will not corrode internal gun surfaces or plug the cooling passages with mineral scale.

5.7.0 FCAW-S Guns

FCAW-S guns are similar to GMAW/FCAW-G guns but are often designed to handle larger diameter wire. The FCAW gun supports and guides the wire electrode. It also contains a remote switch for starting and stopping the welding current. The electrode wire extends through the contact tube located at the end of the gun. The contact tube provides the electrical connection from the welding machine electrode lead to the electrode wire. The contact tube is replaceable and must always be matched to the electrode wire size. If it is too small, the wire will jam; if it is too large, poor contact will cause damaging arcing between the tip and the wire. The contact tube is subject to wear from sliding friction and electrical erosion caused by the wire and must be replaced periodically.

Guns designed for self-shielded FCAW do not have provisions to deliver shielding gas. This makes the gun nozzles smaller in diameter. To protect the contact tube, these guns have a fiber insulator that threads onto the end of the gun nozzle.

FCAW-S generates smoke, which can be removed at the gun tip with a special smoke extraction nozzle. The nozzle is connected to a vacuum pump or blower by a flexible tube to draw off the smoke and thereby improve visibility. *Figure 31* shows an FCAW-S gun without the smoke extractor.

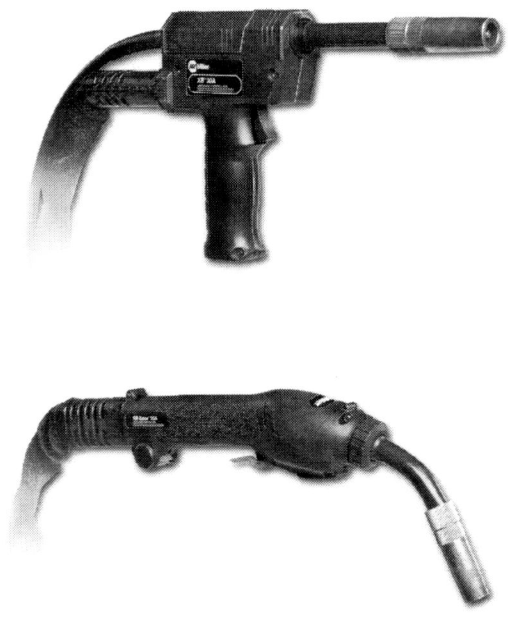

206F28.EPS

Figure 28 ◆ Pull guns for push-pull wire feeders.

206F29.EPS

Figure 29 ◆ Spool gun.

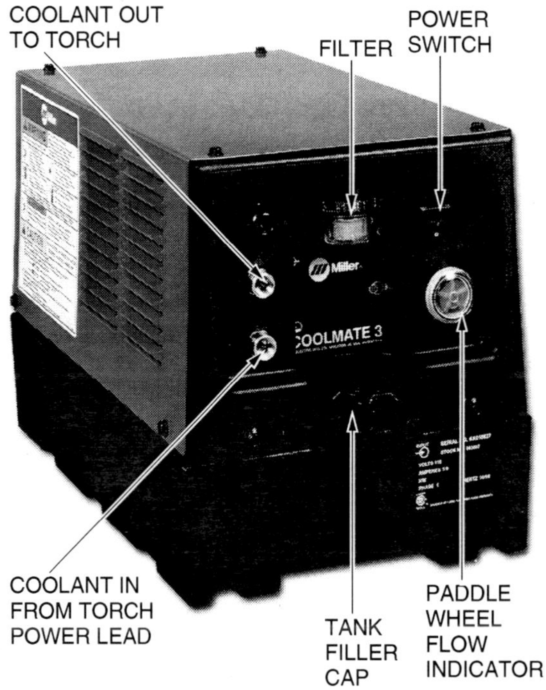

206F30.EPS

Figure 30 ◆ Typical closed-loop cooling unit.

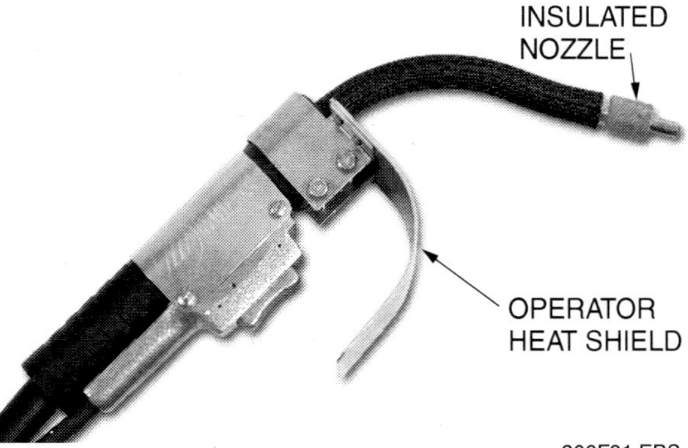

206F31.EPS

Figure 31 ◆ FCAW gun without smoke extractor.

5.8.0 Shielding Gas Supply

The GMAW/FCAW-G process always uses shielding gas to protect the welding electrode (wire) and the weld from contamination. The shielding gas is directed around the electrode end and the weld zone by the nozzle of the gun. The most common shielding gases are argon (Ar), helium (He) and carbon dioxide (CO_2). These may be used alone or mixed. A small amount of oxygen (O_2) is often added to argon to stabilize the arc, increase the filler transfer rate, and improve wetting and bead shape.

Some FCAW core wires require the use of a shielding gas. The shielding gas is directed around the electrode end and the weld zone by the nozzle of the gun. The shielding gas used depends on the filler wire and base metal being welded. The most common shielding gases for FCAW carbon and low-alloy steels are carbon dioxide (CO_2) or an argon-carbon dioxide mix. For FCAW high-alloy or stainless steels, argon or an argon-carbon dioxide mixture is generally used. These gases are available in pressurized cylinders of various sizes and in bulk liquid.

CAUTION

Check your WPS or Site Quality Standards for the type of shielding gas to use. If these are not available, refer to the filler metal supplier's specifications.

5.8.1 Cylinder Safety

Cylinders must always be handled with great care because of the extremely high-pressure gases they contain. When a valve is broken off a high-pressure cylinder, the gas escapes with explosive force and can severely injure personnel with the blast and blown debris. The cylinder itself can become a powerful uncontrolled missile. When transporting and handling cylinders, always observe the following rules:

- The safety cap should always be installed over the valve except when the cylinder is connected for use.

WARNING!

Do not remove the protective cap unless the cylinder is secured. If a cylinder falls over and the nozzle breaks off, the cylinder will shoot like a rocket, causing severe injury or death to anyone in its way.

- When in use, the cylinder must always be secured to prevent it from falling. It should be chained or clamped to the welding machine or to a post, beam, or pipe.
- A cylinder cart should be used to transport a cylinder.
- Cylinders should never be hoisted by slings or magnets. They can slip out of the sling or fall from the magnets. Always use a hoisting basket or similar device.
- Always open the cylinder valve slowly. Once pressure is applied to the system, the valve should be completely opened to prevent leakage from around the valve stem.

5.8.2 Gas Regulators/Flowmeters

A gas regulator/flowmeter is required to meter the shielding gas to the gun at the proper flow rate. A typical regulator/flowmeter consists of a preset pressure regulator with a cylinder valve spud, a flowmetering (needle) valve, and a flow-rate gauge. The pressure regulator is usually equipped with a tank pressure gauge to show cylinder pressure. The metering valve is used to adjust the gas flow to the gun nozzle. The flow gauge indicates the gas flow rate in cubic feet per hour (cfh) or liters per minute as shown by a ball-type flow indicator. *Figure 32* shows a shielding gas regulator/flowmeter.

Gas flow to the gun is started and stopped by an electric solenoid valve that is usually located inside the control unit or the wire feeder unit. The solenoid is controlled by the operator with the trigger switch on the gun. The same trigger switch simultaneously starts and stops the wire feeder and welding current.

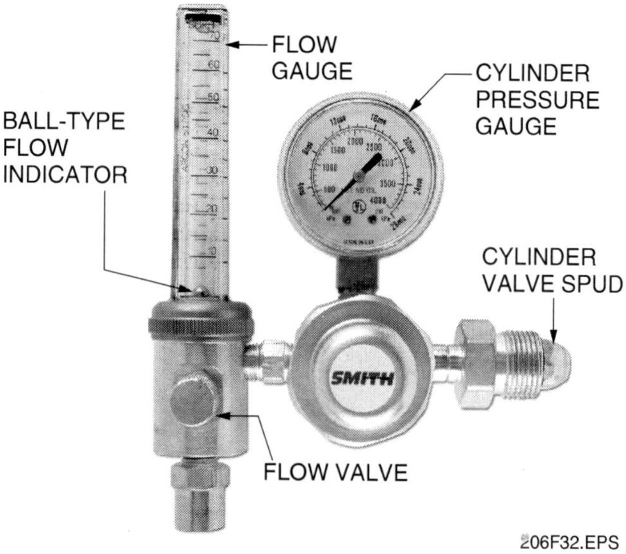

Figure 32 ◆ Shielding gas regulator/flowmeter.

6.0.0 ◆ SHIELDING GASES

Shielding gases are used to displace atmosphere from the weld zone to prevent contamination of the weld puddle by oxygen, nitrogen, and moisture. The three principal shielding gases are argon (Ar), helium (He), and carbon dioxide (CO_2). These gases are often mixed in various proportions for specific applications.

6.1.0 Shielding Gas Characteristics

Each of the shielding gases has distinctive individual performance characteristics. Each welds differently and works differently with different arc modes and base metals. Mixtures of gases often have the best features of the individual constituents. The following sections explain some of the features of the principal gases and their primary uses.

6.1.1 Argon

Argon (Ar), an inexpensive, inert gas, is the most commonly used shielding gas. Because of its current transfer and ionization properties, argon can be used alone or in combination with other shielding gases. It forms a tight arc column with high current density, concentrating the arc in a small area and producing deep penetration.

Pure argon is used with aluminum, nickel-based alloys, copper alloys, and the reactive metals zirconium, titanium, and tantalum. It provides excellent arc stability, penetration, and bead shape. For ferrous metals, argon is usually mixed with oxygen, helium, hydrogen, carbon dioxide, and/or nitrogen.

6.1.2 Helium

Because of thermal conductivity (heat transfer), helium is used when deeper penetration and higher travel speed are required.

Arc stability with helium is not as good as it is with argon. The helium arc column is wider than the argon arc column, resulting in reduced current density. It is better for welding aluminum, magnesium, and copper alloys because of better puddle fluidity and bead wetting. Some mixtures of helium and argon have the beneficial characteristics of both gases. Helium mixtures of 90% helium, 7.5% argon, and 2.5% carbon dioxide are widely used for short-circuit GMAW of stainless steel in all positions.

6.1.3 Carbon Dioxide

Carbon dioxide is commonly used to weld mild steel because of its low cost and weld performance. In the welding arc it is not an inert gas, because it breaks down into carbon monoxide and free oxygen. The oxygen combines with elements in the weld puddle to form slag. Carbon dioxide does not support spray transfer, so short circuiting transfer and globular transfer must be used. Carbon dioxide provides good penetration. However, the weld surface is heavily oxidized and has high weld spatter. Normally, argon is mixed with no more than 25% carbon dioxide to improve the performance characteristics. *Figure 33* shows the effects of shield gas on the welding arc.

6.1.4 Gas Mixtures

The principal shielding gases are often mixed to improve their overall welding characteristics. For

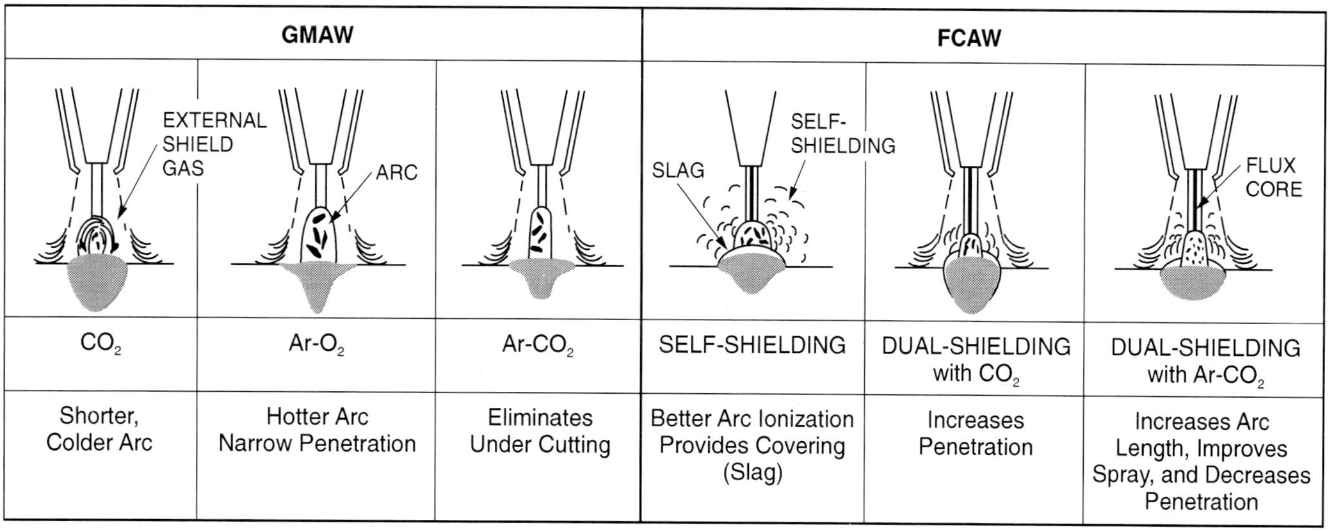

Figure 33 ◆ Effects of shield gas on the welding arc.

steel, small amounts of oxygen added to argon stabilize the arc, increase filler metal deposition rate, and improve wetting and bead shape. Small percentages of oxygen added to carbon dioxide improve arc characteristics. Helium is added to argon to increase the arc voltage and heat for welding nonferrous metals. Small amounts of nitrogen and oxygen are added to argon for welding stainless steel when the austenitic characteristics should be preserved. Helium and carbon dioxide are added to argon to improve arc stability, wetting, and bead profile. *Table 2* (see *Appendix*) shows some examples and uses of gas mixtures. Many other typical gas mixtures are available as shown in *Table 3* (see *Appendix*). The AWS classification for gases and gas mixtures is covered in *ANSI/AWS 5.32/A5.32M*.

6.2.0 Shielding Gas Selection

Selecting the best shielding gas for a particular welding task involves the consideration of many factors. The shielding gas selected will affect arc shape, arc density, arc temperature, arc stability, rate of filler metal transfer, degree of spatter, weld penetration, weld bead shape, weld bead appearance, weld porosity, weld chemistry, and weld quality.

Shielding gases affect the welding process in three ways:

- Gas thermal conductivity affects arc voltage and heat delivered to the weld. The thermal conductivity of both helium and carbon dioxide is much higher than that of argon; both helium and carbon dioxide supply more heat to the weld. Because of this, they also require more voltage and power to maintain a stable arc.
- Some gases can react with the filler and base metals and also affect arc stability. Carbon dioxide and most oxygen-bearing gases cannot be used with aluminum, because they form oxides. However, they provide better fusion and arc stability when used with steels.
- Gases affect the mode of metal transfer and penetration depth. Mixtures containing more than 15% carbon dioxide do not allow true spray transfer.

Table 4 (see *Appendix*) lists common short-circuiting shielding gases and their applications in short-circuiting transfer welding. *Table 5* (see *Appendix*) lists common shielding gases and their applications in spray transfer welding.

6.3.0 Shielding Gas Flow Rate

Gas flow rate from the gun tip is important because it affects the quality and the cost of a weld. Too low a flow rate will not shield the weld zone adequately and will result in a poor-quality weld; excessive gas flow wastes expensive gases and can cause turbulence at the weld. The turbulence can pull atmosphere into the weld zone, causing porosity in the weld. Welding specifications specify shielding gas flow rates. *Table 6* (see *Appendix*) lists typical shielding gas flow rates for various metals and wire sizes.

7.0.0 ◆ FILLER METALS

Filler metals are explained in two sections:

- GMAW filler metals
- FCAW filler metals

7.1.0 GMAW Filler Metals

GMAW uses a continuous solid wire electrode for filler material. This filler wire is drawn from high-grade pure alloys compounded for different applications. During manufacture, the wire is drawn, cleaned and inspected for defects several times. Standard wire diameters are 0.023, 0.030, 0.035, 0.045, 0.052, and 0.062 inches. The finished wire is wound on spools 4, 8, 12, or 14 inches in diameter. The most common spool is 12 inches in diameter and weighs 30 pounds (mild steel wire). Reels up to 30 inches in diameter and weighing up to 1,000 pounds are also available for stationary production welding.

Several industry organizations and the U.S. government publish specification standards for welding wire. The most common are the American Welding Society (AWS) specifications. The purpose of the AWS specifications is to set standards that all manufacturers must follow when manufacturing welding consumables. This ensures consistency for the user regardless of who manufactured the product. The specifications set standards for the following:

- Classification system, identification, and marking
- Chemical composition of the deposited weld metal
- Mechanical properties of the deposited weld metal

Filler Wire and Filler Metal

The terms filler wire and filler metal are both used to describe the filler materials used in GMAW. Filler metal is also referred to as filler rod when used for (GTAW).

The following AWS specifications pertain to gas metal arc welding wires and rods:

- *A5.7, Specification for Copper and Copper Alloy Bare Welding Rods and Electrodes*
- *A5.9, Specification for Bare Stainless Steel Welding Electrodes and Rods*
- *A5.10, Specification for Bare Aluminum and Aluminum Alloy Welding Electrodes and Rods*
- *A5.14, Specification for Nickel and Nickel Alloy Bare Welding Electrodes and Rods*
- *A5.16, Specification for Titanium and Titanium Alloy Welding Electrodes and Rods*
- *A5.18, Specification for Carbon Steel Filler Metals for Gas Shielded Arc Welding*
- *A5.19, Specification for Magnesium Alloy Welding Rods and Electrodes*
- *A5.28, Specification for Low-Alloy Steel Electrodes for Gas Shielded Metal Arc Welding*

> **NOTE**
> The specification number is generally followed by a dash and a two-digit number, such as *A5.18-93* or *A5.28-96*. The dash and the following number indicate the year that the specification was last revised.

Filler wire is graded for three major areas of use:

- *General use* – Wire meets specifications. No record of chemical composition, strength, or similar factors is supplied to the user with the wire purchase.
- *Rigid control fabrication* – A Certificate of Conformance is supplied with the wire at purchase. The stock is identified by code numbers located on the roll package.
- *Critical use* – A Certified Chemical Analysis report is supplied, and records are kept by the fabricator of welds and processes for later reference on aircraft, nuclear reactors, and pressure vessels.

The following factors affect the selection of a GMAW filler wire:

- Base metal chemical composition
- Base metal mechanical properties
- Weld joint design
- Service or specification requirements
- Shielding gas used

Commonly used filler wire types include:

- Carbon steel
- Stainless steel
- Aluminum and aluminum alloy
- Copper and copper alloy
- Low-alloy steel

Table 7 (see *Appendix*) summarizes the GMAW filler metal specifications and the major AWS class covered by each specification.

7.1.1 Carbon Steel Filler Metals

Carbon steel filler metals are identified by specification *ANSI/AWS A5.18*. The filler metal is available in wire reels and rod form. The wire classification (number) is located on a label on the side of the reel or container. *Figure 34* shows a typical AWS classification for carbon steel filler metal.

All carbon steel filler metals contain alloys such as silicon, manganese, aluminum, and carbon. Other alloys such as nickel, chromium, and molybdenum are also often added. The purpose of the alloys is as follows:

- *Silicon (Si)* – Concentrations of 0.40–1.00% are employed to deoxidize the puddle and to strengthen the weld. Silicon above 1% may make the welds crack-sensitive. Silicon also lowers the melting temperature of the wire and promotes wetting.
- *Manganese (Mn)* – Concentrations of 1–2% are also employed as a deoxidizer and to strengthen the weld. Manganese also decreases hot-crack sensitivity.
- *Aluminum (Al), titanium (Ti) and zirconium (Zr)* – One or more may be added in very small amounts for deoxidizing. Strength may also be increased.
- *Carbon (C)* – Concentrations of 0.05–0.12% are employed to add strength without adversely affecting ductility, porosity, or toughness.
- *Nickel (Ni), chromium (Cr), and molybdenum (Mo)* – May be added in small amounts to improve corrosion resistance, strength, and toughness.

> **Hot Tip**
>
> ***Rod and Electrode Designations***
>
> The letter R that follows the E in an AWS classification number indicates that the classification is used as a metal filler rod as well as an electrode. GMAW and GTAW filler metals are both labeled the same way when used for gas tungsten arc welding (GTAW).

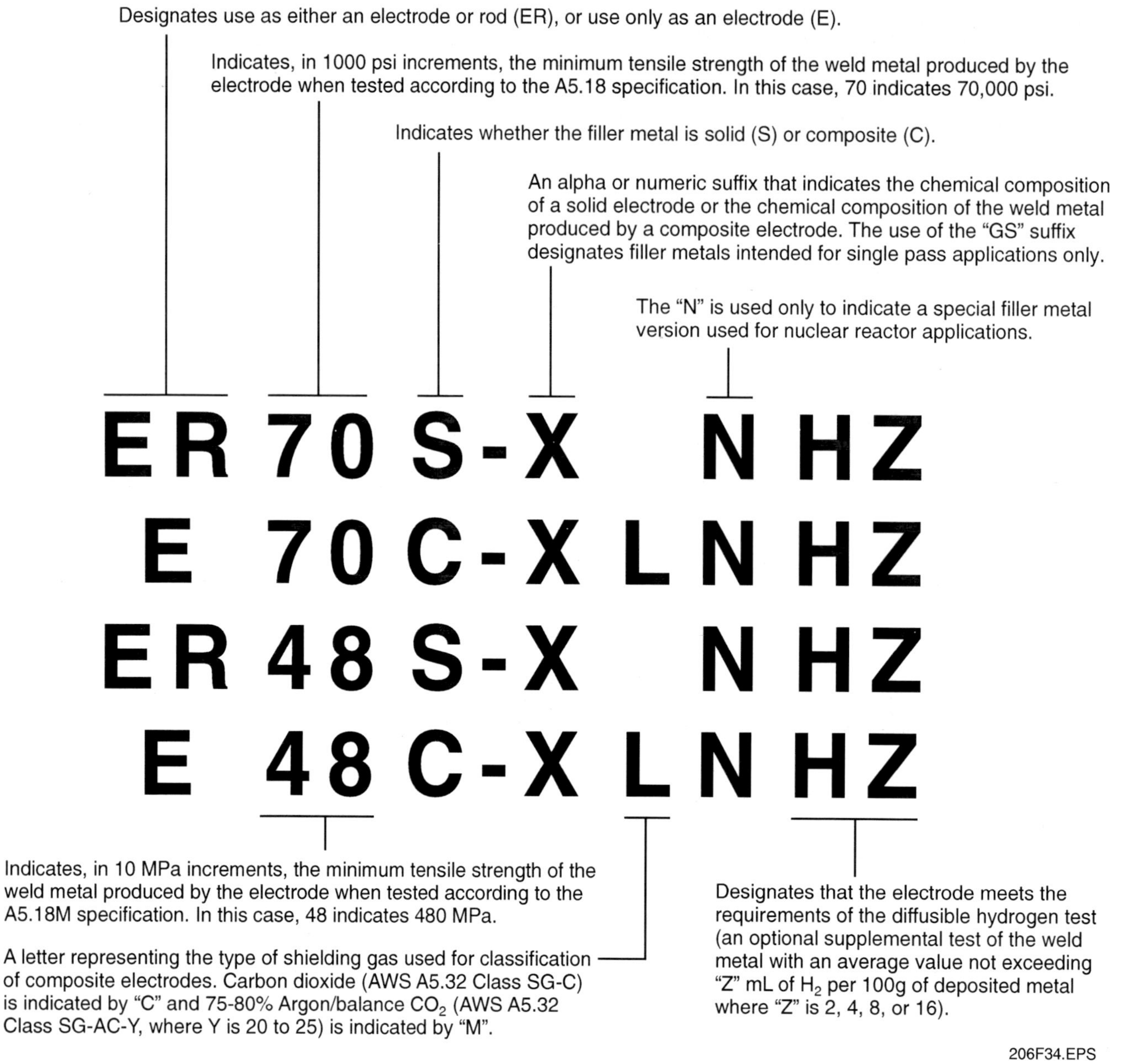

Figure 34 ◆ AWS classification for carbon steel filler metal.

In many cases, carbon steel filler metal wire is copper coated. The copper coating protects the filler metal wire from corrosion and makes electrical contact easier as the wire passes through the copper contact tube at the end of a gun.

Table 8 (see *Appendix*) lists AWS carbon steel filler metal wire classifications and their uses.

7.1.2 Low-Alloy Steel Filler Metals

Low-alloy steel filler metals are identified by *ANSI/AWS Specification A5.28*. The filler metal is available in wire reel and rod form. The wire classification (number) is located on a label on the side of the reel or container. The classification is similar to the carbon steel filler metal classification except that there is a letter in the suffix at the end of the classification. The letter suffix indicates that the filler metal contains alloys (nickel, manganese, molybdenum, etc.) that give the deposited weld metal special properties. Also, the tensile strength of low-alloy filler metals is often higher than that of carbon steel filler metals. For specific information on low-alloy filler metal composition and intended use, refer to the manufacturer's specifications. *Figure 35* shows the AWS classification for low-alloy steel filler metal. *Table 9* (see *Appendix*) provides some uses and characteristics for each AWS class of filler metal.

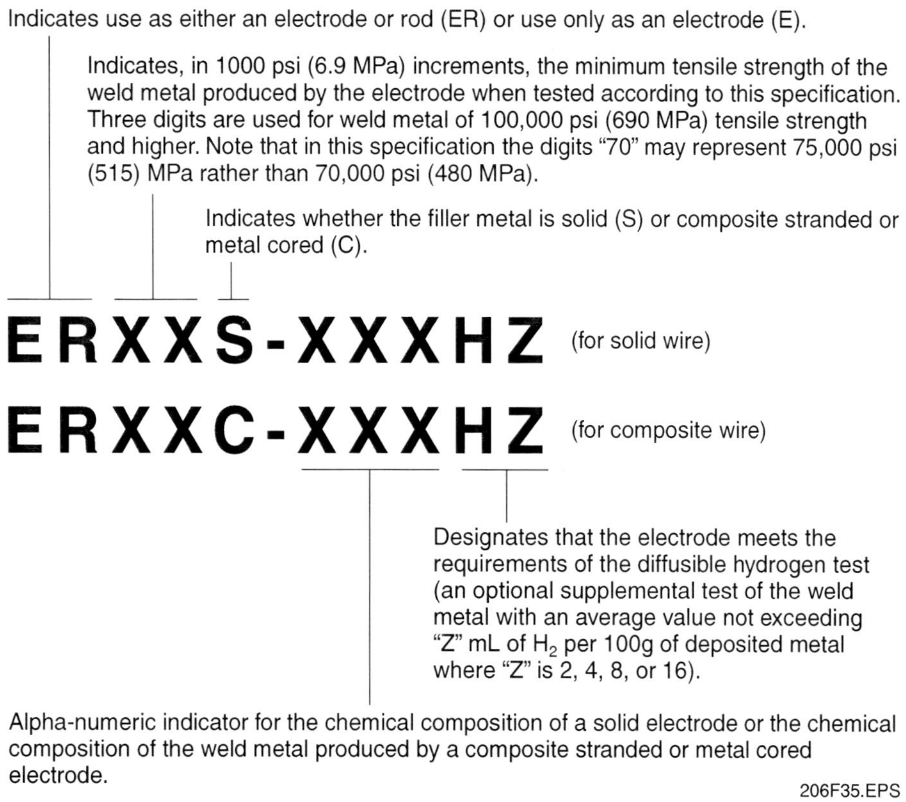

Figure 35 ◆ AWS classification for low-alloy steel filler metal.

7.1.3 Stainless Steel Filler Metals

Stainless steel filler metals are identified by specification *ANSI/AWS A5.9*. The filler metal is available in wire reel, flat strip, and rod form. The wire identifier (number) is located on a label on the side of the reel or container. *Figure 36* shows the AWS classification for stainless steel filler metal.

Stainless steel filler metal should be selected to closely match the alloy composition of the base metal. Stainless steel filler metals also require specific shielding gases or gas mixtures. *Table 10* (see *Appendix*) lists AWS stainless steel filler metal classifications and uses.

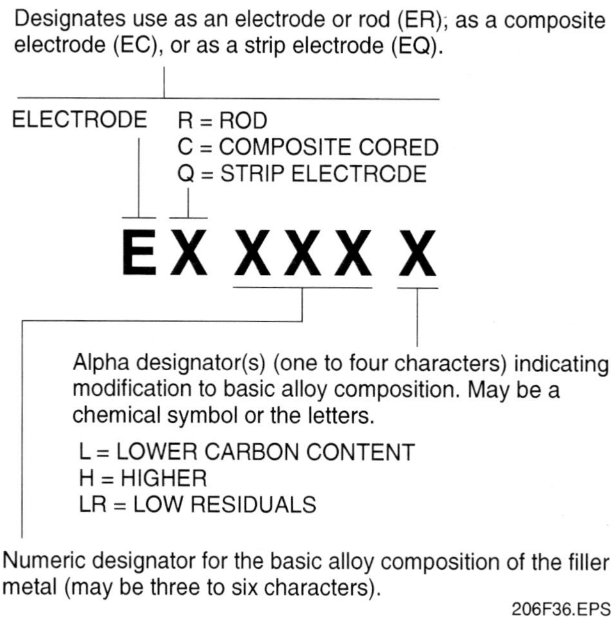

Figure 36 ◆ AWS classification for stainless steel filler metal.

7.1.4 Aluminum and Aluminum Alloy Filler Metals

Aluminum filler metals are covered by *ANSI/AWS Specification A5.10*. Aluminum filler metals usually contain magnesium, manganese, zinc, silicon, or copper for increased strength. Corrosion resistance and weldability are also considerations. Aluminum filler metals are designed to weld specific types of aluminum and should be selected for compatibility. The most widely used aluminum filler metals are ER4043 (contains silicon) and ER5356 (contains magnesium). The filler metal identifier (number) is located on a label on the side of the reel or container. *Figure 37* shows the AWS classification for aluminum filler metal.

Table 11 (see *Appendix*) lists recommended filler metals for common aluminum alloys.

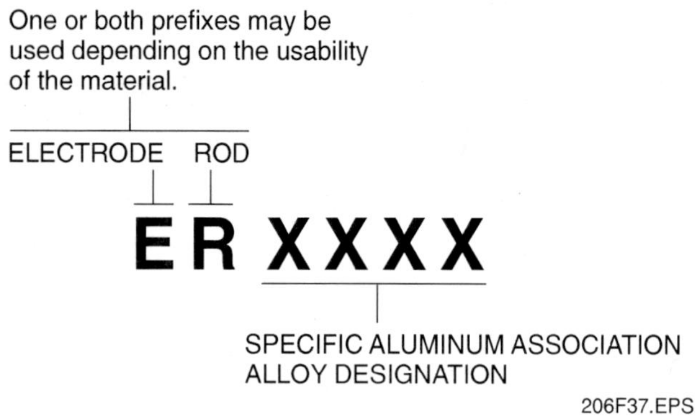

Figure 37 ◆ AWS classification for aluminum filler metal.

7.1.5 Copper and Copper Alloy Filler Metals

Copper filler metals are covered by *ANSI/AWS Specification A5.7*. Most copper filler metals contain other elements that increase strength, deoxidize the weld metal, and match the base metal composition. *Figure 38* shows the AWS classification for copper filler metal.

Table 12 (see *Appendix*) lists AWS copper filler metal classifications and uses.

7.1.6 Nickel and Nickel Alloy Filler Metals

Nickel-based filler metals are covered by *ANSI/AWS Specification A5.14*. These filler metals contain other elements to match base metal applications and to increase the strength and quality of the weld metal. For GTAW, DCEN is used with high-purity argon, helium or both used as a shielding gas. For GMAW, DCEP is used. Argon shielding gas is often used, but mixtures of argon and helium are also used. *Figure 39* shows the AWS classification for nickel filler metals.

Table 13 (see *Appendix*) lists AWS nickel filler metal classifications and uses.

7.1.7 Magnesium Alloy Filler Metals

Magnesium alloy filler metals are covered by *ANSI/AWS Specification A5.19*. These filler metals are usually used with GTAW, GMAW, and PAW processes. Oxyfuel welding should be used only for temporary repair work. For GTAW welding, the techniques and equipment is similar to those for aluminum. Argon, helium, or both are used for shielding. Alternating current is preferred for arc cleaning and penetration. Direct current is also employed with DCEP used for thin materials and DCEN used for mechanized welding with helium for deep penetration. GTAW is recommended for the defect repair of clean magnesium castings. GMAW of magnesium alloys is the same as for other metals. Argon is generally used for shielding; however, mixtures of argon and helium are occasionally used. Pulsed and short-circuit GMAW are used for magnesium alloys. Spray transfer without pulsing is also used. Globular transfer is not suitable. *Figure 40* shows the AWS classification for magnesium filler metals

Table 14 (see *Appendix*) lists recommended filler metals for common magnesium alloys. Note that the prefix ER and R has been removed from the filler metal designations to reduce the size of the table.

7.1.8 Titanium and Titanium Alloy Filler Metals

Titanium and titanium alloy filler metals are covered by *ANSI/AWS Specification A5.16*. These filler metals are generally used with GTAW, GMAW, SAW, and PAW processes. Titanium is sensitive to embrittlement by oxygen, nitrogen, and hydrogen at temperatures above 500°F. Like aluminum, titanium requires weld cleaning and high-purity gas shielding, and especially adequate postflow

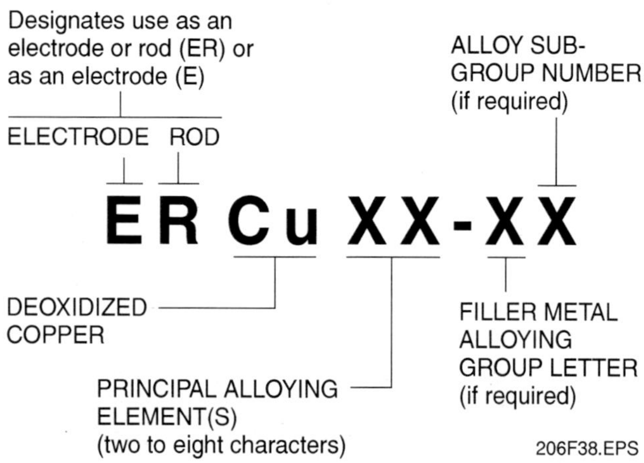

Figure 38 ◆ AWS classification for copper filler metal.

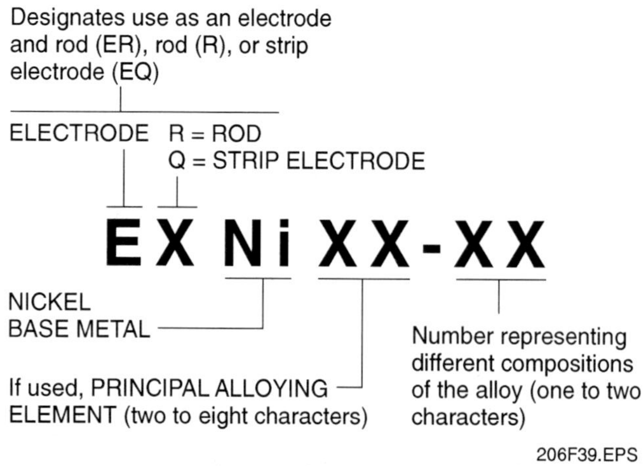

Figure 39 ◆ AWS classification for nickel filler metals.

shielding to avoid embrittlement. Titanium can be successfully fusion welded to zirconium, tantalum, niobium, and vanadium. Titanium should not be fusion welded to copper, iron, nickel, and aluminum. *Figure 41* shows the AWS classification for titanium filler metals.

Table 15 (see *Appendix*) lists AWS titanium filler metal classifications and uses.

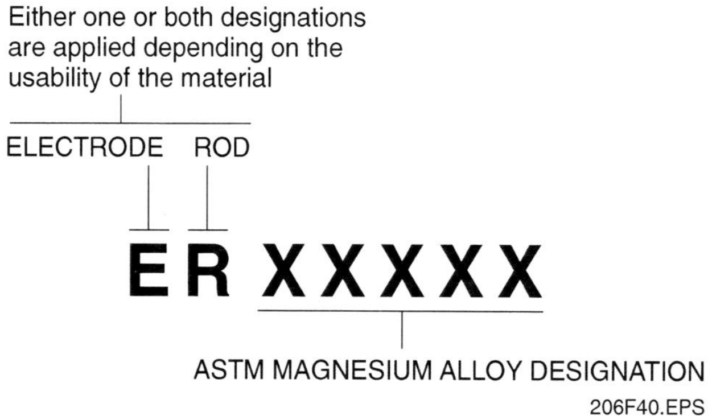

Figure 40 ◈ AWS classification for magnesium filler metals.

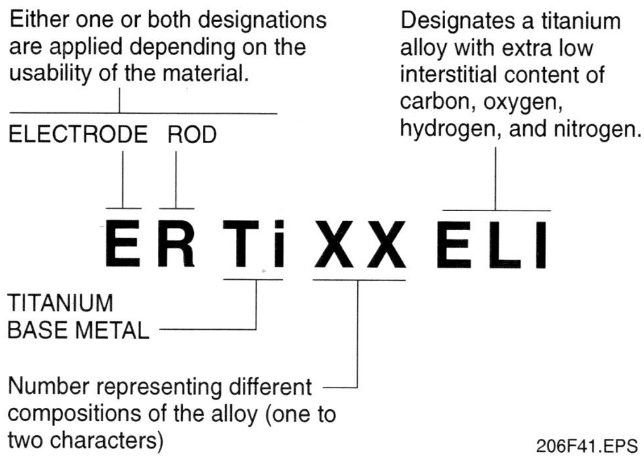

Figure 41 ◈ AWS classification for titanium filler metals.

7.2.0 FCAW Filler Metals

FCAW is presently limited to the welding of ferrous metals. These include low- and medium-carbon steels, some low-alloy steels, cast irons, and some stainless steels.

The FCAW electrode is a continuous flux cored wire electrode. The wire is made by forming a thin filler metal strip into a U shape, filling the U with the flux material, squeezing the strip shut around the flux, and then drawing the tube through dies to size the wire and compact the flux. During manufacture, the wire is carefully inspected for defects. Flux cored wires are manufactured in a range of sizes from 0.030 inch to ⁵⁄₃₂ inch. The finished flux cored wire is coiled on all the standard spool sizes and forms.

NOTE

All flux cored wires are considered to be low hydrogen. For this reason, proper electrode control procedures must be followed to keep the wire dry.

By varying the power settings on the welding machine, smaller size electrode wire (0.045 and ¹⁄₁₆ inch) can be used both for thin metals and for multipass welds on metals of unlimited thickness in all positions.

There are three AWS specifications presently available for flux cored electrodes:

- *ANSI/AWS A5.20 Specification for Carbon Steel Electrodes for Flux Cored Arc Welding*
- *ANSI/AWS A5.22 Specification for Stainless Steel Electrodes for Flux Cored Arc Welding and Stainless Steel Flux Cored Rods for Gas Tungsten Arc Welding*
- *ANSI/AWS A5.29 Specification for Low-Alloy Steel Electrodes for Flux Cored Arc Welding*

Hot Tip

Magnesium Fires

While ignition of magnesium is a very remote possibility while welding, the ignition will cease when the heat source is removed. Ignition of the weld pool is also prevented by the gas shielding used in GTAW, GMAW and PAW processes. Most magnesium fires occur with accumulations of fines from grinding or filing or chips from machining. Accumulations of fines on clothing should be avoided. Graphite-base or salt-base powders recommended for extinguishing magnesium fires should be stored in the work area. If large amounts of fines are produced, they should be collected in a waterwash-type dust collector designed for use with magnesium. Special precautions pertaining to the handling of wet magnesium fines should be followed.

Each specification defines standards for the following:

- Classification, identification, and marking of the electrode wire
- Chemical composition of the deposited weld metal
- Chemical composition of the flux filling
- Mechanical properties of the deposited weld metal

7.2.1 Carbon Steel Flux Cored Electrodes

Figure 42 shows the *ANSI/AWS A5.20* specification format for carbon steel flux cored electrodes.

Table 16 (see *Appendix*) shows the carbon steel flux cored electrode welding characteristics as indicated by the AWS classification.

Table 17 (see *Appendix*) lists carbon steel flux cored electrode classes and uses.

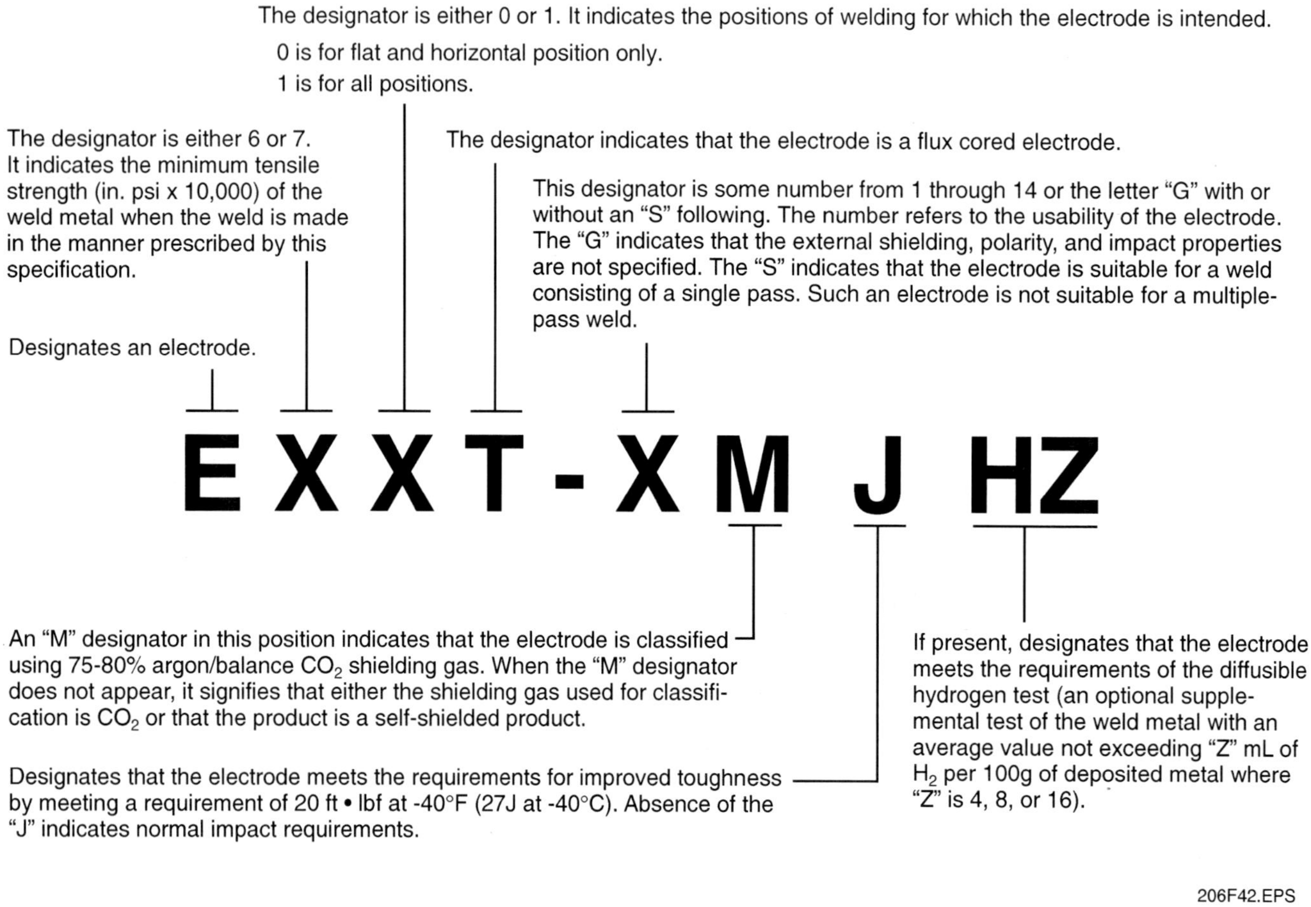

Figure 42 ◆ AWS specification format for carbon steel flux cored electrodes.

FCAW of Structural Steel

For large structural welding projects such as bridges and skyscrapers, many construction companies prefer carbon (or low-alloy) steel FCAW to meet the toughness requirements for new construction using the newer self-shielding E71T-8 or gas-shielded E71T-1 electrodes. Many ironworkers that use SMAW are reluctant to use FCAW, and those that have used FCAW predominantly relied upon E71T-11 for structural welds because it was easy to use in all positions. However, welds with –11 wire sometimes had cracking problems and had to be rewelded. Once welders are properly retrained for FCAW using the newer electrodes, the cracking problems are eliminated and the deposition rates are higher because the newer electrodes run hotter and faster than the –11 electrodes or SMAW electrodes.

7.2.2 Low-Alloy Steel Flux Cored Electrodes

Figure 43 shows the *ANSI/AWS A5.29* specification format for low-alloy steel flux cored electrodes

Table 18 (see *Appendix*) shows the low-alloy steel flux cored electrode welding characteristics as indicated by the AWS classification.M

Table 19 (see *Appendix*) lists the AWS low-alloy steel flux cored electrode classes and uses.

7.2.3 Stainless Steel Flux Cored Electrodes and Rods

Figure 44 shows the *ANSI/AWS A5.22* specification format for stainless steel flux cored electrodes and rods.

Table 20 (see *Appendix*) shows stainless steel flux cored electrode and rod characteristics as indicated by the AWS classification.

Table 21 (see *Appendix*) lists AWS stainless steel flux cored electrode classifications and uses. The table only includes electrodes for FCAW applications because the rods are used for GTAW.

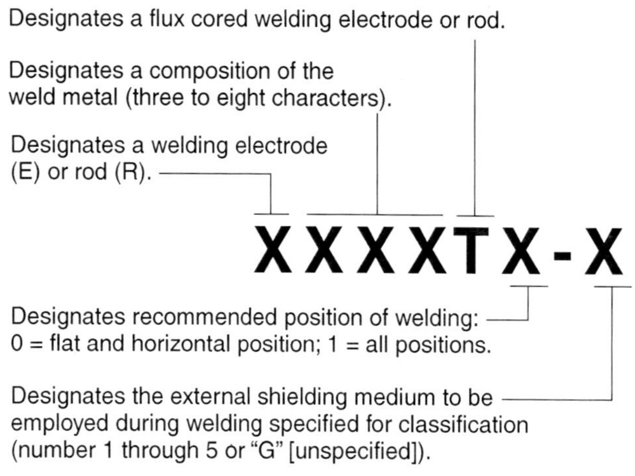

Figure 44 ◆ AWS specification for stainless steel flux cored electrodes and rods.

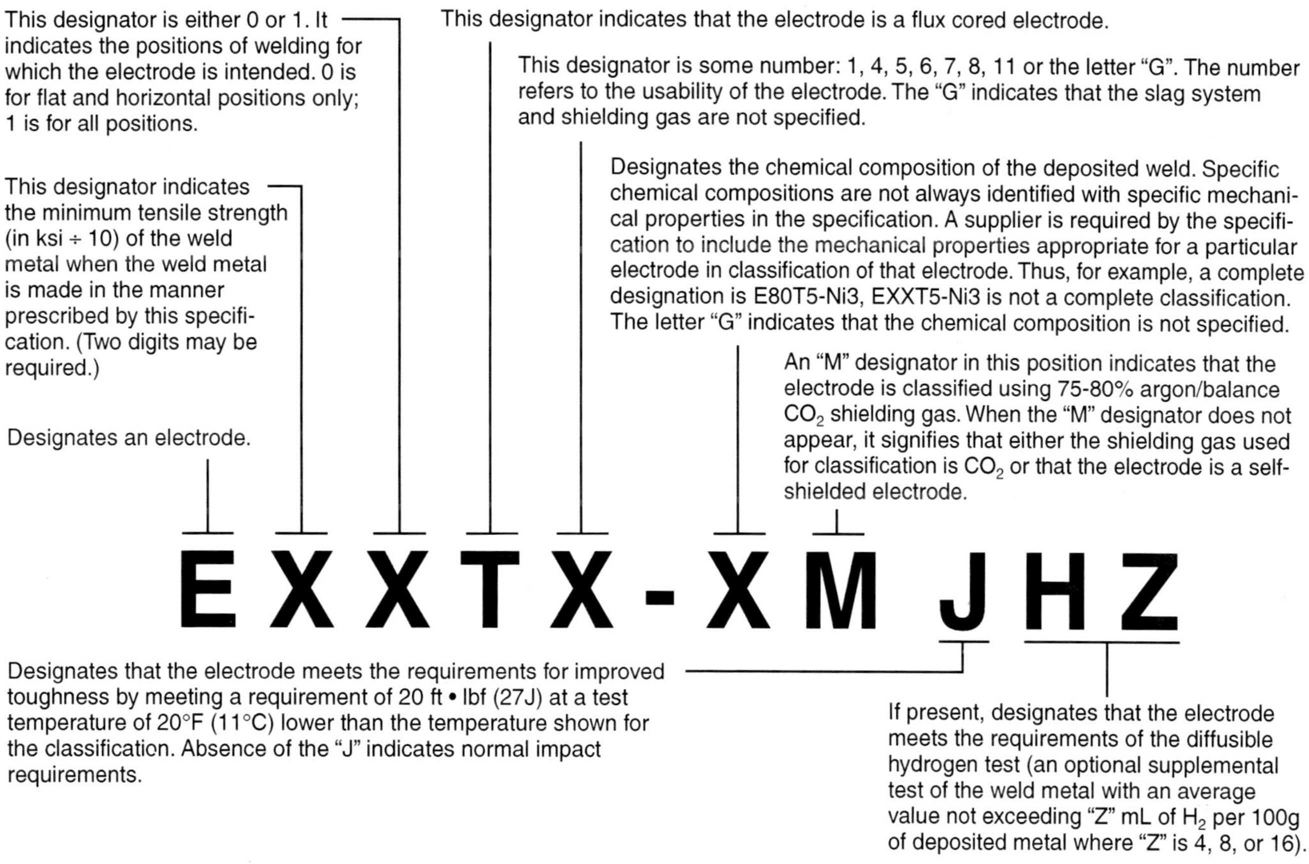

Figure 43 ◆ AWS specification for low-alloy steel flux cored electrodes.

8.0.0 ◆ WELDING EQUIPMENT SETUP

In order to weld safely and efficiently, the welding equipment must be properly set up. The following sections explain the steps for setting up welding equipment. *Figure 45* shows a diagram of a GMAW or FCAW system.

8.1.0 Selecting a GMAW Power Source

To select a power source (welding machine), several factors must be considered.

- GMAW requires a constant-voltage or constant voltage/constant current power source.
- A DCEP-type welding current is required.
- The maximum amperage required must be available.
- Consider the primary power requirements. For example, are there electrical receptacles to plug a welding machine into, or do engine-driven generators/alternators need to be used?

8.2.0 Selecting an FCAW Power Source

To select an FCAW welding machine, the following factors must be considered:

- FCAW requires a constant-voltage power source.
- The welding current required for FCAW is DC.
- The required operating amperage ranges from 150 amps with a 0.045-inch electrode to 650 amps or more with a ⅛-inch electrode; larger electrodes require even higher amperage.
- Consider whether the primary power source is adequate. Are there enough AC electrical receptacles to plug a welding machine into, or do engine-driven generators/alternators need to be used?

8.3.0 Positioning the Equipment

Because of the short length of the gun cables, the wire feeder must be positioned near the work to be performed. The power source must also be located near the wire feeder because of the limited length of the contactor/control cable that runs from the power source to the wire feeder. Normally the wire feeder is located either on top of the power source or on the floor or a cart near the power source.

Select a site where the equipment will not be in the way but will be protected from welding or cutting sparks. There should be good air circulation to keep the machine cool. The environment should be free from explosive or corrosive fumes and as free as possible from dust and dirt. Welding machines have internal cooling fans that will pull these materials into the welding machine if they are present. Also, dust and dirt can collect on the wire, causing weld contamination and clogging of

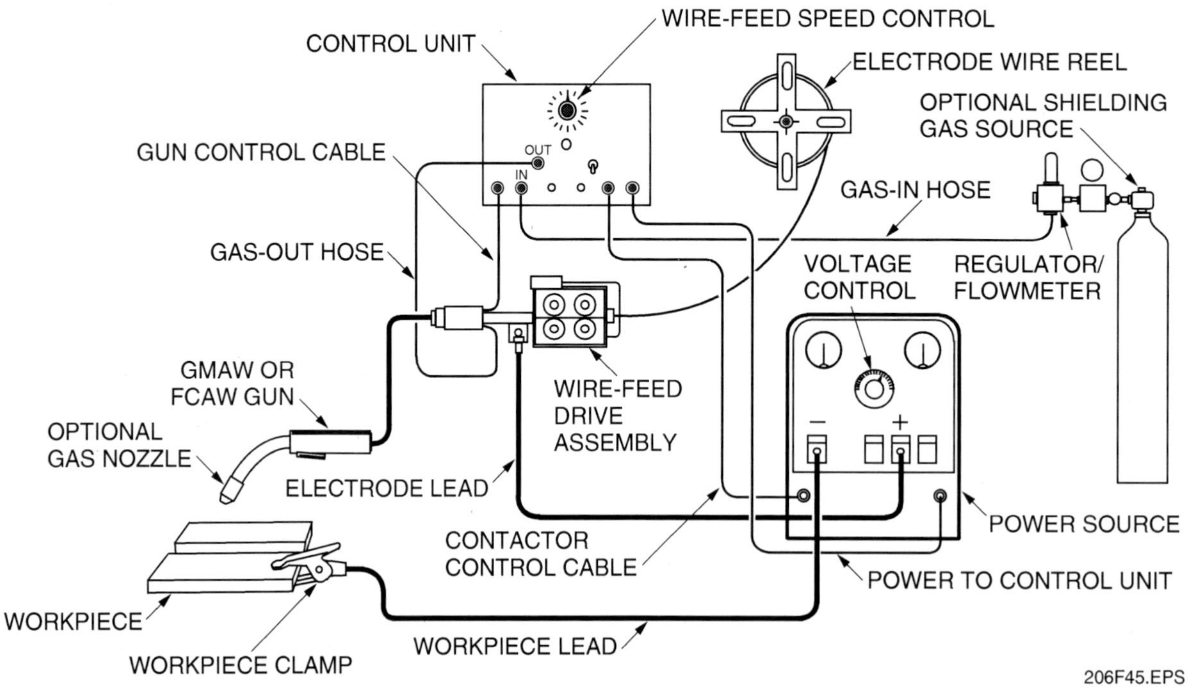

Figure 45 ◆ Diagram of a GMAW/FCAW system.

the liner in the gun cable. The site should also be free of standing water or water leaks. If an engine-driven generator or alternator is used, position it so that it can be easily refueled and serviced.

There should also be easy access to the site so the equipment can be started, stopped, or adjusted as needed. If the machine is to be plugged into an outlet, be sure the outlet has been properly installed by a licensed electrician to ensure it is grounded. Also, be sure to identify the location of the electrical disconnect for the outlet before plugging the welding machine into it.

8.4.0 Moving Welding Power Sources

Large engine-driven generators are mounted on a trailer frame and can easily be moved by a pickup truck or tractor with a trailer hitch. Other types of welding machines may have a skid base or be mounted on steel or rubber wheels. When moving wheel-mounted welding machines by hand, use care. Some machines are top-heavy and could fall over in a tight turn or if the floor or ground is uneven or soft.

CAUTION

Secure or remove the wire feeder and gas cylinder before attempting to move the power source.

WARNING!

If a welding machine starts to fall over, do not attempt to hold it. Welding machines are very heavy, and severe crushing injuries can occur if a welding machine falls on you.

Most welding machines have a lifting eye, which is used to move machines mounted on skids or to lift any machine. Before lifting a welding machine, check the equipment specifications for the weight. Be sure the lifting device and tackle can handle the weight of the machine. Always use a sling and/or a shackle. Never attempt to lift a machine by placing the lifting hook directly in the machine's lifting eye because the safety latch on the hook cannot be closed. Also, before lifting or moving a welding machine, be sure the welding cables are secure. *Figure 46* shows the proper way to lift a welding machine.

8.5.0 Connecting the Shielding Gas

The shielding gas connects to the wire feeder unit or to the power source if the wire feeder is integrated with the power source.

Follow these steps to connect the shielding gas:

Step 1 Identify the required shielding gas by referring to the WPS or Site Quality Standards.

Step 2 Locate a cylinder of the correct gas or mixture, and secure it near the wire feeder.

CAUTION

Be sure to secure the cylinder so that it cannot fall over.

Step 3 Remove the cylinder's protective cap and momentarily crack open the cylinder valve to blow out any dirt, then close it again.

WARNING!

Even though available, do not use an adapter to connect a regulator/flowmeter equipped with one type of CGA (Compressed Gas Association) connection to a gas cylinder that has a different CGA connection. The CGA connections are specific to the types of gas and cylinder pressures permitted for the connection.

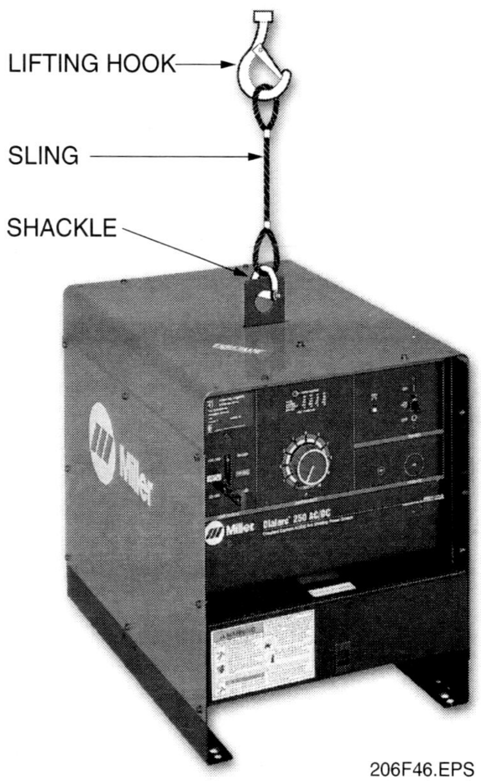

Figure 46 ◆ Lifting a welding machine.

Step 4 Using a regulator/flowmeter with the correct CGA connection for the cylinder, install the pressure regulator/flowmeter on the cylinder.

Step 5 Connect the gas hose to the flowmeter and to the gas connection on the wire feeder (or power source).

Step 6 Slowly crack open the cylinder valve and then open it completely.

Step 7 Adjust the gas flow, using the flowmeter valve, to the specified flow rate.

NOTE

Some flowmeters are equipped with several scales of different calibrations around the same sight tube, for monitoring the flows of different types (densities) of gases. Be sure to rotate the scales or read the correct side for the type of gas in use.

Hot Tip

Adjusting Flowmeters

In most cases, the top of the floating flowmeter ball is used as the reference point for setting the gas flow. However, always check the manufacturer's instructions to determine the correct adjustment method.

206SA05.EPS

8.6.0 Selecting and Installing Filler Wire

Follow these steps to select and install the filler wire:

Step 1 Identify the filler wire required by referring to the WPS or Site Quality Standard.

Step 2 Locate a spool of the wire, and mount it on the wire feeder spool holder. Lock the holder and set the drag brake for a slight drag. The brake control is usually located on the end of the spool axle.

Step 3 Check (and change if necessary) the wire feed drive wheels, liner, and gun contact tube to make sure they are the correct size for the wire selected. Adjust the wheel tension if necessary. For detailed instructions on changing or adjusting the wire feeder drive wheels or changing the gun's contact tube or liner, refer to the specific manufacturer's instructions.

Step 4 Feed the wire into the feeder wheels, and then use the JOG control to feed it through the gun cable and the gun contact tube.

8.7.0 Placing the Workpiece Clamp

The workpiece clamp must be properly placed to prevent damage to surrounding equipment. If the electrical welding current travels through a bearing, seal, valve, or contacting surface, it could cause severe damage through heat and arcing, which would require that these items be replaced. Carefully check the area to be welded, and position the workpiece clamp so the welding current will not pass through any contacting surface. If in doubt, ask your supervisor for assistance before proceeding.

Hot Tip

Gas Flow Rates

More gas flow is not always better. Too much flow can deflect off the workpiece. The resultant swirling gas can create turbulence that pulls in air, which contaminates the weld and creates pinholes. If black flecks occur in the weld puddle, or pinholes occur in the weld, try reducing gas flow.

> **CAUTION**
> Welding current passing through electrical or electronic equipment will cause severe damage. Before welding on any type of mobile equipment, the ground lead at the battery must be disconnected to protect the electrical system. If welding is near the battery, the battery must be removed. Batteries produce hydrogen gas, which is extremely explosive. A welding spark could cause the battery to explode.

> **WARNING!**
> Do not weld near batteries. A welding spark could cause a battery to explode, showering the area with battery acid.

Workpiece clamps must never be connected to pipes carrying flammable or corrosive materials. The welding current could cause overheating or sparks, resulting in an explosion or fire.

The workpiece clamp must make a good electrical contact when it is connected. Dirt and paint will inhibit the connection and cause arcing, resulting in overheating of the workpiece clamp. Dirt and paint also affect the welding current and can cause defects in the weld. Clean the surface before connecting the workpiece clamp. If the workpiece clamp is damaged and does not close securely onto the surface, replace it.

8.8.0 Energizing the Power Source

Electrically powered welding machines are energized by plugging them into an electrical outlet. The electrical requirements (primary current) will be located on the equipment specification tag displayed prominently on the machine. Most machines require single-phase 240-volt power, 208-volt three-phase power, or 460-volt three-phase power. Most machines requiring single-phase 240-volt power have a three-pronged plug. Machines requiring three-phase 208- or 460-volt power will have a four-pronged plug. Both types of plugs are shown in *Figure 47*.

If a welding machine does not have a power plug, an electrician must connect it. The electrician will add a plug or hard-wire the machine directly into an electrical box.

> **WARNING!**
> Never use a welding machine until you identify the electrical disconnect for that machine. In the event of an emergency, you must be able to quickly turn off the power to the welder at the disconnect.

8.9.0 Starting Engine-Driven Generators and Alternators

Before welding can take place with an engine-driven generator or alternator, the engine must be checked and then started. As with a car engine, the engine powering the generator or alternator must also have routine maintenance performed.

8.9.1 Performing Pre-Start Checks

Many sites have pre-start checklists that must be completed and signed prior to starting or operating an engine-driven generator or alternator. Check with your supervisor. If your site has such a checklist, complete and sign it. If your site does

206F47.EPS

Figure 47 ◆ Typical three- and four-pronged plugs.

not have a pre-start checklist, perform the following checks before starting the engine:

- Check the oil, using the engine oil dipstick. If the oil is low, add the appropriate grade of oil for the time of year.
- Check the coolant level in the radiator if the engine is liquid-cooled. If the coolant level is low, add coolant.

CAUTION

Do not add plain water to radiators that contain antifreeze. Antifreeze not only protects radiators from freezing in cold weather, it also has rust inhibitors and additives to aid cooling. If the antifreeze is diluted, it will not function properly. If the weather turns cold, the system may freeze, causing damage to the radiator, engine block, and water pump.

- Check the fuel. The unit may have a fuel gauge or a dipstick. If the fuel is low, add the correct fuel (diesel or gasoline) to the fuel tank. The type of fuel required should be marked on the fuel tank. If it is not marked, contact your supervisor to verify the fuel required and have the tank marked.

CAUTION

Adding gasoline to a diesel engine or diesel to a gasoline engine will cause severe engine problems. It can also cause a fire hazard. Always be sure to add the correct fuel to the fuel tank.

- Check the battery water level unless the battery is sealed. Add demineralized water if the battery water level is low.
- Check the electrode holder to be sure it is not grounded. If the electrode holder is grounded, it will arc and overheat the welding system when the welding machine is started. This is a fire hazard and will cause damage to the equipment.
- Open the fuel shut-off valve if the equipment has one. If there is a fuel shut-off valve, it should be located in the fuel line between the fuel tank and the carburetor.
- Record the hours from the hour meter if the equipment has one. An hour meter records the total number of hours the engine runs. This information is used to determine when the engine needs to be serviced. The hours will be displayed on a gauge similar to an odometer.
- Clean the unit. Use a compressed air hose to blow off the engine and generator or alternator. Use a rag to remove heavier deposits that cannot be removed with the compressed air.

WARNING!

Always wear eye protection when using compressed air to blow dirt and debris from surfaces. Never point the nozzle at yourself or anyone else.

NOTE

Cleaning may not be required on a daily basis. Clean the unit as required.

8.9.2 Starting the Engine

Most engines have an on/off ignition switch and a starter. They may be combined into a key switch similar to the ignition on a car. To start the engine, turn on the ignition switch and press the starter. Release the starter when the engine starts. The engine speed is controlled by the governor. If the governor switch is set for idle, the engine will slow to an idle after a few seconds. If the governor is set to welding speed, the engine will continue to run at the welding speed.

Small engine-driven alternators may have an on/off switch and a pull cord. These are started by turning on the ignition switch and pulling the cord, in much the same way as a lawn mower is started.

Engine-driven generators and alternators should be started about 5 to 10 minutes before they are needed for welding. This will allow the engine to warm up before a welding load is placed on it.

8.9.3 Stopping the Engine

If no welding is required for thirty or more minutes, stop the engine by turning off the ignition switch. If you are finished with the welding machine for the day, close the fuel valve (if there is one).

8.9.4 Performing Preventive Maintenance

Engine-driven generators and alternators require regular preventive maintenance to keep the equipment operating properly. Most sites have a preventive maintenance schedule based on the hours that the engine operates. In severe conditions, such as very dusty or cold weather, maintenance may have to be performed more frequently.

> **CAUTION**
> In order to prevent damage to the equipment, perform preventive maintenance as recommended by the site procedures or by the manufacturer's maintenance schedule in the equipment manual.

The responsibility for performing preventive maintenance varies by site. Check with your supervisor to determine who is responsible for performing preventive maintenance.

When performing preventive maintenance, follow the manufacturer's guidelines in the equipment manual. Typical items to be performed as a part of preventive maintenance include the following:

- Changing the oil
- Changing the gas filter
- Changing the air filter
- Checking/changing the antifreeze
- Greasing the undercarriage
- Repacking the wheel bearings

> **Hot Tip** — *Dust Removal*
> Blow out the welding machine and drive unit with compressed air occasionally to remove dust. A significant buildup of dust inside the machine will provide an alternate path for the electricity and could result in some of the components shorting out.

9.0.0 ◆ HAND TOOLS FOR WELD CLEANING

The tools used for weld cleaning include hand-held tools such as chipping hammers, wire brushes, and pliers, and power tools such as pneumatic weld-flux chippers and needle scalers.

Wire brushes are used to clean welds and to remove paint or surface corrosion. Wire brushing removes light to medium corrosion, but does not remove tight corrosion. Tight corrosion must be removed by filing, grinding, or sandblasting.

Chipping hammers are used to remove cutting and welding slag. The head of a chipping hammer has a point at one end and a chisel at the other. When chipping hammers become dull, they can be sharpened on a grinder. When sharpening chipping hammers, use care not to overheat the head. The head is hardened by tempering, and overheating removes the temper, causing the head to become soft. If the temper is removed from a chipping hammer head, it will mushroom and wear out much faster than a tempered head. Prevent overheating by plunging the head of the chipping hammer into a pail of water every few seconds while grinding.

Welders should also have clamping pliers to handle hot metal or hold and align metal to be welded. Hot metal should not be handled with leather welding gloves. The heat will cause the leather to shrivel and become hard. *Figure 48* shows a wire brush, chipping hammers, various clamping pliers, and a common file. Before use, files should have handles installed on the tang to prevent injury.

Figure 48 ◆ Wire brush, chipping hammers, clamping pliers, and file.

Summary

Both gas metal arc welding (GMAW) and flux cored arc welding (FCAW) can be dangerous if the proper safety precautions are not followed. Before proceeding, be sure you understand the equipment and filler metals required and the safety precautions presented in this module. Before GMAW or FCAW can take place, the appropriate equipment and filler metals must be selected and the equipment must be set up. By following the recommendations in this module, you will be able to select and safely set up the appropriate welding equipment and understand the types of filler metals used for GMAW and FCAW.

Review Questions

1. In gas metal arc welding, the arc is stabilized and the molten filler and base metals protected from oxidation by _____.

a. the application of pressure
b. use of self-fluxing alloy
c. shielding gas
d. use of intense heat

2. Of the two basic FCAW processes, one is self-shielding and one uses ______ shielding gas.

a. nonself-
b. internal
c. external
d. constant

3. When mixed with air or oxygen, _____ will explode at much lower concentrations than any other fuel.

a. propylene
b. propane
c. methane
d. acetylene

4. Depending on the range selected, for each 100 amperes of current the arc voltage of a welding machine is typically _____ volt(s) lower than the open-circuit voltage.

a. 0–1
b. 1–2
c. 2–3
d. 4–5

5. In welding, the current flows through two welding cables; the ground welding leads connect the power source to the ______.

a. base metal
b. wire electrode
c. welding machine
d. work table

6. The advantage of engine-driven generators and alternators is that they are ______.

a. portable and can be used in the field
b. easy to operate
c. less expensive
d. require little maintenance

7. In FCAW, the flux inside the filler core contains ingredients that ______.

a. cause arc flutter
b. oxidize the weld
c. generate shielding gas
d. prevent slag from forming

8. For FCAW, edge preparation is seldom needed for double-welded butt joints on plate up to ⅜-inch thick, because of the ______ of welds made with CO_2 shielding.

a. smooth profile
b. lower temperature
c. deionization
d. deeper penetration

9. When using short-circuiting welding in GMAW, the current passing through the wire electrode heats and then melts the wire in a process called ______.

a. inductance
b. slope
c. pinch effect
d. arc blow

10. GMAW guns are usually rated for amperage at a _____ duty cycle using CO_2.

a. 25%
b. 50%
c. 75%
d. 100%

11. Light-duty GMAW guns are cooled by ______.

a. domestic water
b. special cooling fluid
c. shielding gas flow
d. demineralized water

12. To meter the shielding gas to the welding gun at the proper flow rate, a(n) ______ is required.

a. gate valve
b. gas regulator/flowmeter
c. auxiliary gun control unit
d. cylinder valve

13. Because of its current transfer and ionization properties, ______ can be used alone or in combination with other gases to weld both ferrous and non-ferrous materials.

a. helium
b. carbon dioxide
c. nitrogen
d. argon

14. FCAW is limited to the welding of ______.

a. aluminum
b. copper
c. ferrous metals
d. nonferrous metals

15. To lift a welding machine mounted on skids, the welder should ______ the lifting eye.

a. attach the hook directly to
b. attach a sling and/or a shackle to
c. run a stout rope through
d. attach a heavy-duty chain to

GLOSSARY

Trade Terms Introduced in This Module

Axially: In a straight line along an axis.

Cavitation: The formation of hollow cavities caused by partial vacuums in a liquid flowing across a metal surface. As these cavities collapse, pitting or other damage is caused on the metal surfaces.

Globular transfer: A GMAW process in which the electrode wire does not touch the base metal, and the filler metal is transferred randomly in the arc as large, irregular globules.

Hot shortness: A phenomenon for certain metals in which welding heat causes the metal surrounding a molten weld pool to collapse with no warning. This results in the molten weld pool and underlying base metal falling through the workpiece.

Pulsed transfer: A GMAW welding process that is a variation of the spray transfer process in which the welding power is cycled from a low level (background level) to a higher level (welding level), resulting in lower average voltage and current.

Short-circuiting transfer: A GMAW welding process in which the electrode wire shorts against the base metal. The welding current repeatedly and rapidly melts the wire and establishes an arc, which is extinguished when the wire again touches the base metal.

Slag: The crusty by-product that forms on the weld bead and is produced by flux or by metallic or nonmetallic impurities in the filler metal.

Spray transfer: A GMAW welding process in which the electrode wire does not touch the base metal, and the filler metal is transferred axially (in a straight line) in the arc as fine droplets.

Stickout: The length of electrode that projects from the contact tube, measured from the arc tip to the contact tube.

APPENDIX

Tables

Table 1 Welding Cable Sizes

MACHINE SIZE IN AMPERES	DUTY CYCLE (%)	COPPER CABLE SIZES FOR COMBINED LENGTHS OF ELECTRODES PLUS WORKPIECE LEAD				
		UP TO 50 FEET	50-100 FEET	150-150 FEET	150-200 FEET	200-250 FEET
100	20	#8	#4	#3	#2	#1
180	20	#5	#4	#3	#2	#1
180	30	#4	#4	#3	#2	#1
200	50	#3	#3	#2	#1	#1/0
200	60	#2	#2	#2	#1	#1/0
225	20	#4	#3	#2	#1	#1/0
250	30	#3	#3	#2	#1	#1/0
300	60	#1/0	#1/0	#1/0	#2/0	#3/0
400	60	#2/0	#2/0	#2/0	#3/0	#4/0
500	60	#2/0	#2/0	#3/0	#3/0	#4/0
600	60	#3/0	#3/0	#3/0	#4/0	* * *
650	60	#3/0	#3/0	#4/0	* *	* * *

* *USE DOUBLE STRAND OF #2/0
* * *USE DOUBLE STRAND OF #3/0

206T01.EPS

Table 2 Examples and Uses of Gas Mixtures

GAS MIXTURE	USE
Argon with 1% oxygen	Spray transfer on stainless steels
Argon with 2% oxygen	Spray transfer on carbon steels, low-alloy steels, and stainless steels
Argon with 5% oxygen	General carbon steel welding using highly deoxidized wires
Argon with 8% to 12% oxygen	Single-pass spray transfer welds with high puddle fluidity
Argon with 3% to 10% carbon dioxide	Spray transfer and short-circuiting transfer on carbon steels
Argon with 11% to 20% carbon dioxide	High-speed and out-of-position welds on carbon steels and low-alloy steels
Argon with 21% to 25% carbon dioxide	Short-circuiting transfer on mild steels
Argon with 25% helium	Sometimes used for aluminum where bead appearance is important
Argon with 75% helium	Welding heavy aluminum
Argon with 90% helium	Welding aluminum over 3" thick and for copper over 1" thick
Argon with 5% to 10% carbon dioxide and 1% to 3% oxygen	Welding carbon steel, low-alloy steel, and stainless steel with all transfer modes
Argon with 10% to 30% helium and 5% to 15% carbon dioxide	Pulse spray transfer on heavy out-of-position carbon and low-alloy steel
60% to 70% helium with 20% to 35% argon and 4% to 5% carbon dioxide	Short-circuiting transfer of high-strength steels in all positions
90% helium with 7.5% argon and 2.5% carbon dioxide	Short-circuiting transfer of stainless steels in all positions
Argon, helium, carbon dioxide, and oxygen	High-rate deposition of low-alloy, high-tensile metals, and mild steels

206T02.EPS

Table 3 AWS Classification for Typical Gas Mixtures

AWS Classification	Typical Gas Mixtures (%)	Gas
SG-AC-25	75/25	Argon + Carbon Dioxide
SG-AO-2	98/2	Argon + Oxygen
SG-AHe-10	90/10	Argon + Helium
SG-AH-5	95/5	Argon + Hydrogen
SG-HeA-25	75/25	Helium + Argon
SG-HeAC-7.5/2.5	90/7.5/2.5	Helium + Argon + Carbon Dioxide
SG-ACO-8/2	90/8/2	Argon + Carbon Dioxide + Oxygen
SG-A-G	Special	Argon + Mixture

206T03.EPS

Table 4 Short-Circuiting Shielding Gases and Their Applications

Metal	Shielding Gas	Thickness	Advantages
Carbon steel	75% Argon + 25% Carbon Dioxide	Less than ⅛" (3.2 mm)	High welding speeds without burn-through; minimum distortion and spatter
	75% Argon + 25% Carbon Dioxide	More than ⅛" (3.2 mm)	Minimum spatter; clean weld appearance; good puddle control in vertical and overhead positions
	50% Argon + 50% Carbon Dioxide	—	Deeper penetration; faster welding speeds
Stainless steel	90% Helium + 7.5% Argon + 2.5% Carbon Dioxide	—	No effect on corrosion resistance; small heat-affected zone; no undercutting; minimum distortion
Low alloy steel	60 to 70% Helium + 25 to 35% Argon + 4.5% Carbon Dioxide	—	Minimum reactivity; excellent toughness; excellent arc stability, wetting characteristics, and bead contour; little spatter
	75% Argon + 25% Carbon Dioxide	—	Fair toughness; excellent arc stability, wetting characteristics, and bead contour; little spatter
Aluminum, copper, magnesium, nickel, and their alloys	Argon and Argon + Helium mixtures	Over ⅛" (3.2 mm)	Argon satisfactory on sheet metal; argon-helium preferred base material

206T04.EPS

Table 5 Spray Transfer Shielding Gases and Their Applications

Metal	Shielding Gas	Thickness	Advantages
Aluminum	100% Argon	0 to 1" (0 to 25 mm)	Best metal transfer and arc stability, least spatter
	65% Argon + 35% Helium	1 to 3" (27 to 76 mm)	Higher heat input than straight argon; improved fusion characteristics with 5XXX series Al-Mg alloys
	75% Helium + 25% Argon	Over 3" (76 mm)	Highest heat input; minimizes porosity
Magnesium	100% Argon	—	Excellent cleaning action
Carbon steel	Argon + 3 to 5% Oxygen	—	Improves arc stability; produces a more fluid and controllable weld puddle; good coalescence and bead contour; minimizes undercutting; permits higher speeds than pure argon
	Argon + 8 to 10% Carbon Dioxide	—	High-speed mechanized welding; low-cost manual welding
Low-alloy steel	98% Argon + 2% Oxygen	—	Minimizes undercutting; provides good toughness
Stainless steel	99% Argon + 1% Oxygen	—	Improves arc stability; produces a more fluid and controllable weld puddle; good coalescence and bead contour; minimizes undercutting on heavier stainless steels
	98% Argon + 2% Oxygen	—	Provides better arc stability, coalescence, and welding speed than 1% oxygen mixture for thinner stainless steel materials
Nickel, copper, and their alloys	100% Argon	Up to ⅛" (3.2 mm)	Provides good wetting; decreases fluidity of weld material
	Argon + Helium mixtures	—	Higher heat inputs of 50 and 75% helium mixtures offset high heat dissipation of heavier gauges
Titanium	100% Argon	—	Good arc stability; minimum weld contamination; inert gas backing is required to prevent air contamination on back of weld area

206T05.EPS

Table 6 Typical Shielding Gas Flow Rates for Various Metals and Wire Sizes

PLATE THICKNESS (in.)	WIRE SIZE (in.)	GAS FLOW RATE (cfh)*
	CARBON STEEL (SPRAY-TRANSFER)	
Under 1⁄8	0.030	(short arc length)10 to 15
3⁄16	1⁄16	10 to 15
1⁄2	3⁄32	40 to 50
	ALUMINUM (SPRAY-TRANSFER)	
0.040	0.030	30
0.050	0.030	15
0.093	0.030	15
1⁄4	3⁄64	35 to 40
1⁄2	1⁄16	50
	STAINLESS STEEL (SPRAY-TRANSFER)	
0.063	0.030	(short arc length) 15 to 20
1⁄8	1⁄16	35
1⁄4	1⁄16	35
1⁄2	3⁄32	40 to 50
	COPPER	
Up to 1⁄4	1⁄16 or 3⁄32	50

Factors that may affect the shielding gas flow rate include the following:
DRAFTS: In a drafty location, the flow rate must be increased to maintain the gas shield around the weld zone.
GAS USED: For example, helium usually requires a higher flow rate than argon because it rises much faster due to its low density.
WELDING CURRENT: High welding currents require higher flow rates.
NOZZLE SIZE: Larger nozzles require higher flow rates.
JOINT TYPE: Welds on flat surfaces require higher flow rates than welds in grooves or fillets.
WELDING SPEED: Fast advance speeds require higher flow rates than slower advance speeds.
WELD POSITION: Vertical and horizontal position welds require higher flow rates than flat or overhead welds.
* cfh = Cubic feet per hour

206T06.EPS

Table 7 GMAW Filler Metal Specifications and Classification System

Material	Filler Metal		X Designator Description	Example
	AWS Spec.	AWS Class*		
Carbon steel	A5.18	ERXXS-Y EXXC-Y	Tensile strength × 1,000 (psi)	ER70S-3 E70C-3
Low-alloy steel	A5.28	ERXXS-Y EXXC-Y	Tensile strength × 1,000 (psi)	ER80S-B2 E80C-B2
Stainless steel	A5.9	ERXXXY	Stainless alloy (308, 410, etc.)	ER308L EC308L
Stainless steel (flux cored)	A5.22	RXXXTi-5	Stainless alloy (308, 309, etc.)	R309LT1-5
Aluminum	A5.10	ERXXXX-Y	Aluminum alloy (4043, 5083, etc.)	ER4043
Nickel	A5.14	ERNiXX-Y	Major alloying elements (Cr, Fe, Mo, etc.)	ERNiCr-3
Copper	A5.7	ERCuXX-Y	Major alloying elements (Al, Ni, Si, etc.)	ERCuAl-A2
Magnesium	A5.19	ERXXYYY	Major alloying elements (Al, Zn, etc.)	ERAZ92A
Titanium	A5.16	ERTi-Y		ERTi-5

*Legend
E — Filler metal may be used as an electrode.
R — Filler metal may be used as a rod.
S — Solid filler metal
C — Composite or stranded filler metal
Y — Designator (or combination of designators) that describes specific alloy, shielding gas to be used, diffusible hydrogen limit, etc. Refer to the appropriate AWS Filler Metal Specification shown in table for explanation.

206T07.EPS

Table 8 AWS Carbon Steel Wire Classifications and Uses

WIRE CLASSIFICATION	INTENDED USES
ER70S-2 ER48S-2	Welds all grades of carbon steel, including killed (highly deoxidized by adding aluminum powder or ferrosilicon), semikilled, and rimmed (slightly deoxidized by adding a small amount of aluminum powder). Also welds carbon steel with rusted surfaces. Acceptable shielding gases include Ar-O_2, Ar-CO_2, and CO_2. The viscous weld puddle is well suited to out-of-position welding. Wetting is helped by keeping the oxygen or CO_2 percentage relatively high.
ER70S-3 ER48S-3	Welds killed and semikilled steels (used in appliances, automobiles, and farm equipment) with CO_2, Ar-O_2, or Ar-CO_2 shielding gases.
ER70S-4 ER48S-4	Welds semikilled and rimmed steels (used in boilers and pressure vessels, piping, ship steels, structural steels, and A515 grades 55-70 steels) with CO_2, Ar-O_2, or Ar-CO_2 shielding gases.
ER70S-6 ER48S-6	Welds many carbon steels with CO_2 and high welding current. High-speed welding can be done with Ar-O_2 with 5% or more O_2. Either spray transfer or short-circuiting transfer can be used.
ER70S-7 ER48S-7	Welds a wide variety of steels with superior weld quality and appearance. Porosity and weld defects caused by mill scale or rust are minimized.
ER70S-G ER48S-G E70C-G E48C-G	Filler metals not included in the preceeding classes that only have certain mechanical properties specified. Absence of a C or M suffix means that the manufacturer should be consulted for the recommended shielding gas and intended applications.
E70C-GS E48C-GS	Composite stranded or metal cored electrodes intended for only single pass applications. These electrodes may have higher alloy content and the manufacturer should be consulted for specific applications.
E70C-3 E48C-3 E70C-6 E48C-6	Composite stranded or metal cored electrodes intended for both single and multipass applications on steel. They are characterized by a spray arc and excellent bead wash characteristics.

206T08.EPS

Table 9 AWS Low-Alloy Steel Filler Metal Classifications and Uses

AWS Classification	Intended Uses
ER70S-A1	These filler metals are similar to many carbon steel filler metals except that ½ percent molybdenum has been added to increase the strength of the weld at elevated temperatures. It also provides some increase in corrosion resistance. Typically used for welding ASTM plate and A335-P1 pipe.
ER80S-B2 E80C-B2	These filler metals are used to weld various chromium-molybdenum steels for elevated temperature and corrosive service. Also used for joining dissimilar combinations of chromium-molybdenum and carbon steels. All GMAW transfer modes may be used. Careful control of preheat, interpass temperature, and postweld heat treatment (PWHT) is required to prevent cracking.
ER70S-B2L E70C-B2L	These filler metals are nearly identical to ER80S-B2 and E80C-B2 except for a low carbon content resulting in lower strength. They exhibit greater resistance to cracking if left in the as-welded condition or if PWHT is questionable.
ER90S-B3 E90S-B3	Filler metals used to weld chromium-molybdenum steels for high-temperature/high-pressure piping and pressure vessels. Also used to join chromium-molybdenum steels to carbon steels. All GMAW transfer modes may be used. Careful control of preheat, interpass temperature, and PWHT is required to prevent cracking.
ER80S-B3L E80C-B3L	These fillers are similar to ER90S-B3 and E90C-B3 except for low carbon content resulting in lower strength. They exhibit greater resistance to cracking if left in the as-welded condition or if PWHT is questionable.
ER80-Ni1 E80C-Ni1	Filler metals with 1 percent nickel that deposit weld metal similar to E8018-C3 covered electrodes. They are used for welding low-alloy high-strength steels requiring good toughness at temperatures as low as –50°F.
ER80S-Ni2 E70C-Ni2 E80C-Ni2	These filler metals are used for welding 2½ percent nickel steels. The weld metal is similar to E8018-C1 covered electrodes. These fillers are used in applications where good toughness is required at temperatures as low as –80°F.
ER80S-Ni3 E80C-Ni3	These filler metals deposit weld metal similar to E8018-C2 covered electrodes. They are typically used for welding 3½ percent nickel steels for low-temperature service.
ER80S-D2 ER90S-D2 E90C-D2	These filler metals have the same chemical requirements as E70S-1B for carbon steel. The difference between the ER80 and ER90 fillers is the change in shielding gas and mechanical properties. These fillers contain molybdenum to control porosity when used with CO_2 shielding gas. They are used to weld a variety of carbon and low-alloy high-strength steels in both an as-welded or PWHT conditions using GMAW short circuiting and pulsed arc processes.
ER100S-1 ER110S-1 ER120S-1	These filler metals deposit high-strength, very tough weld metal for critical applications where tensile strength exceeds 100ksi and in-service temperatures as low as –60°F. Mechanical properties of the weld vary depending on the heat input used.
ER80S-B6	These filler metals are used to weld material of similar composition, usually pipe or tubing. Preheat and PWHT is required.
ER80S-B8	These filler metals contain about 9 percent chromium and 1 percent molybdenum. Their use and heat treatment is the same as ER80S-B6.
ER90S-B9	These fillers are a chromium-molybdenum alloy modified with columbium (niobium) and vanadium to provide strength, toughness, fatigue life, and oxidation/corrosion resistance at elevated temperatures. This allows components now fabricated with stainless and ferritic steels to be made from a single alloy, eliminating dissimilar metal welding problems.
ERXXS-G EXXS-G	These are filler metals, not included in the previous classes, for which only certain mechanical properties are specified. The manufacturer should be consulted for composition characteristics and intended uses of filler metals supplied under this classification.

206T09.EPS

Table 10 AWS Stainless Steel Filler Metal Classifications and Uses (1 of 7)

AWS Classification	Intended Uses
ER209	These filler metals are most often used to weld AISI Type 209 (UNS S20910) base metals. The filler metal is a nitrogen-strengthened austenitic stainless steel alloy with high strength and good toughness over a wide range of temperature. Weldments using this filler metal are not subject to carbide precipitation in the as-welded condition. Nitrogen alloying reduces the tendency for intergranular carbide precipitation in the weld area by inhibiting carbon diffusion. Nitrogen alloying coupled with the molybdenum content provides superior resistance to pitting and crevice corrosion in chloride solutions. These filler metals are also used to join dissimilar metals, such as carbon steel to stainless steel, and for direct overlay on carbon steel for corrosion applications.
ER218	These filler metals are most often used to weld AISI Type 218 (UNS S21800) base metals. The filler metal is a nitrogen-strengthened austenitic stainless steel with high strength and good toughness over a wide range of temperature. Nitrogen alloying in this filler metal provides significantly improved in wear resistance in particle-to-metal and metal-to-metal (galling) applications when compared to the conventional austenitic stainless steels such as Type 304. The filler metal can be used for welding dissimilar alloys like mild steel to stainless steel, and also for direct overlay on mild steel for corrosion and wear applications when used with the gas metal arc process. The gas tungsten arc, plasma arc, and electron beam processes are not suggested for direct application of this filler metal on mild steel.
ER219	These filler metals are most often used to weld AISI Type 219 (UNS S21900) base metals and have similar composition, characteristics, and uses to ER209 filler metals.
ER240	These filler metals are most often used to weld AISI Type 240 and 241 base metals. They have similar composition, characteristics, and uses to ER209 filler metals. Additionally, they exhibit an improvement in resistance to wear in particle-to-metal and metal-to-metal (galling) applications compared to the more conventional austenitic stainless steels like Type 304. Weldments in alloys AISI 240 and AISI 241, when compared to Type 304, exhibit improved resistance to transgranular stress corrosion cracking in hot chloride solutions.
ER307	These filler metals are used primarily for moderate strength welds with good crack resistance between dissimilar steels such as austenitic manganese steel and carbon steel forgings or castings.
ER308	These filler metals are most often used to weld base metal of the same or similar composition, such as AISI Types 301, 302, 304, and 305. Commercial specifications for filler, and base metals vary in minimum alloy requirements and names such as 18-8, 19-9 and 20-10 are often associated with this class of filler metals.
ER308H	These filler metals are the same as ER308 except that the allowable carbon content has been increased to the highest portion of the range for ER308. The high carbon content provides higher tensile and creep strengths at elevated temperatures. These filler metals are used for welding Type 304H base metal.
ER308L	These filler metals are the same as ER308, except for restricted carbon content. The low-carbon content of welds made by these filler metals reduces the possibility of intergranular carbide precipitation, which increases the resistance to intergranular corrosion without the use of stabilizers such as columbium (niobium) or titanium. The low carbon welds, however, are not as strong at elevated temperature as the columbium (niobium)-stabilized alloys or 304H.

206T10A.EPS

Table 10 AWS Stainless Steel Filler Metal Classifications and Uses (2 of 7)

AWS Classification	Intended Uses
ER308LSi	These filler metals are the same as ER308L, except for the higher silicon content. This improves the usability of the filler metal in the gas metal arc welding process. If the dilution by the base metal produces a low ferrite or fully austenitic weld, the crack sensitivity of the weld is somewhat higher than that of lower silicon content weld metal.
ER308Mo	These filler metals are the same as ER308 except for the addition of molybdenum. The filler metals are recommended for welding ASTM CF8M stainless steel castings because they match the base metal with regard to chromium, nickel, and molybdenum. They can be used for welding wrought materials such as Type 316 stainless when increased ferrite is desired that cannot be attained with ER316 filler metals.
ER308LMo	These filler metals are recommended for welding ASTM CF3M stainless steel castings, as they match the base metal with regard to chromium, nickel, and molybdenum. The filler metals may also be used for welding wrought materials such as Type 316L stainless when increased ferrite is desired that cannot be attained with ER316L filler metals.
ER308Si	This filler metal is the same as ER308, except for the higher silicon content. This improves the usability of the filler metal in the gas metal arc welding processes. If the dilution by the base metal produces a low ferrite or fully austenitic weld metal, the crack sensitivity of the weld is somewhat higher than that of lower silicon content weld metal.
ER309	These filler metals are commonly used for welding similar alloys in wrought or cast form. They are used for welding dissimilar metals, such as joining Type 304 to carbon steel, welding the clad side of Type 304 clad steels, and applying stainless-steel sheet linings to carbon steel shells. Occasionally, they are used to weld Type 304 and similar base metals where severe corrosion conditions exist requiring higher alloy weld metal.
ER309L	This filler metal is the same as ER309, except for restricted carbon content. The low carbon content of these welds reduces the possibility of intergranular carbide precipitation, which increases the resistance to intergranular corrosion without the use of stabilizers such as columbium (niobium) and titanium. The low-carbon welds are not as strong at elevated temperatures as the columbium (niobium)-stabilized alloys or higher carbon content ER309 welds.
ER309Si	These filler metals are the same as ER309, except for higher silicon content. This improves the usability of the filler metal in gas metal arc welding processes. If the dilution by the base metal produces a low ferrite or fully austenitic weld metal deposit, the crack sensitivity of the weld is somewhat higher than that of a lower silicon content weld metal.
ER309Mo	The weld metal is the same as ER309 filler metals except for the addition of molybdenum to increase pitting corrosion resistance in halide-containing environments. The primary use of these filler metals is the surfacing of base metals to improve their corrosion resistance. It is used as a first layer of multilayer overlays with filler metals such as ER316 or ER317 to reduce chromium and molybdenum dilution from the base metal. Other applications include the welding of molybdenum-containing stainless steel linings to carbon steel shells, joining carbon steel base metals clad with molybdenum-containing stainless steel, and the joining of dissimilar base metals such as carbon steel to Type 304 stainless steel.

206T10B.EPS

Table 10 AWS Stainless Steel Filler Metal Classifications and Uses (3 of 7)

AWS Classification	Intended Uses
ER309LMo	The weld metal is the same as ER309Mo filler metals except for the restricted carbon content. The lower carbon content of the weld metal reduces the possibility of carbon precipitation, which increases resistance to intergranular corrosion, and is used in first layer applications where excessive pickup of carbon from base metal dilution could occur.
ER309LSi	This filler metal is the same as ER309L, except for higher silicon content. This improves the usability of the filler metal in gas metal arc welding processes. If the dilution by the base metal produces a low ferrite or fully austenitic weld, the crack sensitivity of the weld is somewhat higher than that of lower silicon content weld metal.
ER310	These filler metals are most often used to weld base metals of similar composition.
ER312	These filler metals were originally designed to weld cast alloys of similar composition. They have been found to be valuable in welding dissimilar metals, especially if one of them is a stainless steel, high in nickel. This filler metal gives a two-phase weld deposit with substantial amounts of ferrite in an austenitic matrix. Even with considerable dilution by austenite-forming elements, such as nickel, the microstructure remains two-phase and highly resistant to weld metal cracks and fissures.
ER316	These filler metals are used for welding Type 316 and similar alloys. They have been used successfully in certain applications involving special base metals for high-temperature service. The presence of molybdenum in the weld metal provides creep resistance at elevated temperatures. Rapid corrosion of ER316 weld metal may occur when the following three factors co-exist: (1) The presence of a continuous or semi-continuous network of ferrite in the weld metal microstructure (2) A composition balance of the weld metal giving a chromium-to-molybdenum ratio of less than 8.2 to 1 (3) Immersion of the weld metal in a corrosive medium Attempts to classify the media in which accelerated corrosion will take place by attack on the ferrite phase have not been entirely successful. Strong oxidizing and mildly reducing environments have been present where a number of corrosion failures were investigated and documented. The literature should be consulted for the latest recommendations.
ER316H	These filler metals are the same as ER316 except that the allowable carbon content has been increased to the higher portion of the ER316 range. Higher carbon content provides higher tensile and creep strengths at elevated temperatures. These filler metals are used for welding Type 316H base metal.
ER316L	These filler metals are the same as ER316 except for restricted carbon content. The lower carbon content of the weld metal reduces the possibility of intergranular carbide precipitation, which increases the resistance to intergranular corrosion without the use of stabilizers such as columbium (niobium) or titanium. These filler metals are used principally for welding low-carbon, molybdenum-bearing austenitic alloys. Tests have shown that the low-carbon content in the weld metal gives adequate protection against intergranular corrosion in most cases. This low-carbon filler metal is not as strong at elevated temperatures as Type ER316H.

206T10C.EPS

Table 10 AWS Stainless Steel Filler Metal Classifications and Uses (4 of 7)

AWS Classification	Intended Uses
ER316LSi	These filler metals are the same as ER316L, except for the higher silicon content. This improves the usability of the filler metal in the gas metal arc welding process. If the dilution by the base metal produces a low ferrite or fully austenitic weld, the crack sensitivity is somewhat higher than that of lower silicon content weld metal.
ER316Si	These filler metals are the same as ER316, except for the higher silicon content. This improves the usability of the filler metal in the gas metal arc welding process. If the dilution by the base metal produces a low ferrite or fully austenitic weld, the crack sensitivity of the weld is somewhat higher than that of lower silicon content weld metal.
ER317	The alloy content of weld metal deposited by these filler metals is somewhat higher than that of ER316 filler metals, particularly in molybdenum. These filler metals are usually used for welding alloys of similar composition. They are also used in severely corrosive environments (such as those containing halogens) where crevice and pitting corrosion are of concern.
ER317L	The weld metal is the same as for ER317 filler metals except for restricted carbon content. The lower carbon content of weld metal reduces the possibility of intergranular carbide precipitation, which increases the resistance to intergranular corrosion without the use of stabilizers such as columbium (niobium) or titanium. The low-carbon welds are not as strong at elevated temperatures as the columbium (niobium)-stabilized alloys or the standard Type 317 weld metal with higher-carbon content.
ER318	The weld metal is the same as for ER316 filler metals except for the addition of columbium (niobium). Columbium (niobium) provides resistance to intergranular carbide precipitation, which increases resistance to intergranular corrosion. These filler metals are used primarily for welding base metals of similar composition.
ER320	These filler metals have columbium (niobium) added to improve resistance to intergranular corrosion. The filler metals are primarily used to weld base metals of similar composition for applications where resistance to severe corrosion is required for a wide range of chemicals, including sulfuric and sulfurous acids and their salts. These filler metals can be used to weld both castings and wrought alloys of similar composition without postweld heat treatment. A modification of this grade without columbium (niobium) is available for repairing castings that do not contain columbium (niobium). With this modified composition, solution annealing is required after welding.
ER320LR	The LR (Low Residuals) welds of ER320LR filler metals are the same as ER320 welds except that elements C, Si, P, and S are specified at lower maximum levels, and Cb (Nb) and Mn are controlled within narrower ranges. These changes reduce the weld metal fissuring (while maintaining the corrosion resistance) frequently encountered in fully austenitic stainless steel weld metals. Consequently, welding practices typically used for austenitic stainless steel weld metals containing ferrite can be used in bare filler metal welding processes such as GTAW and GMAW can also be used. This filler metal has also been used in submerged arc overlay welding but may be prone to cracking when used to join base metal by this process. Type ER320LR weld metal has a lower minimum tensile strength than ER320 weld metal.

206T10D.EPS

Table 10 AWS Stainless Steel Filler Metal Classifications and Uses (5 of 7)

AWS Classification	Intended Uses
ER321	These filler metals are similar to ER347, except that titanium is added instead of columbium (niobium). The titanium acts in the same way as columbium (niobium) in Type 347 in reducing intergranular chromium carbide precipitation and thus increasing resistance to intergranular corrosion. The filler metals are used for welding chromium-nickel stainless steel base metals of similar composition, using an inert gas shielded process. They are not suitable for use with the submerged arc process because only a small portion of the titanium will be recovered in the weld metal.
ER330	These filler metals are commonly used where heat- and scale-resisting properties above 1800°F are required. However, high-sulfur environments may adversely affect performance at elevated temperature. Repairs of defects in alloy castings and the welding of castings and wrought alloys of similar composition are the most common applications.
ER347	These filler metals have columbium (niobium) or columbium (niobium) and tantalum added as a stabilizer. Either of these additions reduces the possibility of intergranular chromium carbide precipitation, which increases resistance to intergranular corrosion. These filler metals are usually used for welding chromium-nickel alloys that are stabilized with either columbium (niobium) or titanium. Although columbium (niobium) is the stabilizing element usually specified in Type 347 alloys, it should be recognized that tantalum also is present. Tantalum and columbium (niobium) are almost equally effective in stabilizing carbon and in providing high-temperature strength. This specification recognizes the usual commercial practice of reporting columbium (niobium) as the sum of columbium (niobium) plus tantalum. If dilution by the base metal produces a low ferrite or fully austenitic weld metal deposit, crack sensitivity of the weld may increase substantially.
ER347Si	These filler metals are the same as ER347, except for the higher silicon content. This improves the usability of the filler metal in the gas metal arc welding process. If the dilution by the base metal produces a low ferrite or fully austenitic weld, the crack sensitivity of the weld is somewhat higher than that of lower silicon content weld metal.
ER383	These filler metals are used to weld base metal of a similar composition to itself, to UNS N08028, and to other grades of stainless steel. Type ER383 weld metal is recommended for sulfuric and phosphoric acid environments. The elements C, Si, P, and S are specified at low levels to minimize weld metal hot cracking and fissuring (while maintaining the corrosion resistance) frequently encountered in fully austenitic stainless steel weld metals.
ER385	These filler metals are used primarily for welding Type 904L and ASTM B625, B673, B674, and B677 materials for the handling of sulfuric acid and many chloride solutions. The ER385 filler metals are used to join Type 317L material where improved corrosion resistance in specific media is needed. They are also used for joining Type 904L base metal to other grades of stainless. The elements C, Si, P, and S are specified at low levels to minimize weld metal hot cracking and fissuring (while maintaining corrosion resistance) frequently encountered in fully austenitic weld metals.

206T10E.EPS

Table 10 AWS Stainless Steel Filler Metal Classifications and Uses (6 of 7)

AWS Classification	Intended Uses
ER409	These filler metals differ from Type 410 material because they have a ferritic microstructure. A titanium addition forms carbides to improve corrosion resistance, increase strength at high temperature, and promote the ferritic microstructure. The ER409 filler metals may be used to join matching or dissimilar base metals. The greatest usage is for applications where thin stock is fabricated into exhaust system components.
ER 409Cb	These filler metals are the same as ER409 except that columbium (niobium) is used instead of titanium to achieve similar results. Oxidation losses across the arc are generally lower. Applications are the same as those of ER409 filler metals.
ER410	These filler metals are air-hardening steel. Preheat and postheat treatments are required to achieve welds of adequate ductility for many purposes. The most common application of these filler metals is for welding alloys of similar compositions. They are also used for surfacing of carbon steels to resist corrosion, erosion, or abrasion.
ER410NiMo	These filler metals are used for welding ASTM CA6NM castings or similar materials, as well as light-gauge Type 410, 410S, and 405 base metals. The weld metal for these filler metals is modified to contain less chromium and more nickel than the weld metal of ER410 filler metals. This is to eliminate ferrite in the microstructure because ferrite has a bad effect on the mechanical properties of the alloy. Final postweld heat treatment should not exceed 1150°F. Higher temperatures may result in rehardening due to untempered martensite in the microstructure after cooling to room temperature.
ER420	These filler metals are similar to ER410, except for slightly higher chromium and carbon contents. ER420 is used for many surfacing operations requiring corrosion resistance provided by the chromium along with higher hardness than weld metal deposited by ER410 electrodes. This increases wear resistance.
ER430	The composition of these filler metals is balanced by providing sufficient chromium to give adequate corrosion resistance for the usual applications and yet retain sufficient ductility in the heat-treated condition to meet the mechanical requirements of the specification. Excessive chromium will result in lowered ductility. Welding with an ER430 filler metal usually requires preheat and postheat. Optimum mechanical properties and corrosion resistance are obtained only when the weldment is heat-treated following the welding operation.
ER446LMo	These filler metals are used for welding base metal of the same composition with inert gas-shielded welding processes. Due to the high purity of both base metal and filler metal, cleaning of the parts before welding is most important. Complete coverage by shielding gas during welding is extremely important to prevent contamination by oxygen and nitrogen. Non-conventional gas shielding methods (leading, trailing, and back shielding) are often employed.
ER502	These filler metals are used for welding material of similar composition, usually in the form of pipe or tubing. The alloy is an air-hardening material. When welding with this filter metal, preheating and postweld heat treatment is required.
ER505	These filler metals are used for welding base metal of similar composition, usually in the form of pipe or tubing. The alloy is an air-hardening material. When welding with this filler metal, preheating and postweld heat treatment is required.

206T10F.EPS

Table 10 AWS Stainless Steel Filler Metal Classifications and Uses (7 of 7)

AWS Classification	Intended Uses
ER630	These filler metals are primarily designed for welding ASTM A564, Type 630, and some other precipitation-hardening stainless steels. The weld metal is modified to prevent the formation of ferrite networks in the martensite microstructure that could have a bad effect on mechanical properties. Dependent on the application and weld size, the weld may be used in an as-welded, precipitation-hardened, or solution-treated and precipitation-hardened condition.
ER16-8-2	These filler metals are used primarily for welding stainless steel, such as Types 16-8-2, 316, and 347, for high-pressure, high-temperature piping systems. The weld metal usually has a Ferrite Number no higher than 5 FN. The weld metal also has good hot ductility properties, which offer relative freedom from weld or crater cracking even under high-restraint conditions. The weld is usable in either the as-welded or solution-treated condition. These filler metals depend on a very carefully balanced chemical composition to develop their fullest properties. Corrosion tests indicate that Type 16-8-2 weld metal may have less corrosion resistance than Type 316 base metal depending on the corrosive media. Where the weldment is exposed to severe corrodents, surface layers should be deposited with a more corrosion-resistant weld metal.
ER19-10H	These filler metals are similar to ER308H except that the chromium content is lower and there are additional limits on other alloy content. This lower limit of Cr and additional limits on other Cr equivalent elements allows a lower ferrite range to be attained. A lower ferrite level in the weld metal decreases the chance of sigma embrittlement after long-term exposure at temperatures in excess of 1000°F. This filler metal should be used in conjunction with welding processes and other welding consumables that do not deplete or otherwise significantly change the amount of chromium in the weld metal. If used with submerged arc welding, a flux that neither removes nor adds chromium to the weld metal is highly recommended. This filler metal also has a higher carbon level required for improved creep properties in high-temperature service. Actual weld application qualification testing is recommended to assure that an acceptable weld metal carbon level is obtained. If corrosion or scaling is a concern, special testing should be included in application testing.
ER2209	These filler metals are used primarily to weld duplex stainless steels that contain approximately 22% chromium, such as UNS S31803. The weld metal of these filler metals has a duplex microstructure consisting of an austenite-ferrite matrix. ER2209 filler metals combine increased tensile strength with improved resistance to pitting, corrosive attack, and stress-corrosion cracking.
ER2553	These filler metals are used primarily to weld duplex stainless steels that contain approximately 25% chromium. The weld metal of these filler metals has a duplex microstructure consisting of an austenite-ferrite matrix. ER2553 filler metals combine increased tensile strength with improved resistance to pitting, corrosive attack, and stress-corrosion cracking.
ER3556	These filler metals are used for welding ASTM B435, B572, B619, B622, B626, and UNS R30556 base metal to itself, for joining steel to other nickel alloys, and for surfacing steel by the GTAW, GMAW, and PAW processes. The filler metal is resistant to high-temperature corrosive environments containing sulfur.

206T10G.EPS

Table 11 Recommended Filler Metals for Common Aluminum Alloys (1 of 2)

Base Metal to Base Metal	201.0 206.0 224.0	319.0, 333.0, 354.0, 355.0, C355.0	356.0, A356.0 357.0, A357.0 413.0, 443.0, A444.0	511.0, 512.0, 513.0, 514.0, 535.0	7004, 7005, 7039, 710.0, 712.0	6009 6010 6070	6005, 6061, 6063, 6101, 6151, 6201, 6351, 6951	5456	5454
1060, 1070, 1080, 1350	ER4145	ER4145	ER4043[a,b]	ER5356[c,d]	ER5356[c,d]	ER4043[a,b]	ER4043[b]	ER5356[d]	ER4043[b,d]
1100, 3003, Alc 3003	ER4145	ER4145	ER4043[a,b]	ER5356[c,d]	ER5356[c,d]	ER4043[a,b]	ER4043[b]	ER5356[d]	ER4043[b,d]
2014, 2036	ER4145[c]	ER4145[e]	ER4145	—	—	ER4145	ER4145	—	—
2219	ER2319[a]	ER4145[c]	ER4145[b,c]	ER4043	ER4043	ER4043[a,b]	ER4043[a,b]	—	ER4043[b]
3004, Alc 3004	—	ER4043[b]	ER4043[b]	ER5356[f]	ER5356[f]	ER4043[b]	ER4043[b,f]	ER5356[d]	ER5356[f]
5005, 5050	—	ER4043[b]	ER4043[b]	ER5356[f]	ER5356[f]	ER4043[b]	ER4043[b,f]	ER5356[d]	ER5356[f]
5052, 5652[i]	—	ER4043[b]	ER4043[f]	ER5356[f]	ER5356[f]	ER4043[b]	ER5356[c,f]	ER5356[f]	ER5356[f]
5083	—	—	ER5356[c,d]	ER5356[d]	ER5356[d]	—	ER5356[d]	ER5183[d]	ER5356[d]
5086	—	—	ER5356[c,d]	ER5356[d]	ER5356[d]	—	ER5356[d]	ER5356[d]	ER5356[d]
5154, 5254[i]	—	—	ER4043[f]	ER5356[f]	ER5356[f]	—	ER5356[f]	ER5356[f]	ER5356[f]
5454	—	ER4043[b]	ER4043[f]	ER5356[f]	ER5356[f]	ER4043[b]	ER5356[c,f]	ER5356[f]	ER5554[c,f]
5456	—	—	ER5356[c,d]	ER5356[d]	ER5556[d]	—	ER5356[d]	ER5556[d]	
6005, 6061, 6063 6101, 6151, 6201 6351, 6951	ER4145	ER4145[b,c]	ER4043[b,f,g]	ER5356[f]	ER5356[c,f]	ER4043[a,f,g]	ER4043[b,f,g]		
6009, 6010, 6070	ER4145	ER4145[b,c]	ER4043[a,f,g]	ER4043	ER4043	ER4043[a,b,g]			
7004, 7005, 7039 710.0, 712.0	—	ER4043[b]	ER4043[b,f]	ER5356[f]	ER5356[d]				
511.0, 512.0, 513.0 514.0, 535.0	—	—	ER4043[f]	ER5356[f]					
356.0, A356.0, 357.0 A357.0, 413.0 443.0, A444.0	ER4145	ER4145[b,c]	ER4043[b,h]						
319.0, 333.0 354.0, 355.0 C355.0	ER4145[c]	ER4145[b,c,h]							
201.0, 206.0, 224.0	ER2319[a,h]								

206T11A.EPS

Table 11 Recommended Filler Metals for Common Aluminum Alloys (2 of 2)

Base Metal to Base Metal	5154 5254[i]	5086	5083	5052 5652[i]	5005 5050	3004 Alc. 3004	2219	2014 2036	1100 3003 Alc. 3003	1060 1070 1080 1350
1060, 1070,	ER5356[c,d]	ER5356[d]	ER5356[d	ER4043[b,d]	ER1100[b,c]	ER4043[b,d]	ER4145[b,c]	ER4145	ER1100[b,c]	ER1188[b,c,h,j]
1080,1350	ER5356[c,d]	ER5356[d]	ER5356[d]	ER4043[b,d]	ER1100[b,c]	ER4043[b,d]	ER4145[b,c]	ER4145	ER1100[b,c]	
1100, 3003, Alc	—	—	—	—	ER4145	ER4145	ER4145[e]	ER4145[e]		
3003	ER4043	—	—	ER4043[b]	ER4043[a,b]	ER4043[a,b]	ER2319[a]			
2014, 2036	ER5356[f]	ER5356[d]	ER5356[d]	ER5356[c,f]	ER5356[c,f]	ER5356[c,f]				
2219	ER5356[f]	ER5356[d]	ER5356[d]	ER5356[c,d]	ER5356[c,f]					
3004, Alc 3004	ER5356[f]	ER5356[d]	ER5356[d]	ER5654[c,f,i]						
5005, 5050	ER5356[d]	ER5356[d]	ER5183[d]							
5052, 5652[i]	ER5356[d]	ER5356[d]								
5083	ER5356[f,i]									
5086										
5154, 5254[i]										

General Notes:

1. Service conditions such as immersion in fresh or salt water, exposure to specific chemicals, or a sustained high temperature (over 150°F [66°C]) may limit the choice of filler metals. Filler metals ER5183, ER5356, ER5556 are not recommended for sustained elevated temperature service.
2. Recommendations in this table apply to gas shielded arc welding processes. For oxyfuel gas welding, only ER1188, ER1100, ER4043, ER4047, and ER4145 filler metals are ordinarily used.
3. Where no filler metal is listed, the base metal combination is not recommended for welding.

Notes:

a. ER4145 may be used for some applications.
b. ER4047 may be used for some applications.
c. ER4043 may be used for some applications.
d. ER5183, ER5356, or ER5556 may be used.
e. ER2319 may be used for some applications. It can supply high strength when the weldment is postweld solution heat treated and aged.
f. ER5183, ER5356, ER5556, and ER5654 may be used. In some cases, they provide: (1) improved color match after anodizing treatment, (2) highest weld ductility, and (3) higher weld strength.
g. ER4643 will provide higher strength in 1/2 in. (12 mm) and thicker groove welds in 6XXX base alloys when postweld solution heat treated and aged.
h. Filler metal with the same analysis as the base metal is sometimes used. the following wrought filler metals possess the same chemical composition limits as cast filler alloys: ER4009 and R4009 as R-C355.0; ER4010 and R4010 as R-A356.0; and R4011 as R-A357.0
i. Base metal alloys 5254 and 5652 are used for hydrogen peroxide service. ER5654 filler metal is used for welding both alloys for service temperatures below 150°F (66°C).
j. ER1100 may be used for some applications.

206T11B.EPS

Table 12 AWS Copper Filler Metal Classifications and Uses

AWS Classification [a]	Intended Uses [b]
ERCu	These filler metals (and the alloys in the rest of this table) are made of deoxidized copper, but also may contain one or more of the following elements: phosphorus, silicon, tin, manganese, and silver. Phosphorus and silicon are added primarily as deoxidizers. The other elements add either to the ease of welding or to the properties of the final weldment. These filler metals generally are used for the welding of deoxidized and electrolytic tough pitch copper. Reactions with hydrogen in oxygen-free copper, and the segregation of copper oxide in tough pitch copper may detract from joint efficiency. ERCU welding electrodes and rods may be used to weld these base metals when the highest quality is not required. Preheating is desirable on most work, and on thick base metals, it is essential. An external source of preheating generally is not needed when welding base metal ¼" thick or thinner. Preheating in the range of 400 to 1000°F is desirable when welding base metal thicker than ¼" if high-quality welds are to be obtained. For thick base metals, gas metal arc welding is preferred. Conventional joint designs consistent with good welding practice are generally satisfactory.
ERCuSi	These filler metals are copper alloys (silicon bronze) containing approximately 3% silicon; they may also contain small percentages of manganese, tin, or zinc. They are used for GTAW and GMAW of copper-silicon and copper-zinc base metals, to themselves and to steel. When using GMAW with these filler metals, it generally is best to keep the weld pool small and the interpass temperature below 150°F to minimize hot cracking. Using narrow weld passes reduces contraction stresses and permits faster cooling through the **hot-shortness** temperature range. When using GTAW with these filler metals, the best results are obtained by keeping the weld pool small. Preheating is not required. Welding can be done in all positions, but the flat position is preferred.
ERCuSn-A	These filler metals are copper alloys (Phosphor Bronze) containing about 5% tin and up to 0.35% phosphorus added as a deoxidizer. Tin increases wear resistance of the weld metal and slows the rate of solidification by broadening the temperature differential between the liquid and solid states. This slower solidification increases the tendency to hot shortness. To minimize this effect, the weld pool should be kept small and welding time as short as possible. These filler metals can be used to weld bronze and brass. They also can be used to weld copper if the presence of tin in the weld metal is not objectionable. When using GTAW with these filler metals, preheating is desirable. Welding is done in the flat position only.
ERCuNi	In these filler metals, the nickel addition strengthens the weld metal and improves the corrosion resistance, particularly against salt water. The weld metal has good hot and cold ductility. Copper-nickel filler metals are used for welding most copper-nickel alloys. When using GTAW or GMAW with these filler metals, preheating is not required. Welding is done in all positions. The arc should be kept as short as possible to assure adequate shielding gas coverage for minimum porosity.
ERCuAl-A1 [c]	This filler metal is an iron-free aluminum bronze. It is recommended for use as a surfacing metal for wear-resistant surfaces having relatively light loads, for resistance to corrosive media such as salt or brackish water, and for resistance to many commonly used acids in varying concentrations and temperatures. This alloy is not recommended for joining metals.

Notes:

a. Not all available types of copper alloy filler metals have assigned AWS classifications.

b. Argon and helium are the only shielding gases recommended for copper welding.

c. This filler metal is not recommended for oxyfuel welding because of aluminum oxide formation. Depending on the thickness of the base metal, preheat may or may not be required. Flat position welding is preferred; however, other positions can be used with pulsed arc welding equipment.

206T12.EPS

Table 13 AWS Nickel Filler Metal Classifications and Uses (1 of 3)

AWS Classification	Intended Uses
ERNi-1	These filler metals are intended for welding wrought and cast forms of commercially pure nickel alloy (ASTM B160, B161, B162, and B163 having UNS numbers N02200 or N02201) to themselves with GTAW, GMAW, SAW, and PAW processes. The filler metals contain sufficient titanium to control weld-metal porosity with these welding processes.
ERNiCu-7	These filler metals are used for welding nickel-copper alloy (ASTM B127, B163, B164, and B165 having UNS number N04400) to itself with GTAW, GMAW, SAW, and PAW processes. The filler metals contain sufficient titanium to control porosity with these welding processes
ERNiCu-8	These filler metals are used for welding age-hardening nickel-copper alloy (ASTM F467 and F468 having UNS number N05500) to itself with GTAW, GMAW, SAW, and PAW processes. The filler metal will age harden on heat treatment. For specific information concerning age hardening, consult the manufacturer or the manufacturer's technical literature.
ERNiCr-3	These filler metals are used for welding nickel-chromium-iron alloy (ASTM B163, B166, B167, and B168 having UNS number N06600) to itself with GTAW, GMAW, SAW, and PAW processes. They are also used with the same processes for welding the clad side of joints in steel clad with nickel-chromium-iron alloy, for surfacing steel with nickel-chromium-iron weld metal, for dissimilar welding of nickel-base alloys, and for joining steel to nickel-base alloys.
ERNiCr-4	These filler metals are used for GTAW of 50/50 nickel/chromium alloy, cladding nickel/chromium alloy onto nickel-iron-chromium tubing, and casting repair. The filler metals are resistant to high-temperature corrosion, including fuel-ash corrosion in atmospheres containing sulfur and vanadium.
ERNiCr-6	These filler metals are used for welding nickel-chromium-iron alloy (ASTM B163, B166, B167, and B168 having UNS number N06600) to itself with GTAW, GMAW, SAW, and PAW processes. They are also used with the same processes for welding the clad side of joints in steel clad with nickel-chromium-iron alloy, for surfacing steel with nickel-chromium-iron weld metal, and for joining steel to nickel-base alloys.
ERNiCrFe-5	These filler metals are used for welding nickel-chromium-iron alloy (ASTM B163, B166, B167, and B168 having UNS number N06600) to itself with GTAW, GMAW, SAW, and PAW processes. The higher niobium content of the filler metal is intended to minimize cracking where high welding stresses are encountered, as in thick-section base metal.
ERNiCrFe-6	These filler metals are used for cladding steel with nickel-chromium-iron weld metal and for joining steel to nickel-base alloys with GTAW, GMAW, SAW, and PAW processes. The weld metal will age harden on heat treatment. For specific information concerning age hardening, consult the manufacturer or the manufacturer's technical literature.
ERNiCrFe-7	These filler metals are used for welding nickel-chromium-iron alloy (ASTM B163, B166, B167, and B168 having UNS number N06690) to itself with GTAW, GMAW, SAW, and PAW processes. They are also used to weld nickel-chromium-iron alloy with the same processes to steels, as an overlay on steels, and to weld steels clad with the nickel-chromium-iron alloys.
ERNiCrFe-8	These filler metals are used for cladding steel with nickel-chromium-iron weld metal and for joining steel to nickel-base alloys using the GTAW, GMAW, SAW, and PAW processes. The weld metal will age harden on heat treatment. For specific information concerning age hardening, consult the manufacturer or the manufacturer's technical literature.
ERNiCrFe-11	These filler metals are used for welding nickel-chromium-iron-aluminum alloy (ASTM B166, B167, and B168 having UNS number N06601) to itself and to other high-temperature compositions using the GTAW process. It is used for severe applications where the exposure temperature can exceed 2100°F.

206T13A.EPS

Table 13 AWS Nickel Filler Metal Classifications and Uses (2 of 3)

AWS Classification	Intended Uses
ERNiFeCr-1	These filler metals are used for welding nickel-iron-chromium-molybdenum-copper alloy (ASTM B423 having UNS number N08825) to itself using the GTAW and GMAW processes.
ERNiFeCr-2	These filler metals are used for welding nickel-chromium-niobium-molybdenum alloy (ASTM B637 and AMS 5589 having UNS number N07718) to itself using the GTAW processes. The weld metal will age harden on heat treatment. For specific information concerning age hardening, consult the manufacturer or the manufacturer's technical literature.
ERNiMo-1	These filler metals are used for welding nickel-molybdenum alloy (ASTM B333 having UNS number N10001) to itself with GTAW and GMAW processes.
ERNiMo-2	These filler metals are used for welding nickel-molybdenum alloy (ASTM B366, B434, and B573 having UNS number N10003) to itself with GTAW and GMAW processes. They are also used to weld nickel-molybdenum alloy with the same processes to steel, to other nickel-base alloys, and for cladding steel with nickel-molybdenum weld metal.
ERNiMo-3	These filler metals are used for weld repair of various nickel-, cobalt-, and iron-base alloys and for dissimilar joining applications of nickel-, cobalt-, and iron-base alloys.
ERNiMo-7	These filler metals are used for welding nickel-molybdenum alloy (ASTM B333 and B335 having UNS number N10665) to itself with GTAW and GMAW processes and for cladding steel with nickel-molybdenum weld metal using the same processes.
ERNiMo-8 ERNiMo-9	These filler metals are used for welding 9% nickel steel (ASTM A333, A334, A353, A553 having UNS number K81340) to itself using the GTAW and SAW processes.
ERNiMo-10	These filler metals are used for welding nickel-molybdenum alloy (ASTM B333, B335, B366, B564, B619, B622, and B626 having UNS number N10675) to itself with GTAW, GMAW, and PAW processes. They are also used for welding the clad side of joints in steel clad with nickel- molybdenum alloy and for welding nickel-molybdenum alloys to steel and to other nickel-base alloys with the same processes.
ERNiCrMo-1	These filler metals are used for welding nickel-chromium-molybdenum alloy (ASTM B581 and B582 having UNS number N06007) to itself with GTAW, GMAW, and PAW processes. They are also used with the same processes to weld nickel-chromium-molybdenum alloy to steel, to other nickel-base alloys, and for cladding steel with nickel-chromium-molybdenum weld metal.
ERNiCrMo-2	These filler metals are used for welding nickel-chromium-molybdenum alloy (ASTM B366, B435, and B572 having UNS number N06002) to itself with GTAW, GMAW, and PAW processes. They are also used with the same processes to weld nickel-chromium-molybdenum alloy to steel, to other nickel-base alloys, and for cladding steel with nickel-chromium-molybdenum weld metal.
ERNiCrMo-3	These filler metals are used for welding nickel-chromium-molybdenum alloy (ASTM B443, B444, and B446 having UNS number N06625) to itself with GTAW, GMAW, SAW, and PAW processes. They are also used with the same processes to weld nickel-chromium-molybdenum alloy to steel, to other nickel-base alloys, for cladding steel with nickel-chromium-molybdenum weld metal, and for welding the clad side of joints in steel with nickel-chromium-molybdenum alloy. These filler metals are recommended for applications where the operating temperature ranges from cryogenic to 1000°F.

206T13B.EPS

Table 13 AWS Nickel Filler Metal Classifications and Uses (3 of 3)

AWS Classification	Intended Uses
ERNiCrMo-4	These filler metals are used for welding nickel-chromium-molybdenum alloy (ASTM B574, B575, B619, B622, and B628 having UNS number N10276) to itself with GTAW and GMAW processes. They are also used with the same processes to weld nickel-chromium-molybdenum alloy to steel, to other nickel-base alloys, and for cladding steel with nickel-chromium-molybdenum weld metal.
ERNiCrMo-7	These filler metals are used for welding nickel-chromium-molybdenum alloy (ASTM B574, B575, B619, B622, and B628 having UNS number N06455) to itself with GTAW, GMAW, and PAW processes. They are also used with the same processes to weld nickel-chromium-molybdenum alloy to steel, to other nickel-base alloys, and for cladding steel with nickel-chromium-molybdenum weld metal.
ERNiCrMo-8	These filler metals are used for welding nickel-chromium-molybdenum alloy (ASTM B581, B582, B619, B622 and B626 having UNS number N06975) to itself with GTAW, GMAW, and PAW processes. They are also used with the same processes to weld nickel-chromium-molybdenum alloy to steel, to other nickel-base alloys, and for cladding steel with nickel-chromium-molybdenum weld metal.
ERNiCrMo-9	These filler metals are used for welding nickel-chromium-molybdenum alloy (ASTM B581, B582, B619, B622, and B626 having UNS number N06007 or N06985) to itself with GTAW, GMAW, and PAW processes. They are also used with the same processes to weld nickel-chromium-molybdenum alloy to steel, to other nickel-base alloys, and for cladding steel with nickel-chromium-molybdenum weld metal.
ERNiCrMo-10	These filler metals are used for welding nickel-chromium-molybdenum alloy (ASTM B574, B575, B619, B622 and B628 having UNS number N06022) to itself with GTAW, GMAW, and PAW processes. They are also used with the same processes to weld nickel-chromium-molybdenum alloy to steel, to other nickel-base alloys, and for cladding steel with nickel-chromium-molybdenum weld metal.
ERNiCrMo-11	These filler metals are used for welding nickel-chromium-molybdenum alloy (ASTM B581, B582, B619, B622, and B626 having UNS number N06030) to itself with GTAW, GMAW, and PAW processes. They are also used with the same processes to weld nickel-chromium-molybdenum alloy to steel, to other nickel-base alloys, and for cladding steel with nickel-chromium-molybdenum weld metal.
ERNiCrMo-13	These filler metals are used for welding nickel-chromium-molybdenum alloy (B564, B574, B575, B619, B622, and B626 having UNS number N06059) to itself with GTAW, GMAW, SAW, and PAW processes. They are also used with the same processes to weld nickel-chromium-molybdenum alloy to steel, to other nickel-base alloys, and for cladding steel with nickel-chromium-molybdenum weld metal.
ERNiCrMo-14	These filler metals are used for welding nickel-chromium-molybdenum alloy (ASTM B574, B575, B619, B622, and B628 having UNS number N06686) to itself with GTAW, GMAW, and SAW processes. They are also used with the same processes to weld nickel-chromium-molybdenum alloy to steel, to other nickel-base alloys, and for cladding steel with nickel-chromium-molybdenum weld metal.
ERNiCrMo-15	These filler metals are used for welding nickel-chromium-molybdenum alloy (ASTM B805, UNS number N07725) to itself and for cladding steel with GMAW and GTAW processes. The weld metal age-hardens on heat treatment. For specific information concerning age hardening, consult the manufacturer or the manufacturer's technical literature.
ERNiCrCoMo-1	These filler metals are used for welding nickel-chromium-cobalt-molybdenum alloy (UNS N06617) to itself with GTAW and GMAW processes.
ERNiCrWMo-1	These filler metals are used for welding nickel-chromium-tungsten-molybdenum-lanthanum alloy (ASTM B366, B435, B564, and B572 having UNS number N06230) to itself with GTAW and GMAW processes.

206T13C.EPS

Table 14 Recommended Filler Metals for Common Magnesium Alloys (1 of 2)

Base Metal	Base Metal											
	AM100A	**AZ10A**	**AZ31B AZ31C**	**AZ61A**	**AZ63A**	**AZ80A**	**AZ81A**	**AZ91C**	**AZ92A**	**EK41A**	**EZ33A**	**HK31A**
	Filler Metal[a,b]											
AM100A	AZ92A											
AZ10A	AZ92A	AZ61A AZ92A										
AZ31B AZ31C	AZ92A	AZ61A AZ92A	AZ61A AZ92A									
AZ61A	AZ92A	AZ61A AZ92A	AZ61A AZ92A	AZ61A AZ92A								
AZ63A	c	c	c	c	AZ101A AZ92A							
AZ80A	AZ92A	AZ61A AZ92A	AZ61A AZ92A	AZ61A AZ92A	c	AZ61A AZ92A						
AZ81A	AZ92A	AZ92A	AZ92A	AZ92A	c	AZ92A	AZ101A AZ92A					
AZ91C	AZ92A	AZ92A	AZ92A	AZ92A	c	AZ92A	AZ92A	AZ101A AZ92A				
AZ92A	AZ92A	AZ92A	AZ92A	AZ92A	c	AZ92A	AZ92A	AZ92A	AZ101A			
EK41A	AZ92A	AZ92A	AZ92A	AZ92A	c	AZ92A	AZ92A	AZ92A	AZ92A	EZ33A		
EK33A	AZ92A	AZ92A	AZ92A	AZ92A	c	AZ92A	AZ92A	AZ92A	AZ92A	EZ33A	EZ33A	
HK31A	AZ92A	AZ92A	AZ92A	AZ92A	c	AZ92A	AZ92A	AZ92A	AZ92A	EZ33A	EZ33A	EZ33A
HM21A	AZ92A	AZ92A	AZ92A	AZ92A	c	AZ92A	AZ92A	AZ92A	AZ92A	EZ33A	EZ33A	EZ33A
HM31A	AZ92A	AZ92A	AZ92A	AZ92A	c	AZ92A	AZ92A	AZ92A	AZ92A	EZ33A	EZ33A	EZ33A
HM32A	AZ92A	AZ92A	AZ92A	AZ92A	c	AZ92A	AZ92A	AZ92A	AZ92A	EZ33A	EZ33A	EZ33A
K1A	AZ92A	AZ92A	AZ92A	AZ92A	c	AZ92A	AZ92A	AZ92A	AZ92A	EZ33A	EZ33A	EZ33A
LA141A					c	c	c	c	c	d	d	d
M1A MG1	AZ92A	AZ61A AZ92A	AZ61A AZ92A	AZ61A AZ92A	c	AZ61A AZ92A	AZ92A	AZ92A	AZ92A	AZ92A	AZ92A	AZ92A
QE22A	d	d	AZ92A	d	c	d	d	d	d	EZ33A	EZ33A	EZ33A
ZE10A	AZ92A	AZ61A AZ92A	AZ61A AZ92A	AZ61A AZ92A	c	AZ61A AZ92A	AZ92A	AZ92A	AZ92A	EZ33A AZ92A	EZ33A AZ92A	EZ33A AZ92A
ZE41A	d	d	d	d	c	d	d	d	d	EZ33A	EZ33A	EZ33A
ZK21A	AZ92A	AZ61A AZ92A	AZ61A AZ92A	AZ61A AZ92A	c	AZ61A AZ92A	AZ92A	AZ92A	AZ92A	AZ92A	AZ92A	AZ92A
ZH62A ZK51A ZK60A ZK61A	c	c	c	c	c	c	c	c	c	c	c	c

Notes:

a. When more than one filler metal is given, they are listed in order of preference.

b. The letter prefix (ER or R), designating usability of the filler metal, has been deleted to reduce clutter in the table.

c. Welding not recommended

d. No data available

206T14A.EPS

Table 14 Recommended Filler Metals for Common Magnesium Alloys (2 of 2)

Base Metal	HM21A	HM31A	HZ32A	K1A	LA141A	M1A MG1	QE22A	ZE10A	ZE41A	ZK21A	ZH62A ZK51A ZK60A ZK61A
	Filler Metal[a,b]										
HM21A	EZ33A										
HM31A	EZ33A	EZ33A									
HZ32A	EZ33A	EZ33A									
K1A	EZ33A	EZ33A	EZ33A	EZ33A							
LA141A	EZ33A	d	d	d	EZ33A						
M1A MG1	AZ92A	AZ92A	AZ92A	AZ92A	d	AZ61A AZ92A					
QE22A	EZ33A	EZ33A	EZ33A	EZ33A	EZ33A	c	EZ33A				
ZE10A	EZ33A AZ92A	EZ33A AZ92A	EZ33A AZ92A	EZ33A AZ92A	EZ33A	AZ61A AZ92A	EZ33A AZ92A	AZ61A AZ92A			
ZE41A	EZ33A	EZ33A	EZ33A	EZ33A	d	d	EZ33A	d	EZ33A		
ZK21A	AZ92A	AZ92A	AZ92A	AZ92A	d	AZ61A AZ92A	AZ92A	AZ61A AZ92A	AZ92A	AZ61A AZ92A	
ZH62A ZK51A ZK60A ZK61A	c	c	c	c	c	c	c	c	c	c	EZ33A

Notes:
a. When more than one filler metal is given, they are listed in order of preference.
b. The letter prefix (ER or R), designating usability of the filler metal, has been deleted to reduce clutter in the table.
c. Welding not recommended
d. No data available

206T14B.EPS

Table 15 AWS Titanium Filler Metal Classifications and Uses

AWS Classification	Intended Uses
ERTi-1 ERTi-2 ERTi-3 ERTi-4	The alloys used for these filler metals are commonly referred to as commercially pure (C.P.) grades (1 through 4) of titanium with the level of impurities and mechanical properties increasing slightly from ERTi-1 to ERTi-4. C.P. Grade 2 (equivalent to ERTi-2) is the most widely used titanium alloy for industrial applications because of its good balance of strength, formability, and weldability. Typical uses are in seawater and brackish water heat exchangers, chemical process heat exchangers, pressure vessels and piping systems, pulp bleaching systems, air pollution control scrubbers, and electrochemical and chemical storage tanks. These grades also have some uses in the aerospace industry.
ERTi-5	The alloy used for these filler metals commonly referred to as 6-4 titanium and is probably the most widely used titanium alloy. Its high strength, ability to be heat treated, weldability, excellent fatigue strength, and hardness make the alloy excellent for industrial fans, pressure vessels, aircraft components, compressor blades, and automotive and jet engine parts.
ERTi-5ELI	This filler metal is a slightly purer version of ERTi-5 with ELI. With special processing, this alloy can develop high fracture toughness. Primary uses are in surgical implants, cryogenic vessels, and airframe components.
ERTi-6	The alloy used for these filler metals has good weldability, oxidation resistance, stability, and strength at elevated temperature. Typical uses of the alloy include gas turbine engine casings, aerospace structural members at elevated temperatures near engines and wing leading edges, and chemical process equipment where high elevated temperature strength is required.
ERTi-6ELI	These filler metals are a slightly purer version of ERTi-6 with ELI. They are used to fabricate pressure vessels for liquefied gases and other high pressure cryogenic vessels where better ductility and toughness with slightly lower strength are required.
ERTi-7	Welds made with these filler metals probably are the most corrosion-resistant titanium welds used in industrial applications. Mechanical and physical properties are similar to those of ERTi-2. This alloy extends the use of titanium into mildly reducing media, into much higher chloride levels, or where the environment fluctuates between oxidizing and reducing.
ERTi-9	These filler metals often are referred to as half 6-4 because the major components are roughly half that found in ERTi-5. The primary use has been in welding hydraulic tubing and fittings for aircraft. This alloy is desirable when high strength and the ability to maintain strength at elevated temperatures is required. Corrosion resistance, in most environments, appears to be similar to, or slightly less than, that of weld metal from ERTi-2 electrodes.
ERTi-9ELI	The reduced oxygen content of the ERTi-9ELI alloy results in slightly lower strength and improved toughness in comparison with weld metal from ERTi-9 filler metals.
ERTi-12	Welds made with these filler metals offer improved corrosion resistance, especially to crevice corrosion in hot brines, and higher strength levels compared to similar welds made using ERTi-2. Uses in industrial applications are similar to those of ERTi-2, but can be extended to less oxidizing conditions.
ERTi-15	Welds made with these filler metals have excellent resistance to salt water corrosion combined with good toughness and moderate strength. Typical uses are the fabrication of submersible hulls and pressure vessels using base material of a matching composition.

206T15.EPS

Table 16 Carbon Steel Flux Cored Electrode Welding Position, Shielding, Polarity, and Application Requirements (1 of 2)

AWS Classification	Position of Welding[a,b]	External Shielding[c]	Polarity[d]	Application[e]
E70T-1	H and F	CO_2	DCEP	M
E70T-1M	H and F	75-80% Ar/bal CO_2	DCEP	M
E71T-1	H, F, VU, OH	CO_2	DCEP	M
E71T-1M	H, F, VU, OH	75-80% Ar/bal CO_2	DCEP	M
E70T-2	H and F	CO_2	DCEP	S
E70T-2M	H and F	75-80% Ar/bal CO_2	DCEP	S
E71T-2	H, F, VU, OH	CO_2	DCEP	S
E71T-2M	H, F, VU, OH	75-80% Ar/bal CO_2	DCEP	S
E70T-3	H and F	None	DCEP	S
E70T-4	H and F	None	DCEP	M
E70T-5	H and F	CO_2	DCEP	M
E70T-5M	H and F	75-80% Ar/bal CO_2	DCEP	M
E71T-5	H, F, VU, OH	CO_2	DCEP or DCEN[f]	M
E71T-5M	H, F, VU, OH	75-80% Ar/bal CO_2	DCEP or DCEN[f]	M
E70T-6	H and F	None	DCEP	M
E70T-7	H and F	None	DCEN	M
E71T-7	H, F, VU, OH	None	DCEN	M
E70T-8	H and F	None	DCEN	M
E71T-8	H, F, VU, OH	None	DCEN	M
E70T-9	H and F	CO_2	DCEP	M

206T16A.EPS

Table 16 Carbon Steel Flux Cored Electrode Welding Position, Shielding, Polarity, and Application Requirements (2 of 2)

AWS Classification	Position of Welding[a,b]	External Shielding[c]	Polarity[d]	Application[e]
E70T-9M	H and F	75-80% Ar/bal CO_2	DCEP	M
E71T-9	H, F, VU, OH	CO_2	DCEP	M
E71T-9M	H, F, VU, OH	75-80% Ar/bal CO_2	DCEP	M
E70T-10	H and F	None	DCEN	S
E70T-11	H and F	None	DCEN	M
E71T-11	H, F, VD, OH	None	DCEN	M
E70T-12	H and F	CO_2	DCEP	M
E70T-12M	H and F	75-80% Ar/bal CO_2	DCEP	M
E71T-12	H, F, VU, OH	CO_2	DCEP	M
E71T-12M	H, F, VU, OH	75-80% Ar/bal CO_2	DCEP	M
E61T-13	H, F, VD, OH	None	DCEN	S
E71T-13	H, F, VD, OH	None	DCEN	S
E71T-14	H, F, VD, OH	None	DCEN	S
EX0T-G[g]	H and F	Not Specified	Not Specified	M
EX1T-G[g]	H, F, VD or VU, OH	Not Specified	Not Specified	M
EX0T-GS[g]	H and F	Not Specified	Not Specified	S
EX1T-GS[g]	H, F, VD or VU, OH	Not Specified	Not Specified	S

Notes:

a. H = horizontal position; F = flat position; OH = overhead position; VD = vertical position with downward progression; VU = vertical position with upward progression

b. Electrode sizes suitable for out-of-position welding, i.e., welding position other than flat or horizontal, usually are those sizes that are smaller than the 3⁄32" (2.4 mm) size. For that reason, electrodes meeting the requirements for the groove weld tests and the fillet weld tests may be classified as EX1T-X or EX1T-XM (where X represents the tensile strength and usability designators) regardless of their size.

c. Properties of weld metal from electrodes that are used with external gas shielding (EXXT-1, EXXT-1M, EXXT-2, EXXT-2M, EXXT-5, EXXT-5M, EXXT-9, EXXT-9M, EXXT-12, and EXXT-12M) vary according to the shielding gas employed. Electrodes classified with the specified shielding gas should not be used with other shielding gases without first consulting the manufacturer of the electrode.

d. The term "DCEP" refers to direct current electrode positive (dc, reverse polarity). The term "DCEN" refers to direct current electrode negative (dc, straight polarity).

e. M = single or multiple pass; S = single pass only. (Single-pass only classes are highly oxidizing and are used on metals that are rusty or have mill scale.)

f. Some E71T-5 and E71T-5M electrodes may be recommended for use on DCEN for improved out-of-position welding. Consult the manufacturer for the recommended polarity.

g. The letter "X" can be replaced with either a "6" or "7" to designate the weld metal's tensile strength.

206T16B.EPS

Table 17 Carbon Steel Flux Cored Electrode Classes and Uses (1 of 2)

AWS Classification	Intended Uses
EXXT-1 EXXT-1M	Electrodes designated as –1 use CO_2 shielding. However, other gas mixtures can be used to improve arc stability if recommended by the manufacturer. Electrodes designated as –1M use an argon-CO_2 mixture to increase manganese and silicon in the weld metal to increase yield and tensile strengths. Reducing argon in a gas mixture or using CO_2 shielding only with –1M electrodes may cause deterioration of arc and out-of-position welding characteristics and loss of strength. Both electrodes have spray transfer characteristics and low spatter. Most electrodes of this class have rutile-base slag and produce high deposition rates.
EXXT-2 EXXT-2M	These electrodes are similar to –1 and –1M types with higher manganese or silicon or both. The higher levels of deoxidizers allow single-pass welding of heavily oxidized or rimmed steel. The –2M electrodes have an even higher manganese content and can be used on metals with heavier mill scale, rust, or foreign matter that cannot be tolerated by some –1 or –1M electrodes.
EXXT-3	These self-shielding electrodes are designed for single-pass, very high speed welding and use spray transfer. Besides the listed positions, they can be used on 20° inclined downhill welds of sheet metal. They are not recommended for T or lap joints in materials thicker than 3⁄16" or for butt, edge, or corner joints in materials thicker than 1⁄4".
EXXT-4	These self-shielding electrodes use globular-type transfer and are designed for very high deposition rates. They produce low-sulfur welds that resist hot cracking. They are also designed for low penetration beyond the root of the weld because of poor joint fitup.
EXXT-5 EXXT-5M	These electrodes have a globular transfer and a lime-fluoride base slag that produces welds that have better impact strength and hot and/or cold cracking resistance than other similar classes with rutile-based slags. The –5 electrodes are used with a CO_2 shielding gas, however, argon-CO_2 mixes can be used to reduce spatter if recommended by the manufacturer. Reducing argon in a gas mixture or using CO_2 shielding only with –5M electrodes may cause deterioration of arc and out-of-position welding characteristics and loss of strength. Operator appeal for both types of electrodes is not as good as other similar classes with rutile slags.
EXXT-6	These self-shielding electrodes use spray-type transfer. The slag system is designed for good penetration into the root of the weld and excellent slag removal, even in deep grooves.
EXXT-7	These self-shielding electrodes use a droplet/spray-type transfer and produce a low-sulfur weld that is very resistant to cracking. The slag system is designed to produce high deposition rates for larger size electrodes in the flat or horizontal positions.
EXXT-8	These self-shielded electrodes have small droplet/spray-type transfer. They are suitable for all positions. The weld metal has good toughness, low temperature performance, and crack resistance.
EXXT-9 EXXT-9M	These electrodes are essentially the same as the –1 and –1M electrodes except they deposit weld metal with improved impact properties. Some electrodes in this class require joints that are relatively clean of oil, excess oxide, and scale.
EXXT-10	These self-shielding electrodes have small droplet transfer. They are used for high speed, single-pass welds on any thickness material in the flat, horizontal, and 20° inclined vertical positions.
EXXT-11	These self-shielding electrodes have a smooth spray-type transfer and are general-purpose electrodes for all positions. Their use on material thicknesses greater than 3⁄4" is not generally recommended unless preheat and interpass temperatures are maintained. Consult the manufacturer for specific recommendations.

206T17A.EPS

Table 17 Carbon Steel Flux Cored Electrode Classes and Uses (2 of 2)

AWS Classification	Intended Uses
EXXT-12 EXXT-12M	These electrodes are essentially the same as –1 and –1M electrodes except they are modified to improve impact toughness and to lower manganese requirements of the A-1 Analysis Group in the *ASME Boiler and Pressure Code* Section IX. As a result, tensile strength and hardness will be decreased from –1 and –1 electrodes. Users should check hardness when a required hardness level is specified or required.
EXXT-13	These self-shielding electrodes are usually welded with a short circuiting transfer. The slag system allows an all-position root pass on pipe welds. They are recommended only for the root pass and are not generally recommended for multiple passes.
EXXT-14	These self-shielding electrodes have a smooth spray-type transfer. The slag system allows high speed welds in all positions. They are used to make welds on sheet metal up to 3⁄16" thick and are often designed especially for galvanized, aluminized or other coated steels. They are generally not recommended for T or lap joints in materials thicker than 3⁄16" or butt, edge, or corner joints in materials thicker than 1⁄4". The manufacturer should be consulted for specific recommendations.
EXXT-G	These are multipass electrodes not covered by other classifications. Except for requirements to assure carbon steel deposits and tensile strengths, other requirements are agreed upon by the purchaser and manufacturer.
EXXT-GS	These are single-pass electrodes not covered by other classifications. Except for tensile strength, other requirements are agreed upon by the purchaser and the manufacturer.

206T17B.EPS

Table 18 Low-Alloy Steel Flux Cored Electrode Welding Position, Shielding, Polarity, and Application Requirements

AWS Classification[a]	Position[b,c] of Welding	External Shielding[d]	Polarity[e]	Application[f]
EX0T1-X	H, F	CO_2	DCEP	M
EX0T1-XM	H, F	75-80% Ar/bal CO_2	DCEP	M
EX1T1-X	H, F, VU, OH	CO_2	DCEP	M
EX1T1-XM	H, F, VU, OH	75-80% Ar/bal CO_2	DCEP	M
EX0T4-X	H, F	None	DCEP	M
EX0T5-X	H, F	CO_2	DCEP	M
EX0T5-XM	H, F	75-80% Ar/bal CO_2	DCEP	M
EX1T5-X	H, F, VU, OH	CO_2	DCEP or DCEN[g]	M
EX1T5-XM	H, F, VU, OH	75-80% Ar/bal CO_2	DCEP or DCEN[g]	M
EX0T6-X	H, F	None	DCEP	M
EX0T7-X	H, F	None	DCEN	M
EX1T7-X	H, F, VU, OH	None	DCEN	M
EX0T8-X	H, F	None	DCEN	M
EX1T8-X	H, F, VU or VD, OH[h]	None	DCEN	M
EXXT1-K9	VU, H, F, OH	CO_2	DCEP	M
EXXT1-K9M	VU, H, F, OH	75-80% Ar/bal CO_2	DCEP	M
EX0T11-X	H, F	None	DCEN	M
EX1T11-X	H, F, VD, OH	None	DCEN	M
EX0TG-X	H, F	—	Not Specified	—
EX1TG-X	H, F, VU or VD, OH	—	Not Specified	—

Notes:

a. The "X" indicates the tensile strength and chemical composition.

b. H = horizontal position; F = flat position; OH = overhead position; VD = vertical position with downward progression; VU = vertical position with upward progression

c. Electrode sizes suitable for welding in all positions usually are those sizes that are smaller than the 3/32" size. For that reason, electrodes meeting the requirements for the groove weld tests and the fillet weld tests may be classified as EX1TX-X or EX1TX-XM (where X represents the tensile strength and usability designator) regardless of their size.

d. Properties of weld metal from electrodes that are used with external gas shielding (EXXT1-X, EXXT1-XM, EXXT5-X, and EXXT5-XM) vary according to the shielding gas employed. Electrodes classified with the specified shielding gas should not be used with other shielding gases without first consulting the manufacturer of the electrode.

e. The term DCEP refers to direct current electrode positive (dc, reverse polarity). The term DCEN refers to direct current electrode negative (dc, straight polarity).

f. M – single and multipass

g. Some EX1T5-X and EX1T5-XM electrodes may be recommended for use on DCEN for improved out-of-position welding. Consult the manufacturer for the recommended polarity.

h. Per manufacturer's recommendations.

206T18.EPS

Table 19 AWS Low-Alloy Steel Flux Cored Electrode Classifications and Uses (1 of 3)

AWS Classification	Intended Uses
EXXT1-X EXXT1-XM	Electrodes designated as –X use CO_2 shielding. However, other gas mixtures can be used to improve arc stability if recommended by the manufacturer. Electrodes designated as –XM use an argon-CO_2 mixture to increase manganese and silicon in the weld metal to increase yield and tensile strengths. Reducing argon in a gas mixture or using CO_2 shielding only with –XM electrodes may cause deterioration of arc and out-of-position welding characteristics and loss of strength. Both electrodes have spray transfer characteristics and low spatter. Most electrodes of this class have rutile-base slag and produce high deposition rates.
EXXT4-X	These self-shielding electrodes use globular-type transfer and are designed for very high deposition rates. They produce low-sulfur welds that resist hot cracking. They are also designed for low penetration beyond the root of the weld because of poor joint fitup.
EXXT5-X EXXT5-XM	These electrodes have a globular transfer and a lime-fluoride base slag that produces welds that have better impact strength and hot and/or cold cracking resistance than other similar classes with rutile based slags. The –X electrodes are used with a CO_2 shielding gas, however, argon-CO_2 mixes can be used to reduce spatter if recommended by the manufacturer. Reducing argon in a gas mixture or using CO_2 shielding only with –5M electrodes may cause deterioration of arc and out-of-position welding characteristics and loss of strength. Operator appeal for both types of electrodes is not as good as other similar classes with rutile slags.
EXXT6-X	These self-shielding electrodes use spray-type transfer. The slag system is designed for good penetration into the root of the weld and excellent slag removal, even in deep grooves.
EXXT7-X	These self-shielding electrodes use a droplet/spray-type transfer and produce a low-sulfur weld that is very resistant to cracking. The slag system is designed to produce high deposition rates for larger size electrodes in the flat or horizontal positions.
EXXT8-X	These self-shielded electrodes have small droplet/spray-type transfer. They are suitable for all positions. The weld metal has good toughness, low temperature performance, and crack resistance.
EXXT11-X	These self-shielding electrodes have a smooth spray-type transfer and are general-purpose electrodes for all positions. Consult the manufacturer for any material thickness limitations or other specific recommendations.
EXXTX-G EXXTG-X EXXTG-G	These are multipass electrodes not covered by other classifications. The mechanical properties can be anything within this specification. Requirements are established by the digits chosen to complete the classification. Placement of the G designates that the alloy requirements, shielding gas/slag system, or both are not defined and are agreed upon by the purchaser and manufacturer.
EXXTX-AL	These electrodes are similar to the E7XT-X carbon-steel electrodes, except that ½ percent molybdenum has been added to increase the strength of the weld metal at elevated temperatures and provide an increase in corrosion resistance. They are commonly used in the fabrication and erection of boilers and pressure vessels. Typical applications include the welding of C-Mo steel base metals such as ASTM A161, A204, and A302 Gr. A plate and A335-PI pipe.

206T19A.EPS

Table 19 AWS Low-Alloy Steel Flux Cored Electrode Classifications and Uses (2 of 3)

AWS Classification	Intended Uses
EXXTX-BX EXXTX-BXL EXXTX-BXH	These electrodes produce welds that contain between ½ percent and 9 percent chromium, and between ½ percent to 1 percent molybdenum. They are designed to produce welds for high-temperature service and that match the properties of typical base metals as follows: EXXTX-B1 ASTM A335-P2 pipe ASTM A387 Gr. 2 plate EXXTX-B2 ASTM A335-P11 pipe ASTM A387 Gr. 11 plate EXXTX-B2L Thin-wall A335-P11 pipe or tube for use in the as-welded condition or for applications where low hardness is a primary concern. EXXTX-B3 ASTM A335-P22 pipe ASTM A387 Gr. 22 plate EXXTX-B3L Thin-wall ASTM A335-P22 pipe for use in the as-welded condition or for applications where lower hardness is of primary concern. EXXTX-B6 ASTM A213-TS tube ASTM A335-P5 pipe EXXTX-B8 ASTM A213-M tube ASTM A335-P9 pipe For two of these Cr-Mo electrodes, low-carbon EXXTX-BXL classifications have been established. While regular Cr-Mo electrodes produce welds with 0.05 percent to 0.12 percent carbon, the L-Grades are limited to a maximum of 0.05 percent carbon. While the lower carbon in the weld will improve ductility and lower hardness, it will also reduce the high-temperature strength and creep resistance of the weld metal. Several of these grades also have had high-carbon grades (EXXTX-BXH) established. In these cases, the electrode produces welds with 0.10 percent to 0.15 percent carbon, which may be required for high-temperature strength in some applications. Since all Cr-Mo electrodes produce weld metal that will harden in still air, both preheat and PWHT are required for most applications. No minimum notch toughness requirements have been established for any of the Cr-Mo electrode classifications. While it is possible to obtain Cr-Mo electrodes with minimum toughness values at ambient temperatures down to 32°F, specific values and testing must be agreed to by the manufacturer and the purchaser
EXXTX-DX	These electrodes produce welds that contain about 1½% to 2% manganese and between ⅓% and ⅔% molybdenum. This weld metal provides higher strength and better notch toughness than EXXTX-A1 and EXXTX-KX electrodes. However, the weld metal from these electrodes is quite air-hardenable and usually requires preheat and PWHT. The individual electrodes classified under this electrode group have been designed to match the mechanical properties and corrosion resistance of the high-strength, low-alloy pressure vessel steels, such as ASTM A302 Gr. B and HSLA steels and manganese molybdenum castings such as ASTM A49, A291 and A735.
EXXTX-K(X) (Listed below)	This group of electrodes (K1 through K9) produces weld metal of several different chemical compositions. These electrodes are primarily intended for as-welded applications.
EXXTX-K1	These electrodes produce welds with nominally 1% nickel and ½% molybdenum. They can be used for long-term stress-relieved applications or for welding low-alloy, high-strength steels, in particular 1% nickel steels.

206T19B.EPS

Table 19 AWS Low-Alloy Steel Flux Cored Electrode Classifications and Uses (3 of 3)

AWS Classification	Intended Uses
EXXTX-K2	These electrodes produce welds that have a chemical composition of 1½% nickel and up to 0.35% molybdenum. They are used on many high-strength applications ranging from 80 to 110 ksi minimum yield strength. Typical applications include the welding of submarines, aircraft carriers, and many structural applications where excellent low-temperature toughness is required. Steels welded include HY-80, HY-100, ASTM A710, A514, and other similar high-strength steels.
EXXTX-K3	Electrodes of this type produce weld deposits with higher levels of Mn, Ni, and Mo than the EXXTX-K2 types and are usually higher strength than the –K1 and -K2 types. Typical applications include the welding of HY-100 and A514 steels.
EXXTX-K4	Electrodes of this type are similar to that of the -K3 electrodes, with the addition of approximately 0.5% chromium. This provides the higher weld strength needed for many applications needing in excess of 120,000 psi tensile, such as armor plate.
EXXTX-K5	These electrodes produce welds designed to match the mechanical properties of steels such as SAE 4130 and 8630 after the weldment is quenched and tempered. They are only required to meet as-welded mechanical properties; therefore, the user is encouraged to perform qualification testing.
EXXTX-K6	These electrodes produce welds that use less than 1% nickel to achieve toughness in the 60,000 and 70,000 tensile-strength ranges. Applications include structural applications, offshore construction, and circumferential pipe welding.
EXXTX-K7	These electrodes produce welds that have similarities to welds produced with EXXTX-Ni2 and EXXTX-Ni3 electrodes. This weld metal has approximately 1½% manganese and 2½% nickel.
EXXTX-KG	This classification was designed for electrodes intended for use in circumferential girth welding of line pipe. The weld contains approximately 1½% manganese, 1% nickel, and small amounts of other alloys. It is especially intended for use on API 5LX80 pipe steels.
EXXTI-K9	This electrode produces welds similar to those of the -K2 and -K3 type electrodes, but is intended to be similar to the military requirements of MIL-101TM and 101TC electrodes in MIL-E-24403/2C. The electrode is designed for welding HY-80 steel.
EXXTX-NiX	These electrodes have been designed to produce welds with increased strength without being air-hardenable or with increased toughness at temperatures as low as –100°F. They have nickel content that fall into three nominal levels of 1, 2¼, and 3¼% Ni in steel. With carbon levels of up to 0.12%, strength increases and permits some of these Ni-steel electrodes to be classified as E8XTX-NiX and E9XTX-NiX. However, some classifications may produce low-temperature toughness to match the base-metal properties of nickel steels, such as ASTM A203 Gr. A, ASTM A352 Grade LC1 and LC2. Typical base metals would also include ASTM A302, A572, A575, and A734. Electrodes of the EXXTX-Ni(X) type are often used in structural applications where excellent toughness (Charpy V-notch or CTOD) is required. Many low-alloy steels require PWHT heat treatment to stress relieve the weld or temper the weld metal and heat-affected zone (HAZ) to achieve increased ductility. However, for many applications, nickel-steel weld metal can be used without PWHT. If PWHT is specified for a nickel-steel weldment, the holding temperature should not exceed the maximum temperature for the classification considered, since nickel steels can be embrittled at higher temperatures.
EXXTX-WX	These electrodes have been designed to produce welds that match the corrosion resistance and the coloring of the ASTM weathering-type structural steels. These special properties are achieved by the addition of about ½% copper to the weld metal. To achieve strength, ductility, and toughness in the weld metal, some chromium and nickel additions are also made. These electrodes are used to weld typical weathering steel, such as ASTM A242 and A588.
EXXTX-G	These electrodes may be either modifications of other electrodes or totally new classifications. The purchaser and user should determine the description and intended use of the electrode from the manufacturer.

206T19C.EPS

Table 20 Stainless Steel Flux Cored Electrode and Rod Welding Process, Shielding, and Polarity Requirements

AWS Classification[a]	External Shielding Medium[b]	Welding Polarity	Welding Process
EXXXTX-1	CO_2	DCEP	FCAW
EXXXTX-3	none (self shielded)	DCEP	FCAW
EXXXTX-4	70-80% Ar/bal CO_2	DCEP	FCAW
RXXXT1-5	100% Argon	DCEN	GTAW
EXXXTX-G	Not Specified	Not Specified	FCAW
RXXXT1-G	Not Specified	Not Specified	GTAW

Notes:

a. The letters XXX stand for the designation of the chemical composition. The X after the T designates the position of operation. A 0 indicates flat or horizontal operations; a 1 indicates all position operation.

b. The requirement for the use of specified external shielding medium shall not be construed to restrict the use of any other medium for which the electrodes are found suitable, for any application other than the classification tests.

206T20.EPS

Table 21 AWS Stainless Steel Alloy Flux Cored Electrode Rod Classifications and Uses (1 of 4)

AWS Classification	Intended Uses
E307TX-X	These electrodes are used primarily for moderate strength welds with good crack resistance between dissimilar steels. This includes welding austenitic manganese steel to carbon steel forgings or castings.
E308TX-X	These electrodes are most often used to weld base metal of similar composition including AISI Types 301, 302, 304, 305, and 308.
E308LTX-X	The weld metal of these electrodes is the same as that of E308TX-X, except for lower (L) carbon content. By specifying low carbon in this alloy, it is possible to obtain resistance to intergranular corrosion due to carbide precipitation without the use of stabilizers such as columbium (niobium) or titanium. This low-carbon alloy, however, is not as strong at elevated temperature as the E308 and columbium (niobium)-stabilized alloys.
E308HTX-X	The weld metal of these electrodes is the same as that of E308TX-X except for higher carbon content. Higher carbon content provides higher tensile and creep strength at elevated temperatures. These electrodes are used primarily for welding type 304H base metal.
E308MoTX-X	The weld metal of these electrodes is the same as that of E308TX-X weld metal, except for the addition of molybdenum. These electrodes are recommended for welding ASTM CF8M stainless steel castings because they match the base metal with regard to chromium, nickel, and molybdenum. These electrodes may also be used for welding wrought metals such as Type 316L when a ferrite content higher than that attainable with E316LTX-X electrodes is desired.
E308HmoT0-3	The weld metal of these electrodes is the same as that of E308MoTX-X, except for higher carbon content. The higher carbon content provides higher strength at elevated temperatures.
E308LMoTX-X	The weld metal of these electrodes is the same as that of E308MoTX-X weld metal, except for lower carbon content. These electrodes are recommended for welding ASTM CF3M stainless steel castings to match the base metal with regard to chromium, nickel, and molybdenum. These electrodes also may be used for welding wrought metals such as Type 316L stainless when ferrite content higher than that attainable with E316LTX-X electrodes is desired.
E309TX-X	These electrodes are commonly used for welding similar alloys in wrought or cast forms. They are used in welding dissimilar metals, including joining Type 304 to mild steel, welding the stainless steel side of Type 304 clad steels, and applying stainless steel sheet linings to carbon steel sheets. Occasionally, they are used to weld Type 304 base metals where severe corrosion conditions exist that require higher alloy content weld metal.

206T21A.EPS

Table 21 AWS Stainless Steel Alloy Flux Cored Electrode Rod Classifications and Uses (2 of 4)

AWS Classification	Intended Uses
E309LTX-X	The weld metal of these electrodes is the same as E309TX-X, except for the lower carbon content. By specifying low carbon in this alloy, it is possible to obtain resistance to intergranular corrosion due to carbide precipitation without the use of stabilizers such as columbium (niobium) or titanium. This low carbon alloy is not as strong at elevated temperature as Type 309 or the columbium-stabilized (niobium) modification. A primary application of this alloy is the first layer cladding of carbon steel if no columbium (niobium) additions are required.
E309MoTX-X	The weld metal of these electrodes is the same as that of E309TX-X weld metal, except for the addition of molybdenum. These electrodes are used to join stainless steel to carbon and low-alloy steels for service below 600°F and for overlaying of carbon and low-alloy steels. The presence of molybdenum provides pitting resistance in a halide environment and helps provide high temperature ductility in dissimilar joints.
E309LMoTX-X	These electrodes are the same as E309MoTX-X except for lower carbon content. They are used to join stainless steel to carbon and low-alloy steels for service below 600°F and for overlaying of carbon and low-alloy steels. The presence of molybdenum provides pitting resistance in a halide environment and helps provide high temperature ductility in dissimilar joints.
E309LNMoTX-X	These electrodes are essentially the same as E309LMoTX-X except for lower chromium and ferrite content and higher nickel content. The purpose of this modification is to achieve a lower deposit ferrite content when compared to E309LMoTX-X. This chemistry is required by the pulp and paper industry for joining applications. The lower ferrite content leads to better corrosion resistance due to a decreased potential for chromium-nitride precipitation.
E309LCbTX-X	The weld metal of these electrodes is the same as E309LTX-X weld metal except for the addition of columbium (niobium). These electrodes are used to overlay carbon and low-alloy steels and produce a columbium (niobium)-stabilized first layer on such overlays.
E310TX-X	These electrodes are most often used to weld base metals of similar compositions.
E312TX-X	These electrodes most often are used to weld dissimilar metals that are high in nickel. This alloy gives a two-phase weld deposit with substantial amounts of ferrite in an austenitic matrix. Even with considerable dilution by austenite-forming elements, such as nickel, the microstructure remains two-phase and thus highly resistant to weld metal cracks and fissures.
E316TX-X	These electrodes are used for welding similar base metal alloys. They have been used successfully in applications involving special alloys for high-temperature service. The presence of molybdenum provides increased creep resistance at elevated temperatures and pitting resistance in a halide environment.
E316LTX-X	The weld metal of these electrodes is the same as E316TX-X electrodes except for lower carbon content. By specifying low carbon in this alloy, it is possible to obtain resistance to intergranular corrosion due to carbide precipitation without the use of stabilizers such as columbium (niobium) or titanium. However, this low-carbon alloy is not as strong at elevated temperatures as the columbium (niobium)-stabilized alloys.

206T21B.EPS

Table 21 AWS Stainless Steel Alloy Flux Cored Electrode Rod Classifications and Uses (3 of 4)

AWS Classification	Intended Uses
E316LKTO-3	The weld metal of these electrodes is the same as E316LTX-X. These electrodes, however, are self-shielding and are used primarily for welding stainless steels for cryogenic service as designated by the K in the classification. Although the nominal chromium, nickel, and molybdenum content of E316LKTO-3 filler metal is essentially the same as the other E316 grades, special attention is given to it in order to maximize low-temperature toughness. Minimizing the content of carbon and nitrogen improves the toughness. Low nitrogen content is achieved by providing a more efficient slag system than is employed with EXXXTO-3 self-shielding electrodes. Delta ferrite in the weld metal has an adverse effect on toughness; therefore, the chemical composition of the weld metal is balanced to provide a low maximum ferrite content. Fully austenitic weld metal is preferred from a toughness standpoint; however, it is recognized that the tendency for weld metal fissuring is greater in fully austenitic weld metals.
E317LTX-X	These electrodes usually are used for welding alloys of similar composition and are usually limited to severe corrosion applications. Low carbon in this filler metal reduces the possibility of intergranular carbide precipitation and thereby increases the resistance to intergranular corrosion without the use of stabilizers such as columbium (niobium) or titanium. This low-carbon alloy, however, may not be as strong at elevated temperatures as the columbium-stabilized alloys or Type 317.
E347TX-X	The alloy is often referred to as a columbium (niobium)-stabilized Type 308 alloy, indicating that it normally is not subject to intergranular corrosion from carbide precipitation. These electrodes are usually used for welding chromium-nickel steel base metals of similar composition that are stabilized either with columbium (niobium) or titanium. Although columbium (niobium) is the stabilizing element usually specified in 347 alloys, it should be recognized that tantalum also might be present. Tantalum and columbium (niobium) are almost equally effective in stabilizing carbon and in providing high-temperature strength. The usual commercial practice is to report columbium (niobium) as the sum of the columbium plus tantalum. Crack sensitivity of the weld may increase substantially, if dilution by the base metal produces a low ferrite or fully austenitic weld metal deposit.
E409TX-X	These electrodes, which produce a ferritic microstructure, often are used to weld base metal of similar composition.
E410TX-X	The alloy of these electrodes is an air-hardening steel, and therefore, requires preheat and postheat treatments in order to achieve welds of adequate ductility for most purposes. The most common application of these electrodes is for welding alloys of similar composition. They also are used for surfacing carbon steels to resist corrosion, erosion, or abrasion, such as occurs in valve seats and other valve parts.
E410NiMoTX-X	These electrodes generally are used to weld ASTM CA6NM castings or similar materials. The electrodes are modified to contain less chromium and more nickel to eliminate ferrite in the microstructure because ferrite has a bad effect on mechanical properties. Postweld heat treatment in excess of 1150°F may result in rehardening due to untempered martensite in the microstructure after cooling to room temperature.
E410NiTiTX-X	These electrodes contain a titanium stabilizer. They generally are used to weld similar base metals.

206T21C.EPS

Table 21 AWS Stainless Steel Alloy Flux Cored Electrode Rod Classifications and Uses (4 of 4)

AWS Classification	Intended Uses
E430TX-X	The weld metal of these electrodes is balanced by providing sufficient chromium to give adequate corrosion resistance for the usual applications and yet retain sufficient ductility in the heat-treated condition. (Excessive chromium will result in lower ductility.) Welding with these electrodes may produce a partially hardened microstructure that requires preheating and postweld heat treatment. Optimum mechanical properties and corrosion resistance are obtained only when the weldment is heat treated following the welding operation.
E502TX-X	These electrodes are used for welding base metal of similar composition, usually in the form of a pipe or tube. Because this alloy is air-hardening, preheating and postweld heat treatment is strongly recommended.
E505TX-X	These electrodes are used for welding base metal of similar composition, usually in the form of a pipe or tube. Because the alloy is air-hardening, preheating and postweld heat treatment is strongly recommended.
E2209TX-X	These electrodes are used to join duplex stainless-steel base metals containing chromium. The microstructure of the weld deposit consists of a mixture of austenite and ferrite. Because of the two-phase structure, the alloy is one of the family of duplex stainless-steel alloys. The alloy has good resistance to stress corrosion cracking and pitting corrosion attack.
E2209TX-X	These electrodes are used to join duplex stainless-steel base metals containing approximately 22 wt.% chromium. The microstructure of the weld deposit consists of a mixture of austenite and ferrite. Because of the two-phase structure, the alloy is one of the family of duplex stainless-steel alloys. The alloy has good resistance to stress corrosion cracking and pitting corrosion attack.

206T21D.EPS

Additional Resources

This module is intended to be a thorough resource for task training. The following reference works are suggested for further study. These are optional materials for continued education rather than for task training.

Gas Metal Arc Welding Handbook, 2000. William H. Minnick. Tinley Park, IL: The Goodheart-Willcox Co. Inc.

Modern Welding Technology. Howard B. Cary. Englewood Cliffs, NJ: Prentice Hall, Inc.

Protection from Step and Touch Potentials – 1910.269 App C, (Standards – 29 CFR), Current Edition. Washington, DC: Occupational Safety and Health Administration (OSHA).

The Procedure Handbook of Arc Welding, Current Edition. Cleveland, OH: The Lincoln Electric Company.

Welding Skills. R.T. Miller. Homewood, IL: American Technical Publishers.

Figure Credits

Miller Electric Mfg. Co.	206F04, 206F05, 206F06, 206F07, 206F08, 206F23, 206F24, 206F25, 206F27, 206F28, 206F29, 206F30, 206F31, 206F46, 206SA01, 206SA03, 206SA04
Charles Rogers	206F21, 206F22 (upper), 206F48
Gerald Shannon	206F22 (middle), 206SA02, 206F47, 206F48 (pliers)
Sumner Manufacturing	206F22 (lower)
Smith Equipment Co.	206F32
Rex Ball	206SA05

Figure 19 courtesy of American Welding Society

AWS Committee on Arc Welding and Cutting, 1994, *Recommended Practices for Gas Metal Arc Welding*, ANSI/AWS C5.6-94, Miami: American Welding Society, Figure 6.

Figure 34 courtesy of American Welding Society

AWS A5 Committee on Filler Metals and Allied Materials, 2001, *Specification for Carbon Steel Electrodes and Rods for Gas Shielded Arc Welding*, AWS A5.18/A5.18M: 2001, Miami: American Welding Society, Figure A1.

Figure 42 courtesy of American Welding Society

Committee on Filler Metals, 1995, *Specification for Carbon Steel Electrodes for Flux Cored Arc Welding*, Miami: American Welding Society, Figure A1.

Figure 43 courtesy of American Welding Society

AWS Committee on Filler Metals, 1998, *Specification for Low-Alloy Steel Electrodes for Flux Cored Arc Welding*, ANSI/AWS A5.29:1998, Miami: American Welding Society, Figure A1.

Table 3 courtesy of American Welding Society

AWS Committee on Filler Metals, 1997, *Specification for Welding Shielding Gases*, ANSI/AWS A5.32/A5.32M-97, Miami: American Welding Society, Table 4.

Table 4 courtesy of American Welding Society

Myers, L.W., ed., 1997, *The Everyday Pocket Handbook for Gas Metal Arc Welding (GMAW) and Flux Cored Arc Welding (FCAW)*, Miami: American Welding Society, p. 9.

Table 5 courtesy of American Welding Society

Myers, L.W., ed., 1997, *The Everyday Pocket Handbook for Gas Metal Arc Welding (GMAW) and Flux Cored Arc Welding (FCAW)*, Miami: American Welding Society, p. 7-8.

Table 7 courtesy of American Welding Society

Myers, L.W., ed., 1997, *The Everyday Pocket Handbook for Gas Metal Arc Welding (GMAW) and Flux Cored Arc Welding (FCAW)*, Miami: American Welding Society, p. 6.

Table 11 courtesy of American Welding Society

AWS Committee on Filler Metals, 1999, *Specification for Bare Aluminum and Aluminum Alloy Welding Electrodes and Rods*, AWS A5.10/A5.10M:1999, Miami: American Welding Society, Table A2.

Table 14 courtesy of American Welding Society

AWS Committee on Filler Metal, 1992, *Specification for Magnesium Alloy Welding Electrodes and Rods*, ANSI/AWS A5.19-92, Miami: American Welding Society, Table A1.

Table 16 courtesy of American Welding Society

AWS Committee on Filler Metal, 1995, *Specification for Carbon Steel Electrodes for Flux Cored Arc Welding*, Miami: American Welding Society, Table 2.

Table 18 courtesy of American Welding Society

AWS Committee on Filler Metals, 1998, *Specification for Low-Alloy Steel Electrodes for Flux Cored Arc Welding*, ANSI/AWS A5.29:1998, Miami: American Welding Society, Table 3.

Table 20 courtesy of American Welding Society

Adapted from AWS Committee on Filler Metal, 1995, *Specification for Stainless Steel*

CONTREN™ LEARNING SERIES — USER UPDATES

The NCCER makes every effort to keep these textbooks up-to-date and free of technical errors. We appreciate your help in this process. If you have an idea for improving this textbook, or if you find an error, a typographical mistake, or an inaccuracy in NCCER's Contren™ textbooks, please write us, using this form or a photocopy. Be sure to include the exact module number, page number, a detailed description, and the correction, if applicable. Your input will be brought to the attention of the Technical Review Committee. Thank you for your assistance.

Instructors – If you found that additional materials were necessary in order to teach this module effectively, please let us know so that we may include them in the Equipment/Materials list in the Instructor's Guide.

Write: Curriculum Revision and Development Department
National Center for Construction Education and Research
P.O. Box 141104, Gainesville, FL 32614-1104

Fax: 352-334-0932

E-mail: curriculum@nccer.org

Craft ____________________ Module Name ____________________

Copyright Date __________ Module Number __________ Page Number(s) __________

Description

____

____

____

(Optional) Correction

____

____

(Optional) Your Name and Address

____

____

Module 29207-03

GMAW and FCAW – Plate

COURSE MAP

This course map shows all of the modules in the second level of the Welding curriculum. The suggested training order begins at the bottom and proceeds up. Skill levels increase as you advance on the course map. The local Training Program Sponsor may adjust the training order.

WELDING LEVEL TWO

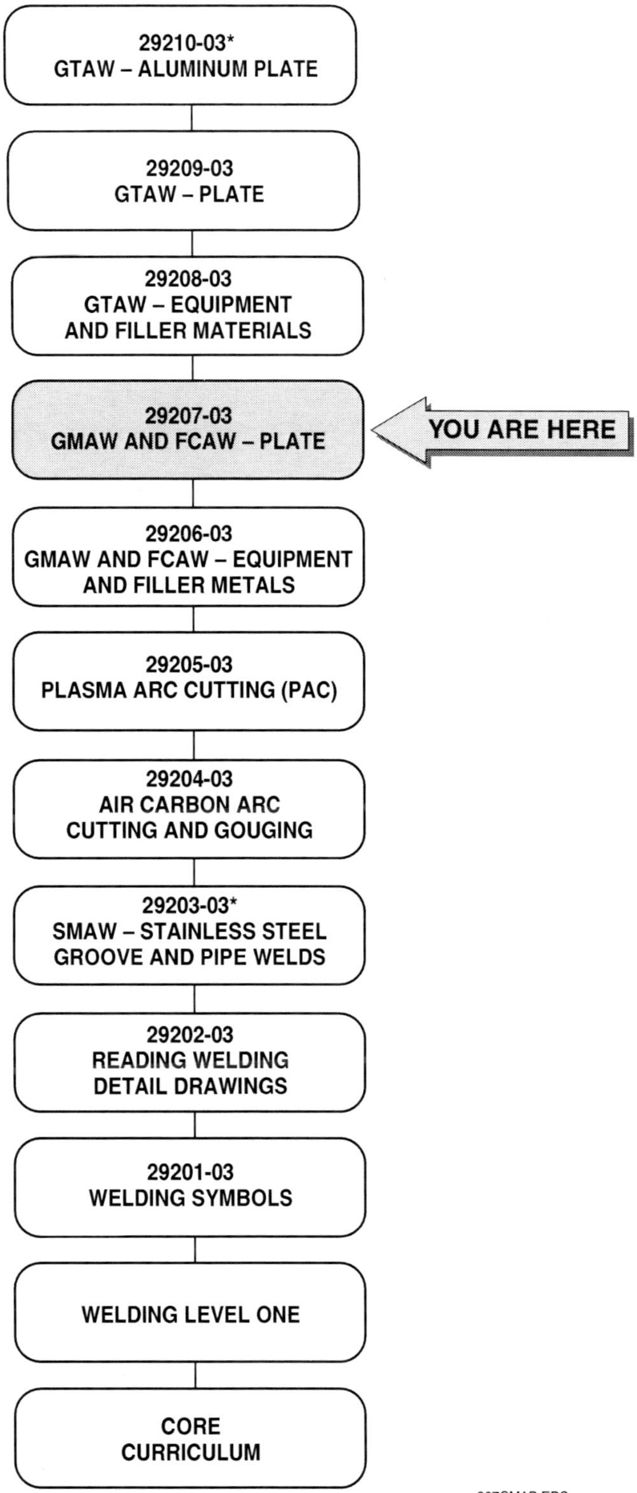

207CMAP.EPS

Please note that Modules 29203-03 and 29210-03 are electives for *Welding Level Two* but are requirements for *AWS Entry Level Welder.

MODULE 29207-03 CONTENTS

MODULE 29207-03 CONTENTS (Continued)

Figures

MODULE 29207-03

GMAW and FCAW – Plate

Objectives

When you have completed this module, you will be able to do the following:

1. Perform GMAW multipass fillet welds on plate, using solid or composite wire and shielding gas, in multiple positions.
2. Perform GMAW multipass groove welds on plate, using solid or composite wire and shielding gas, in multiple positions.
3. Perform GMAW spray fillet and groove welds on plate, using solid or composite wire and shielding gas, in flat and horizontal positions.
4. Perform FCAW multipass fillet welds on plate in multiple positions using flux cored wire and, if required, shielding gas.
5. Perform FCAW multipass groove welds on plate in multiple positions using flux cored wire and, if required, shielding gas.

Prerequisites

Before you begin this module, it is recommended that you successfully complete the following: Core Curriculum; Welding Level One; Welding Level Two, Modules 29201-03 through 29206-03.

Required Trainee Materials

1. Pencil and paper
2. Appropriate personal protective equipment

1.0.0 ◆ INTRODUCTION

This module teaches the trainee how to make fillet and open-root V-groove welds on carbon steel plate using the gas metal arc welding (GMAW) process and/or the flux cored arc welding (FCAW) process in all positions. The procedures for practicing the various welds should be performed using one of the processes and repeated using the other process, or as directed by the instructor.

1.1.0 The GMAW Process

GMAW is an arc welding process that uses a continuous, consumable solid wire or composite cored wire electrode for the filler metal and shielding gas to protect the weld zone. It is a fast and effective method of making high-quality welds. *Figure 1* shows the GMAW process. The GMAW process is commonly used to make welds on carbon steel plate as well as aluminum and other metals.

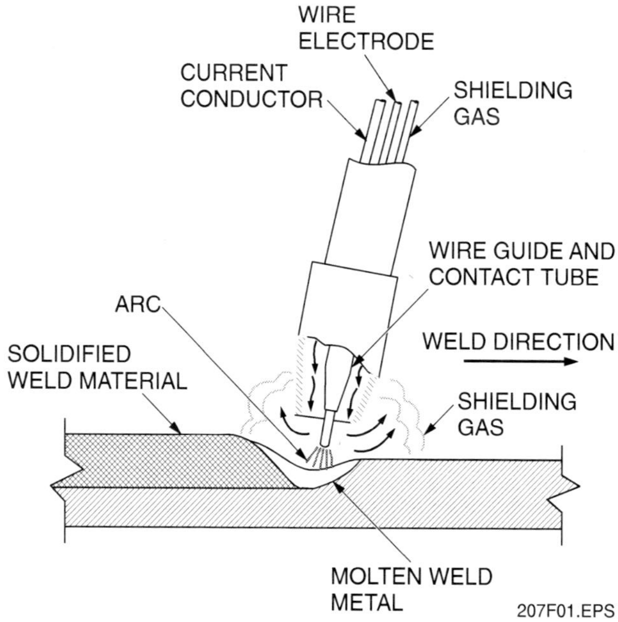

Figure 1 ◆ GMAW process.

> **NOTE**
> Composite cored wire electrodes are a subset of GMAW solid wire carbon steel electrodes as defined in *AWS Specification A5.18.*

1.2.0 The FCAW Process

FCAW is an arc welding process that uses a continuous, consumable flux cored wire as a filler metal to form a weld (see *Figure 2*). Shielding gas protects the molten weld metal from contamination by the atmosphere. The shielding gas can be generated entirely from the flux in the wire core or by both the flux in the wire core and an external shielding gas. In all types of flux cored wires, the flux contains agents for **degassing** the weld, stabilizing the arc, and forming a protective slag over the weld bead. Other elements are often added to the flux cored wire to alloy with the base metal and improve its chemical and/or mechanical characteristics. The FCAW process is basically limited to welding low-carbon steels, some low-alloy steels, and some stainless steels. FCAW is an excellent process for making fillet and groove welds on carbon steel plate.

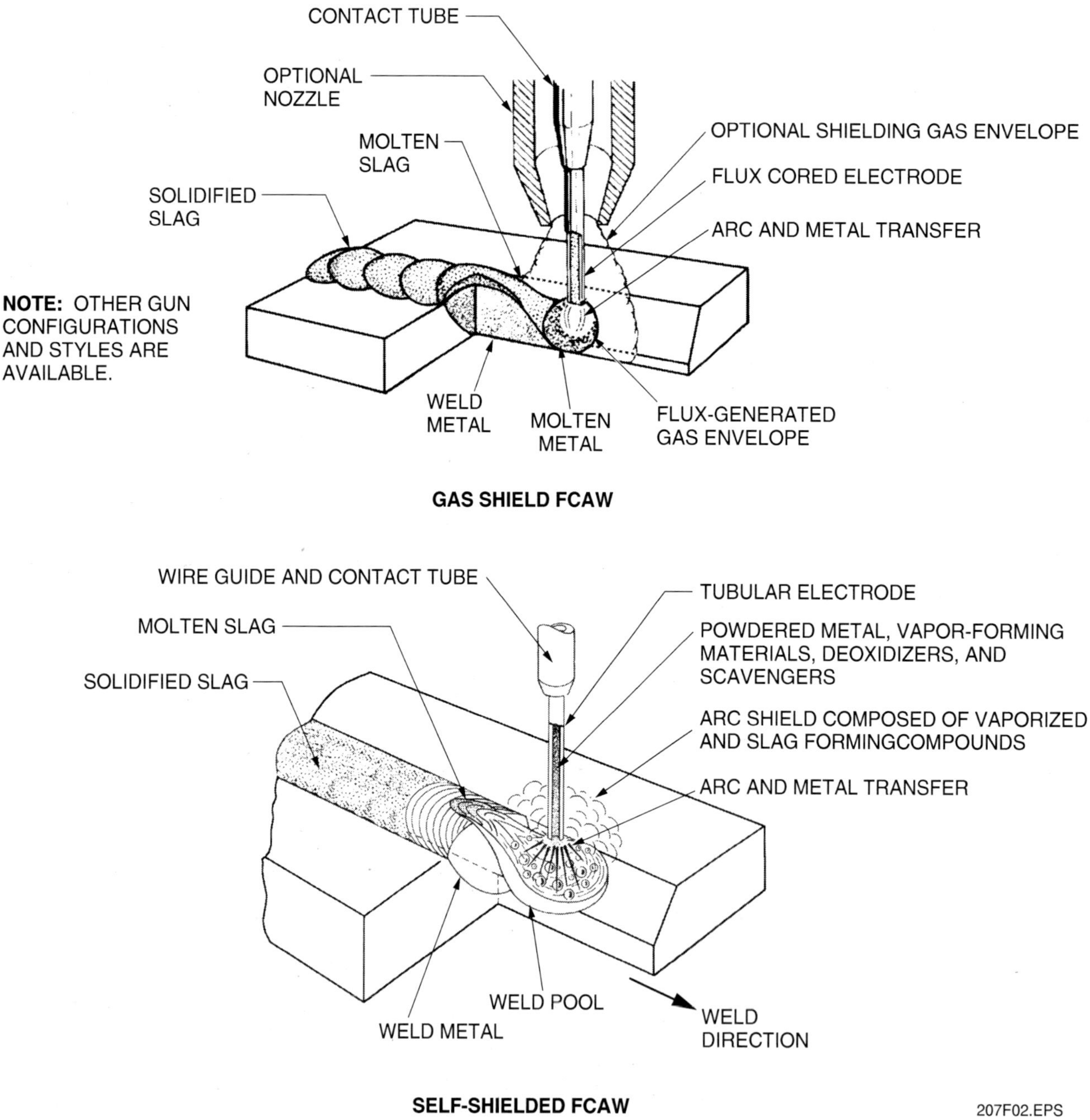

Figure 2 ◆ FCAW process.

1.3.0 GMAW and FCAW Equipment

The equipment used for gas-shielded GMAW and FCAW or self-shielded FCAW is basically the same except for the components needed for the gas used by the gas-shield process. This additional equipment includes guns equipped with a gas nozzle, gas hoses, a gas control solenoid valve, a gas regulator/flowmeter, and a shielding gas cylinder (gas source). *Figures 3* and *4* are illustrations of typical GMAW and FCAW equipment.

Figure 3 ◆ Gas-shielded GMAW equipment.

2.0.0 ◆ SAFETY SUMMARY

The following is a summary of safety procedures and practices that must be observed while cutting or welding. Keep in mind that this is just a summary. Complete safety coverage is provided in the Level One module, *Welding Safety*. If you have not completed that module, do so before continuing. Above all, be sure to wear appropriate protective clothing and equipment when welding or cutting.

2.1.0 Protective Clothing and Equipment

- Always use safety goggles with a full face shield or a helmet. The goggles, face shield, or helmet lens must have the proper light-reducing tint for the type of welding or cutting to be performed. Never directly or indirectly view an electric arc without using a properly tinted lens.

Wind Resistance

Self-shielded FCAW is particularly useful in situations with high winds, because the shielding gas is created inside the welding arc as the flux vaporizes. The molten puddle has a constant supply of shield gas displacing the air from the arc outward and is immediately covered by the residual slag. An external shield gas, in contrast, attempts to force gas down from the nozzle to displace the air and can be blown away from the weld puddle by strong winds.

207SA01.EPS

Figure 4 ◆ Self-shielded FCAW equipment.

- Wear proper protective leather and/or flame-retardant clothing along with welding gloves that will protect you from flying sparks and molten metal as well as heat.
- Wear 8" or taller high-top safety shoes or boots. Make sure that the tongue and lace area of the footwear will be covered by a pant leg. If the tongue and lace area is exposed or the footwear must be protected from burn marks, wear leather spats under the pants or chaps and over the front top of the footwear.
- Wear a solid material (non-mesh) hat with a bill pointing to the rear or, if much overhead cutting or welding is required, a full leather hood with a welding face plate and the correct tinted lens. If a hard hat is required, use a hard hat that allows the attachment of rear deflector material and a face shield.
- If a full leather hood is not worn, wear a face shield and snug-fitting welding goggles over safety glasses for gas welding or cutting. Either the face shield or the lenses of the welding goggles must be an approved shade 5 or 6 filter. For electric arc welding or cutting, wear safety goggles or a welding hood with the correct tinted lens (shade 9 to 14).
- If a full leather hood is not worn, wear earmuffs or at least earplugs to protect your ear canals from sparks.

2.2.0 Fire/Explosion Prevention

- Never carry matches or gas-filled lighters in your pockets. Sparks can cause the matches to ignite or the lighter to explode, causing serious injury.
- Never perform any type of heating, cutting, or welding until a hot-work permit is obtained and an approved fire watch is established. Most work-site fires caused by these types of operations are started by cutting torches.
- Never use oxygen to blow off clothing. The oxygen can remain trapped in the fabric for a time. If a spark hits the clothing during this time, the clothing can burn rapidly and violently out of control.
- Make sure that any flammable material in the work area is moved or shielded by a fire-resistant covering. Approved fire extinguishers must be available before attempting any heating, welding, or cutting operations.
- Never release a large amount of oxygen or use oxygen as compressed air. Its presence around flammable materials or sparks can cause rapid and uncontrolled combustion. Keep oxygen away from oil, grease, and other petroleum products.
- Never release a large amount of fuel gas, especially acetylene. Methane and propane tend to concentrate in and along low areas and can ignite at a considerable distance from the release point. Acetylene is lighter than air but is even more dangerous than methane. When mixed with air or oxygen, it will explode at much lower concentrations than any other fuel.
- To prevent fires, maintain a neat and clean work area and make sure that any metal scrap or slag is cold before disposal.
- Before cutting or welding containers such as tanks or barrels, check to see if they have contained any explosive, hazardous, or flammable materials including petroleum products, citrus products, or chemicals that decompose into toxic fumes when heated. As a standard practice, always clean and then fill any tanks or barrels with water or purge them with a flow of inert gas to displace any oxygen.

2.3.0 Work Area Ventilation

- Make sure confined space procedures are followed before conducting any welding or cutting in the confined space.
- Never use oxygen in confined spaces for ventilation purposes.
- Always perform cutting or welding operations in a well-ventilated area. Cutting or welding operations involving zinc or cadmium materials or coatings result in toxic fumes. For long-term cutting or welding of such materials, always

wear an approved full-face, supplied-air respirator (SAR) that uses breathing air supplied outside of the work area. For occasional, very short-term exposure, a high-efficiency particulate arresting (HEPA)-rated or metal-fume filter may be used on a standard respirator.

- Make sure confined spaces are ventilated properly for cutting or welding purposes.

3.0.0 ◆ WELDING EQUIPMENT SETUP

Before welding can take place, the area has to be made ready, the welding equipment must be set up, and the metal to be welded must be prepared. The following sections will explain how to set up equipment for welding.

3.1.0 Preparing the Welding Area

To practice welding, a welding table, bench, or stand is needed. The welding surface must be steel, and provisions must be made for placing weld coupons out of position. *Figure 5* shows a typical welding station.

207F05.EPS

Figure 5 ◆ Welding station.

To set up the area for welding, follow these steps:

Step 1 Check to be sure the area is properly ventilated. Make use of doors, windows, and fans.

Step 2 Check the area for fire hazards. Remove any flammable materials before proceeding.

Step 3 Check the location of the nearest fire extinguisher. Do not proceed unless the extinguisher is charged and you know how to use it.

Step 4 Position a welding table near the welding machine.

Step 5 Set up flash shields around the welding area.

3.2.0 Preparing the Practice Weld Coupons

If possible, the practice weld coupons should be ⅜" thick carbon steel. If this size is not readily available, ¼" to ¾" thick steel can be used. Use a wire brush or grinder to remove heavy mill scale or corrosion. Prepare weld coupons to practice the welds indicated as follows:

- *Running beads* – The coupons can be any size or shape that can be easily handled.
- *Overlapping beads* – The coupons can be any size or shape that can be easily handled.
- *Fillet welds* – Cut the metal into 4" × 5" rectangles for the base and 3" × 5" rectangles for the web. *Figure 6* shows the weld coupons for fillet welding.

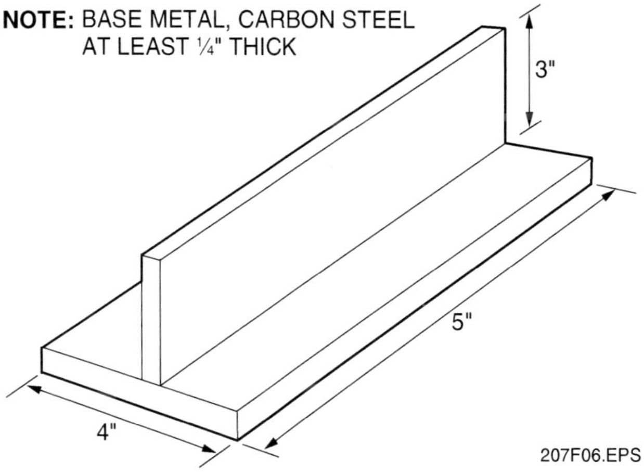

207F06.EPS

Figure 6 ◆ Fillet weld coupons.

- *Open-root V-groove welds* – As shown in *Figure 7*, cut the metal into 3" × 5" rectangles with one (or both) of the 5" lengths beveled at 22½° for stainless steel or 30° for carbon steel or low-alloy plate. Grind a 0" to ⅛" root face on the bevel as directed by your instructor.

> **NOTE**
> Check with your instructor to determine which bevel angles should be used.

Hot Tip: Machine Bevels and Root Faces for GMAW or FCAW

For V-groove joints made in thick materials, machined bevels and root faces provide the cleanest and most uniform method of preparing materials for satisfactory GMAW or FCAW welds. While SMAW, FCAW, or GTAW can tolerate some poor fitup for root passes, controlling penetration of root passes with poor fitup using GMAW is difficult.

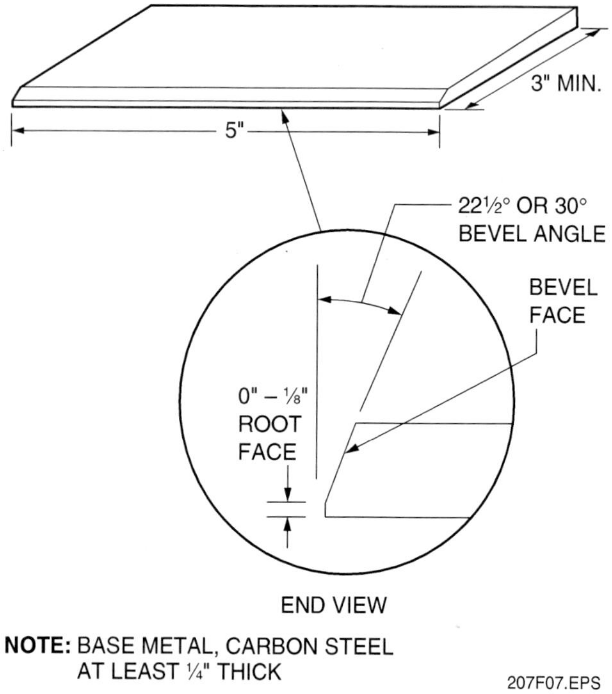

Figure 7 ◆ Metal cut for open-root V-groove weld coupons.

> **NOTE**
> The welding codes allow the root face and opening on open root welds to be 0" to ⅛". Adjust the root face and root opening as needed when you start your welding practices.

Follow these steps to prepare each V-groove weld coupon:

Step 1 Check the bevel face. There should be no dross and a 0" to ⅛" root face. The bevel angle should be 22½° or 30°. An example of this is shown in *Figure 8*.

Step 2 Center the beveled strips with a root opening determined by your instructor and tack weld them in place. Place the tack welds on the back side of the joint. Use three to four ¼" tack welds.

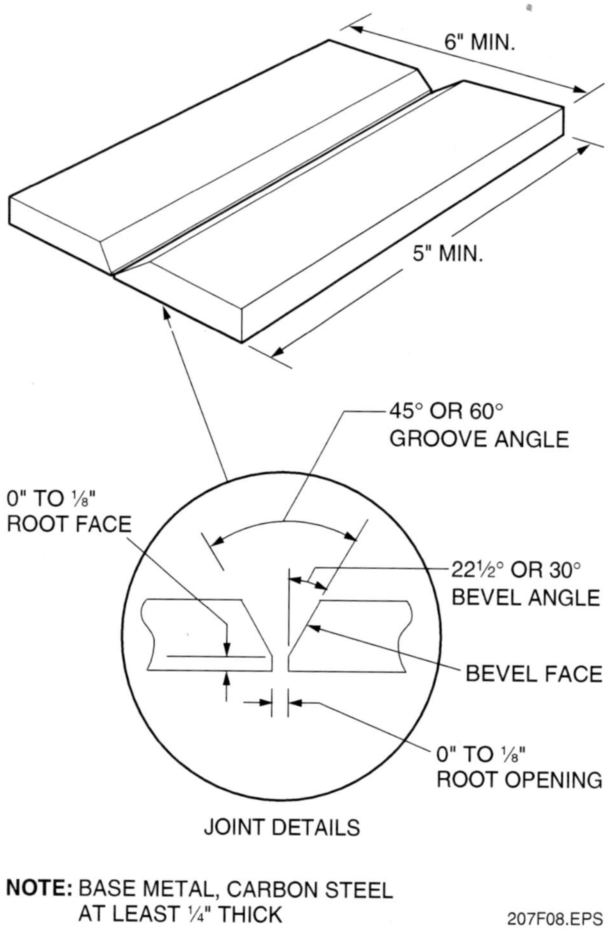

Figure 8 ◆ Open-root V-groove weld coupon.

> **NOTE**
> When tack welding open-root V-groove practice coupons together, use a piece of wire or rod of the correct diameter to space the coupons for the desired root opening.

3.3.0 Welding Machine

Identify a proper welding machine for GMAW/FCAW use and follow these steps to set it up for welding.

Hot Tip

Conserving Material

Steel for practice welding is expensive and difficult to obtain. Every effort should be made to conserve the material that is available. Reuse weld coupons until all surfaces and joints have been welded on by cutting the weld coupon apart and reusing the pieces. Use material that cannot be cut into weld coupons to practice running beads.

Step 1 Verify that the welding machine can be used for GMAW/FCAW with or without gas shielding for FCAW as required.

Step 2 Check to be sure that the welding machine is properly grounded through the primary current receptacle.

Step 3 Verify the location of the primary disconnect.

Step 4 Configure the welding machine for gas-shielded GMAW/FCAW or self-shielding FCAW as directed by your instructor (*Figure 9*). Configure the gun polarity and equip the gun with the correct liner material and contact tube for the diameter of the wire being used and the correct nozzle for the application.

Step 5 In accordance with the manufacturer's instructions, load the wire feeder and gun with a solid, composite, or flux cored electrode wire of the proper diameter as directed by your instructor.

Step 6 If required, connect the proper shielding gas for the application as described in the previous module, as specified by the wire electrode manufacturer, welding procedure specifications (WPS), site quality standards, or your instructor.

Step 7 Connect the clamp of the workpiece lead to the workpiece.

Step 8 Turn on the welding machine, and if necessary, purge the gun as directed by the gun manufacturer's instructions.

Step 9 Set the initial welding voltage, and wire feed speed for the type and size of electrode wire as recommended by the manufacturer.

3.4.0 Welding Voltage

Arc length is determined by voltage, which is set at the power source. Arc length is the distance from the wire electrode tip to the base metal or molten pool at the base metal. If voltage is set too high, the arc will be too long and could cause the wire to fuse to the tip (burnback). Too high a voltage also

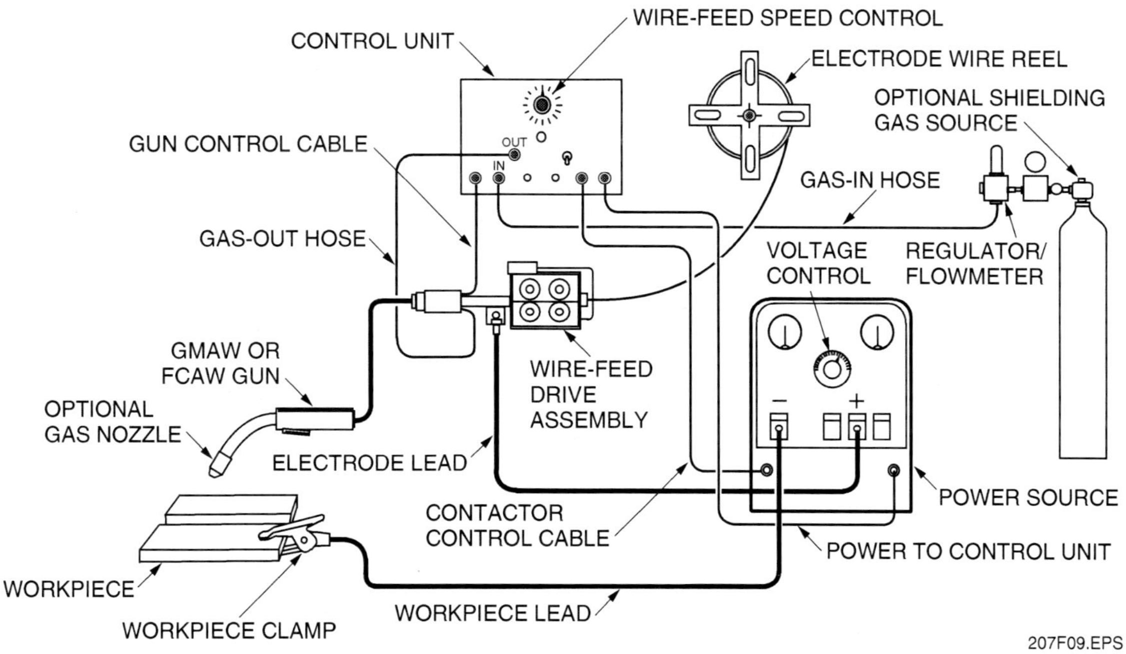

Figure 9 ◆ Configuration diagram of typical GMAW/FCAW welding machine.

Hot Tip

Wire Electrode Manufacturer's Recommendations

Always obtain and follow the manufacturer's recommendations for shielding gas (if required) and for initially setting the welding voltage, current (if variable), and wire feed speed parameters. These balanced parameters are critical and are based on the welding position, size, and composition of the solid, composite, or flux cored wired being used.

Arc Voltage

A minimum arc voltage is needed to maintain spray transfer. However, penetration is not directly related to voltage. Penetration will increase with voltage for a time, but beyond an optimum voltage, it will actually decrease with further voltage increases.

causes porosity and excessive spatter. Voltage must be increased or decreased as wire feed speed is increased or decreased. Set the voltage so that the arc length is just above the surface of the base metal. *Figure 10* shows the arc length.

3.5.0 Welding Amperage

With a standard constant-voltage power source, the electrode feed speed controls the welding amperage after the initial recommended setting. The welding power source provides the amperage necessary to melt the wire electrode while maintaining the selected welding voltage. Within limits, when the wire electrode feed speed is increased, the welding amperage and deposition rate also increase. This results in higher welding heat, deeper penetration, and higher beads. When the wire electrode feed speed is decreased, the welding amperage automatically decreases. With lower welding amperage and less heat, the deposition rate drops and the weld beads are smaller with less penetration.

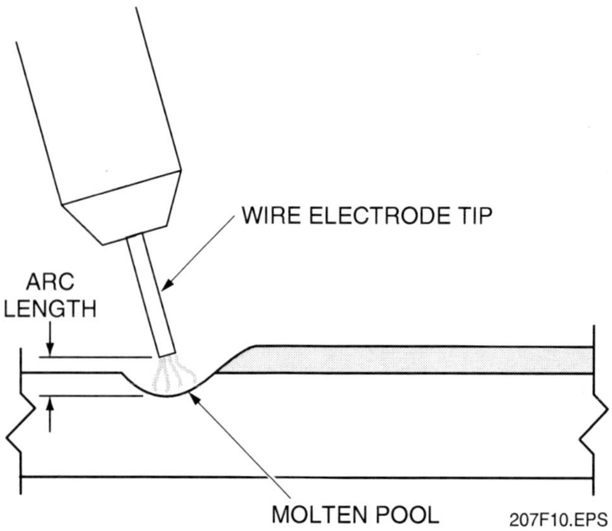

Figure 10 ◆ Arc length.

Note that some constant-voltage power sources used for GMAW/FCAW provide varying degrees of current modification, such as slope or induction adjustments. Power sources with a slope adjustment allow varying the amount of amperage change in relation to voltage range of the unit. This allows the correct short-circuiting current to be established for certain welding applications and/or wire electrode requirements. Standard constant-voltage GMAW/FCAW units have current slope that is fixed by the manufacturer for general welding applications and conditions. Other units allow adjustment of the response time for short-circuit current using an induction control. Instead of an instantaneous short-circuit current draw, the current draw can be allowed to ramp up over a varying amount of time by the use of adjustable induction in the power source circuits. This smoothes the arc characteristics during short-circuiting transfer. Pulse transfer power sources allow the adjustment of the peak current for the pulse and the background current between pulses to match specific welding applications and/or wire electrode requirements. In other cases, where a voltage-sensing wire feeder is used with a constant-current power source, the current is adjustable and the wire feeder varies the wire feed speed to maintain proper arc voltage at the wire electrode tip.

3.6.0 Welding Travel Speed

Weld travel speed is the speed of the electrode tip across the base metal in the direction of the weld, measured in inches per minute (ipm). Travel speed has a great effect on penetration and bead size. Slower travel speeds build higher beads with deeper penetration. Faster travel speeds build smaller beads with less penetration. Ideally, the welding parameters should be adjusted so that the electrode tip is positioned at the leading edge of the weld puddle during travel.

Hot Tip

Travel Speed and Wire Feed Speed

Inexperienced beginners are tempted to turn down the wire feed speed if they experience difficulty controlling the weld puddle. Wire electrodes must be run at certain balanced parameters that cannot be changed individually. Voltage, wirefeed speed, and travel speed are adjusted and balanced together to control the weld puddle.

3.7.0 Gun Position

Gun position for GMAW/FCAW has an effect on weld penetration along with voltage and travel speed. The gun position in relation to the direction of the weld is defined the same as for SMAW welding.

3.7.1 Work Angle

The gun work angle (*Figure 11*) is defined as an angle less than 90° between a line perpendicular to the major workpiece surface at the point of electrode contact, and a plane determined by the electrode **axis** and the weld axis. For a T-joint or corner joint, the line is perpendicular to the nonbutting member. For pipe, the plane is determined by the electrode axis and a line tangent to the pipe surface at the same point.

3.7.2 Travel Angle

The gun travel angle (*Figure 11*) is defined as an angle less than 90° between the electrode axis and a line perpendicular to the weld axis at the point of electrode contact in a plane determined by the electrode axis and the weld axis. For pipe, the plane is determined by the electrode axis and a line tangent to the pipe surface at the same point.

- *Neutral angle* – The term neutral angle describes the travel angle when the centerline of the gun electrode is exactly perpendicular (90°) to the welding surface. The term is nonstandard for a 0° travel angle. Weld penetration is moderate when the gun electrode is used in this position.
- *Push angle* – A push angle is generally used for welding thin materials or when shallow penetration is required. A push angle is the travel

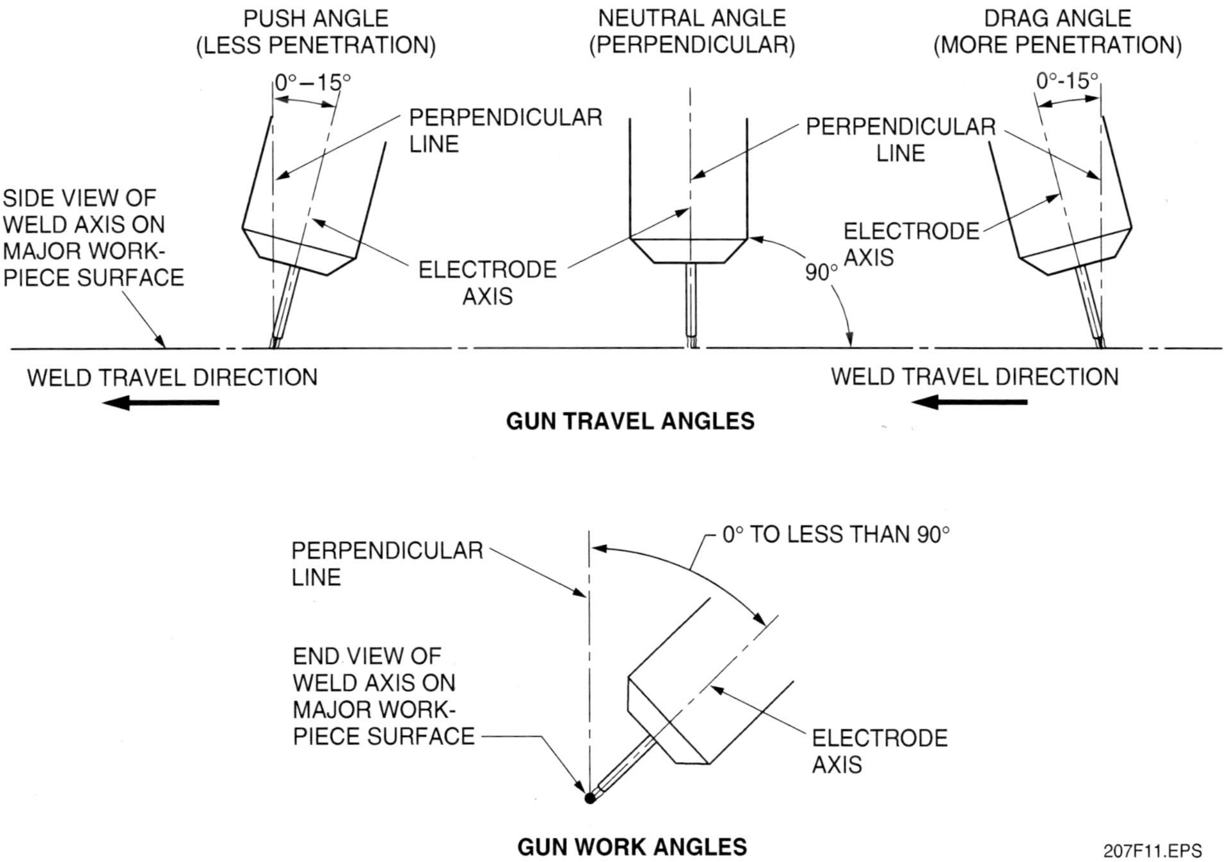

Figure 11 ◆ Gun travel angles.

GMAW of Stainless Steel

For stainless steel, a drag angle is normally used for fillet welds and a push angle for square or grooved butt joints. Stainless steel is usually welded in the flat and horizontal positions with spray transfer. Vertical and overhead welds usually employ short circuiting transfer for at least the root passes, if not the fill and cover passes. Pulsed spray transfer can also be used for fill and cover passes. Remember that all open joints in stainless steel materials must be backed with copper, stainless steel, or inert gas to prevent air from reaching the weld puddle until the puddle solidifies. If practicing welding using stainless steel electrodes and carbon steel coupons, use the travel angles and transfer methods outlined above along with carbon steel backing strips or a backing fixture for open or butt joints, when required. When welding plate that is thicker than ¼", the gun should be moved back and forth in the travel direction and side to side slightly. On thinner materials, only the back and forth motion is necessary.

angle created when the welding gun is tilted back so that the electrode tip precedes the gun in the direction of the weld. In this position, the electrode tip and shielding gas are being directed ahead of the weld bead. Welding with this angle produces a smaller weld bead with less penetration. Push angles greater than 15° are not recommended for manual welding.

- *Drag angle* – A drag angle is the travel angle created when the gun is tilted forward so that the electrode tip trails the gun in the direction of the weld. In this position, the electrode tip and shielding gas are being directed back toward the weld bead. Welding with this angle produces a higher and wider weld bead with more penetration. Maximum penetration is achieved with a drag angle of approximately 25°.

3.8.0 Electrode Extension, Stickout, and Standoff Distance

Electrode extension is the length of the wire that extends beyond the contact tip of the welding gun's contact tube. Changing the extension influences the welding current for low-conductivity metal wires because it changes the preheating of the wire caused by the resistance of the wire to current flow. As the extension increases, preheating of the wire increases. Since the welding power source is self-regulating, it does not have to furnish as much welding current to melt the wire, so the current output automatically decreases. This results in less penetration and increased deposition rates. Increasing extension is useful for bridging gaps and compensating for mismatch but can cause overlap or lack of fusion and a ropy bead appearance. When the extension is decreased, the power source is forced to increase its current output to burn off the wire. Too little extension can cause the wire to weld to the contact tip and develop porosity in the weld. For high-conductivity metal wires, the preheating effect of wire resistance is minimal and wire speed, and voltage settings have a more direct effect.

GMAW and FCAW-G extensions vary between ¼" to 1", depending on the transfer mode. The extension for self-shielding FCAW and an insulating nozzle is ¾" to 1", depending on the welding position and wire size. *Figure 12* shows typical electrode extensions for various GMAW and FCAW gun configurations. In addition, other gun nomenclature is defined. Stickout is the distance from the gas nozzle or insulating nozzle to the end of the electrode.

Arc Blow

When current flows, magnetic fields are created. The magnetic fields tend to concentrate in corners, deep grooves, or the ends of the base metal. When the arc approaches these concentrated magnetic fields, it is deflected from the intended path. Arc blow can cause defects such as excessive weld spatter and porosity. If arc blow occurs, try one or more of these methods to control it:

- Change the position of the workpiece clamp. This will change the flow of welding current, affecting the way the magnetic fields are created.
- Change the angle of the gun. Reducing or even reversing the angle of the gun can compensate for the arc blow.

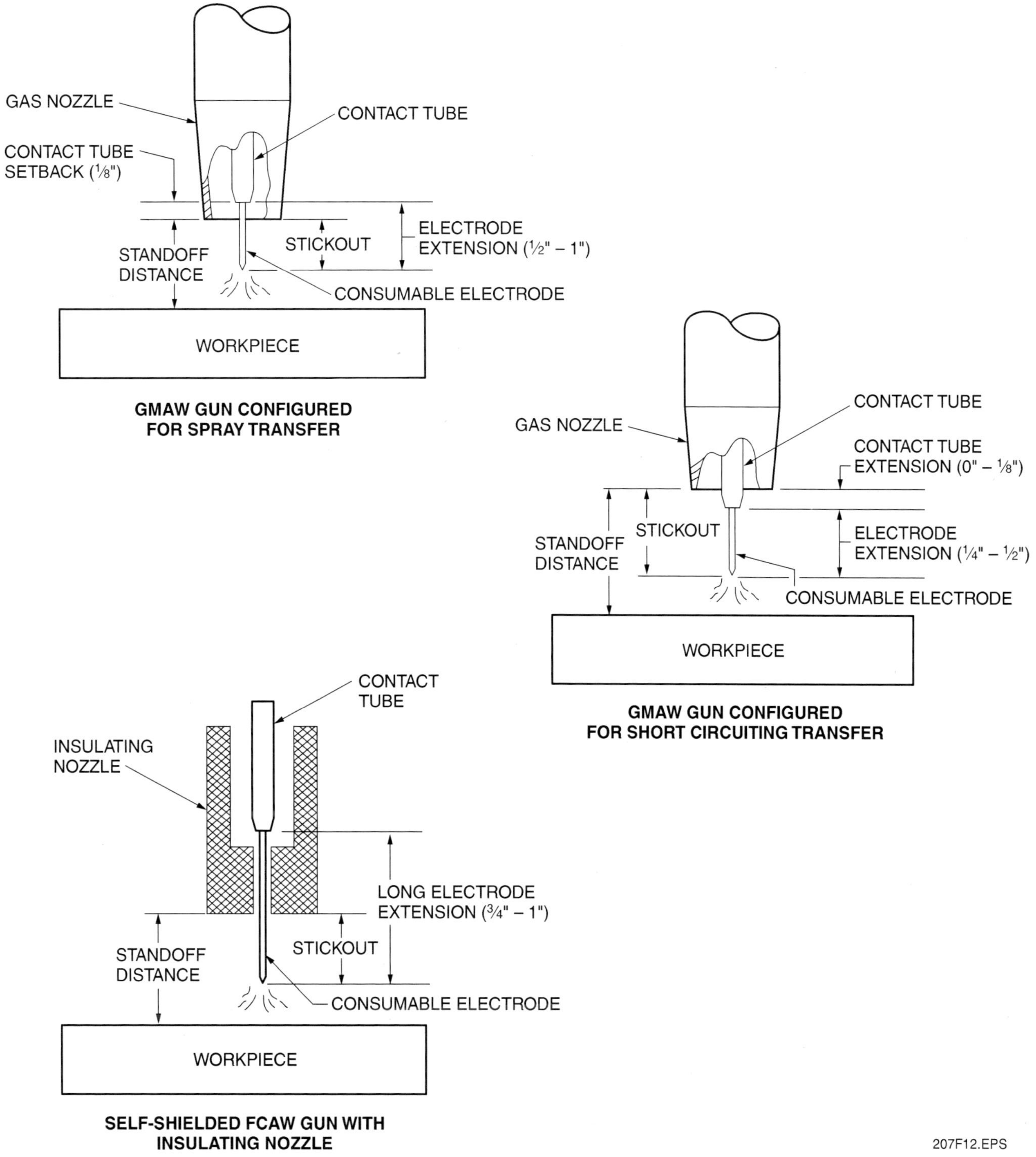

Figure 12 ◆ GMAW and FCAW gun configurations.

Hot Tip

Spatter Control

With excessive stickout, more voltage is lost in the electrode length, the arc voltage is lower, and irregular arc action and spatter may occur. Penetration is decreased as stickout is increased.

With shorter stickout and less voltage drop in the wire, higher arc voltage is available. If the stickout is too short, spatter will increase and rapidly build up in the nozzle and on the contact tip.

Standoff distance is the distance from the gas nozzle or insulating nozzle to the workpiece. Contact tip extension or setback is usually dependent on the transfer mode for the GMAW application.

3.9.0 Gas Nozzle Cleaning

As the welding machine is used, weld spatter accumulates on the gas nozzle and the contact tip. If the gas nozzle is not occasionally cleaned, it will restrict the shielding gas flow and cause porosity in the weld.

Clean the gas nozzle with a reamer or round file or with the tang of a standard file. After cleaning, the nozzle can be sprayed or dipped in a special antispatter compound. The antispatter compound helps prevent the spatter from sticking to the nozzle.

4.0.0 ◆ BEAD TYPES

The two basic bead types are:

- Stringer beads
- Weave beads

Antispatter Materials

Use only antispatter material specifically designed for GMAW and FCAW-G gas nozzles. Water-based products may cause porosity in aluminum and hydrogen embrittlement in hydrogen-sensitive steel. Always refer to material safety data sheets (MSDS) for specific safety considerations.

207SA02.EPS

4.1.0 Stringer Beads

Stringer beads are made with little or no side-to-side motion of the electrode. The width of a stringer bead is generally no more than six times the wire electrode diameter for smaller diameter (0.045" to 1/16")wire or four times the wire electrode diameter for larger diameter (3/32" to 1/8") wire. *Figure 13* shows a stringer bead made with 0.023" diameter wire.

CAUTION

The width of stringer beads is specified in the welding code or WPS being used at your site. Do not exceed the widths specified for your site.

4.1.1 Practicing Stringer Beads

Practice running stringer beads in the flat position. Experiment with different voltages, travel angles, and stickouts. Change the wire feed speed and observe the bead size. Try different travel angles and observe the bead shape and size.

NOTE

Practice stringer bead welds using GMAW and/or FCAW as directed by your instructor. Use solid, composite, or flux cored electrode wire sizes and, if required, appropriate shielding gas as directed by your instructor.

Follow these steps to run stringer beads:

Step 1 Hold the gun at the desired travel angle (drag or push), with the electrode tip directly over the point where the weld is to begin, and pull the gun trigger.

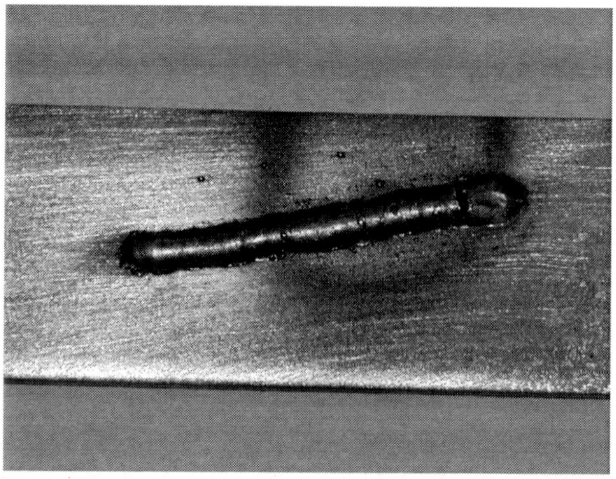

207F13.EPS

Figure 13 ◆ Close-up view of a GMAW stringer bead.

Step 2 Hold the arc in place until the weld puddle begins to form.

Step 3 Slowly advance the arc while maintaining the gun angle.

> **NOTE**
> Stay in the leading edge of the puddle to prevent incomplete fusion.

Step 4 Continue to weld until a bead about 2" to 3" long is formed.

Step 5 Stop, raise the gun to a neutral (0°) travel angle, pause until the crater is filled, and then release the trigger.

Step 6 Inspect the bead for the following:

- Straightness of the bead
- Uniform appearance of the bead face
- Smooth, flat transition with complete fusion at the toes of the weld
- No porosity
- No excessive undercut
- No overlap
- No lack of fusion
- Complete penetration
- No pinholes (fisheyes)
- No inclusions
- Crater filled
- No cracks

Step 7 Continue practicing stringer beads until you can make acceptable welds every time.

4.2.0 Weave Beads

Weave beads are made with wide side-to-side motions of the electrode. The width of a weave bead is determined by the amount of side-to-side motion. *Figure 14* shows a weave bead made with 0.023" diameter wire.

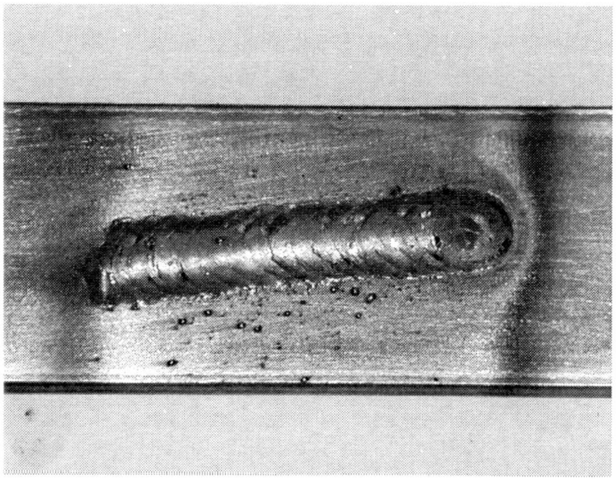

207F14.EPS

Figure 14 ◆ Close-up view of a GMAW weave bead.

When making a weave bead, use care at the toes to ensure proper tie-in to the base metal. To ensure proper tie-in at the toes, slow down or pause slightly at the edges. The pause at the edges will also flatten out the weld, giving it the proper profile.

4.2.1 Practicing Weave Beads

Practice running weave beads about 5/8" wide in the flat position. Experiment with different weave motions, travel angles, and stickouts. Change the wire feed speed and observe the bead size. Try different gun angles and observe the bead shape and size.

> **NOTE**
> Practice weave bead welds using GMAW and/or FCAW as directed by your instructor. Use solid, composite, or flux cored electrode wire sizes and, if required, appropriate shielding gas as directed by your instructor.

Hot Tip

Weave Bead Width

The width of weave beads is often specified in the welding code or WPS being used at your site. Don't try to exceed the widths specified for your site. *ANSI/AWS D1.1-2002* limits the width of a weave bead on prequalified joints to 5/8" in flat, horizontal, and overhead positions and 1" in the vertical positions.

Weave Beads and Stainless Steel

Weave beads are not normally used with stainless steel because of gas shielding problems and excessive heat input to the base metal. However, some fully automatic processes can make reasonable weave beads using pulse transfer techniques.

Follow these steps to run weave beads:

Step 1 Hold the gun at the desired travel angle, with the electrode tip directly over the point where the weld is to begin.

Step 2 Pull the gun trigger.

Step 3 Hold the arc in place until the weld puddle begins to form.

Step 4 Slowly advance the arc in a weaving motion while maintaining the gun angle. *Figure 15* shows the weave motion.

Step 5 Continue to weld until a bead about 2" to 3" long is formed.

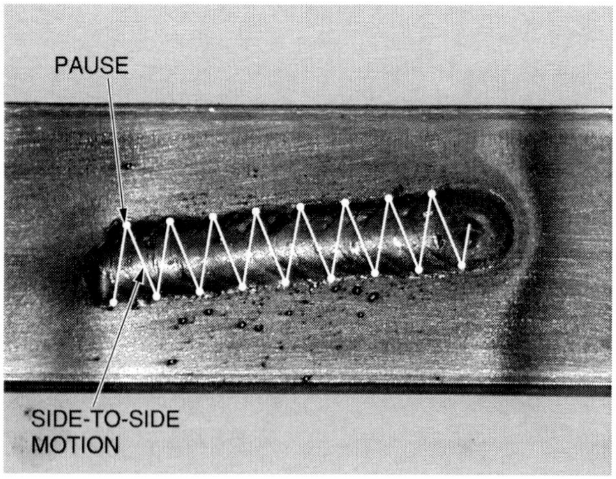

Figure 15 ◆ Weave motion.

Step 6 Stop, raise the gun to a neutral (0°) travel angle, pause until the crater is filled, and then release the trigger.

Step 7 Inspect the bead for the following:

- Straightness of the bead
- Uniform appearance of the bead face
- Smooth, flat transition with complete fusion at the toes of the weld
- No porosity
- No excessive undercut
- No overlap
- No lack of fusion
- Complete penetration
- No pinholes (fisheyes)
- No inclusions
- Crater filled
- No cracks

Step 8 Continue practicing weave beads until you can make acceptable welds every time.

4.3.0 Weld Restarts

A restart is the junction where a new weld connects to continue the bead of a previous weld. Restarts are important because an improperly made restart creates a weld discontinuity. One advantage of the semiautomatic GMAW and FCAW process is that restarts can be minimized. Plan the weld to eliminate restarts, if possible. If a restart has to be made, it must be made so that it blends smoothly with the previous weld and does not stand out. The technique for making a restart

Bead Cleaning

There is no such thing as a self-cleaning electrode or shield gas. Welders must make sure that surfaces to be welded are clean and free of contaminants. It is important that multipass welds be cleaned before each successive bead is laid down. Attempting to lay multipass beads without properly cleaning the previous bead will result in inclusions in the welded joint. A radiograph of a butt weld with two slag lines easily visible in the weld root is shown here.

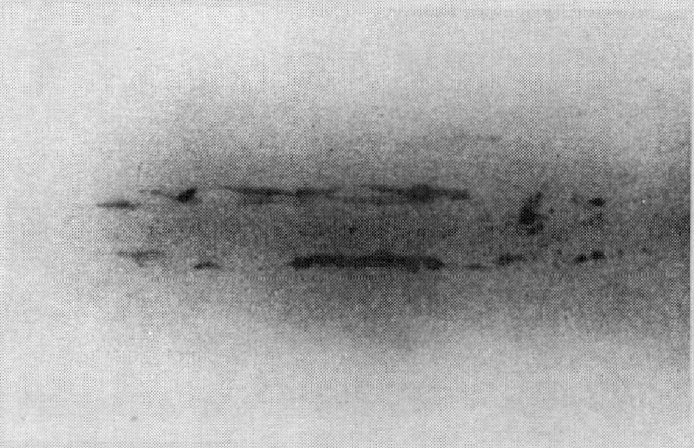

207SA03.EPS

is the same for both stringer and weave beads. Practice weld restarts using GMAW and/or FCAW as directed by your instructor.

Follow these steps to make a restart:

Step 1 With the weld clean, hold the torch at the proper travel angle while restarting the arc directly over the center of the crater.

> **NOTE**
> The welding codes do not allow arc strikes outside the area of permanent welds.

Step 2 Move the electrode tip over the crater in a small circular motion, to fill the crater with a molten puddle.

Step 3 Move to the leading edge of the puddle, and as soon as the puddle fills the crater, continue the stringer or weave bead pattern being used.

Step 4 Inspect the restart.

> **NOTE**
> A properly made restart will blend into the bead, making it hard to detect. If the restart has undercut, not enough time was spent in the crater to fill it. If the undercut is on one side or the other, use more of a side-to-side motion as you move back into the crater. If the restart has a lump, it was overfilled: too much time was spent in the crater before resuming the forward motion.

Step 5 Continue to practice restarts until they are correct.

> **NOTE**
> Use the same techniques for making restarts whenever performing GMAW and FCAW.

4.4.0 Weld Terminations

A weld termination is made at the end of a weld. A termination normally leaves a crater. When making a termination, the welding codes require that the crater must be filled to the full cross section of the weld. This can be difficult, since most terminations are at the edge of a plate, where welding heat tends to build up, making filling the crater more difficult. *Figure 16* shows the proper technique for terminating a weld.

> **NOTE**
> Practice weld terminations using GMAW, FCAW, or both as directed by your instructor.

Follow these steps to make a weld termination:

Step 1 Start to bring the gun up to a 0° travel angle and slow the forward travel as you approach the end of the weld.

Step 2 Stop the forward movement about ⅛" from the end of the plate.

Step 3 With the gun at a 0° travel angle, allow the crater to fill and then move about ⅛" toward the start of the weld.

Step 4 Release the trigger when the crater is completely filled.

> **CAUTION**
> Do not remove the gun from the weld until the puddle has solidified. The gas postflow that continues after the welding stops protects the molten metal. If the gun and shielding gas are removed before the crater has solidified, crater porosity or cracks can occur.

Step 5 Inspect the termination. The crater should be filled to the full cross section of the weld.

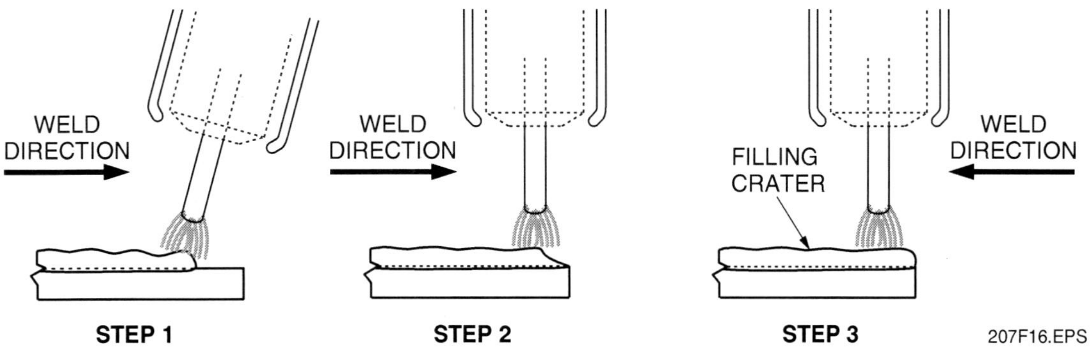

Figure 16 ◆ Terminating a weld.

4.5.0 Overlapping Beads

Overlapping beads are made by depositing connective weld beads parallel to one another. The parallel beads overlap, forming a flat surface. This is also called padding. Overlapping beads are used to build up a surface and for making multipass welds. Both stringer and weave beads can be overlapped. When viewed from the end, properly overlapped beads will have a relatively flat surface. *Figure 17* shows proper and improper overlapping beads.

4.5.1 Practicing Overlapping Stringer or Weave Beads

Follow these steps to weld overlapping stringer or weave beads, using a specific diameter solid, composite, or flux cored electrode wire and the proper shielding gas as directed by your instructor.

NOTE

Practice overlapping beads using GMAW and/or FCAW as directed by your instructor. Avoid making weave beads on stainless steel.

Step 1 Mark out a square on a piece of steel.

Step 2 Weld a stringer or weave bead along one edge at a 0° work angle.

Step 3 Clean the weld.

Step 4 Position the gun at a work angle of 10° to 15° toward the side of the previous bead to get proper tie-in, and pull the trigger. *Figure 18* shows the work angle for overlapping beads.

Step 5 Continue running overlapping stringer or weave beads until the square is covered.

NOTE

Chip or brush each bead before running the new bead.

Step 6 Continue building layers of stringer or weave beads, one on top of the other, until the technique is perfected.

WARNING!

The base metal will get very hot as it is being built up.

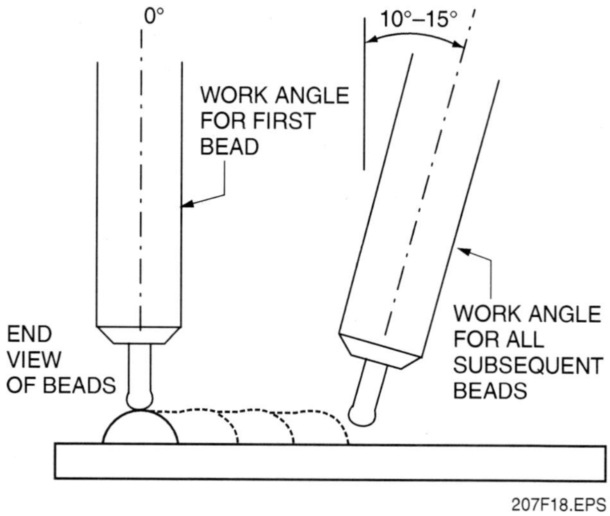

Figure 18 ◆ Work angles for overlapping beads.

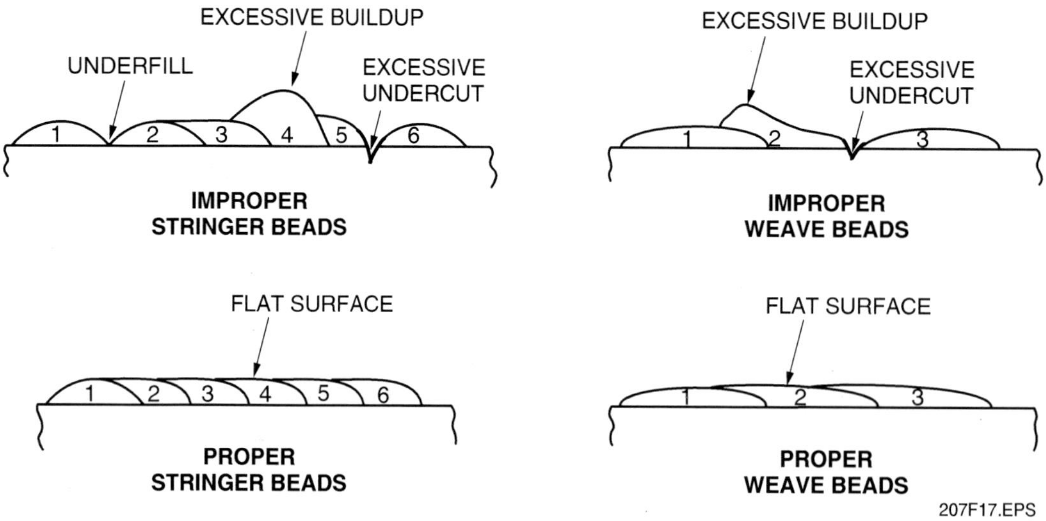

Figure 17 ◆ Proper and improper overlapping beads.

4.5.2 Building a Pad with Stringer or Weave Beads

Build a pad using stringer beads, and then repeat building the pad using weave beads, as shown in *Figure 19.*

5.0.0 ◆ FILLET WELDS

Fillet welds require little base metal preparation, except for the cleaning of the weld area and the removal of any excess material from cut surfaces. Any slag from oxyfuel cutting or dross from plasma arc or carbon arc cutting will cause porosity in the weld. For this reason, the codes require that this material be entirely removed prior to welding.

The most common fillet welds are made in lap and T-joints. The weld position for plate is determined by the weld axis and the orientation of the workpiece. The positions for fillet welding on plate are flat, or 1F (the letter F stands for fillet); horizontal, or 2F; vertical, or 3F; and overhead, or 4F. In the 1F and 2F positions, the weld axis can be inclined up to 15°. Any weld axis inclination for the other positions varies with the rotational position of the weld face as specified in AWS standards. *Figure 20* shows the fillet welding positions for plate.

Fillet welds can be concave or convex, depending on the WPS or site quality standards. The welding codes require a fillet weld to have a uniform concave or convex face, although a slightly nonuniform face is acceptable. Convexity is the distance the weld extends above a line drawn between the toes of the weld. The convexity of a fillet weld

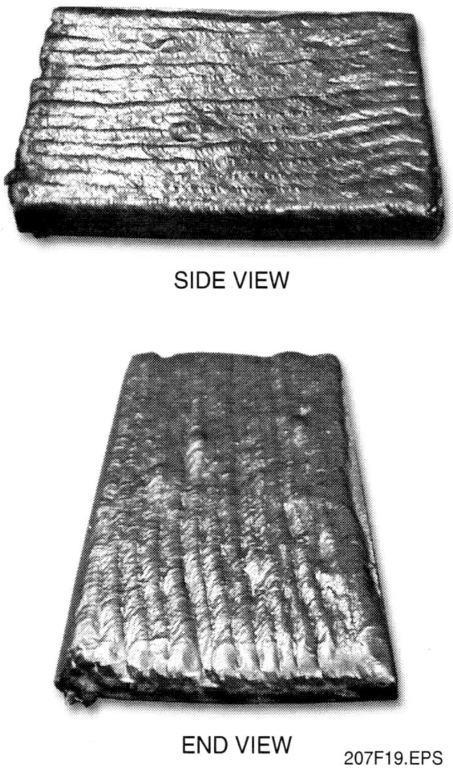

Figure 19 ◆ Pad built with weave beads.

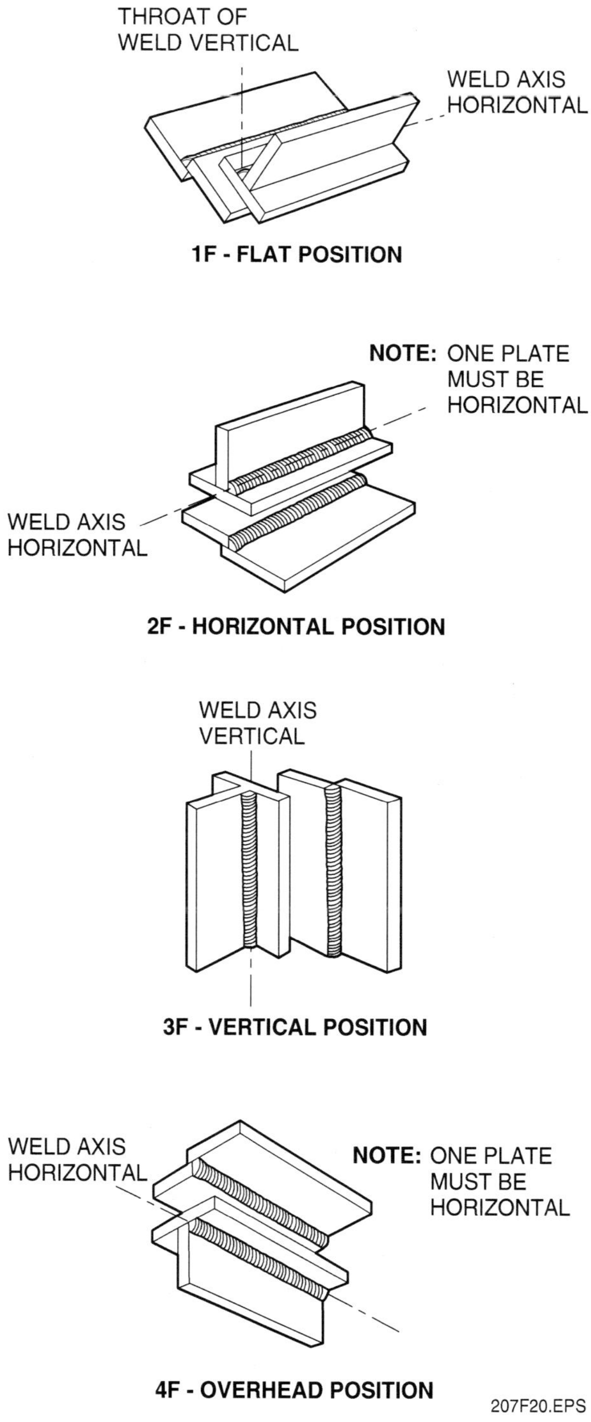

Figure 20 ◆ Fillet weld positions.

Hot Tip

Preferred Filet Weld Contours

In single-pass, weave-bead fillet welds where two workpieces are being joined at an angle (not lap joints), flat or slightly convex faces are usually preferred because weld stresses are more uniformly distributed through the fillet weld and workpieces.

or individual surface bead must not exceed that permitted by the applicable code or standard. A fillet weld must be repaired if the profile has defects.

A fillet weld is unacceptable and must be repaired if the profile has insufficient throat, excessive convexity, excessive undercut, overlap, insufficient leg, or incomplete fusion.

Figure 21 shows acceptable and unacceptable fillet weld profiles.

5.1.0 Practicing Flat (1F) Position Fillet Welds

Practice flat (1F) position fillet welds by welding multipass (six-pass) convex fillet welds in a T-joint. Use an instructor-specified size of solid, composite, or flux cored electrode wire with appropriate shielding gas as required. When making flat fillet welds, pay close attention to the work angle and travel speed. For the first bead, the work angle is vertical (45° to both plate surfaces). The angle is adjusted for all subsequent beads. Increase or decrease the travel speed to control the amount of weld metal buildup. *Figure 22* shows the work angles for 1F position fillet welds.

NOTE

Practice 1F fillet welds using GMAW (spray and short-circuiting transfer) and/or FCAW as directed by your instructor.

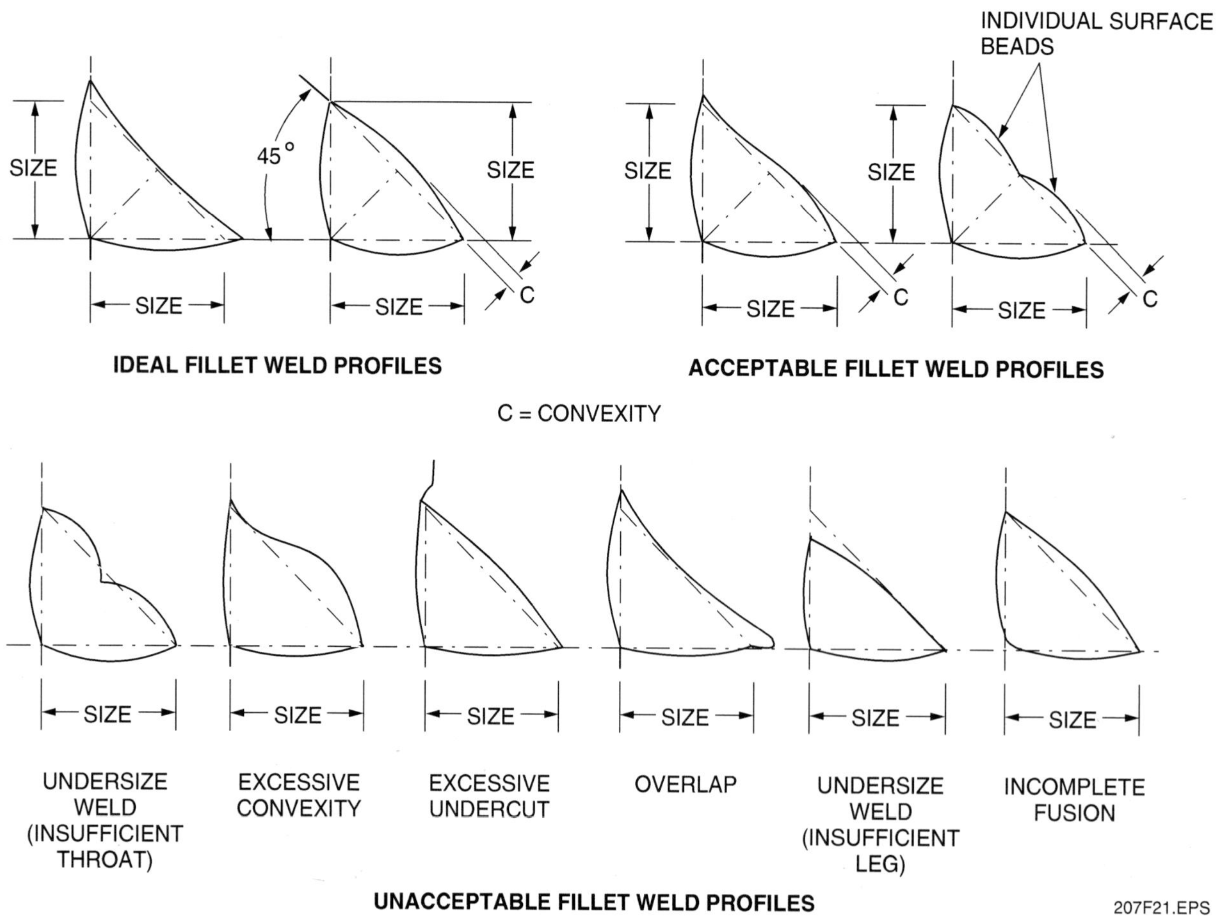

Figure 21 ◆ Acceptable and unacceptable fillet weld profiles.

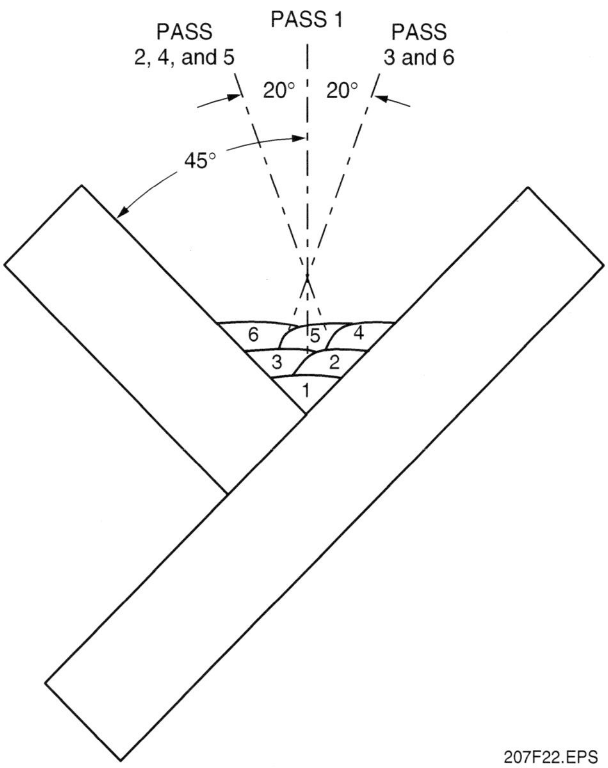

Figure 22 ◆ Multiple-pass 1F weld sequences and work angles.

Follow these steps to make a flat fillet weld. Please note in the following steps to clean all weld beads before beginning the next bead.

Step 1 Tack two plates together to form a T-joint for the fillet weld coupon. Clean the tack welds. *Figure 23* shows the fillet weld coupon.

NOTE: BASE METAL, CARBON STEEL AT LEAST ¼" THICK

Figure 23 ◆ Fillet weld coupon.

Step 2 Clamp or tack-weld the coupon in the flat position.

Step 3 Run the first bead along the root of the joint, using a work angle of 45° with a 5° to 10° drag angle.

> **NOTE**
> Use a slight side-to-side motion (**oscillation**).

> **Hot Tip**
> ### *T-Joint Heat Dissipation*
> In T-joints, the welding heat dissipates more rapidly in the thicker or nonbutting member. On various bead passes, the arc may have to be concentrated slightly more on the thicker or the nonbutting member to compensate for the heat loss.

Step 4 Run the second bead along a toe of the first weld, overlapping about 75% of the first bead. Alter the work angle as shown in *Figure 22* and use a 5° to 10° drag angle with a slight oscillation.

Step 5 Run the third bead along the other toe of the first weld, filling the groove created when the second bead was run. Use the work angle shown in *Figure 22* and a 5° to 10° drag angle with a slight oscillation.

Step 6 Run the fourth bead along the outside toe of the second weld, overlapping about half the second bead. Use the work angle shown in *Figure 22* and a 5° to 10°drag angle with a slight oscillation.

Step 7 Run the fifth bead along the inside toe of the fourth weld, overlapping about half the fourth bead that was run. Use the work angle shown in *Figure 22* and a 5° to 10° drag angle with a slight oscillation.

> **Hot Tip**
> ### *Tacking and Aligning Workpieces*
> When tacking workpieces together, both sides of the workpieces are usually tacked with welds that are about ½" long to position the workpieces and to minimize distortion when the final welds are made. After the first tack weld, use a hammer or other tool to align the workpieces side to side and end to end: then tack the opposite side. Tack the far ends of the workpieces in the same manner. Intermediate tack welds can be made every 5" to 6" as necessary to minimize lengthwise distortions.

Step 8 Run the sixth bead along the toe of the fifth weld, filling the groove created when the fifth bead was run. Use the work angle shown in *Figure 22* and a 5° to 10° drag angle with a slight oscillation.

Step 9 Inspect the weld. The weld is acceptable if it has the following:

- Uniform appearance on the bead face
- Craters and restarts filled to the full cross section of the weld
- Uniform weld size ±1⁄16"
- Acceptable weld profile in accordance with the applicable code or standard
- Smooth transition with complete fusion at the toes of the weld
- No porosity
- No excessive undercut
- No overlap
- No inclusions
- No cracks

5.2.0 Practicing Horizontal (2F) Position Fillet Welds

Practice horizontal (2F) fillet welding by placing multiple-pass fillet welds in a T-joint using solid, composite, or flux cored electrode wire sizes and, if required, appropriate shielding gas as directed by your instructor. When making horizontal fillet welds, pay close attention to the gun angles and travel speed. For the first bead, the electrode work angle is 45°. The work angle is adjusted for all other welds. Increase or decrease the travel speed to control the amount of weld-metal buildup.

NOTE

Practice 2F fillet welds using GMAW (spray and short-circuiting transfer) and/or FCAW as directed by your instructor.

Follow these steps to make a horizontal fillet weld:

Step 1 Tack two plates together to form a T-joint for the fillet weld coupon. Clean the tack welds.

Step 2 Clamp or tack-weld the coupon in the horizontal position.

Step 3 Run the first bead along the root of the joint using a work angle of approximately 45° with a 5° to 10° drag angle and a slight side-to-side oscillation (*Figure 24*).

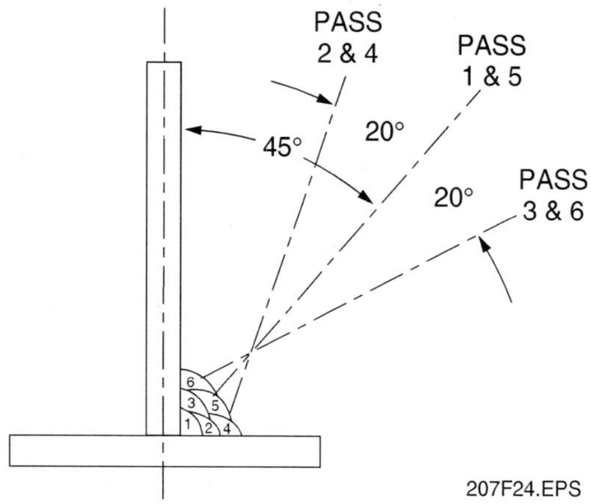

Figure 24 ◆ Multipass 2F weld sequences and work angles.

Step 4 Clean the weld.

Step 5 Run the remaining passes using a 5° to 10° drag angle and a slight oscillation at the appropriate work angles (*Figure 24*). Overlap each previous pass and clean the weld after each pass.

Step 6 Have the instructor inspect the weld. The weld is acceptable if it has:

- Uniform appearance on the bead face
- Craters and restarts filled to the full cross section of the weld
- Uniform weld size ±1⁄16"
- Acceptable weld profile in accordance with the applicable code or standard
- Smooth transition with complete fusion at the toes of the weld
- No porosity
- No excessive undercut
- No overlap
- No inclusions
- No cracks

T-Joint Vertical Plate Undercut

The most common defect for T-joints is undercut on the vertical plate of the joint. If the problem occurs, angling the arc slightly more toward the vertical plate and the bead at the top of the weave will force more metal into the bead at the top edge of the weld.

5.3.0 Practicing Vertical (3F) Position Fillet Welds

Practice vertical (3F) fillet welding by placing multiple-pass fillet welds in a T-joint using solid, composite, or flux cored electrode wire sizes and, if required, appropriate shielding gas as directed by your instructor. Normally, vertical welds are accomplished by welding uphill (also called vertical up welding) from the bottom to the top using a gun push angle (up-angle). However, downhill welding (also called vertical down) can also be accomplished using a slight drag angle. When vertical welding, either stringer or weave beads can usually be used. However, weave beads are avoided on stainless steel or in downhill welding. On the job, the site WPS or quality standard will specify which technique to use.

> **NOTE**
> Practice 3F fillet welds using GMAW and/or FCAW as directed by your instructor. Check with your instructor to see if you should run stringer beads or weave beads, or practice both techniques.

5.3.1 Vertical Weave Bead Fillet Welds

Follow these steps to make an uphill fillet weld:

Step 1 Tack two plates together to form a T-joint for the fillet weld coupon.

> **Hot Tip**
> ***Vertical Fillet Welds***
> When making vertical fillet welds, pay close attention to the gun angles and travel speed. For the first bead, the work angle is approximately 45°. The work angle is adjusted for all other welds. Increase or decrease the travel speed to control the amount of weld metal buildup.

Step 2 Clamp or tack-weld the coupon in the vertical position.

Step 3 Starting at the bottom, run the first bead along the root of the joint, using a work angle of approximately 45° with a 10° to 15° push angle. Pause in the weld puddle to fill the crater.

Step 4 Clean the weld.

Step 5 Run the remaining passes using a 5° to 10° push angle and a side-to-side weave technique with a 45° work angle (*Figure 25*). Use a slow motion across the face of the weld, pausing at each toe for penetration and to fill the crater. Adjust the travel speed across the face of the weld to control the buildup. Overlap each previous pass as required and clean the weld after each pass.

> **Hot Tip**
> ***Vertical Weave Bead***
> To help control undercut, an alternate pattern, such as a small triangular weave, can be used in vertical welds. By pausing the rod at the edges, the previous undercut can be filled. This action also creates undercut at the existing weld pool, but that will be filled in by the next weave.

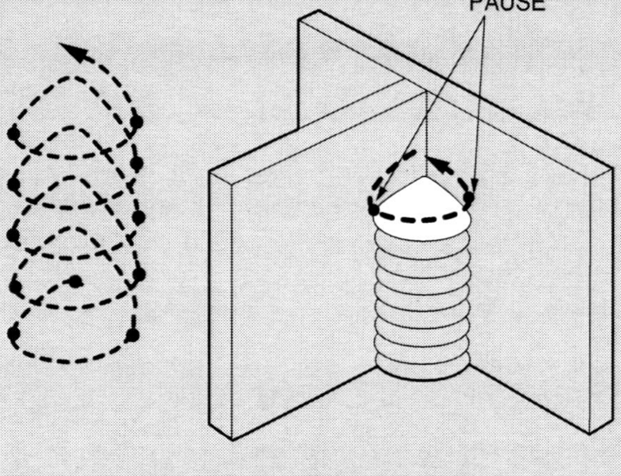

TRIANGULAR WEAVE PATTERN

207SA04.EPS

Step 6 Have the instructor inspect the weld. The weld is acceptable if it has:

- Uniform appearance on the bead face
- Craters and restarts filled to the full cross section of the weld
- Uniform weld size ±1/16"
- Acceptable weld profile in accordance with the applicable code or standard
- Smooth transition with complete fusion at the toes of the weld
- No porosity
- No excessive undercut
- No overlap
- No inclusions
- No cracks

5.3.2 Vertical Stringer Bead Fillet Welds

Repeat vertical fillet (3F) welding using stringer beads. Use a slight oscillation and side-to-side motion, pausing slightly at each toe to prevent undercut. For stringer beads, use a 0° to 10° push angle and the work angles shown in *Figure 25* as required.

5.4.0 Practicing Overhead (4F) Position Fillet Welds

Practice overhead (4F) fillet welding by welding multiple-pass fillet welds in a T-joint using solid, composite, or flux cored electrode wire sizes and, if required, appropriate shielding gas as directed by your instructor. When making overhead fillet welds, pay close attention to the gun angles and travel speed. For the first bead, the work angle is approximately 45°. The work angle is adjusted for all other welds. Increase or decrease the travel speed to control the amount of weld metal buildup.

> **NOTE**
> Practice 4F fillet welds using GMAW and/or FCAW as directed by your instructor.

Follow these steps to make an overhead fillet weld:

Step 1 Tack two plates together to form a T-joint for the fillet weld coupon.

Step 2 Clamp or tack-weld the coupon so it is in the overhead position.

Step 3 Run the first bead along the root of the joint, using a work angle of approximately 45° with a 5° to 10° drag angle. Use a slight oscillation to tie-in the weld at the toes.

Step 4 Clean the weld.

Step 5 Using a slight oscillation, run the remaining passes using a 5° to 10° drag angle and the work angles shown in *Figure 26*. Overlap each previous pass and clean the weld after each pass.

Step 6 Have the instructor inspect the weld. The weld is acceptable if it has:

- Uniform appearance on the bead face
- Craters and restarts filled to the full cross section of the weld
- Uniform weld size ±1/16"
- Acceptable weld profile in accordance with the applicable code or standard
- Smooth transition with complete fusion at the toes of the weld
- No porosity
- No excessive undercut
- No overlap
- No inclusions
- No cracks

6.0.0 ◆ OPEN-ROOT V-GROOVE WELDS

The open-root V-groove weld is a common groove weld normally made on plate and pipe. Practicing the open-root V-groove weld on plate will prepare the welder to make the more difficult pipe welds covered in the next level.

6.1.0 Root Pass

The most difficult part of making an open-root V-groove weld is the root pass. The root pass is made from the V-groove side of the joint and must have complete penetration, but not an excessive amount of burn-through root reinforcement. The penetration is controlled with a side-to-side motion or oscillation.

After the root pass is run, it should be cleaned and inspected. Inspect the face of the weld for excess buildup or undercut, as shown in *Figure 27*, which could trap slag if a FCAW root pass is run. Remove excess buildup or undercut with a hand grinder by grinding the face of the root pass with the edge of a grinding disk. Use care not to grind through the root pass or widen the groove.

6.2.0 Groove Weld Positions

As illustrated in *Figure 28*, groove welds can be made in all positions. The weld position for plate is determined by the axis of the weld and the orientation of the workpiece. Groove weld positions for plate are flat, or 1G (groove); horizontal, or 2G; vertical, or 3G; and overhead, or 4G.

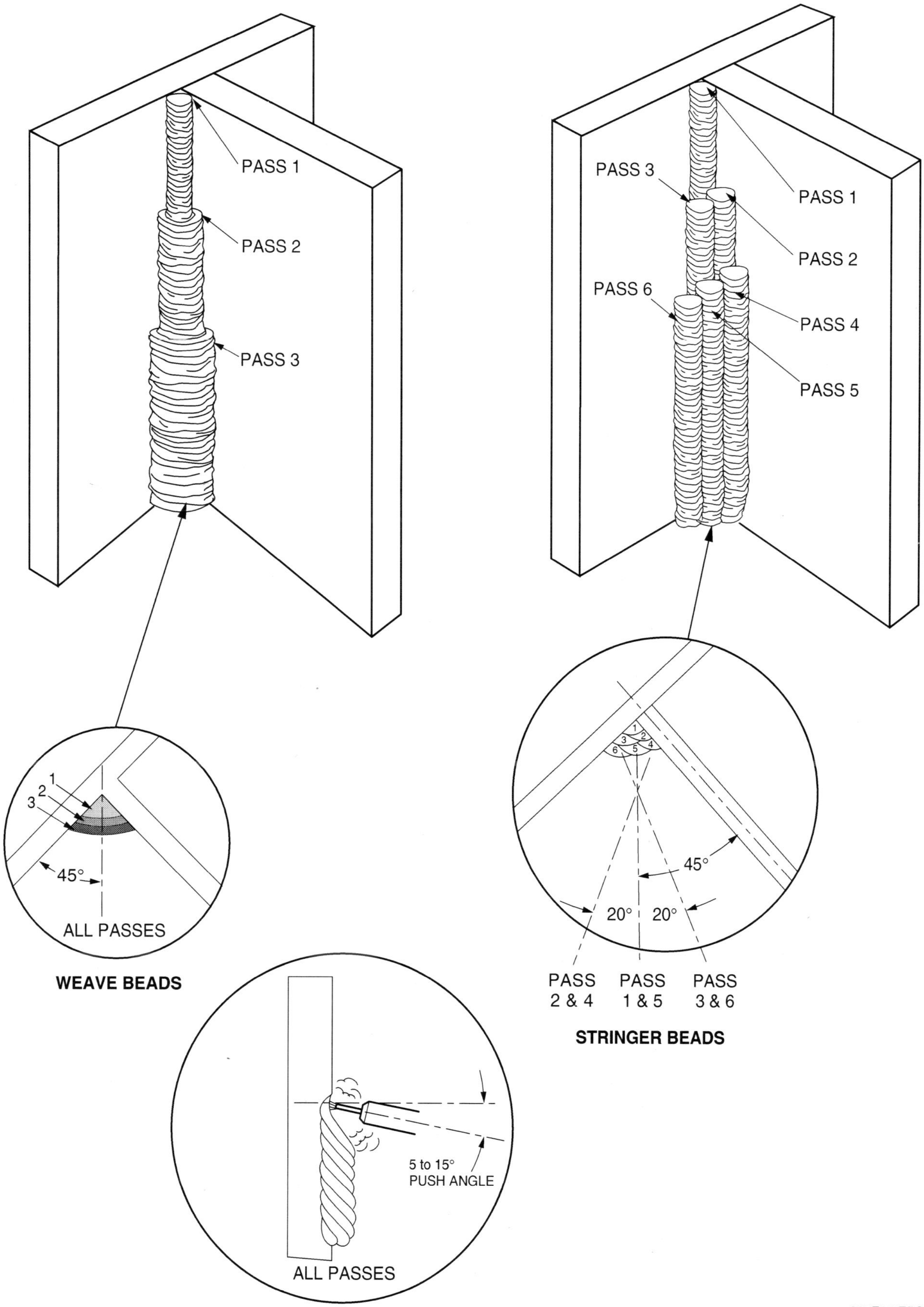

Figure 25 ◆ Multiple-pass 3F weld sequences and work angles for weave beads.

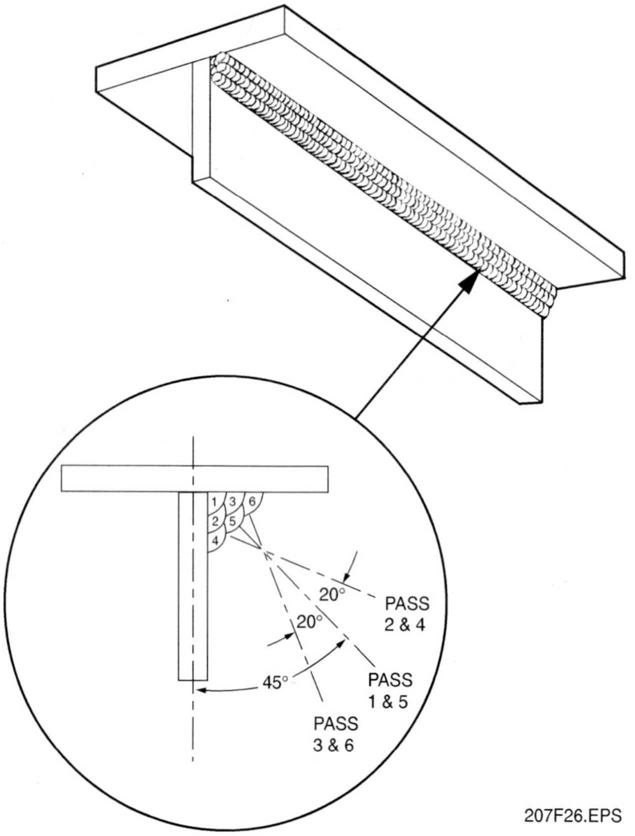

Figure 26 ◆ Multiple-pass 4F weld sequences and work angles.

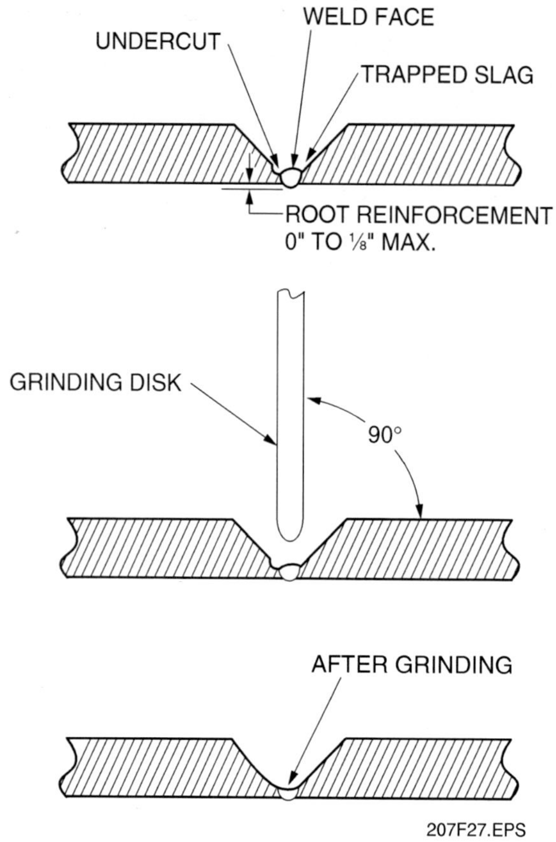

Figure 27 ◆ Grinding a root pass.

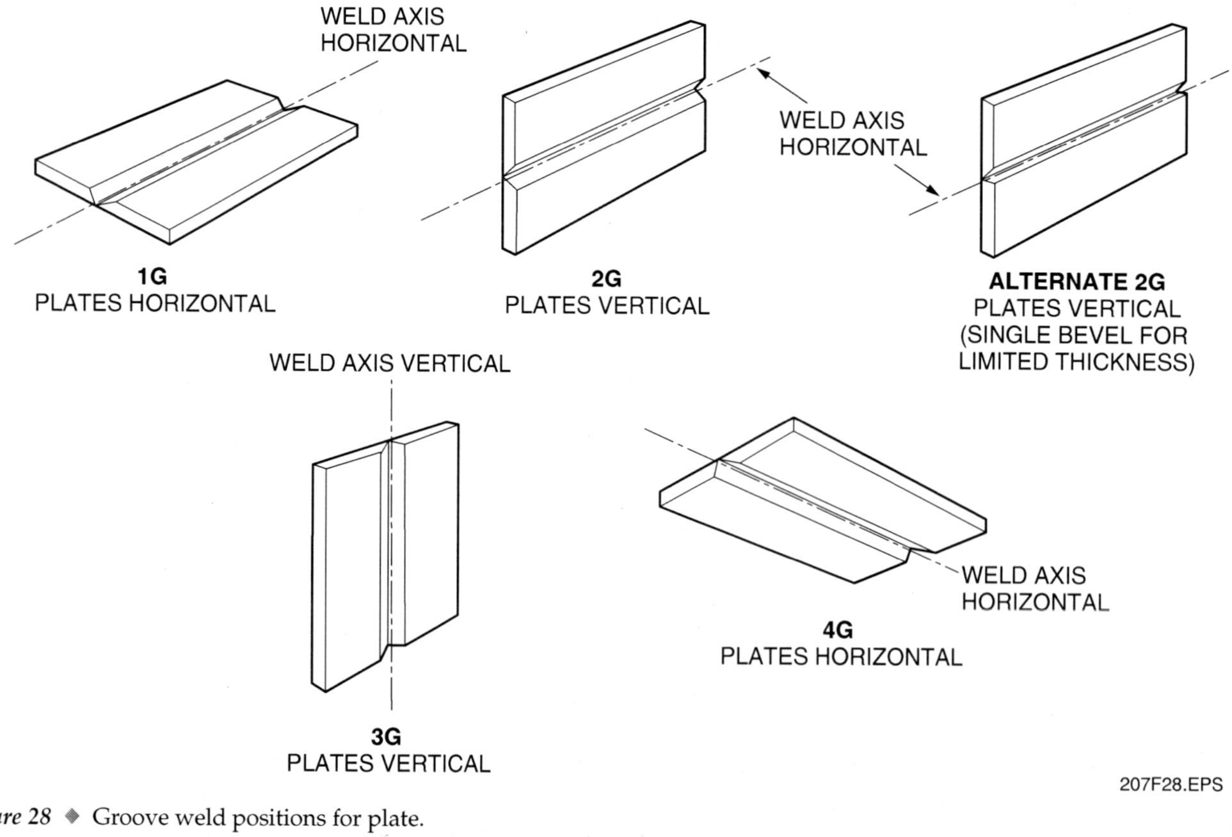

Figure 28 ◆ Groove weld positions for plate.

V-Groove Welds for Stainless Steel

If practicing V-groove welds for stainless steel, remember to practice using a push angle instead of a drag angle. Also practice using a backing for the weld, either with base metal strips or with a backing gas fixture.

207SA05.EPS

In the 1G and 2G positions, the weld axis can be inclined up to 15°. Any weld axis inclination for the other positions varies with the rotational position of the weld face as specified in AWS standards.

6.3.0 Acceptable and Unacceptable Groove Weld Profiles

Groove welds should be made with slight reinforcement and a gradual transition to the base metal at each toe. As shown in *Figure 29*, groove welds must not have excess reinforcement, underfill, excessive undercut, or overlap. If a groove weld has any of these defects it must be repaired.

6.4.0 Practicing Flat (1G) Position Open-Root V-Groove Welds

Practice flat (1G) open-root V-groove welds, as depicted in *Figure 30,* using solid, composite, or flux cored electrode wire sizes and, if required, appropriate shielding gas as directed by your instructor. Use weave beads and keep the gun angle at 90° to the plate surface (0° work angle) with a 5° to 10° drag angle (except for the root pass). When using stringer beads, use a 5° to 10° drag angle (except for the root pass), but adjust the work angle to tie in and fill the bead. Pay particular attention at the termination of the weld to fill the crater. If the practice coupon gets too hot between passes, you may have to cool it in water.

Grinding and Back Welding of FCAW Root Pass

When FCAW is used for the root pass, grinding of the weld face is usually required. This is to remove any trapped slag along the toes of the root pass before the filler passes are run. In addition, the completed V-groove weld must be ground out to the root pass on the back side of the weld with the edge of a grinding wheel. Then a back weld pass is made to achieve proper root reinforcement. In some cases, a backing weld is made first and its root is ground out from the groove side of the workpiece. Then, the root pass and remaining passes are made on the groove side of the material.

Root Pass Welding

In practice, root passes are usually made in various metals using GTAW or SMAW methods. Then the fill and cover passes are made using FCAW or GMAW.

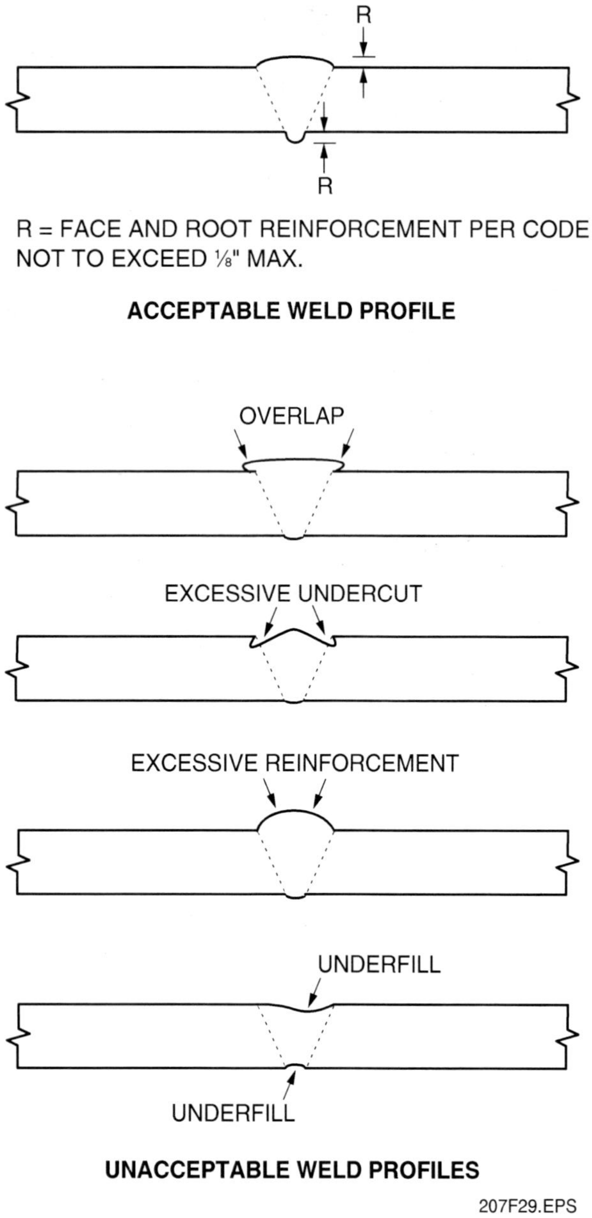

Figure 29 ◆ Acceptable and unacceptable groove weld profiles.

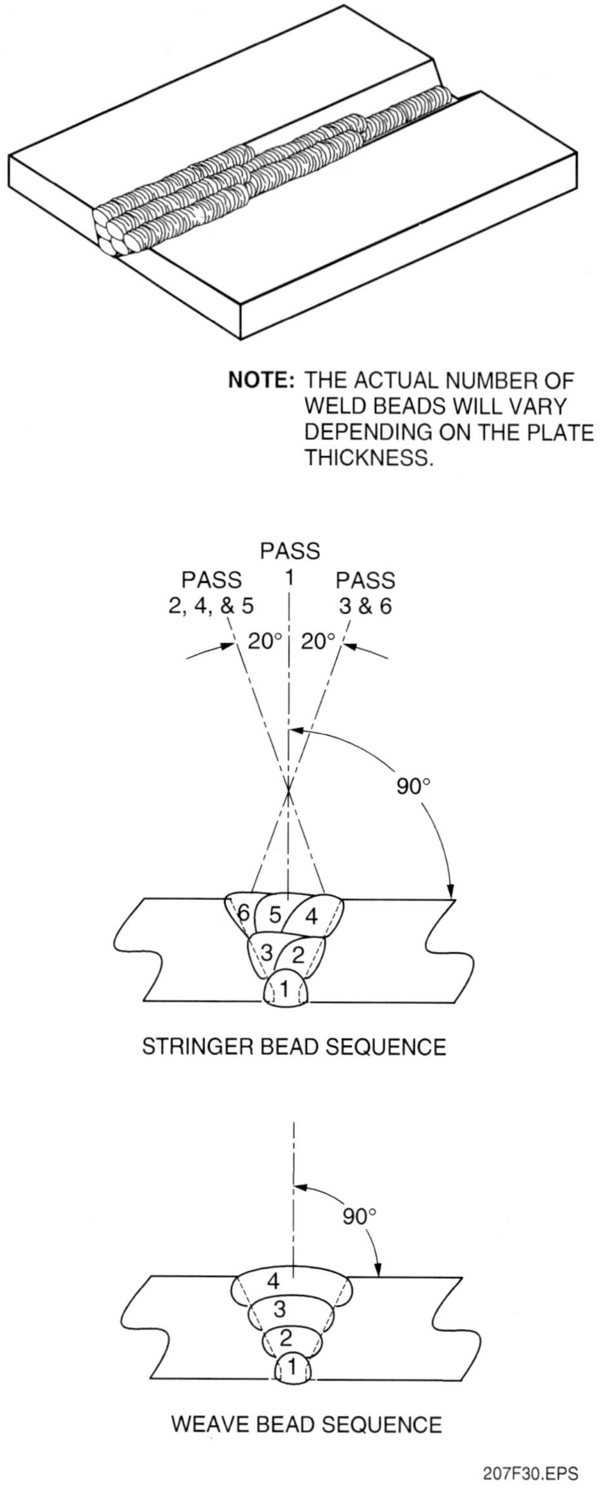

Figure 30 ◆ Multipass 1G weld sequences and work angles.

> **NOTE**
> Practice 1G open-root V-groove welds using GMAW (spray and short circuiting transfer) and/or FCAW as directed by your instructor.

Hot Tip

Whiskers

When making a root pass, make sure that the wire electrode does not get ahead of the leading edge of the weld puddle. If it does, a portion of the wire may penetrate through the joint and be melted off and fused into the back side of the weld. This will cause pieces of wire, sometimes called *whiskers*, to protrude from the back of the weld..

> **WARNING!**
> Use pliers to handle the hot practice coupons. Wear gloves when placing the practice coupon in water. Steam will rise off the coupon and can burn or scald unprotected hands.

> **NOTE**
> Cooling with water is done only on practice coupons. Never cool test coupons or on-the-job welds with water. Cooling with water can cause weld cracks and affect the mechanical properties of the base metal.

Follow these steps to practice open-root V-groove welds in the flat position:

> **NOTE**
> Avoid using weave beads for stainless steel.

Step 1 Tack-weld the practice coupon together as explained earlier.

Step 2 Clamp or tack-weld the weld coupon in the flat position above the welding table surface.

Step 3 Run the root pass using a 5° to 10° drag angle and a slight side-to-side motion to control penetration.

Step 4 Clean the root pass (grind if required).

Step 5 Run the remaining passes using a 5° to 10° drag angle and applicable work angles. Overlap each previous pass and clean the weld after each pass.

Step 6 Inspect the weld. The weld is acceptable if it has the following:

- Uniform appearance on the bead face
- Craters and restarts, filled to the full cross section of the weld
- Uniform weld width ±1/16"
- Acceptable weld profile in accordance with the applicable code or standard
- Complete uniform root reinforcement at least flush with the base metal to a maximum buildup of 1/8"
- Smooth, flat transition with complete fusion at the toes of the weld
- No porosity
- No excessive undercut
- No inclusions
- No cracks

6.5.0 Practicing Horizontal (2G) Position Open-Root V-Groove Welds

A horizontal (2G) open-root V-groove weld is shown in *Figure 31*. Practice horizontal open-root V-groove welds using solid, composite, or flux cored electrode wire sizes and, if required, appropriate shielding gas as directed by your instructor.

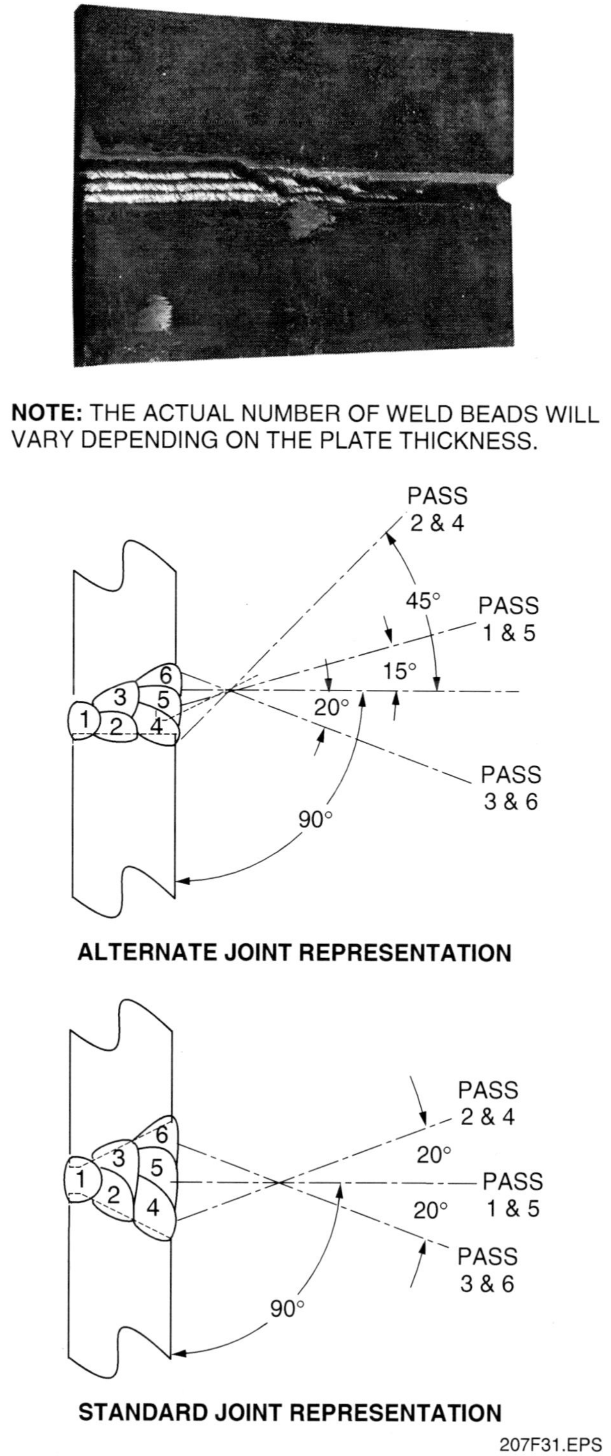

Figure 31 ◆ Multipass 2G weld sequences and work angles.

For the gun angles, use a 5° to 10° drag angle (except for the root pass) and adjust the work angle as required. Pay particular attention to fill the crater at the termination of the weld.

> **NOTE**
> Practice 2G open-root V-groove welds using GMAW and/or FCAW as directed by your instructor.

Follow these steps to practice welding open-root V-groove welds in the horizontal position:

Step 1 Tack-weld the practice coupon together as explained earlier. Use the standard or alternate horizontal weld coupon as directed by your instructor.

Step 2 Tack-weld the flat plate weld coupon into the horizontal position.

Step 3 Run the root pass using a 5° to 15° drag angle and the appropriate work angle. Use a slight side-to-side motion to control penetration.

Step 4 Clean the weld.

Step 5 Run the remaining passes using a 5° to 10° drag angle and appropriate work angles. Overlap each previous pass and clean the weld after each pass.

Step 6 Inspect the weld. The weld is acceptable if it has the following:

- Uniform appearance on the bead face
- Craters and restarts filled to the full cross section of the weld
- Uniform weld width ±1⁄16"
- Acceptable weld profile in accordance with the applicable code or standard
- Complete uniform root reinforcement at least flush with the base metal to a maximum buildup of 1⁄8"
- Smooth, flat transition with complete fusion at the toes of the weld
- No porosity
- No excessive undercut
- No inclusions
- No cracks

6.6.0 Practicing Vertical (3G) Position Open Root V-Groove Welds

Practice vertical (3G) multipass groove welds using stringer beads or weave beads. Set the welding parameters at the lower ends of the suggested ranges. Out-of-position welds are made at lower welding voltages and currents than flat or horizontal welds, to prevent the molten weld metal from sagging or running.

> **NOTE**
> When welding vertically, either stringer or weave beads can be used. However, weave beads are not normally used on stainless steel. On the job, the WPS or site quality standards will specify which technique to use. Check with your instructor to see if you should run stringer beads or weave beads or practice both techniques. The root bead can be run either uphill or downhill. The filler and cover beads can be run uphill or downhill as well. However, downhill weave beads are not normally used. Use the direction preferred at your site or as specified in the WPS or site quality standards.

Run the root bead, using a work angle of 90° (nozzle centerline bisects the V-groove) and a 5° to 15° push angle. The work angle is adjusted for all other welds.

6.6.1 Vertical Open-Root V-Groove Weld with Stringer Beads

Practice 3G open-root V-groove welds using GMAW and/or FCAW as directed by your instructor. Use solid, composite, or flux cored electrode wire sizes and appropriate shielding gas as directed by your instructor.

Follow these steps to make a vertical groove weld with stringer beads:

Step 1 Tack-weld the practice coupon together as explained earlier.

Step 2 Clamp or tack-weld the coupon in the vertical position.

Step 3 Run the root pass uphill with a work angle of 0° (nozzle centerline bisects the V-groove) and a 5° to 15° push angle.

Step 4 Clean the weld.

Step 5 Run the remaining passes at the work angles shown in *Figure 32*, using a 5° to 15° push angle. Overlap each previous pass and use a slight side-to-side oscillation to tie-in the welds at the toes. Clean the weld after each pass.

Step 6 Inspect the weld. The weld is acceptable if it has the following:

- Uniform appearance on the bead face
- Craters and restarts filled to the full cross section of the weld

NOTE: THE ACTUAL NUMBER OF WELD BEADS WILL VARY WITH THE METAL THICKNESS.

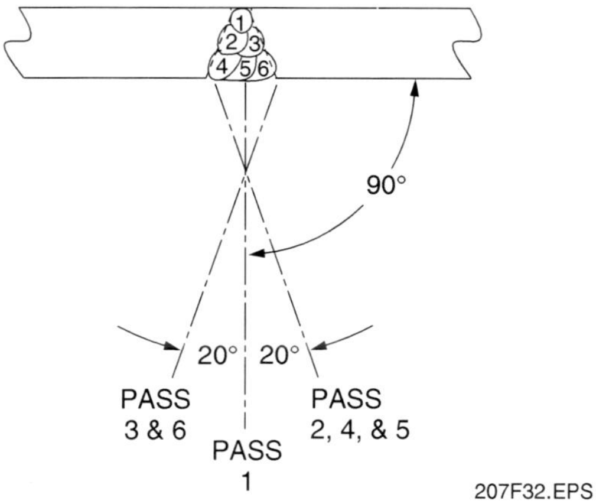

Figure 32 ◆ Multipass 3G weld sequences and work angles for stringer beads.

- Uniform weld width ± 1/16"
- Acceptable weld profile in accordance with the applicable code or standard
- Complete uniform root reinforcement at least flush with the base metal to a maximum buildup of 1/8"
- Smooth, flat transition with complete fusion at the toes of the weld
- No porosity
- No excessive undercut
- No inclusions
- No cracks

6.6.2 Vertical Open-Root V-Groove Weld with Weave Beads

Follow these steps to make a vertical groove weld with weave beads:

> **NOTE**
> Avoid using weave beads for stainless steel.

Step 1 Tack-weld the practice coupon together as explained earlier.

Step 2 Clamp or tack-weld the coupon in the vertical position.

Step 3 Run the root pass uphill with a work angle of 0° (90° from the plate) and a 5° to 15° push angle. Use a slight side-to-side oscillation to control the penetration.

Step 4 Clean the weld.

Step 5 Run the remaining passes as uphill weave beads across the root bead at a 0° work angle. Clean the weld after each pass.

> **NOTE**
> Use a zigzag weave pattern across the root bead and pause at each toe for good fusion. Use a push angle of 5° to 15°. Increase or decrease the travel speed to control the amount of metal buildup.

Figure 33 shows the zigzag pattern and weld sequence for weave beads.

Step 6 Inspect the weld. The weld is acceptable if it has the following:

- Uniform appearance on the bead face
- Craters and restarts filled to the full cross section of the weld
- Uniform weld width ±1/8"
- Acceptable weld profile in accordance with the applicable code or standard
- Complete uniform root reinforcement at least flush with the base metal to a maximum buildup of 1/8"
- Smooth, flat transition with complete fusion at the toes of the weld
- No porosity
- No excessive undercut
- No inclusions
- No cracks

6.6.3 Practicing Overhead (4G) Position Open-Root V-Groove Welds

Practice overhead (4G) open-root V-groove welds using GMAW and/or FCAW as directed by your instructor. Use solid, composite, or flux cored electrode wire sizes and appropriate shielding gas as directed by your instructor.

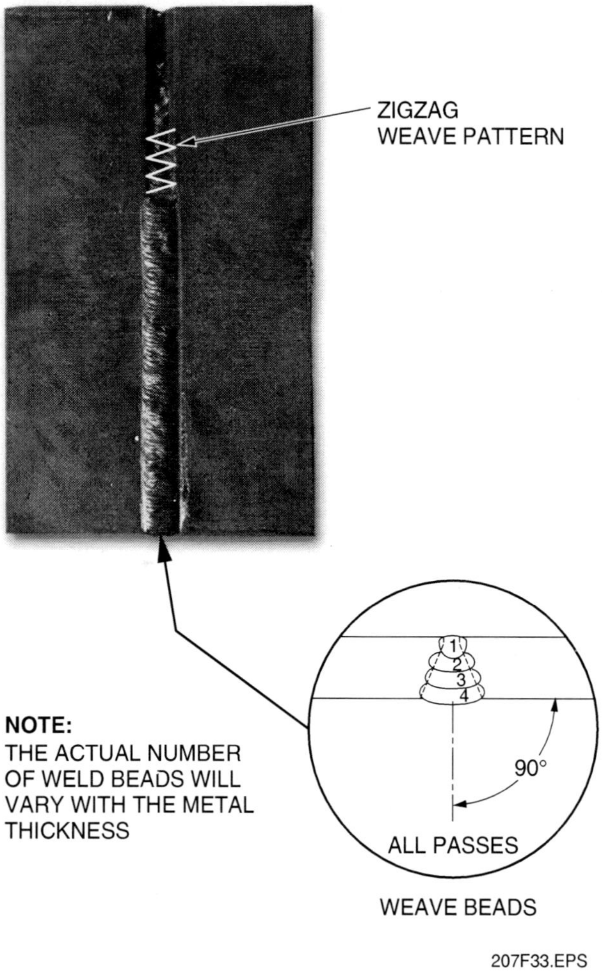

Figure 33 ◆ Multipass 3G weld sequences and work angle for weave beads.

> **NOTE**
> To reduce heat and prevent the molten weld metal from sagging or running, GMAW and FCAW overhead welds are made with lower welding voltages and currents than flat or horizontal welds.

Use a work angle of 0° (90° to plate) and a 5° to 15° drag angle for the root bead.

Follow these steps to make an overhead groove weld with stringer beads:

Step 1 Tack-weld the practice coupon together as explained earlier.

Step 2 Clamp or tack-weld the coupon in the overhead position.

Step 3 Run the root pass, using a work angle of 0° and a 5° to 15° drag angle. Use a slight side-to-side oscillation to control the penetration.

Step 4 Clean the weld.

Step 5 Run the remaining passes at the work angles shown in *Figure 34* using a 5° to 10° drag angle. Overlap each previous pass and use a slight side-to-side motion to tie-in the welds at the toes. Each bead should be cleaned before running the next bead.

Step 6 Inspect the weld. The weld is acceptable if it has the following:

- Uniform appearance on the bead face
- Craters and restarts filled to the full cross section of the weld
- Uniform weld width ±1⁄16"
- Acceptable weld profile in accordance with the applicable code or standard
- Complete uniform root reinforcement at least flush with the base metal to a maximum buildup of 1⁄8"
- Smooth, flat transition with complete fusion at the toes of the weld
- No porosity
- No excessive undercut
- No inclusions
- No cracks

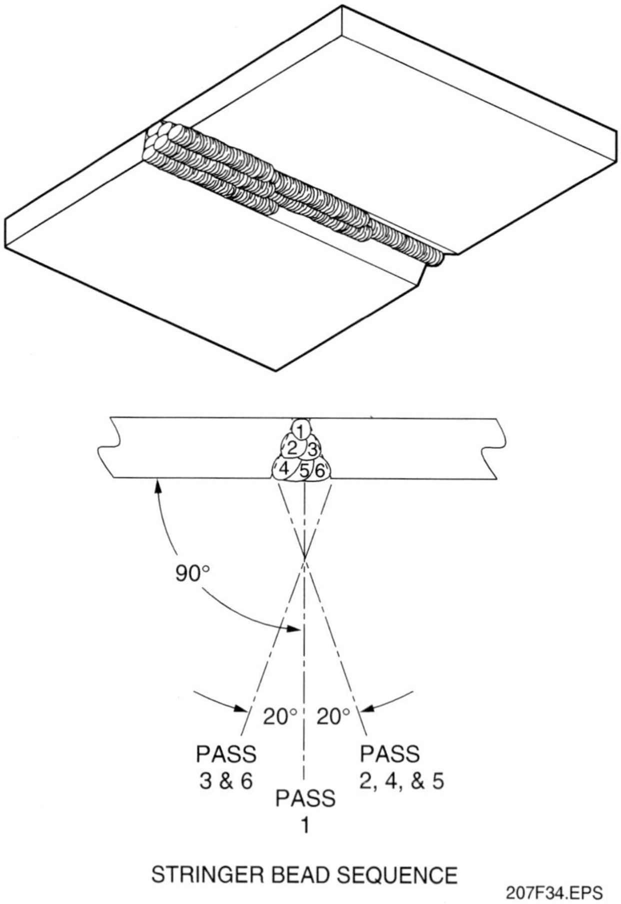

Figure 34 ◆ Multipass 4G weld sequences and work angles.

Summary

Setting up GMAW and/or FCAW equipment, preparing the welding work area, running stringer and weave beads, and making acceptable fillet welds and open-root V-groove welds on plate in all positions are essential skills a welder must have to perform basic welding jobs and to progress to more difficult welding procedures. Practice these welds until you can consistently produce acceptable welds as defined in the criteria for acceptance.

Review Questions

1. In all types of flux cored wires, the flux contains agents for degassing the weld, stabilizing the arc, and _____ the weld bead.

a. removing the slag from
b. forming a protective slag over
c. removing contaminants from
d. strengthening

2. If the welding voltage is set too high, the arc will be too long and could cause _____.

a. the wire to melt in the tip
b. excessive root penetration
c. overlap
d. longitudinal cracking

3. When the wire electrode speed is increased, the result is usually _____.

a. deeper penetration
b. lower welding heat
c. flatter bends
d. lower amperage

4. In GMAW/FCAW, weld penetration is most affected by _____.

a. current
b. voltage
c. travel speed
d. gun position

5. The length of the wire that extends beyond the contact tip of the welding gun's contact tube is the _____.

a. standoff distance
b. contact tip extension
c. electrode extension
d. stickout

6. When running string beads, stay in the leading edge of the puddle to prevent _____.

a. non-fusion
b. cracking
c. porosity
d. undercuts

7. Weave beads are made with _____ motion of the electrode.

a. little or no side-to-side
b. wide side-to-side
c. little or no back-to-back
d. wide back-to-back

8. If a restart has undercut, _____.

a. too much time was spent in the crater
b. not enough time was spent in the crater
c. there were arc strikes outside the weld area
d. there was not enough side-to-side motion

9. A weld termination normally leaves _____.

a. a crack
b. porosity
c. a defect
d. a crater

10. Deposit connective weld beads parallel to one another to make _____ beads.

a. stringer
b. weave
c. solid
d. overlapping

11. When making flat fillet welds, increase or decrease the travel speed to control the amount of _____.

a. porosity
b. base metal fusion
c. weld metal buildup
d. root penetration

12. For the first bead in a fillet weld, the work angle is _____° to both plate surfaces.

a. 15
b. 20
c. 45
d. 55

13. Groove welds can be made in _____.

a. the 1G and 2G positions only
b. all positions
c. the 2G and 3G positions only
d. the 1G, 2G, and 3G positions only

14. To make a vertical groove weld with weave beads, use a zigzag pattern across the root bead and pause at each toe _____.

a. for good fusion
b. to prevent excessive spatter
c. for complete penetration
d. to fill in the crater

15. Compared to flat or horizontal welds, out-of-position welds are made at _____.

a. lower voltages and currents
b. higher voltages and currents
c. lower voltages and higher currents
d. higher voltages and lower currents

GLOSSARY

Trade Terms Introduced in This Module

Axis: The centerline of the weld or prepared base metal.

Degassing: Removing the gases that are formed during the welding process. These gases can cause imperfections in the weld.

Oscillation: Side-to-side movement of the electrode during welding.

APPENDIX

Performance Accreditation Tasks

The Performance Accreditation Tasks (PATCs) correspond to and support the learning objectives in AWS *EG2.0-95, Guide for the Training and Qualification of Welding Personnel: Entry-Level Welder*.

PATCs provide specific acceptable criteria for performance and help to ensure a true competency-based welding program for students.

The following tasks are designed to evaluate your ability to run beads and fillet welds with GMAW or FCAW equipment. Perform each task when you are instructed to do so by your instructor. As you complete each task, show it to your instructor for evaluation. Do not proceed to the next task until instructed to do so by your instructor.

MAKE A FILLET WELD IN THE (1F) FLAT POSITION

As directed by the instructor, use GMAW, FCAW, or both processes with carbon steel solid or flux cored electrode wire sizes, and, if required, appropriate shielding gas to make a six-pass fillet weld using stringer beads on carbon steel plate, as shown.

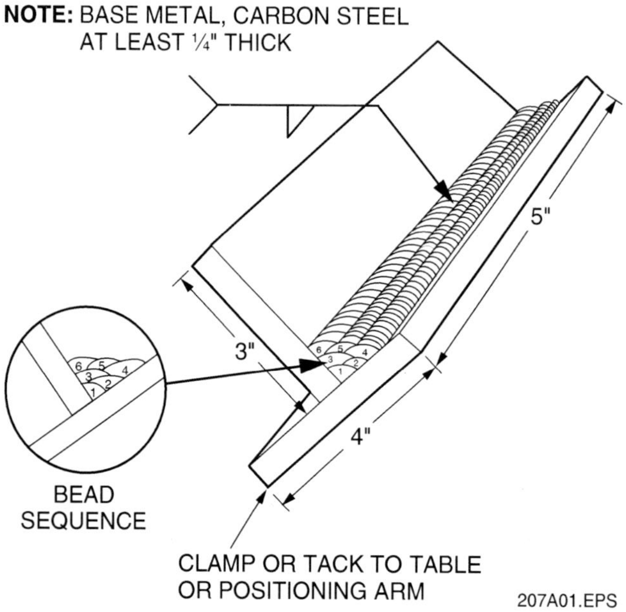

Criteria for Acceptance

- Uniform rippled appearance on the bead face __________
- Craters and restarts filled to the full cross section of the weld __________
- Uniform weld size ±1/16" __________
- Smooth flat transition with complete fusion at the toes of the welds __________
- Acceptable weld profile in accordance with the applicable code or standard __________
- No porosity __________
- No excessive undercut __________
- No overlap __________
- No inclusions __________

MAKE A FILLET WELD IN THE (2F) HORIZONTAL POSITION

As directed by the instructor, use GMAW, FCAW, or both processes with carbon steel solid or flux cored electrode wire sizes, and, if required, appropriate shielding gas to make a six-pass fillet weld using stringer beads on carbon steel plate, as shown.

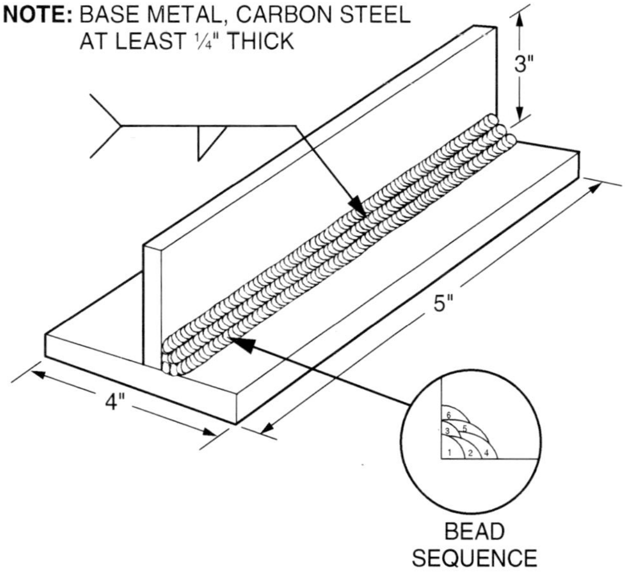

Criteria for Acceptance

- Uniform rippled appearance on the bead face ________
- Craters and restarts filled to the full cross section of the weld ________
- Uniform weld size ±1⁄16" ________
- Smooth flat transition with complete fusion at the toes of the welds ________
- Acceptable weld profile in accordance with the applicable code or standard ________
- No porosity ________
- No excessive undercut ________
- No overlap ________
- No inclusions

MAKE A FILLET WELD IN THE (3F) VERTICAL POSITION

As directed by the instructor, use GMAW, FCAW, or both processes with carbon steel solid or flux cored electrode wire sizes, and, if required, appropriate shielding gas to make a vertical fillet weld on carbon steel plate, as shown.

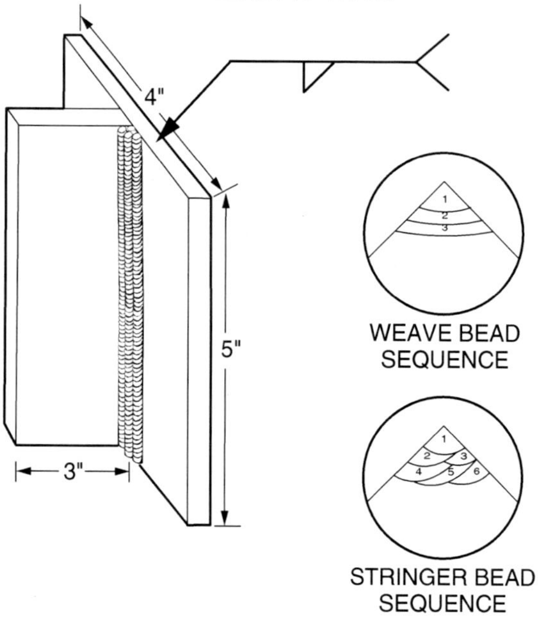

207A03.EPS

Criteria for Acceptance

- Uniform rippled appearance on the bead face ____________
- Craters and restarts filled to the full cross section of the weld ____________
- Uniform weld width ±1⁄16" ____________
- Acceptable weld profile in accordance with the applicable code or standard ____________
- Smooth flat transition with complete fusion at the toes of the weld ____________
- No porosity ____________
- No excessive undercut ____________
- No inclusions ____________
- No cracks

MAKE A FILLET WELD IN THE (4F) OVERHEAD POSITION

As directed by the instructor, use GMAW, FCAW, or both processes with carbon steel solid or flux cored electrode wire sizes, and, if required, appropriate shielding gas to make a six-pass fillet weld using stringer beads on carbon steel plate, as shown.

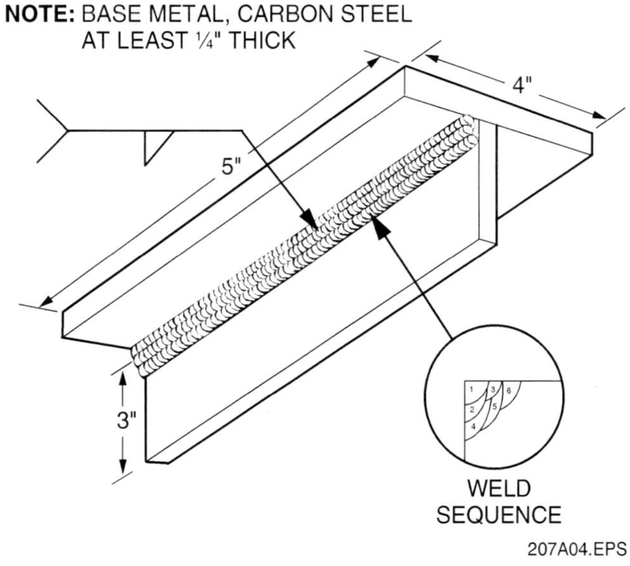

Criteria for Acceptance

- Uniform rippled appearance on the bead face ______
- Craters and restarts filled to the full cross section of the weld ______
- Uniform weld width ±1⁄16" ______
- Acceptable weld profile in accordance with the applicable code or standard ______
- Smooth flat transition with complete fusion at the toes of the weld ______
- No porosity ______
- No excessive undercut ______
- No inclusions ______
- No cracks ______

MAKE A GROOVE WELD IN THE (1G) FLAT POSITION

As directed by the instructor, use GMAW, FCAW, or both processes with carbon steel solid or flux cored electrode wire sizes, and, if required, appropriate shielding gas to make a multipass groove weld using stringer beads on carbon steel plate, as shown.

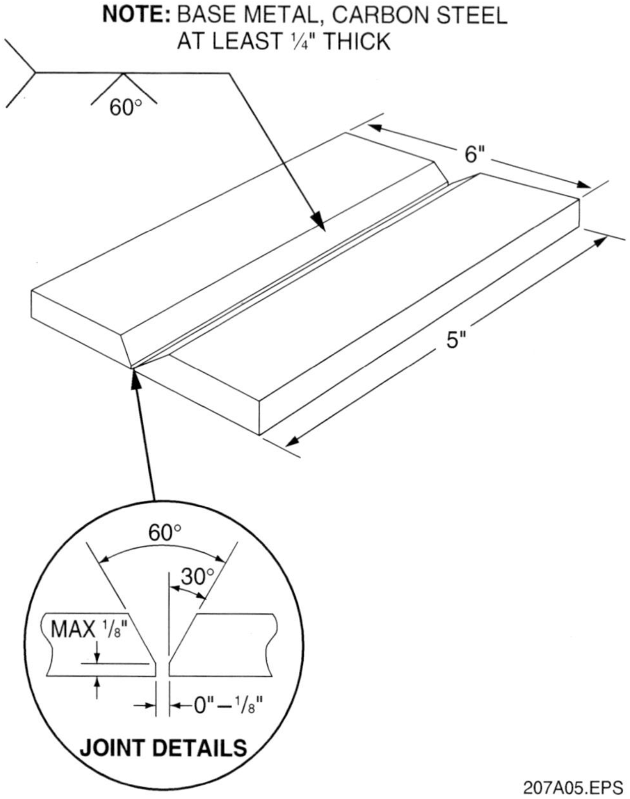

207A05.EPS

Criteria for Acceptance

- Uniform rippled appearance on the bead face ______
- Craters and restarts filled to the full cross section of the weld ______
- Uniform weld width ±1⁄16" ______
- Acceptable weld profile in accordance with the applicable code or standard ______
- Complete uniform root reinforcement at least flush with the base metal to a maximum buildup of 1⁄16" ______
- Smooth flat transition with complete fusion at the toes of the weld ______
- No porosity ______
- No excessive undercut ______
- No inclusions ______
- No cracks ______

MAKE A GROOVE WELD IN THE (2G) HORIZONTAL POSITION

As directed by the instructor, use GMAW, FCAW, or both processes with carbon steel solid or flux cored electrode wire sizes, and, if required, appropriate shielding gas to make a multipass groove weld on carbon steel plate, as shown.

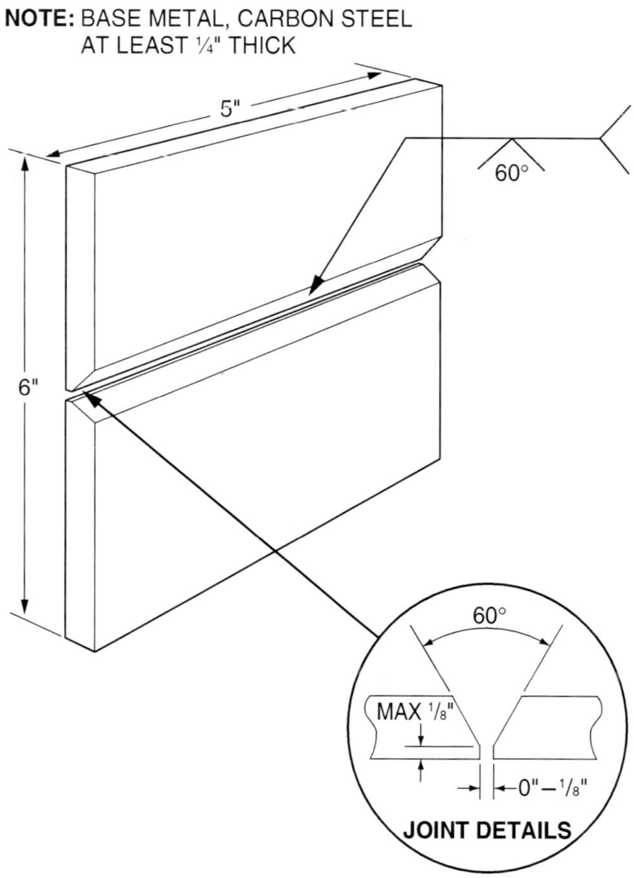

207A06.EPS

Criteria for Acceptance

- Uniform rippled appearance on the bead face ______
- Craters and restarts filled to the full cross section of the weld ______
- Uniform weld width ±1⁄16" ______
- Acceptable weld profile in accordance with the applicable code or standard ______
- Complete uniform root reinforcement at least flush with the base metal to a maximum buildup of 1⁄16" ______
- Smooth flat transition with complete fusion at the toes of the weld ______
- No porosity ______
- No excessive undercut ______
- No inclusions ______
- No cracks ______

MAKE A GROOVE WELD IN THE (3G) VERTICAL POSITION

As directed by the instructor, use GMAW, FCAW, or both processes with carbon steel solid or flux cored electrode wire sizes, and, if required, appropriate shielding gas to make a multipass groove weld on carbon steel plate, as shown.

Note: Run the root vertical up or vertical down as specified by your instructor.

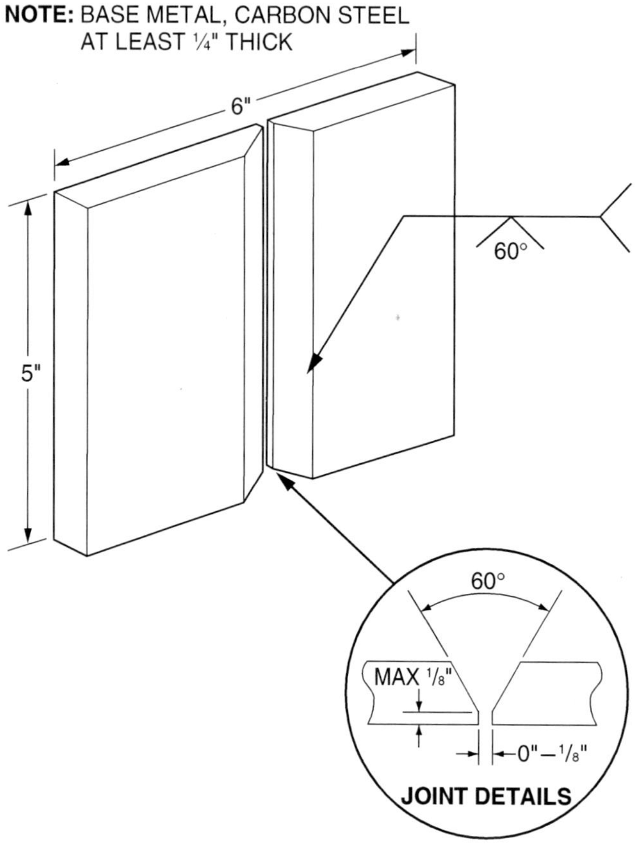

207A07.EPS

Criteria for Acceptance

- Uniform rippled appearance on the bead face ________
- Craters and restarts filled to the full cross section of the weld ________
- Uniform weld width ±1/16" ________
- Acceptable weld profile in accordance with the applicable code or standard ________
- Complete uniform root reinforcement at least flush with the base metal to a maximum buildup of 1/16" ________
- Smooth flat transition with complete fusion at the toes of the weld ________
- No porosity ________
- No excessive undercut ________
- No inclusions ________
- No cracks ________

MAKE A GROOVE WELD IN THE (4G) OVERHEAD POSITION

As directed by the instructor, use GMAW, FCAW, or both processes with carbon steel solid or flux cored electrode wire sizes, and, if required, appropriate shielding gas to make a multipass groove weld on carbon steel plate, as shown.

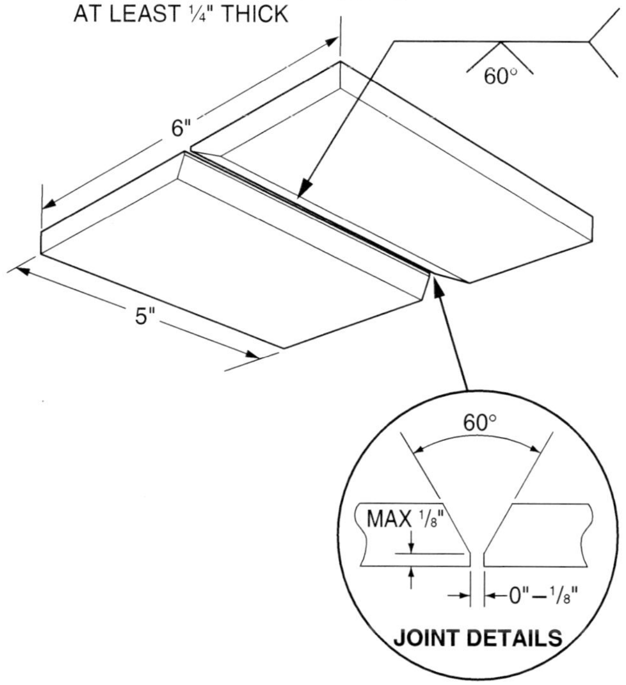

207A08.EPS

Criteria for Acceptance

- Uniform rippled appearance on the bead face ________
- Craters and restarts filled to the full cross section of the weld ________
- Uniform weld width ±1⁄16" ________
- Acceptable weld profile in accordance with the applicable code or standard ________
- Complete uniform root reinforcement at least flush with the base metal to a maximum buildup of 1⁄16" ________
- Smooth flat transition with complete fusion at the toes of the weld ________
- No porosity ________
- No excessive undercut ________
- No inclusions ________
- No cracks ________

Additional Resources

This module is intended to be a thorough resource for task training. The following reference works are suggested for further study. These are optional materials for continued education rather than for task training.

ANSI/AWS C5.6 Recommended Practices for Gas Metal Arc Welding. Miami, FL: The American Welding Society

Protection from Step and Touch Potentials – 1910.269 App C (Standards – 29 CFR), Current Edition. Washington, DC: Occupational Safety and Health Administration (OSHA).

Welding Handbook, Current Edition. Miami, FL: The American Welding Society.

Welding Skills, Current Edition. R.T. Miller. Homewood, IL: American Technical Publishers, Inc.

Welding Technology, Current Edition. J.W. Giachino, W.R. Weeks, and G.S. Johnson. Homewood, IL: American Technical Publishers, Inc.

Welding Technology, Current Edition. Gower A. Kennedy. Indianapolis, IN: Macmillan Publishing Company, Inc.

Figure Credits

Miller Electric Manufacturing Company	207SA01, 207F03, 207F04
Gerald Shannon	207F05, 207SA02, 207F12, 207F13, 207F14, 207F15, 207F23, 207F24
Copyright 2002, TWI, Ltd.	207SA03
Rex Ball	207F19
John Yochum	207SA05
Charles Rogers	207F31, 207F32, 207F33

CONTRENT™ LEARNING SERIES — USER UPDATES

The NCCER makes every effort to keep these textbooks up-to-date and free of technical errors. We appreciate your help in this process. If you have an idea for improving this textbook, or if you find an error, a typographical mistake, or an inaccuracy in NCCER's Contren™ textbooks, please write us, using this form or a photocopy. Be sure to include the exact module number, page number, a detailed description, and the correction, if applicable. Your input will be brought to the attention of the Technical Review Committee. Thank you for your assistance.

Instructors – If you found that additional materials were necessary in order to teach this module effectively, please let us know so that we may include them in the Equipment/Materials list in the Instructor's Guide.

Write: Curriculum Revision and Development Department
National Center for Construction Education and Research
P.O. Box 141104, Gainesville, FL 32614-1104

Fax: 352-334-0932

E-mail: curriculum@nccer.org

Craft ______________ Module Name ______________

Copyright Date ______________ Module Number ______________ Page Number(s) ______________

Description ______________

____

____

____

(Optional) Correction ______________

____

____

(Optional) Your Name and Address ______________

____

____

Module 29208-03

GTAW – Equipment and Filler Materials

COURSE MAP

This course map shows all of the modules in the second level of the Welding curriculum. The suggested training order begins at the bottom and proceeds up. Skill levels increase as you advance on the course map. The local Training Program Sponsor may adjust the training order.

WELDING LEVEL TWO

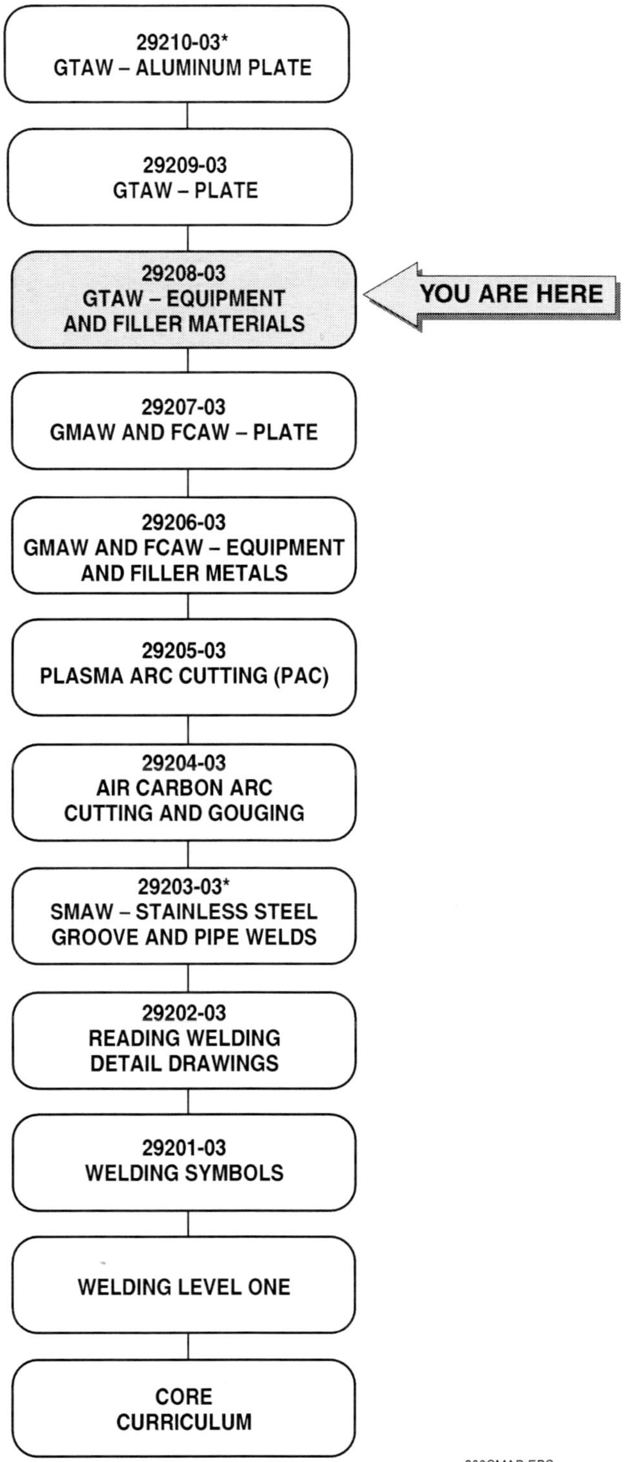

208CMAP.EPS

Please note that Modules 29203-03 and 29210-03 are electives for *Welding Level Two* but are requirements for *AWS Entry Level Welder.

MODULE 29208-03 CONTENTS

MODULE 29208-03 CONTENTS (Continued)

Figures

Tables (*Appendix*)

GTAW – Equipment and Filler Materials

Objectives

When you have completed this module, you will be able to do the following:

1. Explain gas tungsten arc welding (GTAW) safety.
2. Identify and explain the use of GTAW equipment.
3. Identify and explain the use of GTAW filler metals.
4. Identify and explain the use of GTAW shielding gases.
5. Set up GTAW equipment.

Prerequisites

Before you begin this module, it is recommended that you successfully complete the following: Core Curriculum; Welding Level One; Welding Level Two, Modules 29201-03 through 29207-03.

Required Trainee Materials

1. Pencil and paper
2. Appropriate personal protective equipment

1.0.0 ◆ Introduction

Gas **tungsten** arc welding (GTAW) is an arc welding process. It uses a welding power source to produce an electric arc between a tungsten electrode and the base metal. The arc melts and fuses the base metal to form the weld. If required, a filler metal can be added. In manual welding, the filler metal is usually a hand-held metal rod or wire that is manually fed into the leading edge of the weld metal pool. An inert gas or gas mixture is used to shield the electrode and molten weld metals to prevent oxidation and contamination by the atmosphere. The finished weld has no slag because no flux is used. Also, since no filler metal is transported within the arc, the process produces little or no spatter. GTAW is used to weld a variety of metals, especially aluminum. The process is similar to oxyfuel welding and is slower than GMAW; however, it produces high quality welds. GTAW has been commonly used for manual root pass welds on pipe where the subsequent filler and cover passes are performed with manual GMAW or FCAW. In the past, GTAW was called tungsten inert gas (TIG) welding. Today, the nonstandard term *TIG* is still used, especially in the control nomenclature marked on some welding machines. *Figure 1* shows manual GTAW being performed.

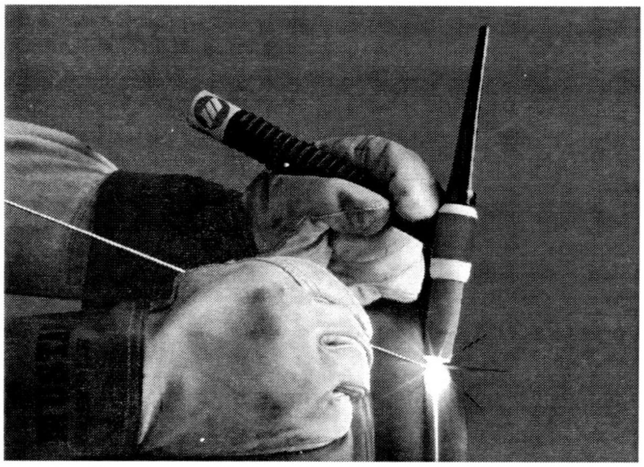

208F01.EPS

Figure 1 ◆ Performing manual GTAW.

This module provides an overview of the different items of equipment used for GTAW. Topics include safety and safety equipment, GTAW equipment, electrode types and preparation, filler metal types and classifications, shielding gas characteristics and gas systems, and equipment setup. Upon completion of this module, trainees will be aware of the safety concerns associated with GTAW and will be able to select and set up GTAW equipment safely and efficiently.

1.1.0 Safety Practices

The following is a summary of safety procedures and practices that must be observed while cutting or welding. Keep in mind that this is just a summary. Complete safety coverage is provided in the Level One module, *Welding Safety*. If you have not completed that module, do so before continuing. Above all, be sure to wear appropriate protective clothing and equipment when welding or cutting.

1.1.1 Protective Clothing and Equipment

- Always use safety goggles with a full face shield or a helmet. The goggles, face shield, or helmet lens must have the proper light-reducing tint for the type of welding or cutting to be performed. Never directly or indirectly view an electric arc without using a properly tinted lens.
- Wear proper protective leather and/or flame retardant clothing along with welding gloves that will protect you from flying sparks and molten metal, as well as heat.
- Wear 8" or taller high-top safety shoes or boots. Make sure that the tongue and lace area of the footwear will be covered by a pant leg. If the tongue and lace area is exposed or the footwear must be protected from burn marks, wear leather spats under the pants or chaps and over the front top of the footwear.
- Wear a solid material (non-mesh) hat with a bill pointing to the rear or, if much overhead cutting or welding is required, a full leather hood with a welding face plate and the correct tinted lens. If a hard hat is required, use a hard hat that allows the attachment of rear deflector material and a face shield.
- If a full leather hood is not worn, wear a face shield and snug fitting welding goggles over safety glasses for gas welding or cutting. Either the face shield or the lenses of the welding goggles must be an approved shade five or six filter. For electric arc welding or cutting, wear safety goggles or a welding hood with the correct tinted lens (shade 5 to 14).
- If a full leather hood is not worn, wear earplugs to protect your ear canals from sparks.

1.1.2 Fire/Explosion Prevention

- Never carry matches or gas-filled lighters in your pockets. Sparks can cause the matches to ignite or the lighter to explode, causing serious injury.
- Never perform any type of heating, cutting, or welding until a hot-work permit is obtained and an approved fire watch is established. Most work-site fires caused by these types of operations are started by cutting torches.
- Never use oxygen to blow off clothing. The oxygen can remain trapped in the fabric for a time. If a spark hits the clothing during this time, the clothing can burn rapidly and violently out of control.
- Make sure that any flammable material in the work area is moved or shielded by a fire-resistant covering. Approved fire extinguishers must be available before attempting any heating, welding, or cutting operations.
- Always comply with any site requirement for a hot-work permit and/or a fire watch.
- Never release a large amount of oxygen or use oxygen as compressed air. Its presence around flammable materials or sparks can cause rapid and uncontrolled combustion. Keep oxygen away from oil, grease, and other petroleum products.
- Never release a large amount of fuel gas, especially acetylene. Methane and propane tend to concentrate in and along low areas and can ignite at a considerable distance from the release point. Acetylene is lighter than air but is even more dangerous than methane. When mixed with air or oxygen, it will explode at much lower concentrations than any other fuel.
- To prevent fires, maintain a neat and clean work area and make sure that any metal scrap or slag is cold before disposal.
- Before cutting or welding containers such as tanks or barrels, check to see if they have contained any explosive, hazardous, or flammable materials including petroleum products, citrus products, or chemicals that decompose into toxic fumes when heated. As a standard practice, always clean and then fill any tanks or barrels with water or purge them with a flow of inert gas to displace any oxygen.

1.1.3 Work Area Ventilation

- Make sure confined space procedures are followed before conducting any welding or cutting in the confined space.
- Never use oxygen in confined spaces for ventilation purposes.
- Always perform cutting or welding operations in a well-ventilated area. Cutting or welding operations involving zinc or cadmium materials or coatings result in toxic fumes. For long-term cutting or welding of such materials, always wear an approved full-face, supplied-air respirator (SAR) that uses breathing air supplied outside of the work area. For occasional, very short-term exposure, a high efficiency particulate arresting (HEPA)-rated or metal-fume filter may be used on a standard respirator.
- Make sure confined spaces are ventilated properly for cutting or welding purposes.

2.0.0 ◆ WELDING CURRENT

The welding arc required for GTAW is produced by electrical current from a constant-current welding power source. Electrical current is the flow of electrons along a conductor. The current flows through two welding cables, one connected to the base metal (work) and the other connected to the torch, which holds a tungsten electrode. In welding, an arc is established in the gap between the end of the tungsten electrode and the work. The gap has a high resistance. When the electrical current meets resistance, heat is generated. When the arc is struck across the gap, the resistance generates intense heat of 6,000 to 10,000 degrees, melting the base metal and forming a weld. The area surrounding the arc is flooded with a shielding gas to prevent oxidation of the filler rod and base metal surface during the welding process.

Figure 2 diagrams the GTAW welding process.

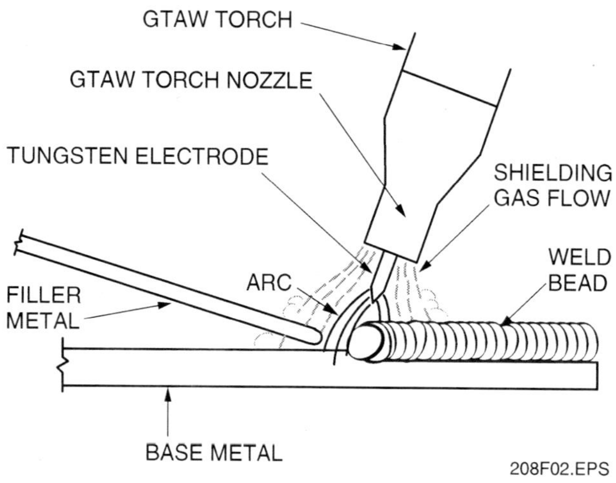

Figure 2 ◆ Diagram of GTAW welding process.

2.1.0 Characteristics of Welding Current

The current produced by a welding machine to perform welding has different characteristics than the current flowing through utility power lines. Welding current has low voltage and high amperage, while the power line current has high voltage and low amperage.

2.1.1 Voltage

Voltage is the measure of the electromotive force or pressure that causes current to flow in a circuit. There are two types of voltage associated with welding current: open-circuit voltage and operating voltage. Open-circuit voltage is the voltage present when the machine is on but no welding is taking place. For GTAW, there are usually ranges of open-circuit voltages that can be selected, up to about 80V. Operating voltage, or arc voltage, is the voltage after the arc is struck. With GTAW, this voltage is generally much lower than the open-circuit voltage. The arc voltage drops rapidly with increasing current flow.

2.1.2 Amperage

Amperage is the electrical current flow in a circuit. The unit of measurement for amperage is the ampere (amp). The number of amperes produced by the welding machine determines the intensity of the arc and the amount of heat generated. The higher the amperage, the hotter the arc. Usually, the welding machine current used in GTAW is controlled by a finger control or foot pedal.

2.2.0 Types of Welding Current

There are two types of welding current: direct current (DC) and alternating current (AC). The type of current selected depends on the equipment used and the type of welding to be done.

2.2.1 DC Welding Current

DC is electrical current that flows in one direction only. The direction the current is flowing is called the polarity.

2.2.2 DC Polarity

Polarity only applies to DC. The direction in which the current flows in a circuit determines polarity. Direct current electrode negative (DCEN) is almost always used for performing GTAW.

When the welding circuit is set up as DCEN, the tungsten electrode of the GTAW torch is connected to the negative terminal of the welding machine and the workpiece (lead) is connected to the positive terminal. DCEN produces the greatest amount of heat in the base metal. About two-thirds of the arc heat goes to the base metal and one-third goes to the tungsten electrode. Because of this, a small-size tungsten electrode with a pointed end that is **truncated** (blunted) is usually used. The weld bead produced is narrow with deep penetration. DCEN is used to weld steels, stainless steels, nickel, and titanium. Because this type of current has no cleaning action, the base metal surfaces must be cleaned of any contaminants such as rust, mill scale, or oxides.

Direct current electrode positive (DCEP) produces the least amount of heat in the base metal. About one-third of the arc heat goes to the base metal and two-thirds goes to the tungsten electrode. Because of this, a large tungsten electrode with a rounded end is usually required. The weld bead produced is wide with shallow penetration. This current type also has a strong cleaning action on the surface of oxidized base metals. DCEP is only used for special applications when shallow penetration is required. *Figure 3* shows DCEN and DCEP welding circuits.

> **Hot Tip**
>
> ***Polarity Switches***
>
> Many welding machines have a polarity switch. Instead of physically having to disconnect the welding leads and reconnect them to change polarity, polarity is changed by turning a switch.

2.2.3 AC Welding Current

AC is electrical current that alternates (cycles) between positive and negative values. In the positive half of the cycle, the current flows in one direction; during the negative half, the current reverses itself. The number of cycles completed in one second is called frequency. In the United States, AC is almost always 60 cycles per second. *Figure 4* shows one cycle of AC.

Alternating current is DCEN for about half the time and DCEP for the remainder of the time. For this reason, its characteristics fall in between those of DCEN and DCEP. About half of the heat goes to the base metal and the other half to the tungsten electrode. The weld bead size and penetration are midway between the beads produced by DCEN and DCEP. Some cleaning action occurs with AC. The AC tungsten has a small, rounded tip. AC is used with GTAW to weld aluminum, magnesium, and their alloys.

2.2.4 High Frequency

High frequency is electrical current with a very high voltage (3,000V to 5,000V) and very low amperage. It is called high frequency because it cycles at millions of times per second compared to standard current, which has a frequency of 60 cycles per second. This high-voltage, high-frequency current creates a very stable arc able to jump a gap of about ½". Because high-frequency current has such low amperage, it creates very little heat.

High frequency has two purposes: to stabilize AC and allow a DC arc to be started without touching the tungsten electrode to the work.

When AC is used with GTAW, each time the arc cycles, it goes out and must be reestablished in the opposite direction. Even though this happens very fast (60 cycles per second), it can cause the arc to extinguish on the reverse polarity part of the AC cycle when the electron flow is weakest. If this happens, oxides are absorbed into the weld, causing porosity, which weakens the weld. To prevent the AC arc from extinguishing, a high-frequency current is added to the alternating welding current. The high-frequency current, with its high voltage and low amperage, establishes a path for the alternating welding current to travel on, preventing it from extinguishing as it cycles. When non-squarewave AC is being used, the high frequency must be on continuously.

When using DC for GTAW without high frequency, the tungsten must be touched to the work to establish the arc. Except for newer GTAW machines, this may not be desirable since it can cause contamination of the electrode and defects in the weld. To prevent this, high-frequency current can be added to the direct welding current. When the tungsten electrode is brought close to the work, the high-voltage, high-frequency arc jumps the gap, establishing a path for the direct welding current to travel without touching the tungsten to the work. Typically, a timer will automatically shut off the high frequency about two to three seconds after the welding arc is established. Since DC flows in only one direction, it is very stable. The high frequency is only needed for starting the arc.

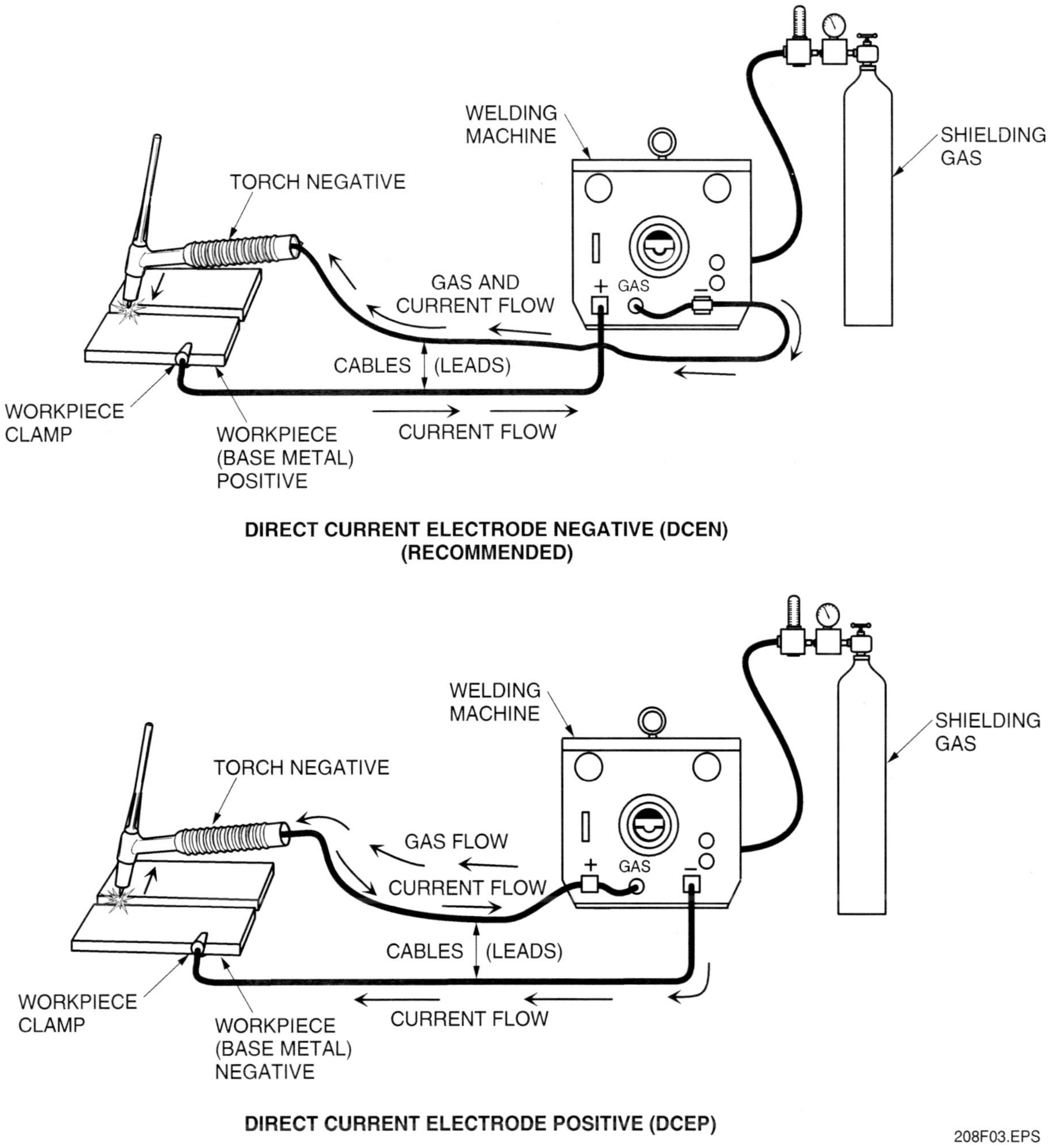

Figure 3 ◆ Representative DCEN and DCEP welding circuits.

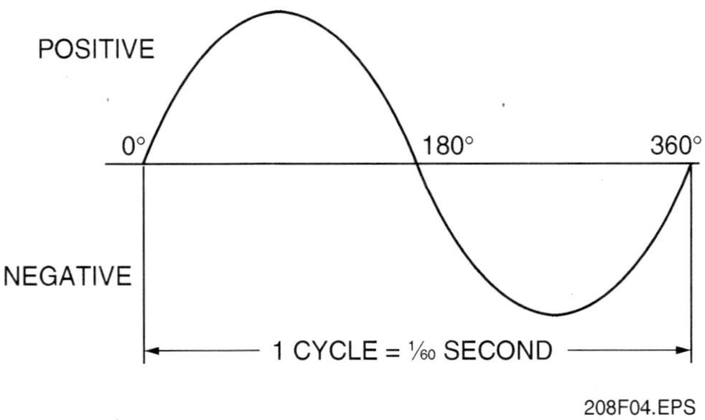

Figure 4 ◆ AC cycle.

Welding power sources specifically designed for GTAW have a high-frequency generator built into them. If conventional power sources are used for GTAW, a separate auxiliary high-frequency generator can be attached if required. The generators require 110VAC to operate. Some auxiliary high-frequency generators only have high-frequency controls; others may have a combination of high-frequency controls and gas and cooling water controls. Most will have a torch switch that plugs into the console. The switch attaches to the torch and can be used to turn the power, shielding

Starting DC GTAW Without High Frequency

DC GTAW is often performed without a high-frequency start for noncritical welds. The arc is touch-started in the weld zone or on a piece of copper. Copper is less likely to contaminate the tungsten. Some machines have a built-in selectable no high-frequency start mode.

gas, and cooling water (if used) on or off. The auxiliary high-frequency generator is connected to the power source by the electrode cable. The GTAW torch is connected to the auxiliary high-frequency generator.

Figure 5 shows an auxiliary high-frequency generator with a non-adjustable gas postflow timer.

3.0.0 ◆ WELDING POWER SOURCE TYPES

GTAW is performed using constant-current welding machines. The type of welding current (DCEN, DCEP) depends on the base metal type and thickness.

Welding current for GTAW may be supplied by many types of welding machines. Usually, any

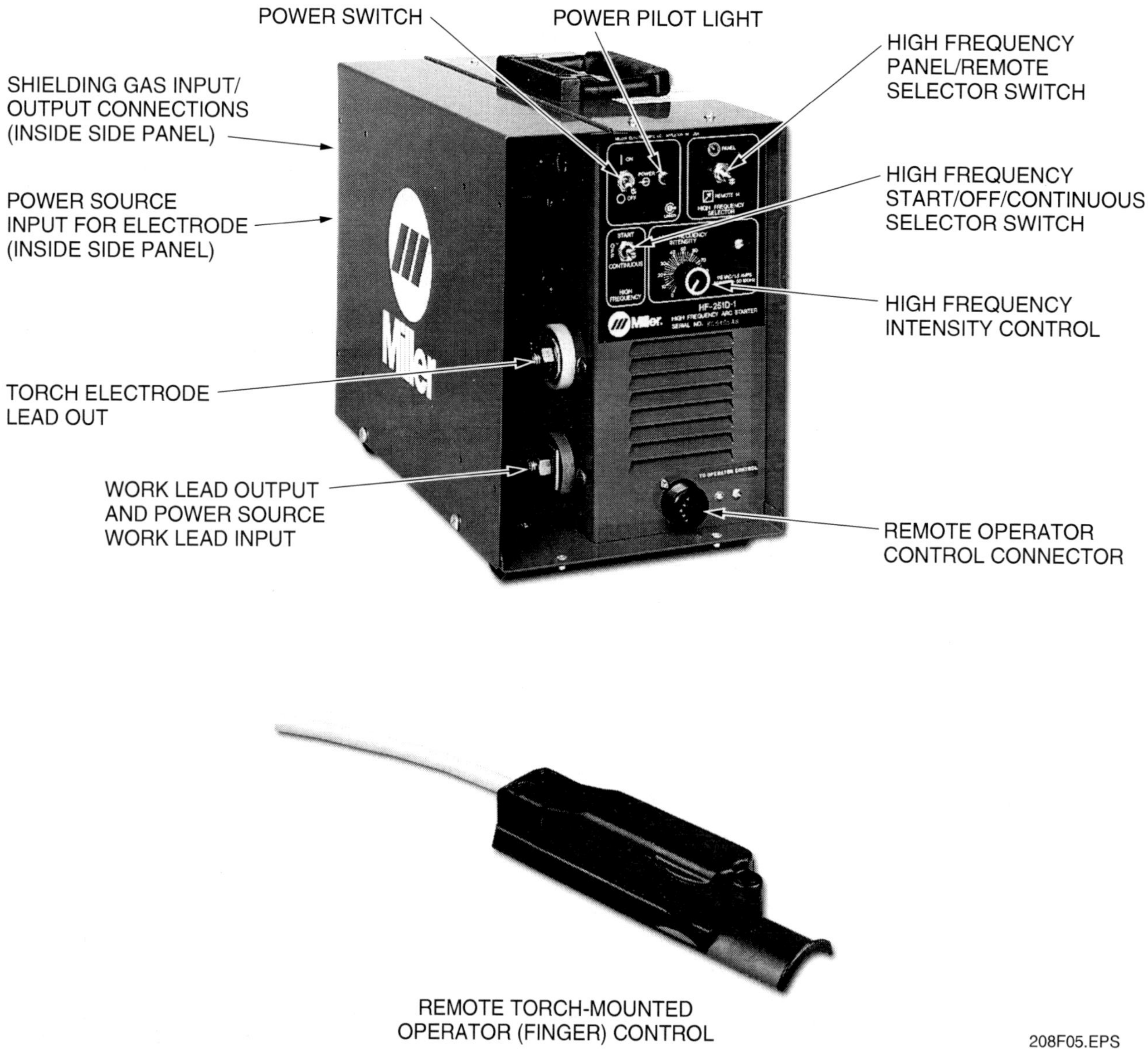

Figure 5 ◆ Auxiliary high-frequency generator.

welding machine that produces constant current, such as those used for shielded metal arc welding (SMAW), can be used. In addition, there are welding machines specially designed for GTAW. These special machines have built-in high frequency and controls for shielding gas and cooling water. Types of welding machines that can be used for GTAW include:

- Transformer welding machines
- Transformer-rectifier welding machines
- Engine-driven generator and alternator welding machines
- Inverter power sources

DC welding machines are usually designed to produce either constant-current or constant-voltage (constant-potential) direct welding current. Constant current means that welding current is produced over a wide voltage range regardless of arc length. This is typical of a SMAW or GTAW welding machine. A constant-voltage machine maintains a constant voltage as the output current varies. This is typical of a gas metal arc welding (GMAW) or flux cored arc welding (FCAW) machine. The output voltages and output currents of a welding machine can be plotted on a graph to form a curve. These curves show how the output voltage relates to the output current as either changes. *Figure 6* shows constant-current and constant-voltage output curves.

3.1.0 Transformer Welding Machines

Transformer welding machines produce alternating welding current only. They use a voltage step-down transformer, which converts high-voltage, low-amperage current from commercial power lines to low-voltage, high-amperage welding current. The primary current required for a transformer welder can be 120V or 230V single-phase or 460V three-phase. Special light-duty transformer welding machines used for sheet metal work are designed to be plugged into a 120V outlet. However, most light-duty transformer welding machines require 230V primary current. Heavy-duty industrial transformer welders require 460V three-phase primary current.

WARNING!

Coming in contact with the primary current of a welding machine can cause death by electrocution. Ensure that welding machines are properly grounded to prevent injury.

Transformer welders are not as common as other types of welding machines on the job site but are used for special jobs. A transformer welder has an on/off switch, amperage control, and terminals for connecting the electrode lead and the workpiece lead. Transformer welding machines used for GTAW must have high frequency built in or must use an auxiliary high-frequency generator.

3.2.0 Transformer-Rectifier Welding Machines

The transformer-rectifier welding machine uses a transformer to transform the high-voltage, low-amperage primary current to low-voltage, high-amperage welding current and a rectifier to convert the AC to DC. Transformer-rectifier

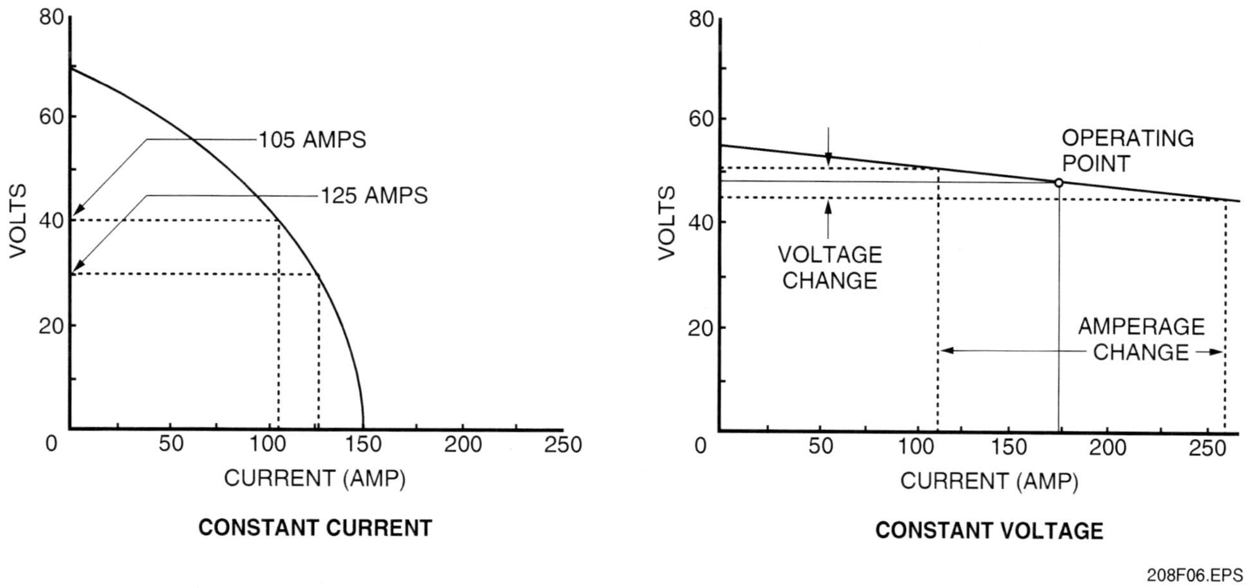

Figure 6 ◆ Constant-current and constant-voltage output curves.

Hot Tip

High-Frequency Operation Precaution

When operating in SMAW mode, the high-frequency generators for most transformer and transformer-rectifier machines must be turned off.

welding machines are designed to produce both alternating and direct welding currents or direct welding current only. Small transformer-rectifier machines that produce both alternating and direct welding currents are usually lighter-duty than those that produce DC only. Transformer-rectifier welding machines that produce direct welding current only are sometimes called rectifiers. Depending on the size, transformer-rectifier welding machines may require 230V single-phase power, 230V three-phase power, or 460V three-phase power.

WARNING!

Coming in contact with the primary current of a welding machine can cause death by electrocution. Ensure that welding machines are properly grounded to prevent injury.

Transformer rectifiers used for GTAW have an on/off switch and an amperage control. The welding cables (electrode lead and workpiece lead) are connected to terminals. They often have selector switches to select DCEN or DCEP. If there is no selector switch, the cables must be manually changed on the machine terminals to select the type of current desired. Most high-amperage machines designed specifically for GTAW are the transformer-rectifier type. Transformer rectifiers designed specifically for GTAW will generally have:

- High-frequency generator
- High-frequency selector switch for DC START, CONTINUOUS (for AC), or OFF (no high frequency)
- Shielding gas preflow timer
- Control to automatically start cooling water flow when an arc is struck
- Shielding gas postflow timer control to prevent weld contamination (keeps shielding gas flowing for a set time after arc is terminated)
- Hand- or foot-operated remote current control
- Remote/local selector switch for remote control
- Start amperage control

GTAW is often used for critical welds on difficult-to-weld materials. For this reason, many special advanced features and controls have been developed and are available on power sources designed for GTAW. These advanced features and controls include:

- High-frequency intensity control for better arc starting
- High-frequency stabilizer control to control tungsten spitting
- Crater fill timer to automatically taper the welding current from the selected setting to a minimum current at the end of a weld
- Balanced wave adjustment for penetration/cleaning control (sets percentage of time AC remains on the straight or reverse side of the cycle)

NOTE

Refer to the machine's operating instruction manual for specific operating information on the controls for the welding machine being used.

Figure 7 shows a typical industrial, high-amperage, heavy-duty transformer-rectifier welding machine designed for GTAW.

3.3.0 Inverter Power Sources

Inverter power sources increase the frequency of the incoming primary power. This provides a smaller, lighter power source with a faster response time and much more waveform control for pulse welding. Inverter power sources are used where limited space and portability are important factors. The controls on these machines vary according to size and application. *Figure 8* is a typical basic GTAW inverter power source rated at 175 amps. This particular welding machine automatically adapts to 120 to 480VAC, single-phase, or three-phase primary power without removing covers or changing internal jumpers. It is very portable at about 40 pounds and provides the following functions:

- *Voltage meter* – This meter displays the open or welding circuit voltage when the welding machine is operating.
- *Parameter display* – This display indicates welding amperage or other functions when appropriate controls are pressed. When the machine is in use, it displays welding amperage.

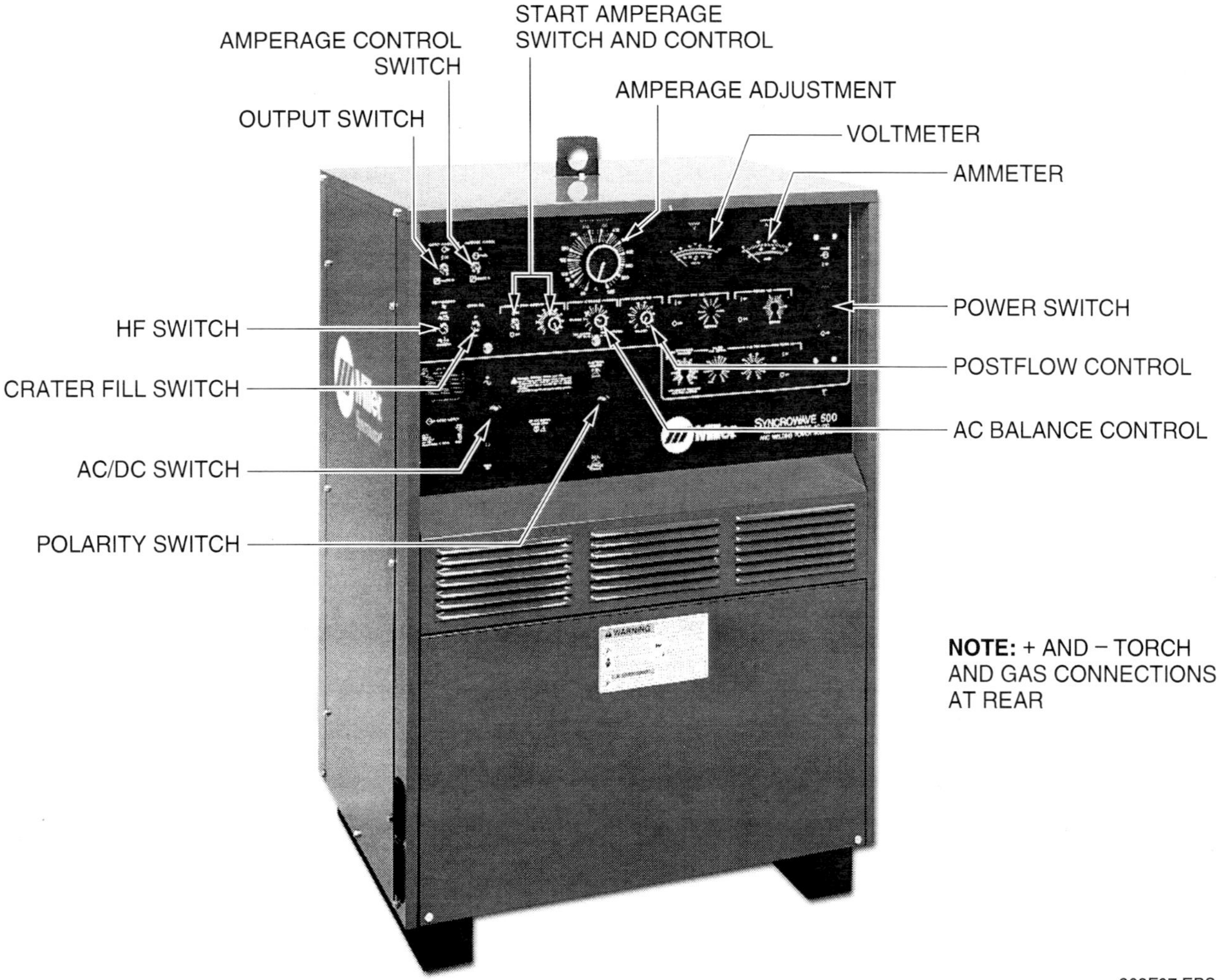

Figure 7 ◆ Transformer-rectifier welding machine designed for GTAW.

- *Encoder control* – This control is used to adjust values on the display for functions selected by other panel controls.
- *Process control* – This control selects a STICK (SMAW DCEP) mode, TIG (GTAW DCEN) HF IMPULSE (high frequency) start mode, or TIG LIFT-ARC™ start mode. The momentary high frequency start enables an electrode non-contact arc start. The LIFT-ARC™ start is a start that does not use high frequency and allows the electrode to heat at a low voltage and high current with a gas preflow to avoid contamination. When the electrode is lifted after several seconds, the arc occurs. The LIFT-ARC™ mode can be used for DCEN or AC TIG starts. STICK (SMAW) starts must be made by the scratch method because only a low sensing voltage is present until the electrode is in contact with the work surface for more than one or two seconds.
- *Output control* – This control selects a constantly ON mode for SMAW welding or lift-arc starting without a remote control. If a remote finger or foot pedal current control is to be used, then the RMT (Remote) STD mode is selected. When long welds are to be made, the RMT 2T HOLD mode can be selected to reduce operator fatigue. In this mode, only a torch trigger is active. It can be pressed and released to start the weld and then pressed and released at the end to stop the weld. Current control is provided only at the welding machine.
- *Amperage control (A)* – When pressed, this control allows the desired output current to be set using the encoder control as indicated on the parameter display with the A light illuminated.
- *Adjust control* – When POSTFLOW is selected, the encoder control is used to set the post flow time in seconds as indicated on the parameter display with the S light illuminated. When the stick mode is active and DIG is selected, the encoder control is used to set the percentage of short circuit current (arc force) available at low arc voltage as indicated on the parameter display with the % light illuminated.

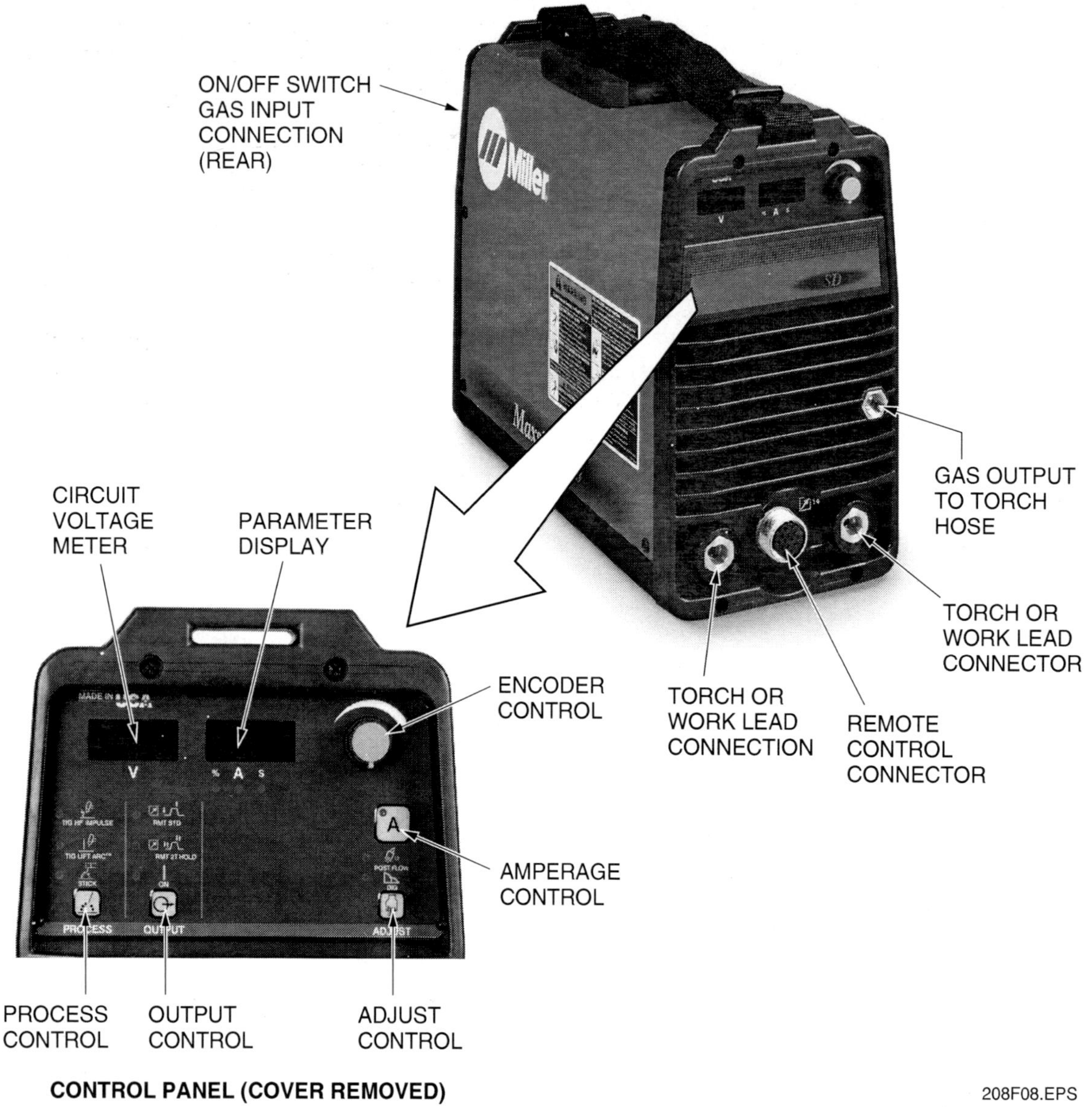

Figure 8 ◆ Basic inverter power supply.

Many production inverter power sources used today are referred to as advanced inverter power sources. They usually include all the controls listed for transformer-rectifier machines and basic inverter power sources plus the following:

- Pulse arc control (allows the welding current to be pulsed between a high-current setting and a low-current setting for puddle control)
- Hot-start current control to enable a controlled surge of welding current to establish a puddle quickly
- Slope-up control for direct hot-start current (controls how quickly the current rises to the hot-start setting)
- Optional spot arc timer control for GTAW spot welding

Figure 9 shows an advanced inverter power source with a liquid cooling unit for torch cooling. Similar inverter power sources are available in capacities of up to 500 amps that weigh approximately 100 pounds.

3.3.1 Advanced Function and Waveform Control Descriptions

The following summarizes the description and application of some of the basic functions and programmable waveform controls provided on the advanced inverter power source control panel

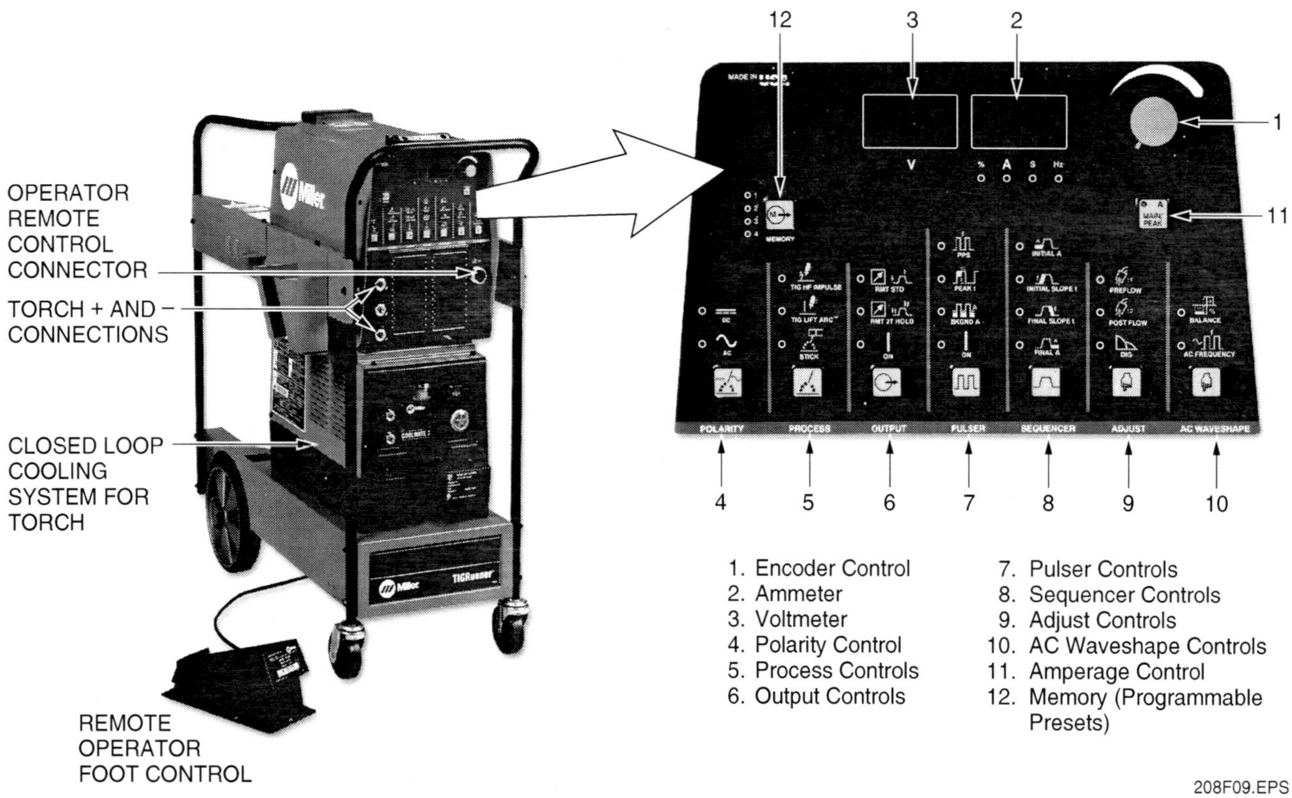

Figure 9 ◆ Advanced inverter power source for GTAW.

shown in *Figure 9*. These descriptions are provided so that the trainee will be aware of the great amount of process and arc control provided by an advanced inverter power source. Similar functions and controls exist on other manufacturers' advanced welding machines.

> **NOTE**
>
> Except for items in parentheses, the capitalized text below represents control panel marking. The panel-mounted ammeter functions not only as an ammeter but also as a readout for parameters that are changed using the panel-mounted encoder control.

- *PROCESS controls* – On this machine, like the basic inverter shown in *Figure 8*, two start modes of TIG (GTAW) welding can be selected: TIG HF IMPULSE or TIG LIFT ARC™. The TIG HF IMPULSE mode allows a short burst of high-frequency energy so that the arc can be started without touching the workpiece. It is used for the DCEN TIG process if a non-contact start is required. The TIG HF IMPULSE amperage, time, and polarity can be adjusted using the ammeter, MAIN/PEAK amperage control, and encoder control when the TIG HF IMPULSE mode is selected. The TIG LIFT ARC™mode reduces the open-circuit voltage so that only a very small sensing voltage exists at the torch tip. When the torch tip is held against the workpiece momentarily, a low voltage and a starting current is established that heats the tungsten, and the preflow shielding gas starts to flow. The instant the torch tip is lifted, the arc is established with normal circuit voltage. This mode is used for a DCEN or AC TIG process if high frequency is prohibited or to replace a scratch start. The remaining selection under *Process* is the conventional STICK (SMAW) mode. However, because of the initial lack of open-circuit voltage, stick electrodes must be scratch started. In all TIG modes, the start is always automatically made with the electrode positive, regardless of polarity selection. This applies maximum heat to the tungsten so that it quickly becomes a better emitter.
- *POLARITY controls* – AC can be selected when either a TIG or STICK mode is selected in *Process*. When DC is selected, the machine is set to electrode negative if a TIG mode is selected or electrode positive if the STICK mode is selected.
- *OUTPUT controls* – On this machine, two standard modes of remote trigger control can be selected for either foot or finger controls at the torch. If the RMT (Remote) STD is selected, an on/off switch, when pressed and held, will start the gas preflow and the program selected

for the SEQUENCER. As programmed, the current will ramp up/down to the selected welding level. When the on/off switch is released, the program selected for the SEQUENCER will, if programmed, ramp the current up/down to a final current, then to off, followed by a cool-down gas post flow. If the RMT 2T HOLD is selected, the same thing will occur, but the on/off switch will be just momentarily activated to start the program. After welding is completed, the switch is again momentarily activated to ramp the current down and off and allow a gas post flow. The advantage of the 2T mode is that the remote on/off switch does not have to be held on during the entire welding time. Other remote trigger modes (4T and spot) can be individually programmed as desired.

- *PULSER controls* – Pulsing is used only during TIG modes and is the periodic raising and lowering of welding amperage. The raised periods are controlled in width (PEAK T), height (peak amperage) and frequency (PPS), producing pulses of weld output. The peak amperage periods, including the low-amperage periods between them (called the background amperage), alternately heat and cool the weld puddle. Precise heat control is extremely important when welding thin gauge steel or stainless steel where burn-through or distortion is a problem. The pulser peak amperage is set on the ammeter using the MAIN/PEAK amperage control and encoder control. Background amperage is set using the ammeter and encoder control when the BKGND A function is selected. The PPS and PEAK T, when selected, are also adjusted using the ammeter and encoder control.
- *SEQUENCER controls* – The sequencer controls are only active when a TIG mode is selected. Each parameter defined by the bolded portion of the waveform in each sequencer selection can be adjusted using the ammeter and encoder control. The INITIAL A selection allows the beginning amperage to be set to assist in preheating cold material prior to depositing filler or to ensure a soft start. The INITIAL SLOPE T selection allows setting a ramp up/down time from the initial amperage to the welding amperage. The FINAL SLOPE T selection allows setting a ramp up/down time to the final amperage to assist when welding crack-sensitive materials or to eliminate the crater at the end of a weld. The FINAL A selection permits setting the final amperage.
- *ADJUST controls* – The PREFLOW and POST FLOW functions are only active if a TIG process has been selected. When selected, shielding gas PREFLOW time can be set using the ammeter and encoder control. The PREFLOW function is used to purge the weld area before the arc starts. The shielding gas POST FLOW function is adjusted in the same manner and is used to cool the tungsten electrode and weld to prevent contamination. The DIG (arc force) function is only active if the STICK process has been selected. When DIG is selected, the amount of short circuit amperage available at low voltage can be increased from normal welding amperage to a higher value. This helps arc starting or allows welding with a very short arc without sticking the electrode in vertical or overhead welds. The value (% increase) is set using the encoder control and ammeter.
- *AC WAVESHAPE controls* – The AC waveshape controls are only active when both an AC mode and TIG mode are selected. When the AC squarewave BALANCE is selected, the ammeter and encoder control can adjust the percentage of time that the torch electrode polarity is negative. This function is used when welding on oxide-forming materials, such as aluminum or magnesium, when minimal cleaning of the work piece has occurred. By increasing the time the electrode is positive, surface oxides on the workpiece are removed from the weld puddle, and a shallower, wider bead is produced. If the weld puddle looks scummy or black flecks are floating on the surface, it is an indication that the electrode positive time should be increased. Increasing electrode negative time creates a deeper, narrower weld bead with better penetration and a smaller etched zone. This helps on thick materials or when appearance is important. Joint configuration, set-up, process variables, and oxide thickness will affect the BALANCE settings. When AC FREQUENCY is selected, the ammeter and encoder are used to adjust the frequency of the AC squarewave. Increasing the frequency focuses the arc, decreases bead and puddle width, and increases travel speed, all of which help when welding in corners, root passes, and fillet welds. Increasing frequency also allows a new aluminum AC welding technique that, combined with increased electrode negative time, uses pointed thoriated or ceriated tungsten electrodes instead of balled electrodes. Decreasing frequency produces wider beads, helps remove impurities from the metal surface, and speeds up heavy metal deposition required for a fill pass.
- *MEMORY controls* – After the machine has been manually set up with parameters that result in a successful weld application, the settings for the application can be retained and recalled

from memory. On this machine, each of four main selections can store four subprograms, resulting in 16 stored application programs.

3.4.0 Engine-Driven Generator and Alternator Welding Machines

Welding machines can also be powered by gasoline or diesel engines. The engine can be connected to a generator or to an alternator. Engine-driven generators produce direct welding current. Engine-driven alternators produce AC that is fed through a rectifier to produce direct welding current.

The size and type of engine used depend on the size of the welding machine. Single-cylinder engines are used to power small rectifier alternators, and six-cylinder engines are used to power larger generators.

To produce welding current, the generator or alternator must turn at a required number of revolutions per minute (rpm). The engines powering alternators and generators have governors that control the engine speed. Most governors have a welding speed switch that can be set to idle the engine when no welding is taking place. When the electrode is touched to the base metal, the governor automatically increases the speed of the engine to the required rpm for welding. After about 15 seconds of no welding, the engine automatically returns to idle. The switch can also be set for the engine to run continuously at the welding speed.

Figure 10 shows an engine-driven generator welding machine.

208F10.EPS

Figure 10 ◆ Engine-driven generator welding machine.

Engine-driven generators and alternators often have an auxiliary power unit to produce 115VAC current for lighting, power tools, and other electrical equipment. When 115VAC current is required, the engine-driven generator or alternator must run continuously at the welding speed.

Engine-driven generators and alternators have engine controls and welding current controls. The engine controls vary with the type and size but normally include:

- Starter
- Voltage gauge
- Temperature gauge
- Fuel gauge
- Hour meter

The welding current controls include:

- Amperage control
- Current range switch
- Amperage and voltage gauge
- Polarity switch

The advantage of engine-driven generators and alternators is that they are portable and can be used in the field where electricity is not available to power other types of welding machines. The disadvantage is that engine-driven generators and alternators are costly to purchase, operate, and maintain.

3.5.0 Power Source Ratings

The rating (size) of a welding machine is determined by the amperage output of the machine at a given duty cycle. The duty cycle of a welding machine is based on a ten-minute period. It is the percentage of ten minutes that the machine can continuously produce its rated amperage without overheating. For example, a machine with a rated output of 300 amps at 60 percent duty cycle can deliver 300 amps of welding current for six minutes out of every ten without overheating.

The duty cycle of a welding machine will be 20%, 30%, 40%, 60%, or 100%. A welding machine having a duty cycle of 20% to 40% is considered a light- to medium-duty machine. Most industrial, heavy-duty machines for manual welding have a 60% or 100% duty cycle. Machines designed for automatic welding operations are 100% duty cycle.

The maximum amperage that a welding machine will produce is always higher than its rated capacity. A welding machine rated 300 amps at 60% duty cycle generally puts out a maximum of 375 to 400 amps. However, since the duty cycle is a function of its rated capacity, the duty cycle will decrease as the amperage is raised over 300 amps. Welding at 375 amps with a welding machine rated 300 amps at 60% duty cycle will lower the duty cycle to about 30%. If welding continues for more than three out of ten minutes, the machine will overheat.

> **NOTE**
> Most welding machines have a heat-activated circuit breaker that will shut off the machine automatically when it overheats. The machine cannot be turned back on until it has cooled.

If the amperage is set below the rated amperage, the duty cycle increases. Setting the amperage at 200 amps for a welding machine rated 300 amps at 60% duty cycle will increase the duty cycle to 100%.

Figure 11 shows the relationship between amperage and duty cycle.

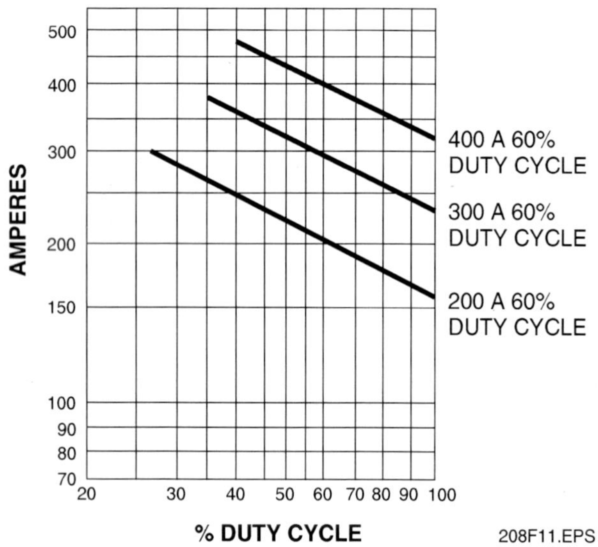

Figure 11 ◆ Relationship between amperage and duty cycle.

4.0.0 ◆ WELDING CABLE

Cables used to carry welding current are designed for maximum strength and flexibility. The conductors inside the cable are made of fine strands of copper wire. The copper strands are covered with layers of rubber reinforced with nylon or dacron cord. *Figure 12* shows the construction of a welding cable.

The size of a welding cable is based on the number of copper strands it contains. Large-diameter cable has more copper strands and can carry more welding current. Typically, the smallest cable size is number 4 and the largest is number 3/0 (3 aught).

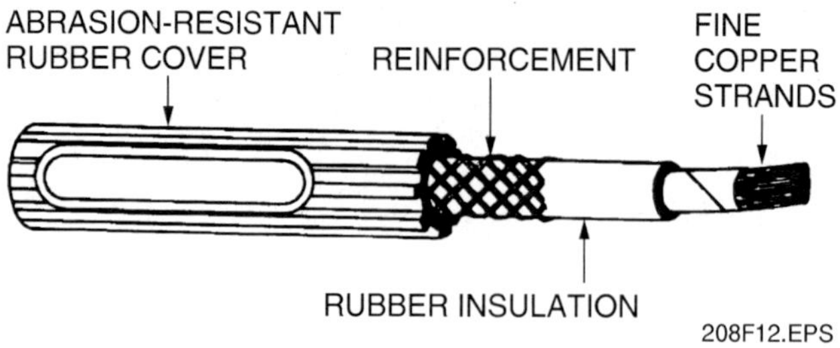

Figure 12 ◆ Construction of a welding cable.

When selecting a welding cable size, consider the amperage load as well as the distance the current will travel. The longer the distance the current has to travel, the larger the cable must be to reduce voltage drop and heating caused by the electrical resistance in the welding cable. When selecting welding cable, use the rated capacity of the welding machine for the cable amperage requirement. For the distance, measure both the electrode and workpiece leads, and add the two lengths together. To identify the welding cable size required, refer to a table of recommended welding cable sizes. These tables are furnished by most welding cable manufacturers.

Table 1 (*Appendix*) shows a typical welding cable size table.

5.0.0 ◆ GTAW EQUIPMENT

In addition to a constant-current power source (and high-frequency generator for AC welding) GTAW requires:

- GTAW torch
- Shielding gas
- Optional torch cooling system

In addition, a remote amperage control is often used with GTAW. *Figure 13* shows a basic GTAW welding system diagram.

5.1.0 GTAW Torches

The GTAW torch provides an electrical path for the welding current between the welding current lead and the tungsten electrode. There are many different manufacturers of torches, but all are basically the same (although parts are not interchangeable). All torches have a collet holder (built in or replaceable) and electrode collet. The electrode collet comes in a variety of sizes to match the size of the electrode being used. The electrode collet holds the electrode in place and is secured by the electrode back cap. The electrode back cap, which is threaded onto the torch, is loosened to change the electrode or to adjust its stickout. The reserve length of electrode

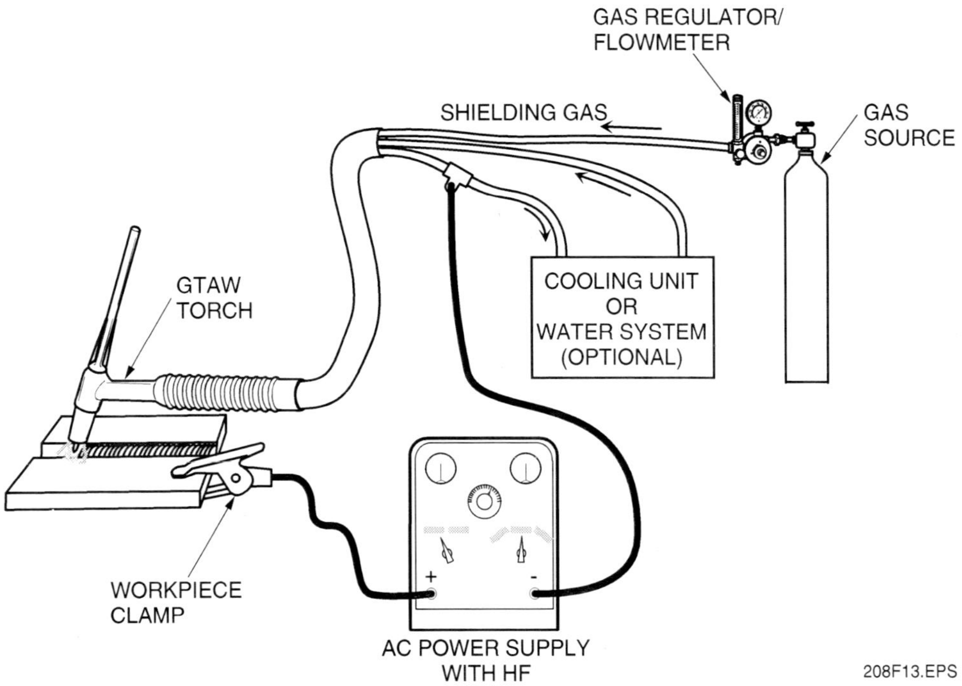

Figure 13 ◆ Basic GTAW welding system diagram.

that extends from the back of the collet is insulated and protected by the electrode back cap. Back caps are made in several lengths to cover tungstens of different lengths. The shorter back caps are used with short tungstens in confined spaces where torch clearance is inadequate. The torch also contains passages for the shielding gas and cooling fluid or water, if used. The shielding gas exits the torch through an insulated gas nozzle, sometimes called a cup, that surrounds the tungsten. *Figure 14* shows GTAW torch nomenclature and *Figure 15* is a disassembled GTAW torch.

Heat is removed from GTAW torches by air cooling or water cooling the torches with either the shielding gas or water. Two styles of air-cooled torches are common. One style uses a single plastic hose that has a power cable inside it. The shielding gas flows around the power cable, cooling it as well as the torch before being discharged to shield the electrode and weld. A special power cable lug attaches to the power terminal on the power source. The single hose connects to one side of the lug. This connection also makes the electrical connection. The shielding gas supply hose connects to the other side of the lug.

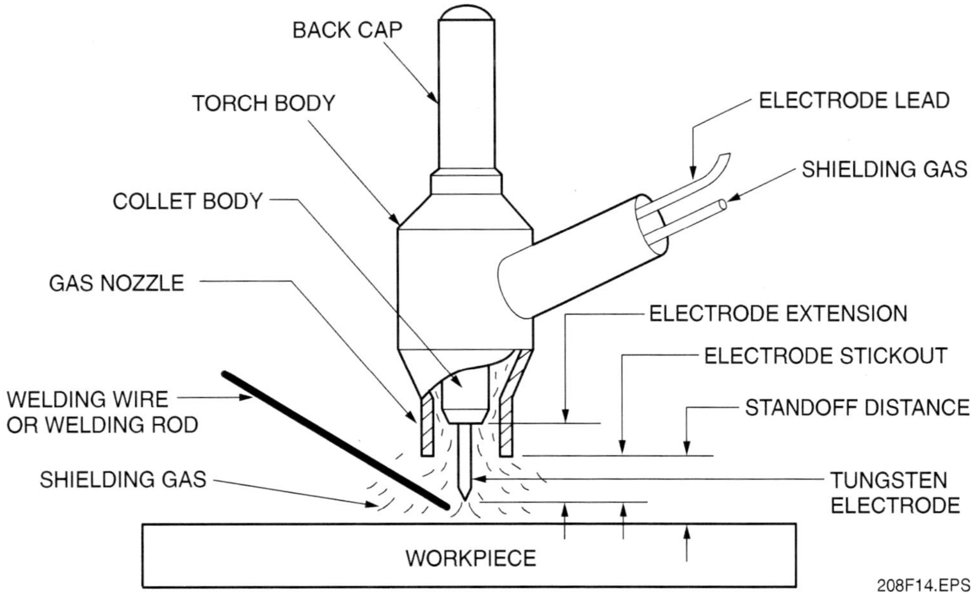

Figure 14 ◆ GTAW torch nomenclature.

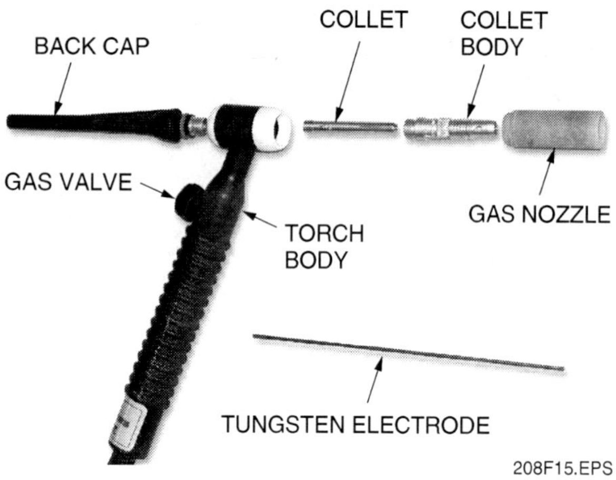

Figure 15 ◆ Disassembled GTAW torch.

The second type of air-cooled torch has a separate power cable and shielding gas hose. Either the single- or double-hose torch can be purchased with a gas control valve in the torch to manually control the shielding gas flow. This is necessary when the power source being used does not have a solenoid valve to automatically start and stop the shielding gas flow.

Figure 16 shows an air-cooled GTAW torch.

Water-cooled torches have three hoses. One hose supplies shielding gas, one hose supplies cooling water to the torch, and the third hose contains the power cable and the cooling water return or discharge hose. The cooling water may be supplied

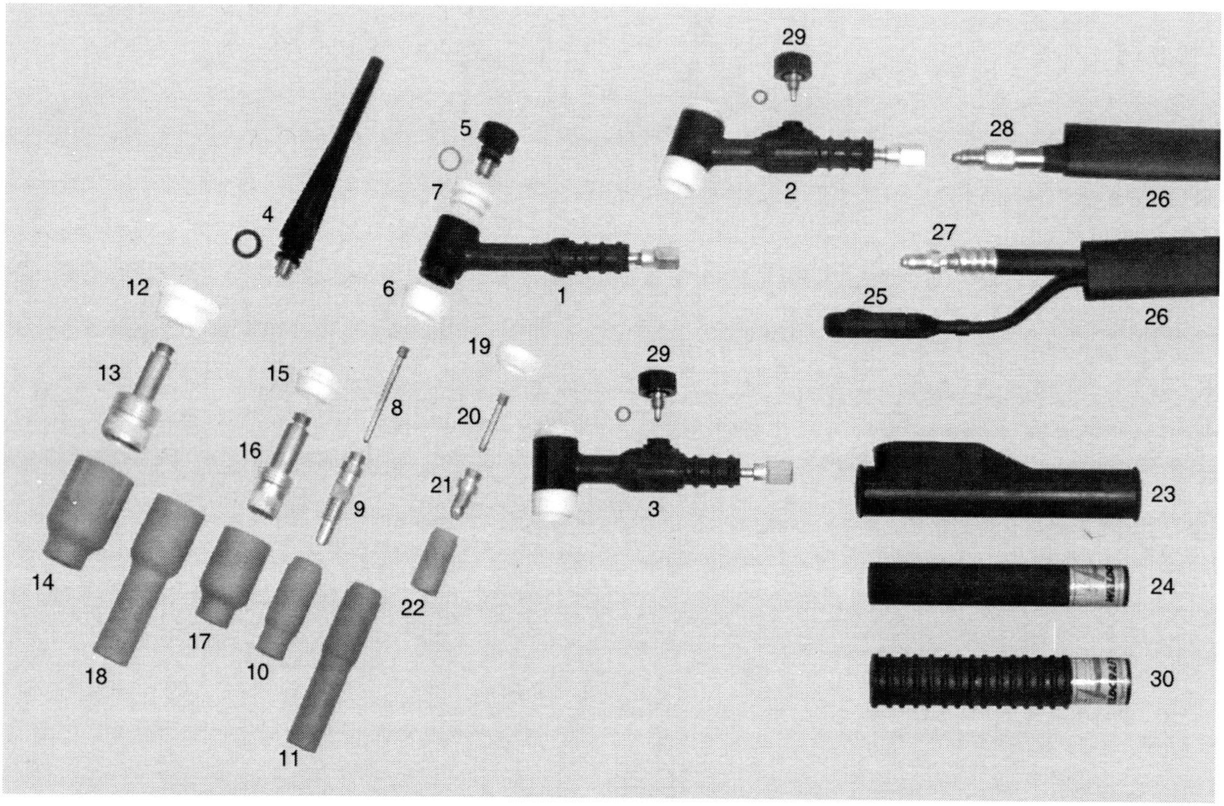

1 – 70° TORCH BODY
2 – 70° TORCH BODY WITH VALVE
3 – 90° TORCH BODY
4 – LONG BACK CAP
5 – SHORT BACK CAP
6 – CUP GASKET
7 – TEFLON INSERT
8 – COLLET
9 – COLLET BODY
10 – ALUMINA NOZZLE
11 – LONG ALUMINA NOZZLE
12 – LARGE GAS LENS INSULATOR
13 – LARGE GAS LENS COLLET BODY
14 – LARGE GAS LENS ALUMINA NOZZLE
15 – STANDARD GAS LENS INSULATOR
16 – STANDARD GAS LENS COLLET BODY
17 – STANDARD GAS LENS ALUMINA NOZZLE
18 – STANDARD GAS LENS LONG ALUMINA NOZZLE
19 – NOZZLE GASKET
20 – SHORT COLLET
21 – SHORT COLLET BODY
22 – SHORT ALUMINA NOZZLE
23 – RUBBER SWITCH COVER
24 – KNURLED PUSH-ON HANDLE
25 – FLAT BUTTON SWITCH
26 – TORCH CABLE COVER
27 – POWER AND GAS CABLE
28 – POWER AND RETURN WATER CABLE
29 – VALVE KNOB AND STEM
30 – RIBBED HANDLE

208F16.EPS

Figure 16 ◆ Air-cooled GTAW torch.

from the domestic water system or from a closed-loop cooling system. Regardless of the system, it should include a fusible link in the torch cooling line. A fusible link is a safety device that will open and stop the welding current flow if the coolant flow to the torch is interrupted. *Figure 17* shows a water-cooled GTAW torch.

Closed-loop cooling systems recirculate demineralized water or special cooling fluids that will not corrode internal torch surfaces or plug the cooling passages with mineral scale. Cooled water is supplied to the torch through a hose in the torch cable assembly. After cooling the torch, the water is typically returned to the system through a hose containing the torch power cable. A cooling system usually consists of a reservoir (tank), circulating pump, and connecting lines. The reservoir may also serve as the unit's base and be equipped with wheels for easy transport. Heavy-duty cooling systems have large reservoirs and may also be equipped with a heat exchanger (radiator) and cooling fan to cool the fluid in the cooling system.

Figure 18 shows a closed-loop cooling unit with an internal radiator and cooling fan.

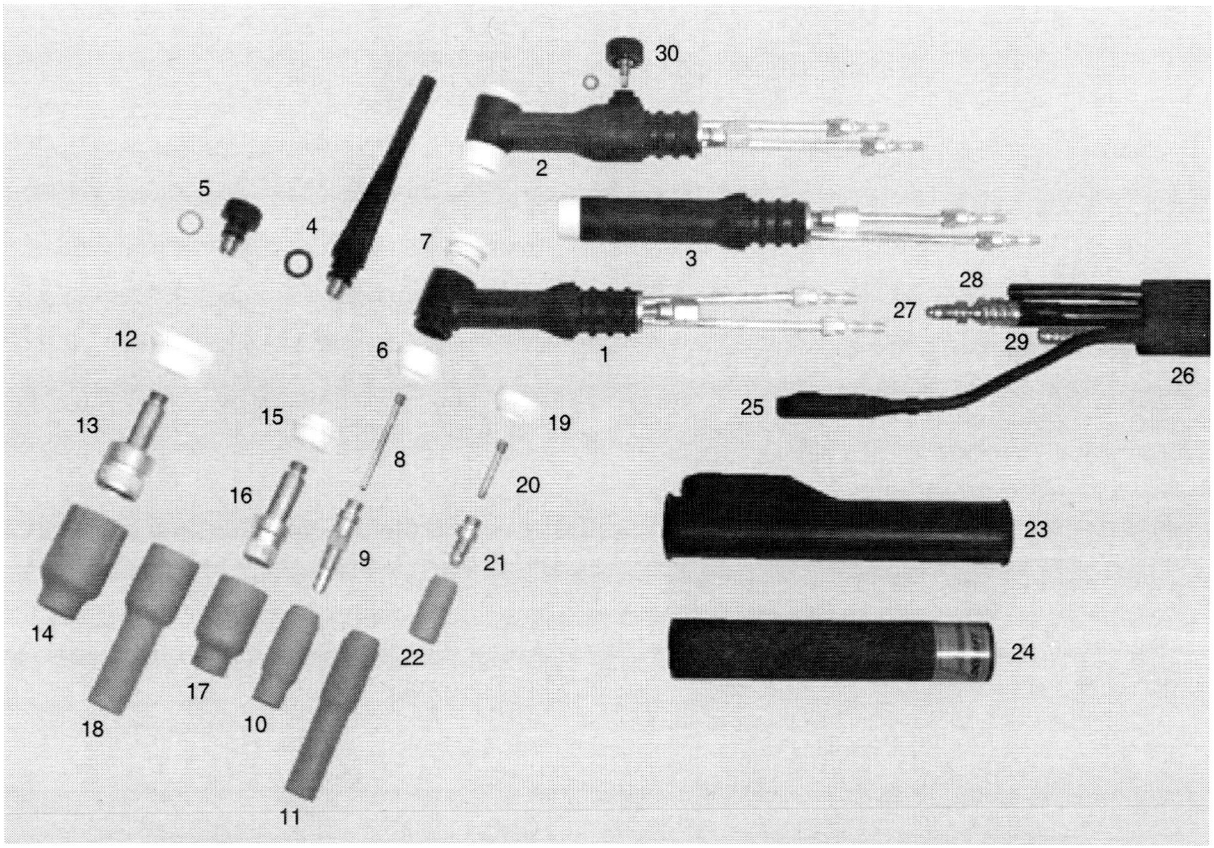

1 – 70° TORCH BODY
2 – 70° TORCH BODY WITH GAS CONTROL VALVE
3 – 180° TORCH BODY (FRONT LOADING)
4 – LONG BACK CAP
5 – SHORT BACK CAP
6 – TEFLON CUP GASKET
7 – TEFLON INSERT (BACK CAP)
8 – COLLET
9 – COLLET BODY
10 – ALUMINA NOZZLE
11 – LONG ALUMINA NOZZLE
12 – LARGE GAS LENS TEFLON INSULATOR
13 – LARGE GAS LENS COLLET BODY
14 – LARGE GAS LENS ALUMINA NOZZLE
15 – STANDARD GAS LENS TEFLON INSULATOR
16 – STANDARD GAS LENS COLLET BODY
17 – STANDARD GAS LENS ALUMINA NOZZLE
18 – STANDARD GAS LENS LONG ALUMINA NOZZLE
19 – NOZZLE GASKET
20 – SHORT COLLET
21 – SHORT COLLET BODY
22 – SHORT ALUMINA NOZZLE
23 – RUBBER SWITCH COVER
24 – KNURLED PUSH-ON HANDLE
25 – FLAT BUTTON SWITCH
26 – TORCH CABLE COVER
27 – POWER AND GAS CABLE
28 – GAS HOSE
29 – WATER SUPPLY HOSE
30 – GAS CONTROL VALVE, KNOB AND STEM

208F17.EPS

Figure 17 ◆ Water-cooled GTAW torch.

Figure 18 ◆ Typical closed-loop cooling unit.

5.2.0 Gas Nozzles

The GTAW gas nozzle shapes and directs the shielding gas flow as it exits the torch. Nozzles, sometimes called cups, are usually threaded onto the torch to form a gas-tight joint. They are made of ceramic material, chrome-plated steel, pyrex glass, or plastic. Ceramic nozzles can be used up to about 300 amperes. Above that, water-cooled, metal-coated ceramic nozzles or water-cooled ceramic nozzles must be used.

> **CAUTION**
> Most welding machines have a heat-activated circuit breaker that will shut off the machine automatically when it overheats. The machine cannot be turned back on until it has cooled.

Gas nozzles are available in different lengths and diameters (see *Figures 16* and *17*). Nozzle length is determined by the job requirements, such as limited clearance or deep grooves. Nozzle size is the exit orifice inside diameter and depends on the torch type and size and the electrode diameter. Nozzle size may be specified in fractions of an inch (1/4", 3/8", 7/16", 1/2", 3/4") or in millimeters (6 mm, 10 mm, 11 mm, 13 mm, 19 mm). Sometimes the diameter is given as a size number, such as 4, 6, 7, 8, or 12. These numbers are the nozzle opening diameters in multiples of 1/16". For example, a number 6 nozzle is six times 1/16 or 6/16 (3/8") in diameter.

A GTAW torch may also be equipped with a gas lens nozzle and a gas lens collet body. A gas lens collet body contains an assembly of fine screens that straightens the gas flow from the nozzle to eliminate turbulence and causes the gas to flow smoothly past the tungsten electrode. This reduces the chance that the turbulence could pull atmosphere into the weld zone and cause contamination and weld defects. *Figure 19* shows a typical gas lens collet body and torch with a gas lens nozzle.

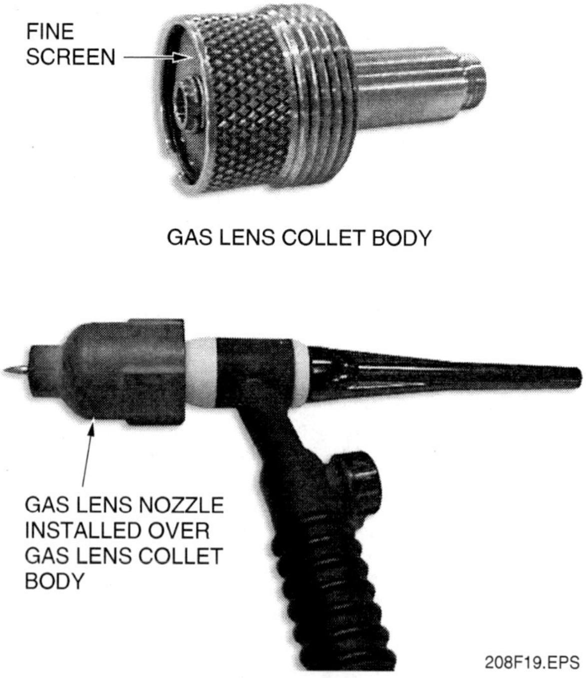

Figure 19 ◆ Typical gas lens collet body and torch with gas lens nozzle.

Coolant and Shielding Gas Flow Inside a Water-Cooled Torch

Shielding gas and cooling flow in a typical water-cooled torch is shown here.

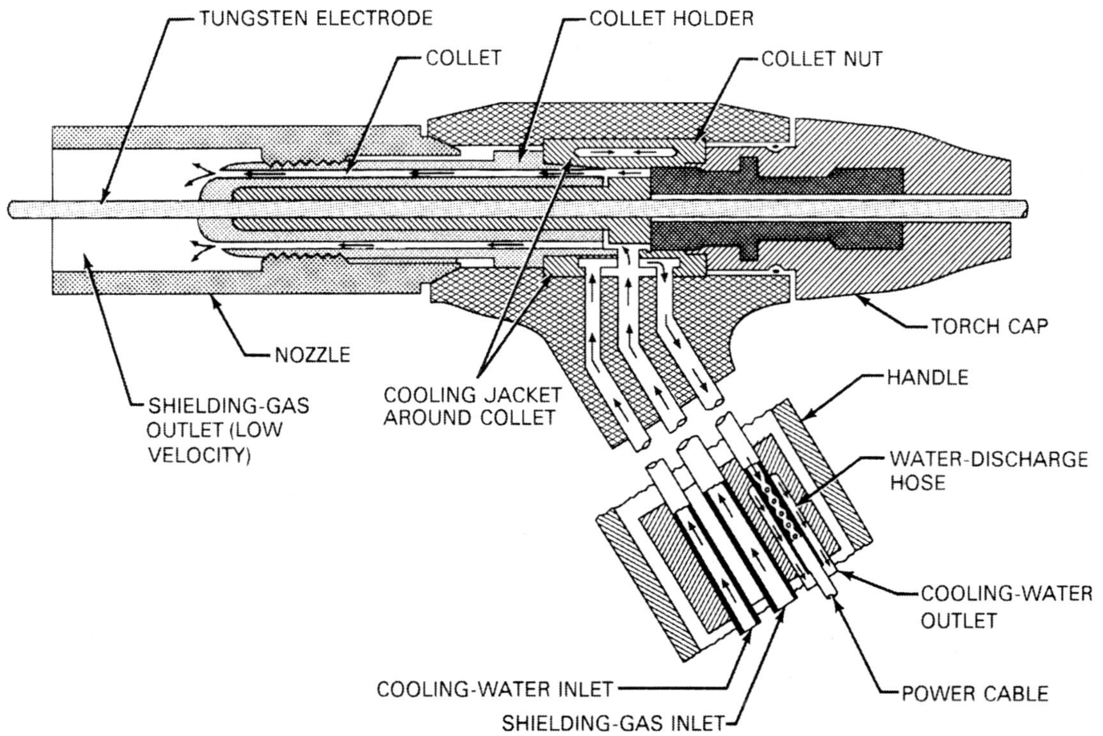

208SA01.EPS

Cable Covers

To prevent damage caused by hot or molten metal to a GTAW torch cable containing hoses for gas or water, leather or fire-resistant cable covers can be added to protect the torch cable. These covers zip or snap around the torch cable for a distance of 10' to 20' behind the torch handle.

SNAP-TYPE CABLE COVER

208SA02.EPS

5.3.0 Tungsten Electrodes

Tungsten electrodes used for GTAW are manufactured in different formulations and sizes to meet the requirements of *American Welding Society specification A5.12*. Electrode diameters range from 0.01" up to ¼" for current ratings of 5 to 1,000 amperes. They are made in lengths of 3", 6", 7", 12", 18", and 24". Manual welding is done with 7" or shorter electrodes.

> **Hot Tip**
>
> ### *Gas Nozzles*
>
> Some manufacturers recommend that the inside diameter of the nozzle should be a minimum of three times the electrode diameter.

> **Think About It**
>
> ### *Nozzle Size*
>
> What size nozzle would a welder select if a ½" diameter opening was needed?

Electrodes are manufactured with two different finishes—chemically cleaned or ground. The ground finish is the more expensive.

A number of types of tungsten electrodes are manufactured. They are:

- Pure tungsten
- Zirconiated tungsten
- Thoriated tungsten
- Ceriated tungsten
- Lanthanated tungsten

Pure tungsten electrodes (commonly called tungstens) are used for AC welding. They provide good arc stability and good resistance against contamination. If used for DC welding, they are easily contaminated. Although expensive, pure tungsten is still the least expensive of the common types of GTAW electrodes.

Zirconiated tungstens are also used for AC welding. They have a small percentage of zirconium added to the tungsten. They are used for AC welding where tungsten inclusions cannot be tolerated. The zirconium also gives them easy arc-starting characteristics. Zirconiated tungstens are the most expensive type of tungsten electrode.

Thoriated tungsten electrodes are used for DC welding. They have a small percentage of thorium added to:

- Make the arc easier to start
- Increase the current range
- Help prevent tip melt
- Reduce the tendency to stick or freeze to the work
- Increase resistance to contamination when properly used

Thoriated tungsten electrodes are manufactured in EWTh-1 (approximately 1% thoria) and EWTh-2 (approximately 2% thoria) concentrations of thorium oxide (thoria).

Ceriated and lanthanated tungsten electrodes are somewhat less effective than thoriated electrodes, but have the advantage of not being radioactive. Ceriated tungsten electrodes are also used for AC welding.

Tungsten electrodes are identified by a color band at one end. *Table 2* (*Appendix*) shows GTAW electrode color codes and AWS classifications.

> **Hot Tip**
>
> ### *Gas Lens*
>
> The use of a gas lens allows longer tungsten stickout. This is useful for welding in tight quarters.
>
> ### *Avoiding Tungsten Electrode Contamination*
>
> Handle tungsten electrodes with clean gloves to prevent contamination from body oils and other materials.
>
> ### *Thoriated Tungsten*
>
> Thoriated tungsten electrodes contain low-level, radioactive thorium. The grinding dust from these electrodes is considered a safety risk. If possible, always use tungsten electrodes containing cerium or lanthanum instead of thorium.

Hot Tip: *Tungsten Electrode Packaging*

Most tungsten electrodes are packaged as shown to prevent contamination or damage to the electrodes. An inner divider containing the electrodes slides out of the package. Note the color band at the end of the electrodes. The color band end is not used when sharpening an electrode.

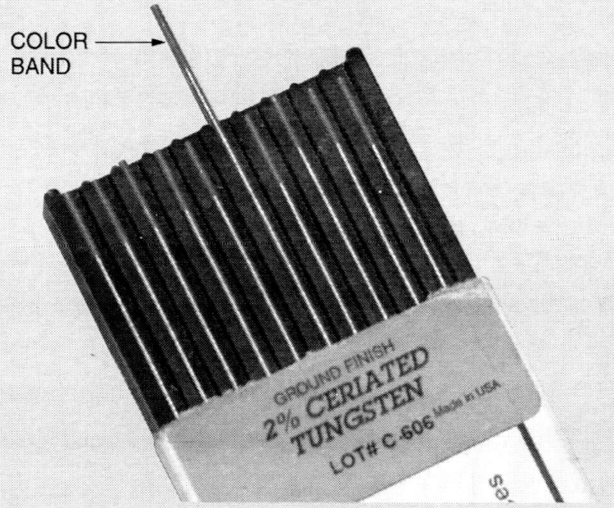

208SA03.EPS

5.4.0 Shielding Gas

The GTAW process always uses inert shielding gases to displace atmosphere from the weld zone to prevent oxidation and contamination of the tungsten electrode, weld puddle, and filler metal. The two principal shielding gases used for GTAW are argon (Ar) and helium (He). These gases are used alone or as mixtures; occasionally, small percentages of other gases are added for special conditions.

Each shielding gas and shielding gas mixture has distinct performance characteristics. Each affects the arc differently and produces different weld characteristics. Mixtures of gases often have the best features of the individual gases. The following sections explain shielding gas characteristics and their principal uses.

Hot Tip: *Always Use the Correct Gas*

Make sure to always use the correct gas or gas mixtures. Using an incorrect gas mixture, like the 75% argon/25% CO_2 mix used for GMAW, will immediately cause the tungsten electrode to be consumed and deposited in the weld as a contaminant.

5.4.1 Argon

Argon is the most common shielding gas used with GTAW. It provides a smooth, quiet arc that requires a lower arc voltage than other shielding gases for a given arc length. This means that it gives the welder the greatest tolerance for arc gap variation. Argon also works well with AC, and provides better base metal cleaning (electrical cathode base metal oxide removal) than helium. Argon is ten times heavier than helium, so it forms a better gas shield than helium, which tends to float (rise) at the same flow rates. Argon is used both alone and in combination with other shielding gases.

When compared with helium, argon has the following advantages:

- A smoother, quieter arc
- Lower arc voltage for a given arc length
- Easier arc starting
- Better cathode cleaning on aluminum and magnesium with AC
- Lower shielding flow rate
- Better shielding in cross drafts

Small amounts of hydrogen are sometimes mixed with argon for deeper penetration in welding stainless steel. Hydrogen cannot be used with aluminum or carbon steels because it produces porosity and underbead cracking. The most common argon-hydrogen mixture is 85% argon and 15% hydrogen.

Nitrogen is sometimes added to argon to stabilize austenitic stainless steel and to increase penetration in welding copper.

5.4.2 Helium

Helium is used when deeper penetration and higher travel speed are required. Arc stability is not as good with helium as it is with argon. However, helium is capable of delivering more heat on the base metal than is possible with argon, although it does require higher voltage. This makes helium better for welding thick sections of high heat conductivity metals such as aluminum and copper.

When compared with argon, helium has the following advantages:

- Deeper penetration into the weld joint
- Increased welding speed
- Welding of high heat conductivity metals

Helium is often mixed with argon. A common mixture is helium-argon mixed at 75% helium and 25% argon. This mixture gives good cathode cleaning action and deep weld penetration, which are good characteristics of both gases.

5.4.3 Cylinder Safety

Shielding gases may be supplied in liquid bulk tanks, in liquid cylinders, or in high-pressure cylinders of various sizes. The most common container is the high-pressure cylinder, which is portable and can be easily moved as needed.

When transporting and handling cylinders, always observe the following rules:

- Always install the safety cap over the valve, except when the cylinder has been secured for use.
- Secure the cylinder to prevent it from falling when it is in use. Chain or clamp it to the welding machine or to a post, beam, or pipe.
- Always use a cylinder cart to transport cylinders.
- Never hoist cylinders with a sling. Cylinders can slip out of the sling. Always use a hoisting basket or similar device.
- Open the cylinder valve slowly, and open it completely to prevent leakage around the valve stem.

WARNING!

Do not remove the protective cap until the cylinder is secured. If the cylinder falls over and the valve breaks off, the cylinder will shoot like a rocket and will cause severe injury or death to anyone in its way.

5.4.4 Gas Regulators/Flowmeters

A gas regulator/flowmeter is required to supply the shielding gas to the torch at the proper pressure and flow rate. A typical regulator/flowmeter consists of a preset pressure regulator with a cylinder valve spud, a flowmetering (needle) valve, and a flow-rate gauge. The pressure regulator is usually equipped with a cylinder pressure gauge to indicate the cylinder gas pressure. The metering valve is used to adjust the gas flow to the torch nozzle. The flow gauge indicates the gas flow rate in cubic feet per hour or liters per minute as shown by a ball-type flow indicator.

Gas flow to the torch is started and stopped either by a manually operated valve or by an electric solenoid valve. The manual valve may be on the torch, or it may be in the gas line to the torch. The solenoid is controlled by welding current flow or a manual switch. *Figure 20* shows a shielding gas regulator/ flowmeter.

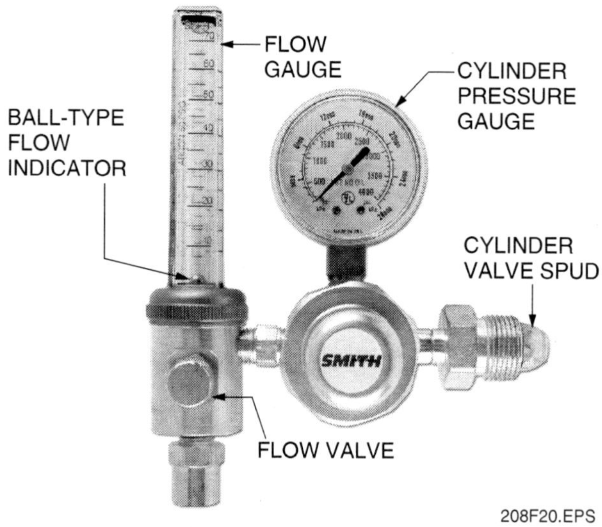

Figure 20 ◆ Shielding gas regulator/flowmeter.

Hot Tip

Gas Cylinders

Gas cylinders must always be handled with great care because of the high-pressure gases they contain. If the valve is broken off a high-pressure gas cylinder, the gas will escape with explosive force, which can cause severe injury from the blast and blown debris. The cylinder itself may become a deadly missile. The tanks must always be chained in an upright position when being transported or positioned in their storage or use locations.

5.5.0 Remote Current Control

With GTAW, control of the welding current is often performed with remote controls such as foot- or hand-operated switches and **potentiometers**. Remote controls are generally designed for use with a specific welding machine, so they cannot be used universally on all machines. Remote controls with only a switch can control (turn on and off) one or more of the following:

- High frequency
- Shielding gas
- Cooling water
- Welding current (on and off only)

Remote controls with only a potentiometer can be used to continuously vary the welding current (up or down) as needed, from zero to the maximum current set on the power source.

The most useful remote control has both switches and a potentiometer. These control:

- High frequency
- Shielding gas
- Cooling water
- Welding current (on and off)
- Welding current up or down as needed (zero to maximum setting on the power source)

Figure 21 shows typical foot- and finger-operated remote controls.

FOOT-OPERATED REMOTE CONTROL

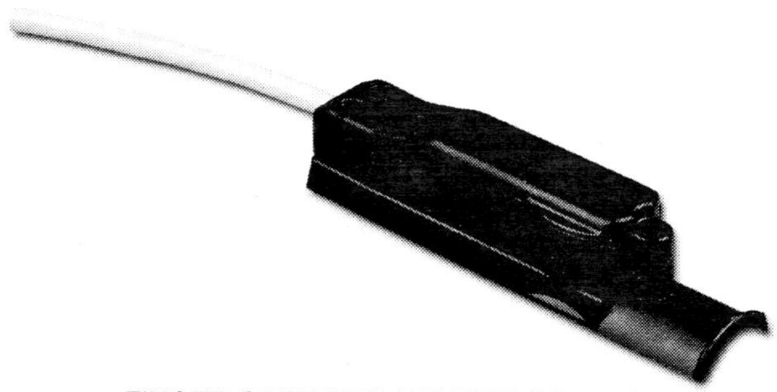

FINGER-OPERATED REMOTE CONTROL
(FASTENED TO TORCH HANDLE)

208F21.EPS

Figure 21 ◆ Typical foot- and finger-operated remote controls.

6.0.0 ◆ GTAW FILLER METALS

Filler metal for manual GTAW is generally supplied in 36" lengths from 1/16" to 1/4" diameters. There are some automatic and manually operated wire feeders used with manual GTAW, but they are usually found only in high-production facilities. Filler metals for GTAW are drawn from high-grade pure alloys compounded for specific applications. Generally, the rods are not coated

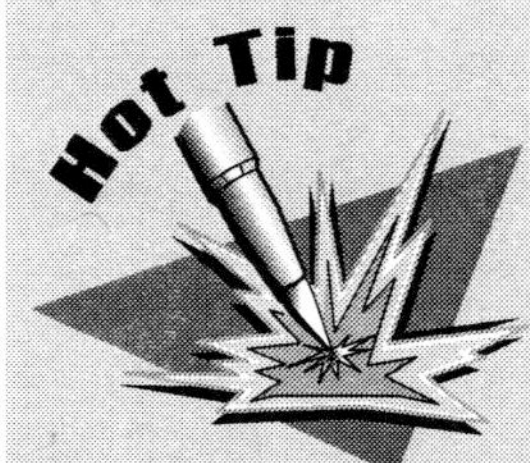

Filler Rod Marking

The AWS classification of metal filler rod is identified on its container. In addition, each rod is usually marked either by stamping the AWS classification into the rod or by a paper or plastic tab wrapped around the rod.

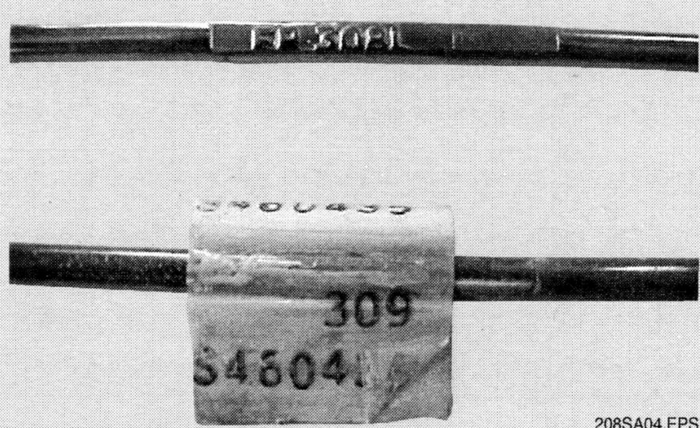

208SA04.EPS

(bare without flux) except for a corrosion-resistant copper electroplating on some carbon steel rods. However, some special-purpose rods may have flux cores or coatings.

Several industry organizations and the U.S. Government publish specification standards for filler metal. The most common standards are those published by the American Welding Society (AWS). The purpose of the AWS specifications is to set standards that all manufacturers must follow when manufacturing welding consumables. This ensures consistency for the user regardless of who manufactured the product. The specifications set standards for the:

- Classification system, identification, and marking
- Chemical composition of the deposited weld metal
- Mechanical properties of the deposited weld metal

Examples of AWS specifications that pertain to GTAW filler metals are:

- *AWS A5.7, Specification for Copper and Copper Alloy Bare Welding Rods and Electrodes*
- *AWS A5.9, Specification for Corrosion Resisting Chromium and Chromium-Nickel Steel Bare and Composite Metal Cored and Stranded Welding Electrodes and Welding Rods*
- *AWS A5.10, Specification for Bare Aluminum and Aluminum Alloy Welding Electrodes and Rods*
- *AWS A5.14, Specification for Nickel and Nickel Alloy Base Welding Electrodes and Rods*
- *AWS A5.16, Specification for Titanium and Titanium Alloy Electrodes and Rods*
- *AWS A5.18, Specification for Carbon Steel Filler Metals for Gas Shielded Arc Welding*
- *AWS A5.19, Specification for Magnesium Alloy Welding Rods and Bare Electrodes*
- *AWS A5.22, Specification for Stainless Steel Electrodes for Flux Cored Arc Welding and Stainless Steel Flux Cored Rods for Gas Tungsten Arc Welding*
- *AWS A5.28, Specification for Low Alloy Steel Filler Metals for Gas Shielded Arc Welding*

NOTE

The specification number is generally followed by a dash and a two-digit number, such as *A5.18-01* or *A5.28-96.* The dash and following number indicate the year that the specification was last revised. These documents are on a five-year revision cycle. Always use the latest version.

Filler metals and wire are graded for three major areas of use:

- *General use* – Wire or rod meets specifications. No record of chemical composition, strength, etc., is supplied to the user with the wire purchase.
- *Rigid control fabrication* – A Certificate of Conformance is supplied with the wire or rod at purchase. The stock is identified by heat numbers or code numbers located on the roll package.
- *Critical use* – A Certified Chemical Analysis report is supplied, and records are kept by the fabricator of welds and processes for later reference on aircraft, nuclear reactors, and pressure vessels.

Factors that affect the selection of a GTAW filler metal include:

- Base metal chemical composition
- Base metal mechanical properties
- Weld joint design
- Service or specification requirements
- Shielding gas used

Commonly used filler metal types include:

- Carbon steel filler metals
- Stainless steel filler metals
- Aluminum and aluminum alloy filler metals
- Copper and copper alloy filler metals
- Low-alloy steel filler metals

Table 3 (*Appendix*) summarizes the GTAW filler metal specifications and the major AWS class covered by each specification.

6.1.0 Carbon Steel and Low-Alloy Steel Filler Metals

Carbon steel filler metals are identified by *AWS Specification A5.18*. Low-alloy steel filler metals are identified by *AWS Specification A5.28*. The rod classification (number) is located on a label on the packaging.

Figures 22 and 23 show the AWS classifications for carbon steel and low-alloy steel filler metals.

All steel filler metals contain alloys such as silicon, manganese, aluminum, and carbon. Other alloys such as nickel, chromium, and molybdenum are also often added. The purpose of the alloys is as follows:

- *Silicon (Si)* – Concentrations of 0.40% to 1.00% are used to deoxidize the puddle and to strengthen the weld. Silicon above 1% may make the welds crack-sensitive.

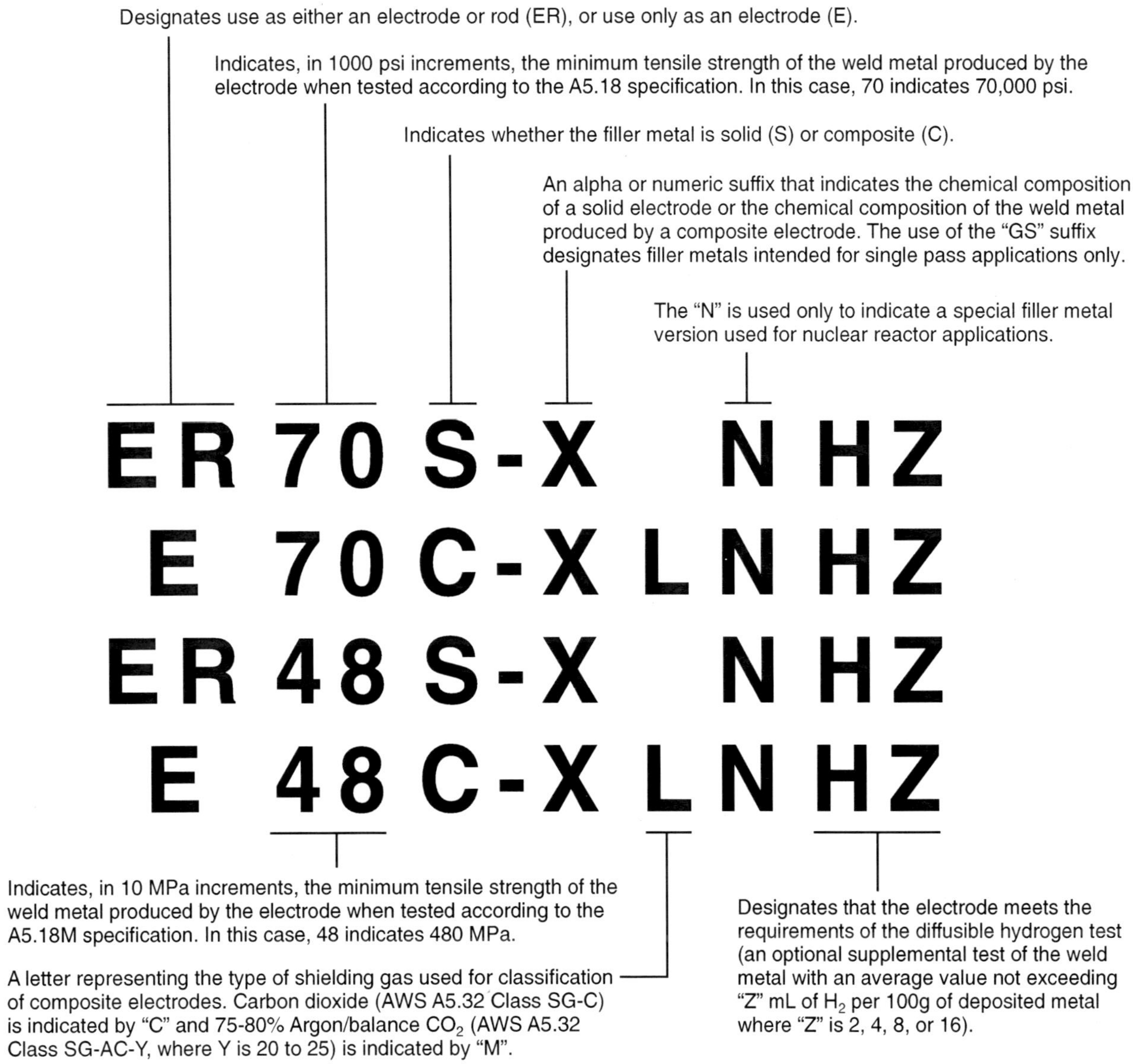

Figure 22 ◆ AWS classification for carbon steel filler metals.

- *Manganese (Mn)* – Concentrations of 1% to 2% are also used as a deoxidizer and to strengthen the weld. Manganese also decreases hot-crack sensitivity.
- *Aluminum (Al), titanium (Ti), and zirconium (Zr)* – One or more may be added in very small amounts for deoxidizing. Strength may also be increased.
- *Carbon (C)* – Concentrations of 0.05% to 0.12% are used to add strength without adversely affecting ductility, porosity, or toughness.
- *Nickel (Ni), Chromium (Cr), and Molybdenum (Mo)* – These elements may be added in small amounts to improve corrosion resistance, strength, and toughness.

Tables 4 and *5* (*Appendix*) list examples of AWS carbon steel and low-alloy electrode rod classifications and their uses.

Hot Tip

Rod and Electrode Designations

The letter R that follows the E in an AWS classification number indicates that the classification is used as a metal filler rod as well as an electrode. Electrode wire is used in GMAW. Filler metal rods are used in oxyfuel welding and in GTAW.

6.2.0 Stainless Steel Filler Metals

Stainless steel filler metals are identified by *AWS Specification A5.9*. *Figure 24* shows a typical AWS stainless steel electrode rod classification.

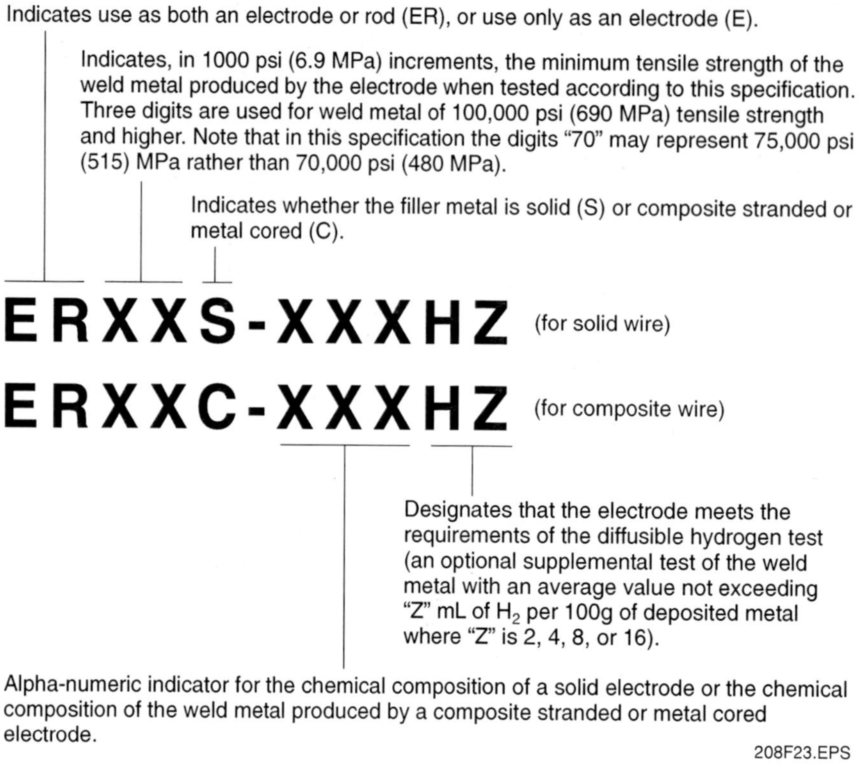

Figure 23 ◆ AWS classification for low-alloy steel filler metals.

Stainless steel filler metal should be selected to closely match the alloy composition of the base metal.

Table 6 (*Appendix*) lists examples of AWS stainless steel electrode rod classifications and uses. The complete list of these classifications and a more extensive description of their intended usage is contained in a previous module.

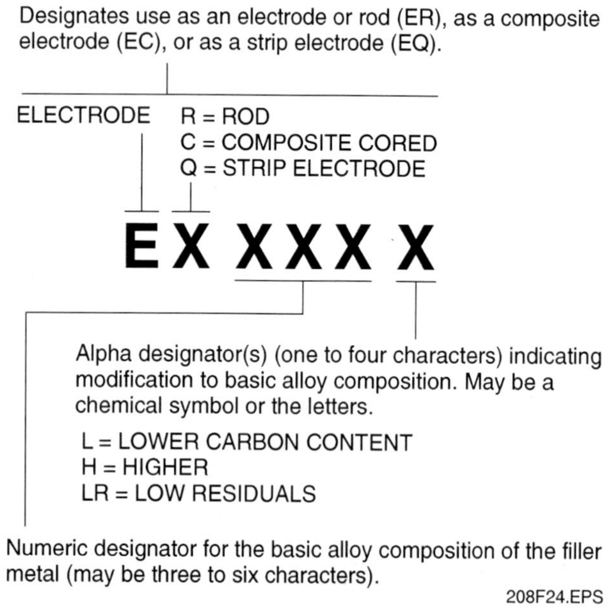

Figure 24 ◆ Typical AWS stainless steel rod classification.

6.3.0 Aluminum and Aluminum Alloy Filler Metals

Aluminum electrode rods are covered by *AWS Specification A5.10*. Aluminum filler metals usually contain magnesium, manganese, zinc, silicon, or copper for increased strength. Corrosion resistance and weldability are also considerations. Aluminum filler metals are designed for specific types of aluminum and should be selected to be compatible. The most widely used aluminum electrode rods are ER4043 (contains silicon) and ER5356 (contains magnesium). *Figure 25* shows a typical AWS aluminum filler metal classification.

Table 7 (*Appendix*) lists recommended filler metals for common aluminum alloys.

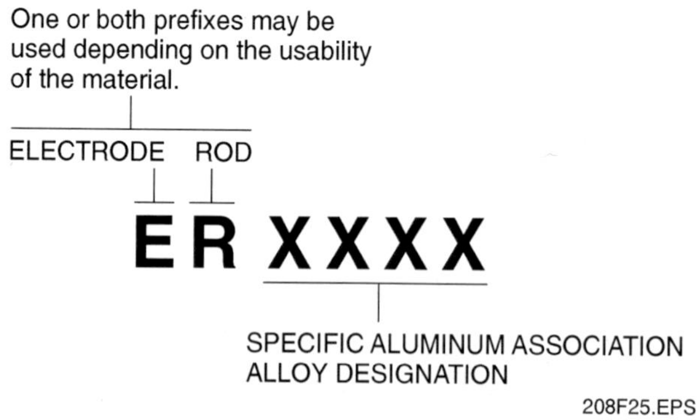

Figure 25 ◆ Typical AWS aluminum filler metal classification.

6.4.0 Copper and Copper Alloy Filler Metals

Copper electrode rods are covered by *AWS Specification A5.6*. Most copper filler metals contain other elements to increase strength, deoxidize the weld metal, and match the base metal composition. *Figure 26* shows a typical AWS copper filler metal classification.

Table 8 (*Appendix*) lists some AWS copper electrode rod identifiers and their principal uses. A more extensive list and description are in a previous module.

6.5.0 Nickel and Nickel Alloy Filler Metals

Nickel-based filler metals are covered by *AWS Specification A5.14*. These filler metals contain other elements to match base metal applications and to increase the strength and quality of the weld metal. For GTAW, DCEN is used with high-purity argon, helium or both used as a shielding gas. *Figure 27* shows the AWS classification for nickel filler metals. *Table 9* (*Appendix*) lists AWS nickel filler metal classifications and uses.

6.6.0 Magnesium Alloy Filler Metals

Magnesium alloy filler metals are covered by *AWS Specification A5.19*. These filler metals are usually used with GTAW, GMAW, and PAW processes. Oxyfuel welding should be used only for temporary repair work. For GTAW welding, the techniques and equipment are similar to those for aluminum. Argon, helium, or both are used for shielding. Alternating current is preferred for arc cleaning and penetration. Direct current is also employed with DCEP used for thin materials and DCEN used for mechanized welding with helium for deep penetration. GTAW is recommended for the defect repair of clean magnesium castings. *Figure 28* shows the AWS classification for magnesium filler metals.

Table 10 (*Appendix*) lists recommended filler metals for common magnesium alloys. Note that the prefix ER and R has been removed from the filler metal designations to reduce the size of the table.

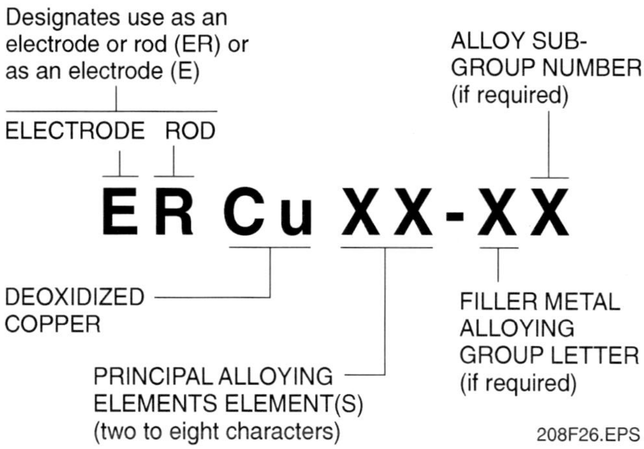

Figure 26 ◆ Typical AWS copper filler metal classification.

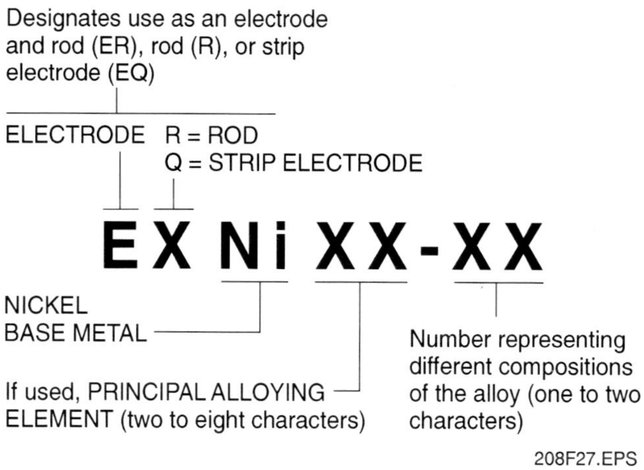

Figure 27 ◆ AWS classification for nickel filler metals.

Hot Tip

Magnesium Fires

Ignition of magnesium is a very remote possibility while welding; however, the fire will cease when the heat source is removed. Ignition of the weld pool is prevented by the gas shielding used in GTAW, GMAW, and PAW processes. Most magnesium fires occur with accumulations of fines from grinding or filing or chips from machining. Accumulations of fines on clothing should be avoided. Graphite-base or salt-base powders recommended for extinguishing magnesium fires should be stored in the work area. If large amounts of fines are produced, they should be collected in a waterwash-type dust collector designed for use with magnesium. Special precautions pertaining to the handling of wet magnesium fines should be followed.

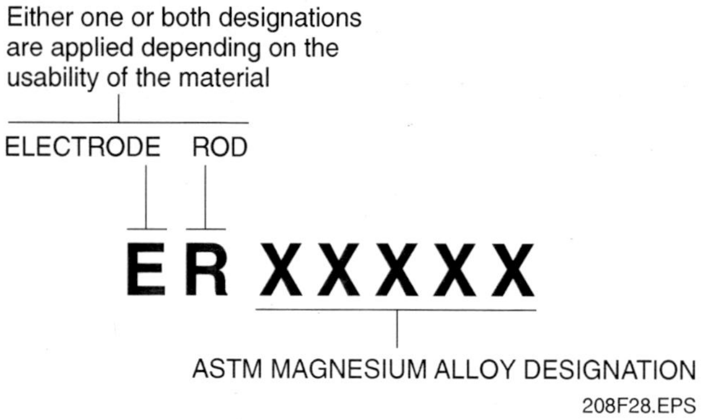

Figure 28 ◆ AWS classification for magnesium filler metals.

6.7.0 Titanium and Titanium Alloy Filler Metals

Titanium and titanium alloy filler metals are covered by *AWS Specification A5.16*. These filler metals are generally used with GTAW, GMAW, SAW, and PAW processes. Titanium is sensitive to embrittlement by oxygen, nitrogen, and hydrogen at temperatures above 500°F. Like aluminum, titanium requires weld cleaning and high-purity gas shielding, and especially adequate postflow shielding, to avoid embrittlement. Titanium can be successfully fusion welded to zirconium, tantalum, niobium, and vanadium. Titanium should not be fusion welded to copper, iron, nickel, and aluminum. *Figure 29* shows the AWS classification for titanium filler metals. *Table 11* (*Appendix*) lists AWS titanium filler metal classifications and uses.

6.8.0 Stainless Steel Flux Cored Electrodes and Rods

Figure 30 shows the *ANSI/AWS A5.22* specification format for stainless steel flux cored electrodes and rods. *Table 12* (*Appendix*) shows stainless steel flux cored rod characteristics as indicated by the AWS classification.

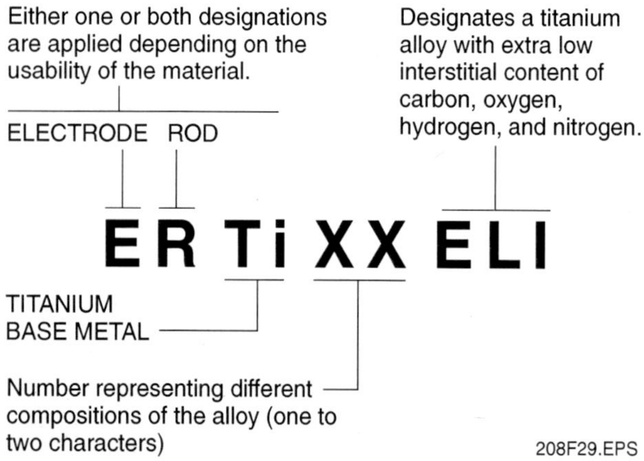

Figure 29 ◆ AWS classification for titanium filler metals.

Table 13 (*Appendix*) lists AWS stainless steel flux cored rod classifications and uses. The table lists only the rods for GTAW applications.

7.0.0 ◆ WELDING EQUIPMENT SETUP

In order to weld safely and efficiently, the welding equipment must be properly set up. The following sections will explain how to properly set up GTAW equipment.

7.1.0 Selecting a GTAW Power Source

Select the power source and current type for the base metal to be welded. To select a power source (welding machine), consider the following factors:

- GTAW requires a constant-current power source. SMAW power sources and welding machines are commonly used for GTAW.
- The welding current type required, AC or DC. Generally, carbon steels, stainless steels, and alloy steels are welded with DC, and aluminum and magnesium are welded with AC and high frequency.
- The maximum amperage required.
- The primary power requirements. Are there electrical receptacles to plug a welding machine into, or does an engine-driven generator/alternator need to be used?

Table 14 (*Appendix*) shows welding current requirements for different base metals.

7.2.0 Locating the GTAW Equipment

Because of the limited length of GTAW torch cables, the shielding gas supply must be located reasonably close to the welding site. The welding machine can be some distance away, but unless

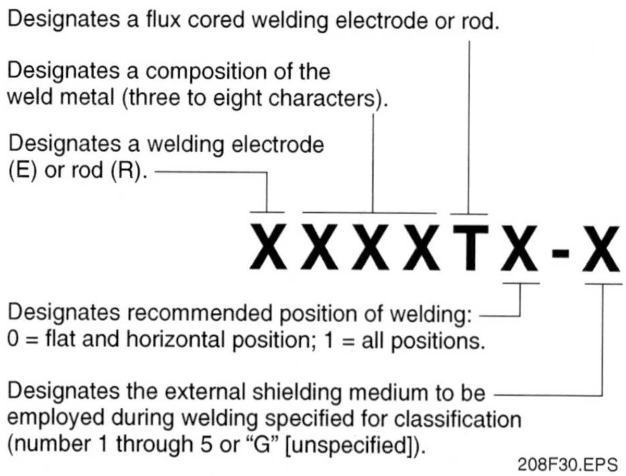

Figure 30 ◆ AWS classification for stainless steel flux cored electrodes and rods.

remote controls are installed, it will not be convenient to operate.

Select a site where the GTAW equipment will not be in the way but will be protected from welding or cutting sparks. There should be good air circulation to keep the welding machine cool. The environment should be free from explosive or corrosive fumes, and as free as possible from dust and dirt; welding machines have internal cooling fans that will pull these particles into the machines if they are present. The site should also be free of standing water or water leaks. If an engine-driven generator or alternator is used, locate it so that it can be easily refueled and serviced.

There should also be easy access to the site so the equipment can be started, stopped, or adjusted as needed. If the machine is to be plugged into an outlet, be sure the outlet has been properly installed by a licensed electrician to ensure that it is grounded. Also, be sure to identify the location of the electrical disconnect for the outlet before plugging the welding machine into it.

7.3.0 Connecting the Shielding Gas

The hose from the shielding gas regulator/ flowmeter connects to the welding machine gas solenoid if the welding machine is designed for GTAW. If a standard welding machine is used, the shielding gas hose connects to the torch cable. To connect the shielding gas, perform the following steps:

Step 1 Identify the shielding gas required by referring to the welding procedure specification (WPS) or site quality standard.

Step 2 Locate a cylinder of the correct gas or mixture, and secure it nearby.

> **CAUTION**
> Be sure to secure the cylinder so it cannot fall over.

Step 3 Remove the cylinder's protective cap, momentarily crack open the cylinder valve to blow out any dirt, and then close it again.

Step 4 Using a regulator/flowmeter with the correct CGA connection for the cylinder, install the regulator/flowmeter on the cylinder.

Step 5 Connect the gas hose to the flowmeter and to the gas solenoid on the welding machine or to the end of the torch cable.

> **NOTE**
> If the gas hose is connected to the torch cable and the torch does not contain a gas shutoff valve, install a valve in the line between the torch cable and the regulator/flowmeter.

Step 6 Check to be sure the flowmeter adjusting valve is released (screwed out).

Step 7 Very slowly open the cylinder valve slightly. After the pressure gauge indicates the cylinder pressure, open the valve completely.

Figure 31 shows how to connect the shielding gas.

7.4.0 Setting the Shielding Gas Flow Rate

Gas flow rate is measured in cubic feet per hour (ft^3/hr. or cfh) or liters per minute (L/min.) and represents the volume of gas flowing from the torch nozzle. Flow rate is important because it affects the quality and the cost of a weld. Too low a flow rate will not shield the weld zone adequately and will result in a poor-quality weld. An excessively high gas flow rate wastes expensive gas and can generate turbulence in the gas column. This turbulence can pull atmosphere into the weld zone and cause oxidation and weld contamination. The nozzle must be large enough to gently flood the weld pool with inert gas. As a general rule, larger nozzles require higher flow rates. Welding specifications contain shielding gas flow rates.

Factors that may affect the shielding gas flow rate include:

- *Drafts* – The flow rate must be increased in a drafty location to maintain the gas shield around the weld zone.
- *Specific gas used* – For example, helium usually requires a higher flow rate than argon because it rises much faster due to its low density.
- *Welding current* – High welding currents require higher flow rates.
- *Nozzle size (exit opening)* – Larger nozzles require higher flow rates.
- *Weld joint type* – Welds on flat surfaces require higher flow rates than welds in deep grooves or fillets.
- *Welding speed* – Fast advance speeds require higher flow rates than slower advance speeds.
- *Weld position* – Vertical- and horizontal-position welds require higher flow rates than flat or overhead welds.

> **Hot Tip** — *Gas Flow Rates*
>
> 
>
> More gas flow is not always better. Too much flow can deflect off the workpiece. The resultant swirling gas can create a turbulence that pulls in air, which contaminates the weld and creates pinholes. If black flecks occur in the weld puddle, or pinholes occur in the weld, try reducing gas flow.

Adjust the flow for the type and thickness of the base metal being welded, or per applicable WPS. To set the shielding gas flow rate, perform the following steps:

Step 1 Initiate the gas flow by depressing the remote control foot pedal or hand-operated switch to open the gas solenoid on the welder or manually open the hand valve on the torch.

> **NOTE**
>
> Most flowmeters have the pressure preset for a specific gas. These flowmeters don't require any pressure adjustment but can only be used with the shield gases indicated on the flowmeter.

Step 2 Set the gas flow by turning the flowmetering adjustment knob out until the ball inside the glass tube rises to the desired flow rate. The flow rate is usually read from the top of the floating ball. However, always check the manufacturer's instructions to determine the correct adjustment method. *Figure 32* shows the flowmeter being adjusted.

> **NOTE**
>
> Some flowmeters are equipped with several scales of different calibrations around the same sight tube for monitoring the flows of different types (densities) of gases. Be sure to rotate the scales or read the correct side for the gas type being used.

Step 3 When the proper gas flow has been achieved, release the remote foot pedal or hand switch or manually close the valve on the torch to stop the gas flow.

The following sections provide general recommendations for flow rates for various types of base metals.

7.4.1 Gas Flow Rates for Carbon Steel

Argon flow rates for GTAW on low- and medium-carbon steels and low-alloy steels (*Table 15, Appendix*) typically vary from 15 cubic feet per hour (cfh) for light welding at 60 amperes to 20 cfh for heavier welding at 200 or more amperes.

7.4.2 Gas Flow Rates for Stainless Steel

Argon flow rates for GTAW on stainless steel (*Table 16, Appendix*) typically vary from 10 cfh for light welding at 80 amperes to 15 cfh for heavy welding at 350 or more amperes.

7.4.3 Gas Flow Rates for Aluminum

Argon flow rates for GTAW on aluminum (*Table 17, Appendix*) typically vary from 15 cfh for light welding at 60 amperes to 30 cfh for heavy welding at 350 or more amperes.

7.4.4 Gas Flow Rates for Copper

A typical argon flow rate for GTAW on deoxidized copper up to 1/8" thick is 15 cfh from 110 to 250 amperes (*Table 18, Appendix*). With copper over 1/8" thick, use helium at 30 to 40 cfh from 190 to 525 amperes.

7.5.0 Selecting the Electrode

The GTAW electrode must be selected and the end properly prepared before it can be installed in the torch. Correct preparation of the electrode is absolutely essential. If the tip is improperly shaped, it will not produce the required arc shape and characteristics.

When selecting a GTAW electrode, choose the type recommended for the welding current required by the base metal. The type of electrode will be specified in the WPS or site quality standards. General recommendations are:

- *Pure tungsten (EWP)* – These electrodes are used for welding of aluminum and magnesium with AC and high frequency.
- *Zirconiated tungsten (EWZr)* – These electrodes are used for welding of aluminum and magnesium with AC and high frequency for welds where tungsten inclusions are not tolerated and higher current capacity is desired.
- *Thoriated tungsten (EWTh-1 or EWTh-2), ceriated tungsten (EWCe-2), or lanthanated tungsten (EWLa-1, -1.5, or -2)* – These electrodes are used for mild steels, alloy steels, and stainless steels with DCEN current.

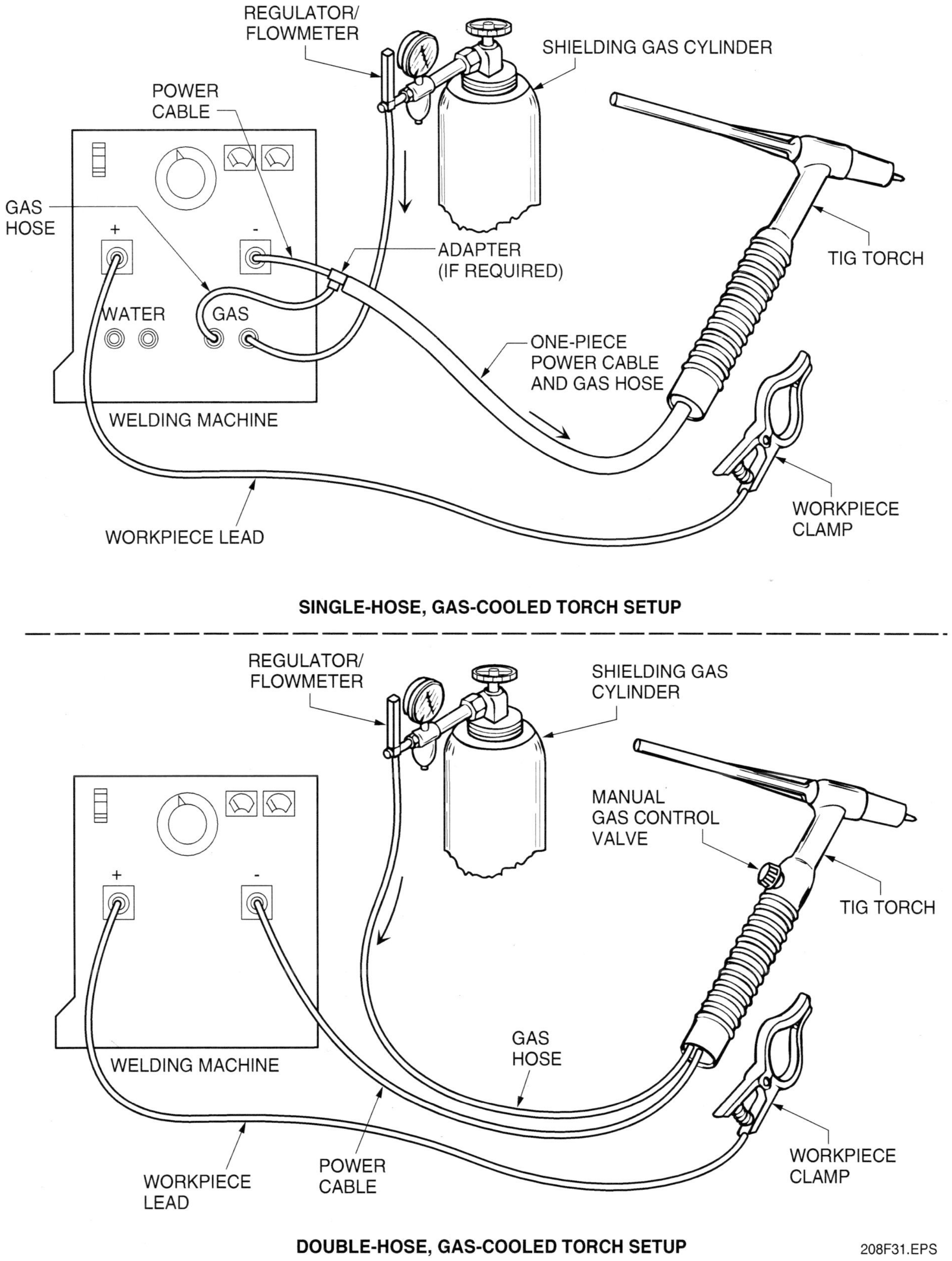

Figure 31 ◈ Connecting the shielding gas.

- Ceriated electrodes also operate successfully with AC current. Ceriated and lanthanated electrodes are used as non-radioactive substitutes for thoriated electrodes.

Select the size electrode rated for the amperage to be used. *Table 19* (*Appendix*) lists examples of electrode current ranges by electrode size and type when used with argon shielding gas under various GTAW modes.

208F32.EPS

Figure 32 ◆ Flowmeter being adjusted.

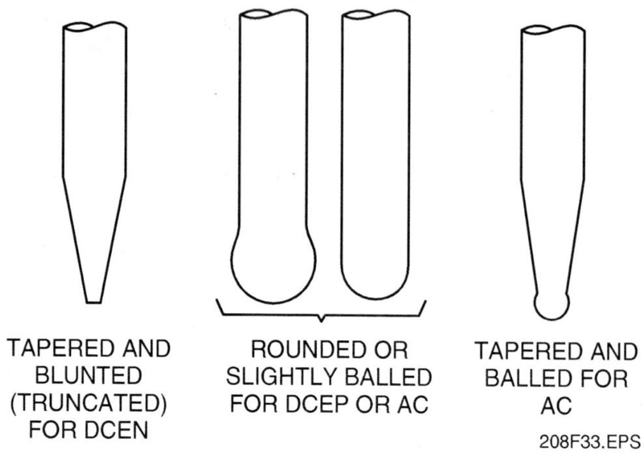

Figure 33 ◆ Various GTAW electrode end shapes for different current types.

7.6.0 Preparing the Electrode

The different types of GTAW welding current require different-shaped electrode ends (tips). For manual DCEN welding, a taper-ground and truncated point is the best shape. For manual DCEP welding, a rounded end is the best shape. For manual AC welding, a tapered and balled end is the best shape. *Figure 33* shows GTAW electrode end shapes for different current types.

Electrode tips must be formed on all new electrodes and on used electrodes with damaged or contaminated tips. On contaminated electrodes, the contaminated section must be cleanly broken off before the new tip is formed. This can be done by striking the electrode with a hammer while it is held against a sharp metal edge or by breaking it between two pairs of pliers.

7.6.1 Pointing the Electrode Tip

For DCEN welding, electrodes can be made pointed by dipping the red-hot end into a special chemical powder, or they can be tapered by grinding or sanding.

Perform the following steps to use a chemical powder to sharpen a tungsten electrode:

Step 1 Place the tungsten electrode in the torch so it extends about 1 inch from the gas nozzle.

Step 2 Heat the tungsten electrode by shorting it on the work or by striking an arc. Be sure shielding gas is flowing to protect the tungsten electrode.

Step 3 When the tip of the tungsten electrode is cherry-red for a distance of about four times the electrode diameter, place the end of the electrode in the chemical powder.

Step 4 Hold the end of the tungsten electrode in the powder for the time recommended on the container for the diameter of the electrode being pointed. The chemical powder will dissolve the tungsten electrode and form a sharp point at the proper angle.

Hot Tip

There's No Universal Electrode Size

Contrary to some popular beliefs, the $\frac{3}{32}$" electrode is not a universal size electrode. Always use the electrode size recommended for the amperage required. Using an electrode that's too big will result in arc starting and stability problems. Using an electrode that's too small can result in tungsten *spitting* (tungsten melting and migration to the weld).

Preserve the Electrode Color Band

Do not prepare the end of the electrode that contains the color-code band or you will not be able to identify the electrode's type in the future.

Handling Tungsten Electrodes

Always wear clean gloves when handling tungsten electrodes to avoid contamination of the electrode.

> **NOTE**
> For an untapered tungsten, the procedure may have to be repeated a number of times to obtain the taper.

Figure 34 shows the typical chemical powder used for sharpening.

More often, tungsten electrodes are sharpened by grinding or sanding. Since tungsten electrodes are very hard, they are best ground with diamond-impregnated silicon carbide or alumina oxide grinding wheels or sanding belts. Use 80 grit for fast shaping and 120 grit for finishing.

Follow these steps to point the end of a tungsten electrode using a grinder:

> **WARNING**
> Observe proper respiratory, hazard, and environmental contamination procedures. Dust from tungsten electrode grinding can be hazardous and may contain radioactive elements. Refer to the manufacturer's MSDS to determine any specific hazards.

Step 1 With the color band end held farthest from the grinder, hold the tungsten electrode against the grinder so that the tungsten electrode is in the plane of the wheel and makes an angle of about 30° with the wheel or belt surface. *Figure 35* shows the proper grinding angle and tungsten electrode orientation.

Step 2 Slowly rotate the tungsten electrode as it is ground to keep the point on the center line of the axis. The grind marks must be oriented toward the point along the long dimension of the tungsten electrode, and not around the point.

208F34.EPS

Figure 34 ◆ Tungsten sharpening powder.

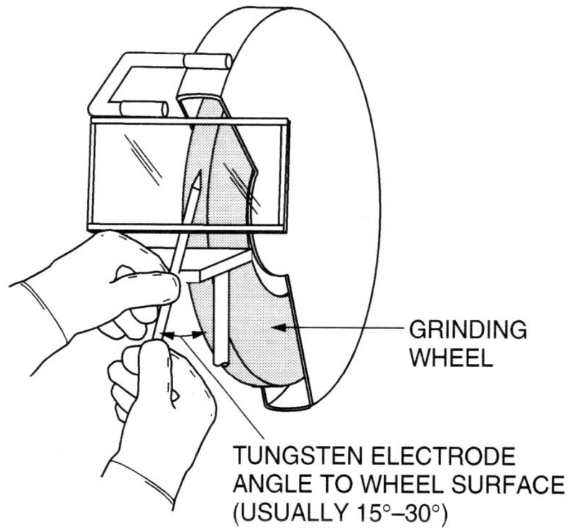

208F35.EPS

Figure 35 ◆ Proper grinding angle and tungsten electrode orientation when hand grinding.

> **NOTE**
> For DCEN electrodes, the taper should be between two and three times the electrode diameter in length and the point should be blunted (truncated) to approximately the tip diameter shown in *Table 20* (*Appendix*).

Hot Tip

Tungsten Grinding Wheels

Use a grinding wheel designated only for tungsten. Be careful to never use the grinding wheel for any other material or contaminants will be deposited on the tungsten electrode and the grinding wheel will have to be discarded. Always wear safety glasses when using a grinder.

Electrode Taper

The degree of taper on an electrode tip affects weld penetration and bead width. A long narrow taper (small angle) produces a wide bead with shallow penetration. A short wide taper (large angle) produces a narrow bead and deep penetration. During any particular welding operation, it is important to maintain the same taper on the electrode during resharpening or replacement so that the bead characteristics remain the same.

Step 3 Check the ground end to make sure the taper length is correct and that the point is in the center of the tungsten electrode and not toward one side (asymmetrical). Correct the shape if necessary.

Figure 36 shows properly and improperly prepared pointed electrode ends.

7.6.2 Balling the Electrode Tip

Only pure tungsten (EWP) or zirconiated tungsten (EWZr) can be balled (rounded). The thorium in thoriated tungstens or other rare earth (except cerium) tungstens inhibits the formation of a ball. If an attempt is made to ball most rare earth tungstens, the tungsten will form a number of small protrusions at the tip instead of a ball. These protrusions cause the arc to be unstable.

If the electrode is to be used for lower-amperage DCEP welding, the electrode end should be tapered about two diameters in length and be quite blunt. A ball will form on the blunt end when welding.

For AC welding, the end is balled with or without tapering. Balling is done by arcing the electrode against clean copper or other clean metal.

CAUTION

Do not arc against carbon, because it will contaminate the electrode.

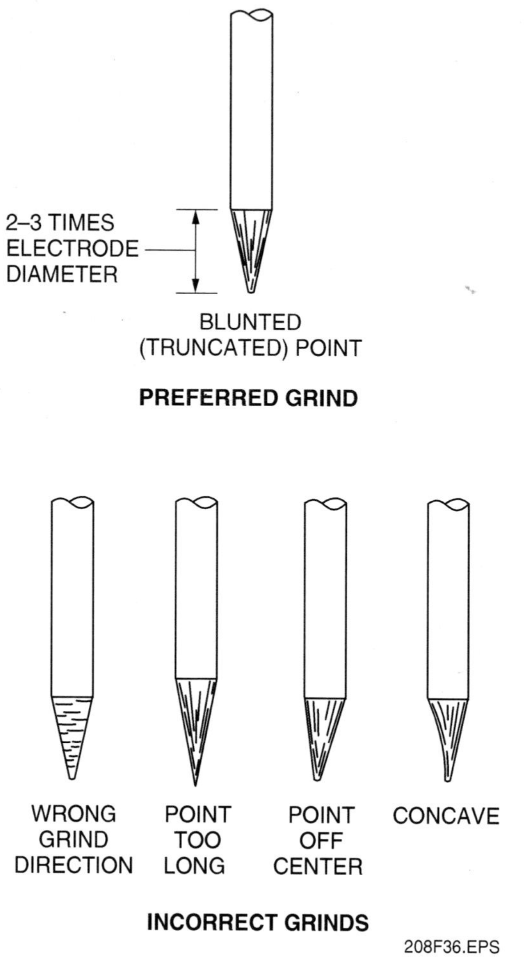

Figure 36 ◆ Properly and improperly prepared pointed electrode ends.

Hot Tip

Diamond-Wheel Grinders

A water-bath diamond-wheel grinder that will properly grind tungsten electrodes is shown here. The water bath captures all the tungsten dust for subsequent disposal. The unit can be set to different electrode grind angles as necessary.

208SA05.EPS

AC or DCEP can be used to ball an electrode. When balling an electrode, start off with a low current and gradually increase it until the end of the electrode starts to melt and form a hemispherical end or slight ball. The ball should be no more than one and one-half times the electrode diameter. Using DCEP will cause a ball to form at a much lower amperage setting than will be formed using AC current.

Figure 37 shows properly and improperly prepared balled electrode ends.

> **NOTE**
> Using DCEP, only about 35 to 45 amps of welding current will be required to ball 3/32" to 1/8" diameter electrodes.

> **CAUTION**
> Using excess amperage will cause the ball to melt and drop off the electrode. Be especially careful when using DCEP as it will generate much more heat on the electrode than the same AC setting.

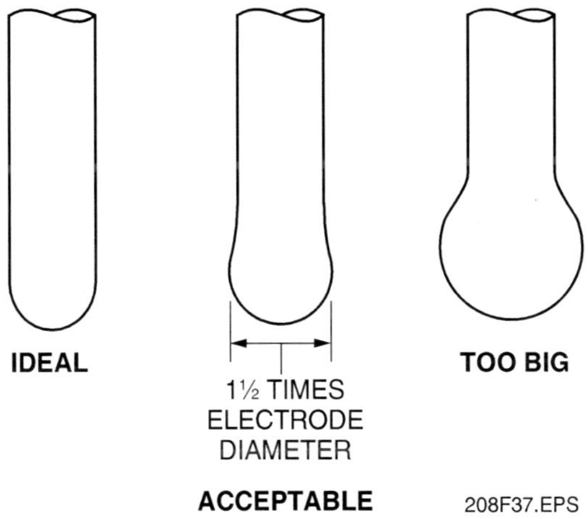

Figure 37 ◆ Properly and improperly prepared balled, non-tapered electrode ends for AC welding.

7.7.0 Selecting and Installing the Nozzle

Gas nozzles are designed to fit specific model torches. Usually, one manufacturer's nozzles will not fit another manufacturer's torches. Typically, a range of nozzle sizes and styles are available for a given torch to accommodate different electrode sizes and welding applications. Nozzles are made for close clearances, for long reaches, for reaching deep into narrow places, and for covering wide beads. Nozzles with chips, cracks, or metal buildup on the end should be discarded because they can affect the gas flow pattern and produce weld defects.

The nozzle must be sized for the electrode diameter to be used and for any unique welding requirement, such as reaching into a V-groove. The nozzle diameter must be large enough to cover the entire weld area with shielding gas. A wide, easy flow of gas is generally preferable over a narrow jet, which is apt to create turbulence and draw atmosphere into the weld zone. Small-diameter nozzles tend to overheat and break easily. However, a small-diameter nozzle is useful for steadying the torch and arc for the root pass in a V-groove weld.

Always refer to the manufacturer's nozzle recommendations for the torch being used to select the proper size and shape nozzle. In general, the nozzle's inside diameter should be about three times the diameter of the electrode being used.

7.8.0 Installing the Electrode

The electrode is usually installed after the nozzle because the electrode stickout (ESO) is measured from the end of the nozzle. The electrode is clamped in place in the torch body by the collet assembly. The collet is a cylindrical clamp that tightens as it is pulled or pushed (depending on the torch design) into a tapered holder as the threaded electrode cap is screwed on and tightened. To adjust the electrode stickout or to remove the electrode, loosen the cap assembly. Collet

> **Hot Tip**
>
> ### *Aluminum Welding with a Balled-Tip Versus a Pointed-Tip Electrode*
>
> Traditional AC welding of aluminum using a balled-tip electrode is required for most applications using conventional power sources. However, for inverter power sources that use square-wave technology with AC balance and variable AC frequency control, welding thinner materials and creating smaller beads is accomplished much easier using the same pointed and blunted 2% thoriated or ceriated electrodes as used for DCEN welding along with increased electrode-negative time.

designs vary with torch manufacturers. *Figure 38* shows the collet assembly of a typical GTAW torch.

Follow these steps to install the electrode:

> **CAUTION**
>
> Handle tungsten electrodes with clean gloves to avoid contamination of the electrode.

Step 1 Remove the electrode cap from the rear of the torch.

Step 2 Check the collet size and collet body. If it is not the correct size for the electrode to be installed, remove and replace it with the correct size.

Step 3 Replace the electrode cap, but do not tighten it. Screw the cap only about one turn to leave the collet clamp loose.

Step 4 Insert the electrode into the collet from the nozzle end of the torch, and adjust it so that it extends from the nozzle about 1 to 1½ times the electrode diameter. Insert the color-coded end first to prevent destroying the electrode markings.

> **CAUTION**
>
> If the electrode cap is overtightened, the threads will be damaged as the torch heats with use. If the cap is too loose, the collet and electrode will overheat and damage the torch.

Step 5 Hand-tighten the electrode cap to lock the electrode in place.

> **Hot Tip**
>
> ### *Electrode Stickout*
>
> Various manufacturers recommend stickout distances that vary from flush at the end of the cup to distances equal to the inside diameter of the cup, depending on the application and task. For instance, a root pass in a deep narrow groove can be accomplished more easily with a narrow cup and a stickout that is about equal to the inside diameter of the cup. However, the use of a gas lens allows a longer tungsten stickout for welding under the same circumstances.

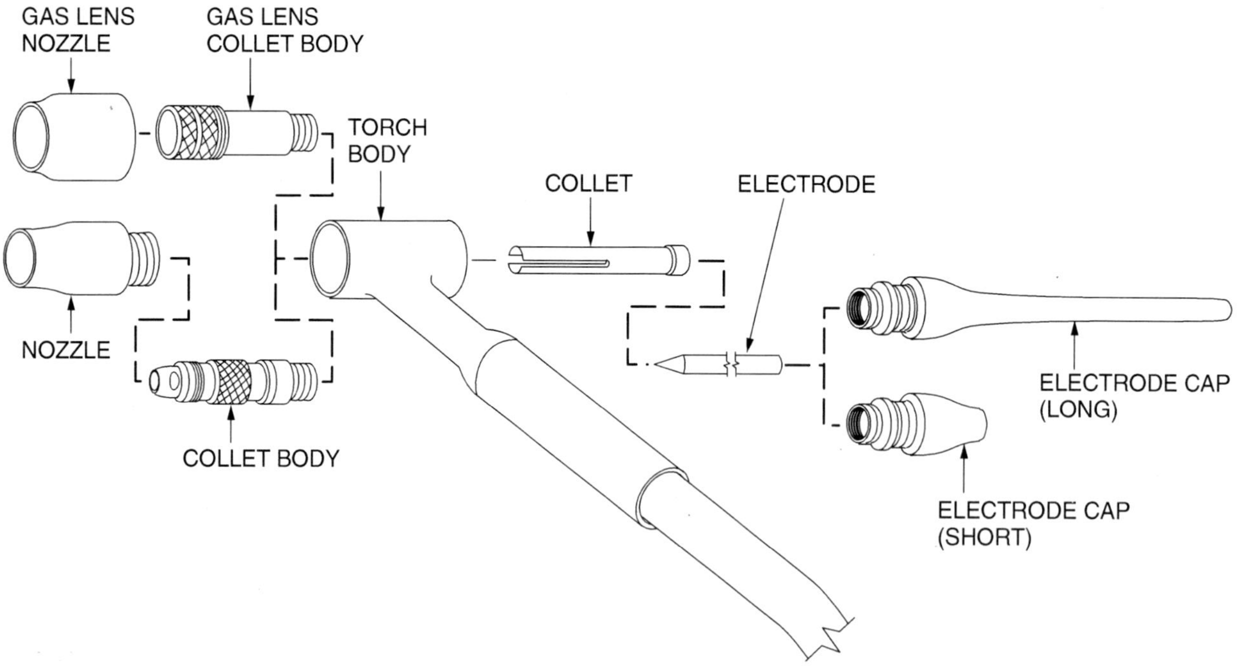

Figure 38 ◆ Collet assembly of a typical GTAW torch.

Summary

Welders must be able to select and set up GTAW equipment safely and efficiently. This module provided an overview of GTAW safety practices, equipment, shielding gas characteristics, electroded types and preparation, filler metal types and classifications, and equipment setup used for GTAW.

Review Questions

1. In GTAW, the torch holds a _____ electrode.
   a. carbon steel
   b. aluminum alloy
   c. tungsten
   d. copper

2. Welding current has _____.
   a. high voltage and low amperage
   b. low voltage and high amperage
   c. low voltage and low amperage
   d. high voltage and high amperage

3. The voltage present when the welding machine is on but no welding is taking place is the _____.
   a. open-circuit voltage
   b. operating voltage
   c. arc voltage
   d. high-frequency voltage

4. There are _____ types of welding current.
   a. two
   b. three
   c. four
   d. five

5. Constant current means that welding current is produced over a(n) _____ voltage range regardless of arc length.
   a. wide
   b. narrow
   c. random
   d. alternating

6. Transformer welding machines produce _____ welding current only.
   a. high-voltage
   b. intermittent
   c. AC
   d. DC

7. Inverter power sources _____ of the incoming primary power.
   a. hold constant the flow
   b. decrease the frequency
   c. increase the frequency
   d. redirect the flow

8. Most industrial, heavy-duty machines for manual welding have a _____ or 100% duty cycle.
   a. 20
   b. 30
   c. 40
   d. 60

9. If the amperage of a welding machine is set below the rated amperage, the duty cycle _____.
   a. does not change
   b. fluctuates slightly
   c. increases
   d. decreases

10. Typically, the smallest welding cable size is number _____.
    a. 1
    b. 2
    c. 4
    d. 3/0

11. Water-cooled torches have _____ hoses.
    a. two
    b. three
    c. four
    d. five

12. The most common shielding gas used with GTAW is _____.
    a. oxygen
    b. helium
    c. argon
    d. hydrogen

13. The widely used aluminum alloy filler metal rod ER4043 also contains _____.
    a. flux
    b. silicon
    c. iron
    d. tungsten

14. Helium shielding gas requires a higher flow rate than argon because it _____.

a. does not work as well
b. is heavier
c. has a higher density
d. has a lower density

15. The electrode will be contaminated if it is arced against _____.

a. aluminum
b. carbon
c. stainless steel
d. copper

GLOSSARY

Trade Terms Introduced in This Module

High frequency: A very high-voltage (several thousand volts), extremely low-amperage current that alternates at a high frequency of several million cycles per second. It is superimposed on the welding current to help establish and maintain the arc.

Potentiometer: A three-terminal resistor with an adjustable center connection, used for variable voltage control.

Truncated: A conical shape having its vertex (point) cut off.

Tungsten: A nonstandard common name for the tungsten electrode used in GTAW and the principal metal from which the electrodes are constructed.

APPENDIX

Tables

Table 1 Welding Cable Size Table for Manual Welding

Machine Size (in amps)	Duty Cycle (%)	Copper Cable Sizes for Combined Lengths of Electrodes Plus Ground Cable				
		Up to 50'	50' to 100'	100' to 150'	150' to 200'	200' to 250'
100	20	#8	#4	#3	#2	#1
180	20	#5	#4	#3	#2	#1
180	30	#4	#4	#3	#2	#1
200	50	#3	#3	#2	#1	#1/0
200	60	#2	#2	#2	#1	#1/0
225	20	#4	#3	#2	#1	#1/0
250	30	#3	#3	#2	#1	#1/0
300	60	#1/0	#1/0	#1/0	#2/0	#3/0
400	60	#2/0	#2/0	#2/0	#3/0	#4/0
500	60	#2/0	#2/0	#3/0	#3/0	#4/0
600	60	#3/0	#3/0	#3/0	#4/0	***
650	60	#3/0	#3/0	#4/0	**	***

208T01.EPS

Table 2 GTAW Electrode Color Codes and AWS Classifications

Electrode	Color Band	Electrode AWS Classification
Pure Tungsten	Green	EWP (Minimum 99.5% Tungsten)
1% Thoriated	Yellow	EWTh-1 (1% Thoria)
2% Thoriated	Red	EWTh-2 (2% Thoria)
Zirconiated	Brown	EWZr (0.15%–0.40% Zirconium Oxide)
Ceriated	Orange	EWCe-2 (2% Ceria)
Lanthanated	Black	EWLa-1.0 (1% Lanthana)
Lanthanated	Gold	EWLa-1.5 (1.5% Lanthana)
Lanthanated	Blue	EWLa-2.0 (2% Lanthana)
EWG	Grey	Undefined

208T02.EPS

Table 3 GTAW Filler Metal Specifications and Classification System

Material	Filler Metal		X Designator Description	Example
	AWS Spec.	AWS Class*		
Carbon steel	A5.18	ERXXS-Y EXXC-Y	Tensile strength × 1,000 (psi)	ER70S-3 E70C-3
Low-alloy steel	A5.28	ERXXS-Y EXXC-Y	Tensile strength × 1,000 (psi)	ER80S-B2 E80C-B2
Stainless steel	A5.9	ERXXXY	Stainless alloy (308, 410, etc.)	ER308L EC308L
Stainless steel (flux cored)	A5.22	RXXXTi-5	Stainless alloy (308, 309, etc.)	R309LT1-5
Aluminum	A5.10	ERXXXX-Y	Aluminum alloy (4043, 5083, etc.)	ER4043
Nickel	A5.14	ERNiXX-Y	Major alloying elements (Cr, Fe, Mo, etc.)	ERNiCr-3
Copper	A5.7	ERCuXX-Y	Major alloying elements (Al, Ni, Si, etc.)	ERCuAl-A2
Magnesium	A5.19	ERXXYYY	Major alloying elements (Al, Zn, etc.)	ERAZ92A
Titanium	A5.16	ERTi-Y		ERTi-5

*Legend
E — Filler metal may be used as an electrode.
R — Filler metal may be used as a rod.
S — Solid filler metal
C — Composite or stranded filler metal
Y — Designator (or combination of designators) that describes specific alloy, shielding gas to be used, diffusible hydrogen limit, etc. Refer to the appropriate AWS Filler Metal Specification shown in table for explanation.

208T03.EPS

Table 4 AWS Carbon Steel Classification and Uses

WIRE CLASSIFICATION	INTENDED USES
ER70S-2 ER48S-2	Welds all grades of carbon steel, including killed (highly deoxidized by adding aluminum powder or ferrosilicon), semikilled, and rimmed (slightly deoxidized by adding a small amount of aluminum powder). Also welds carbon steel with rusted surfaces. The viscous weld puddle is well suited to out-of-position welding.
ER70S-3 ER48S-3	Welds killed and semikilled steels (used in appliances, automobiles, and farm equipment).
ER70S-4 ER48S-4	Welds semikilled and rimmed steels (used in boilers and pressure vessels, piping, ship steels, structural steels, and A515 grades 55-70 steels).
ER70S-6 ER48S-6	Welds many carbon steels.
ER70S-7 ER48S-7	Welds a wide variety of steels with superior weld quality and appearance. Porosity and weld defects caused by mill scale or rust are minimized.
ER70S-G ER48S-G	Filler metals not included in the preceding classes that only have certain mechanical properties specified. Absence of a C or M suffix means that the manufacturer should be consulted for the recommended shielding gas and intended applications.

208T04.EPS

Table 5 AWS Low-Alloy Steel Filler Metal Classifications and Uses

AWS Classification	Intended Uses
ER70S-A1	These filler metals are similar to many carbon steel filler metals except that ½% molybdenum has been added to increase the strength of the weld at elevated temperatures. It also provides some increase in corrosion resistance. Typically used for welding ASTM plate and A335-P1 pipe.
ER80S-B2 E80C-B2	These filler metals are used to weld various chromium-molybdenum steels for elevated temperature and corrosive service. Also used for joining dissimilar combinations of chromium-molybdenum and carbon steels. Careful control of preheat, interpass temperature, and post-weld heat treatment (PWHT) is required to prevent cracking.
ER70S-B2L E70C-B2L	These filler metals are nearly identical to ER80S-B2 and E80C-B2 except for a low carbon content resulting in lower strength. They exhibit greater resistance to cracking if left in the as-welded condition or if PWHT is questionable.
ER90S-B3 E90S-B3	Filler metals used to weld chromium-molybdenum steels for high-temperature/high-pressure piping and pressure vessels. Also used to join chromium-molybdenum steels to carbon steels. Careful control of preheat, interpass temperature, and PWHT is required to prevent cracking.
ER80S-B3L E80C-B3L	These fillers are similar to ER90S-B3 and E90C-B3 except for low carbon content resulting lower strength. They exhibit greater resistance to cracking if left in the as-welded condition or if PWHT is questionable.
ER80-Ni1 E80C-Ni1	Filler metals with 1% nickel that deposit weld metal similar to E8018-C3 covered electrodes. They are used for welding low-alloy, high-strength steels requiring good toughness at temperatures as low as –50°F.
ER80S-Ni2 E70C-Ni2 E80C-Ni2	These filler metals are used for welding 2½% nickel steels. The weld metal is similar to E8018-C1 covered electrodes. These fillers are used in applications where good toughness is required at temperatures as low as –80°F.
ER80S-Ni3 E80C-Ni3	These filler metals deposit weld metal similar to E8018-C2 covered electrodes. They are typically used for welding 3½% nickel steels for low-temperature service.
ER80S-D2 ER90S-D2 E90C-D2	These filler metals have the same chemical requirements as E70S-1B for carbon steel. The difference between the ER80 and ER90 fillers is the change in shielding gas and mechanical properties. These fillers contain molybdenum to control porosity when used with CO_2 shielding gas. They are used to weld a variety of carbon and low-alloy, high-strength steels in both an as-welded or PWHT conditions using GMAW short circuiting and pulsed arc processes.
ER100S-1 ER110S-1 ER120S-1	These filler metals deposit high-strength, very tough weld metal for critical applications where tensile strength exceeds 100ksi and in-service temperatures as low as –60°F. Mechanical properties of the weld vary depending on the heat input used.
ER80S-B6	These filler metals are used to weld material of similar composition, usually pipe or tubing. Preheat and PWHT is required.
ER80S-B8	These filler metals contain about 9% chromium and 1% molybdenum. Their use and heat treatment is the same as ER80S-B6.
ER90S-B9	These fillers are a chromium-molybdenum alloy modified with columbium (niobium) and vanadium to provide strength, toughness, fatigue life, and oxidation/corrosion resistance at elevated temperatures. This allows components now fabricated with stainless and ferritic steels to be made from a single alloy, eliminating dissimilar metal welding problems.
ERXXS-G EXXS-G	These are filler metals, not included in the previous classes, for which only certain mechanical properties are specified. The manufacturer should be consulted for composition characteristics and intended uses of filler metals supplied under this classification.

208T05.EPS

Table 6 AWS Stainless Steel Electrode Rod Classifications and Their Uses

Electrode Rod Classification	Metals Welded
ER308	Welds Type 308 stainless steel and any chromium-nickel alloy containing approximately 19% chromium and 9% nickel.
ER308L	Welds the same metals as ER308 but contains less carbon, reducing the chances of carbon precipitation.
ER308L Si	Welds the same metals as ER308 and ER308L. The higher silicon improves wetting, especially with Ar-1% O_2 shielding gas. The high silicon content can cause greater crack sensitivity with extensive base metal dilution.
ER308H	Welds the same metals as ER308, but the higher range of carbon ensures the creep strength at the high end of the 308 range. Used for welding Type 304H stainless steel.
ER309	Welds Type 309 and 309S stainless steels, Type 304 stainless steel for severe corrosion service, and joins carbon steel to 304 stainless steel.
ER309 ELC	Has extra-low carbon content for resistance to carbide precipitation. Used for Type 309L stainless steel and overlay work.
ER310	Welds Type 310 stainless steel.
ER312	Welds stainless steels to carbon steels. Welds high-strength steels that are difficult to weld with ferritic electrodes. Used to weld carbon steel mounts to stainless steel vessels.
ER316	Welds Type 316 and 319 stainless steels. Used in high-temperature applications for creep resistance.
ER316L	Welds Type 316L stainless steel. More resistant to intergranular corrosion than ER316.
ER316L Si	Contains a higher amount of silicon, which markedly improves wetting characteristics for use in applications demanding a good cosmetic appearance.
ER316H	Welds Type 316H stainless steel. This electrode can be used in place of ER316. The restricted range of carbon ensures that the creep strength of the weld metal is at the high end of the 316 range.
ER320LR	Welds Carpenter 20Cb-3 type stainless steel.
ER347	Welds Type 321 and 347 stainless steel where the service temperature will be below 600°F and maximum resistance to corrosion is required.

208T06.EPS

Table 7 Recommended Filler Metals for Common Aluminum Alloys (1 of 2)

Base Metal to Base Metal	201.0 206.0 224.0	319.0, 333.0, 354.0, 355.0, C355.0	356.0, A356.0 357.0, A357.0 413.0, 443.0, A444.0	511.0, 512.0, 513.0, 514.0, 535.0	7004, 7005, 7039, 710.0, 712.0	6009 6010 6070	6005, 6061, 6063, 6101, 6151, 6201, 6351, 6951	5456	5454
1060, 1070, 1080,1350	ER4145	ER4145	ER4043[a,b]	ER5356[c,d]	ER5356[c,d]	ER4043[a,b]	ER4043[b]	ER5356[d]	ER4043[b,d]
1100, 3003, Alc 3003	ER4145	ER4145	ER4043[a,b]	ER5356[c,d]	ER5356[c,d]	ER4043[a,b]	ER4043[b]	ER5356[d]	ER4043[b,d]
2014, 2036	ER4145[c]	ER4145[e]	ER4145	—	—	ER4145	ER4145	—	—
2219	ER2319[a]	ER4145[c]	ER4145[b,c]	ER4043	ER4043	ER4043[a,b]	ER4043[a,b]	—	ER4043[b]
3004, Alc 3004	—	ER4043[b]	ER4043[b]	ER5356[f]	ER5356[f]	ER4043[b]	ER4043[b,f]	ER5356[d]	ER5356[f]
5005, 5050	—	ER4043[b]	ER4043[b]	ER5356[f]	ER5356[f]	ER4043[b]	ER4043[b,f]	ER5356[d]	ER5356[f]
5052, 5652[i]	—	ER4043[b]	ER4043[f]	ER5356[f]	ER5356[f]	ER4043[b]	ER5356[c,f]	ER5356[f]	ER5356[f]
5083	—	—	ER5356[c,d]	ER5356[d]	ER5356[d]	—	ER5356[d]	ER5183[d]	ER5356[d]
5086	—	—	ER5356[c,d]	ER5356[d]	ER5356[d]	—	ER5356[d]	ER5356[d]	ER5356[d]
5154, 5254[i]	—	—	ER4043[f]	ER5356[f]	ER5356[f]	—	ER5356[f]	ER5356[f]	ER5356[f]
5454	—	ER4043[b]	ER4043[f]	ER5356[f]	ER5356[f]	ER4043[b]	ER5356[c,f]	ER5356[f]	ER5554[c,f]
5456	—	—	ER5356[c,d]	ER5356[d]	ER5556[d]	—	ER5356[d]	ER5556[d]	
6005, 6061, 6063 6101, 6151, 6201 6351, 6951	ER4145	ER4145[b,c]	ER4043[b,f,g]	ER5356[f]	ER5356[c,f]	ER4043[a,f,g]	ER4043[b,f,g]		
6009, 6010, 6070	ER4145	ER4145[b,c]	ER4043[a,f,g]	ER4043	ER4043	ER4043[a,b,g]			
7004, 7005, 7039 710.0, 712.0	—	ER4043[b]	ER4043[b,f]	ER5356[f]	ER5356[d]				
511.0, 512.0, 513.0 514.0, 535.0	—	—	ER4043[f]	ER5356[f]					
356.0, A356.0, 357.0 A357.0, 413.0 443.0, A444.0	ER4145	ER4145[b,c]	ER4043[b,h]						
319.0, 333.0 354.0, 355.0 C355.0	ER4145[c]	ER4145[b,c,h]							
201.0, 206.0, 224.0	ER2319[a,h]								

208T07A.EPS

Table 7 Recommended Filler Metals for Common Aluminum Alloys (2 of 2)

Base Metal to Base Metal	5154 5254[i]	5086	5083	5052 5652[i]	5005 5050	3004 Alc. 3004	2219	2014 2036	1100 3003 Alc. 3003	1060 1070 1080 1350
1060, 1070, 1080,1350	ER5356[c,d]	ER5356[d]	ER5356[d]	ER4043[b,d]	ER1100[b,c]	ER4043[b,d]	ER4145[b,c]	ER4145	ER1100[b,c]	ER1188[b,c,h,j]
1100, 3003, Alc 3003	ER5356[c,d]	ER5356[d]	ER5356[d]	ER4043[b,d]	ER1100[b,c]	ER4043[b,d]	ER4145[b,c]	ER4145	ER1100[b,c]	
2014, 2036	—	—	—	—	ER4145	ER4145	ER4145[e]	ER4145[e]		
2219	ER4043	—	—	ER4043[b]	ER4043[a,b]	ER4043[a,b]	ER2319[a]			
3004, Alc 3004	ER5356[f]	ER5356[d]	ER5356[d]	ER5356[c,f]	ER5356[c,f]	ER5356[c,f]				
5005, 5050	ER5356[f]	ER5356[d]	ER5356[d]	ER5356[c,d]	ER5356[c,f]					
5052, 5652[i]	ER5356[f]	ER5356[d]	ER5356[d]	ER5654[c,f,i]						
5083	ER5356[d]	ER5356[d]	ER5183[d]							
5086	ER5356[d]	ER5356[d]								
5154, 5254[i]	ER5356[f,i]									

General Notes:

1. Service conditions such as immersion in fresh or salt water, exposure to specific chemicals, or a sustained high temperature (over 150°F [66°C]) may limit the choice of filler metals. Filler metals ER5183, ER5356, ER5556 are not recommended for sustained elevated temperature service.
2. Recommendations in this table apply to gas shielded arc welding processes. For oxyfuel gas welding, only ER1188, ER1100, ER4043, ER4047, and ER4145 filler metals are ordinarily used.
3. Where no filler metal is listed, the base metal combination is not recommended for welding.

Notes:

a. ER4145 may be used for some applications.
b. ER4047 may be used for some applications.
c. ER4043 may be used for some applications.
d. ER5183, ER5356, or ER5556 may be used.
e. ER2319 may be used for some applications. It can supply high strength when the weldment is postweld solution heat treated and aged.
f. ER5183, ER5356, ER5556, and ER5654 may be used. In some cases, they provide: (1) improved color match after anodizing treatment, (2) highest weld ductility, and (3) higher weld strength.
g. ER4643 will provide higher strength in 1/2 in. (12 mm) and thicker groove welds in 6XXX base alloys when postweld solution heat treated and aged.
h. Filler metal with the same analysis as the base metal is sometimes used. the following wrought filler metals possess the same chemical composition limits as cast filler alloys: ER4009 and R4009 as R-C355.0; ER4010 and R4010 as R-A356.0; and R4011 as R-A357.0
i. Base metal alloys 5254 and 5652 are used for hydrogen peroxide service. ER5654 filler metal is used for welding both alloys for service temperatures below 150°F(66°C).
j. ER1100 may be used for some applications.

208T07B.EPS

Table 8 AWS Copper Electrode Rod Identifiers and Their Principal Uses

Filler Metals for Fusion Welding Copper Alloys			
AWS Classification			
Covered Electrode	**Bare Wire**	**Common Name**	**Base Metal Applications**
ECu	ERCu	Copper	Coppers
ECuSi	ERCuSi-A	Silicon bronze	Silicon bronzes, brasses
ECuSn-A ECuSn-C	ERCuSn-A	Phosphor bronze	Phosphor bronzes, brasses
ECuNi	ERCuNi	Copper-nickel	Copper-nickel alloys
ECuAl-A2	ERCuAl-A2	Aluminum bronze	Aluminum bronzes, brasses, silicon bronzes, manganese bronzes
ECuAl-B	ERCuAl-A3	Aluminum bronze	Aluminum bronzes
ECuNiAl	ERCuNiAl	—	Nickel-aluminum bronzes
ECuMnNiAl	ERCuMnNiAl	—	Manganese-nickel-aluminum bronzes
	RBCuZn-A	Naval brass	Brasses, copper
	RCuZn-B	Low fuming brass	Brasses, manganese bronzes
	RCuZn-C	Low fuming brass	Brasses, manganese bronzes

208T08.EPS

Table 9 AWS Nickel Filler Metal Classifications and Uses (1 of 3)

AWS Classification	Intended Uses
ERNi-1	These filler metals are intended for welding wrought and cast forms of commercially pure nickel alloy (ASTM B160, B161, B162, and B163 having UNS numbers N02200 or N02201) to themselves with GTAW, GMAW, SAW, and PAW processes. The filler metals contain sufficient titanium to control weld-metal porosity with these welding processes.
ERNiCu-7	These filler metals are used for welding nickel-copper alloy (ASTM B127, B163, B164, and B165 having UNS number N04400) to itself with GTAW, GMAW, SAW, and PAW processes. The filler metals contain sufficient titanium to control porosity with these welding processes.
ERNiCu-8	These filler metals are used for welding age-hardening nickel-copper alloy (ASTM F467 and F468 having UNS number N05500) to itself with GTAW, GMAW, SAW, and PAW processes. The filler metal will age harden on heat treatment. For specific information concerning age hardening, consult the manufacturer or the manufacturer's technical literature.
ERNiCr-3	These filler metals are used for welding nickel-chromium-iron alloy (ASTM B163, B166, B167, and B168 having UNS number N06600) to itself with GTAW, GMAW, SAW, and PAW processes. They are also used with the same processes for welding the clad side of joints in steel clad with nickel-chromium-iron alloy, for surfacing steel with nickel-chromium-iron weld metal, for dissimilar welding of nickel-base alloys, and for joining steel to nickel-base alloys.
ERNiCr-4	These filler metals are used for GTAW of 50/50 nickel/chromium alloy, cladding nickel/chromium alloy onto nickel-iron-chromium tubing, and casting repair. The filler metals are resistant to high-temperature corrosion, including fuel-ash corrosion in atmospheres containing sulfur and vanadium.
ERNiCr-6	These filler metals are used for welding nickel-chromium-iron alloy (ASTM B163, B166, B167, and B168 having UNS number N06600) to itself with GTAW, GMAW, SAW, and PAW processes. They are also used with the same processes for welding the clad side of joints in steel clad with nickel-chromium-iron alloy, for surfacing steel with nickel-chromium-iron weld metal, and for joining steel to nickel-base alloys.
ERNiCrFe-5	These filler metals are used for welding nickel-chromium-iron alloy (ASTM B163, B166, B167, and B168 having UNS number N06600) to itself with GTAW, GMAW, SAW, and PAW processes. The higher niobium content of the filler metal is intended to minimize cracking where high welding stresses are encountered, as in thick-section base metal.
ERNiCrFe-6	These filler metals are used for cladding steel with nickel-chromium-iron weld metal and for joining steel to nickel-base alloys with GTAW, GMAW, SAW, and PAW processes. The weld metal will age harden on heat treatment. For specific information concerning age hardening, consult the manufacturer or the manufacturer's technical literature.
ERNiCrFe-7	These filler metals are used for welding nickel-chromium-iron alloy (ASTM B163, B166, B167, and B168 having UNS number N06690) to itself with GTAW, GMAW, SAW, and PAW processes. They are also used to weld nickel-chromium-iron alloy with the same processes to steels, as an overlay on steels, and to weld steels clad with the nickel-chromium-iron alloys.
ERNiCrFe-8	These filler metals are used for cladding steel with nickel-chromium-iron weld metal and for joining steel to nickel-base alloys using the GTAW, GMAW, SAW, and PAW processes. The weld metal will age harden on heat treatment. For specific information concerning age hardening, consult the manufacturer or the manufacturer's technical literature.
ERNiCrFe-11	These filler metals are used for welding nickel-chromium-iron-aluminum alloy (ASTM B166, B167, and B168 having UNS number N06601) to itself and to other high-temperature compositions using the GTAW process. It is used for severe applications where the exposure temperature can exceed 2100°F.

208T09A.EPS

Table 9 AWS Nickel Filler Metal Classifications and Uses (2 of 3)

AWS Classification	Intended Uses
ERNiFeCr-1	These filler metals are used for welding nickel-iron-chromium-molybdenum-copper alloy (ASTM B423 having UNS number N08825) to itself using the GTAW and GMAW processes.
ERNiFeCr-2	These filler metals are used for welding nickel-chromium-niobium-molybdenum alloy (ASTM B637 and AMS 5589 having UNS number N07718) to itself using the GTAW processes. The weld metal will age harden on heat treatment. For specific information concerning age hardening, consult the manufacturer or the manufacturer's technical literature.
ERNiMo-1	These filler metals are used for welding nickel-molybdenum alloy (ASTM B333 having UNS number N10001) to itself with GTAW and GMAW processes.
ERNiMo-2	These filler metals are used for welding nickel-molybdenum alloy (ASTM B366, B434, and B573 having UNS number N10003) to itself with GTAW and GMAW processes. They are also used to weld nickel-molybdenum alloy with the same processes to steel, to other nickel-base alloys, and for cladding steel with nickel-molybdenum weld metal.
ERNiMo-3	These filler metals are used for weld repair of various nickel-, cobalt-, and iron-base alloys and for dissimilar joining applications of nickel-, cobalt-, and iron-base alloys.
ERNiMo-7	These filler metals are used for welding nickel-molybdenum alloy (ASTM B333 and B335 having UNS number N10665) to itself with GTAW and GMAW processes and for cladding steel with nickel-molybdenum weld metal using the same processes.
ERNiMo-8 ERNiMo-9	These filler metals are used for welding 9% nickel steel (ASTM A333, A334, A353, A553 having UNS number K81340) to itself using the GTAW and SAW processes.
ERNiMo-10	These filler metals are used for welding nickel-molybdenum alloy (ASTM B333, B335, B366, B564, B619, B622, and B626 having UNS number N10675) to itself with GTAW, GMAW, and PAW processes. They are also used for welding the clad side of joints in steel clad with nickel- molybdenum alloy and for welding nickel-molybdenum alloys to steel and to other nickel-base alloys with the same processes.
ERNiCrMo-1	These filler metals are used for welding nickel-chromium-molybdenum alloy (ASTM B581 and B582 having UNS number N06007) to itself with GTAW, GMAW, and PAW processes. They are also used with the same processes to weld nickel-chromium-molybdenum alloy to steel, to other nickel-base alloys, and for cladding steel with nickel-chromium-molybdenum weld metal.
ERNiCrMo-2	These filler metals are used for welding nickel-chromium-molybdenum alloy (ASTM B366, B435, and B572 having UNS number N06002) to itself with GTAW, GMAW, and PAW processes. They are also used with the same processes to weld nickel-chromium-molybdenum alloy to steel, to other nickel-base alloys, and for cladding steel with nickel-chromium-molybdenum weld metal.
ERNiCrMo-3	These filler metals are used for welding nickel-chromium-molybdenum alloy (ASTM B443, B444, and B446 having UNS number N06625) to itself with GTAW, GMAW, SAW, and PAW processes. They are also used with the same processes to weld nickel-chromium-molybdenum alloy to steel, to other nickel-base alloys, for cladding steel with nickel-chromium-molybdenum weld metal, and for welding the clad side of joints in steel with nickel-chromium-molybdenum alloy. These filler metals are recommended for applications where the operating temperature ranges from cryogenic to 1000°F.

208T09B.EPS

Table 9 AWS Nickel Filler Metal Classifications and Uses (3 of 3)

AWS Classification	Intended Uses
ERNiCrMo-4	These filler metals are used for welding nickel-chromium-molybdenum alloy (ASTM B574, B575, B619, B622, and B628 having UNS number N10276) to itself with GTAW and GMAW processes. They are also used with the same processes to weld nickel-chromium-molybdenum alloy to steel, to other nickel-base alloys, and for cladding steel with nickel-chromium-molybdenum weld metal.
ERNiCrMo-7	These filler metals are used for welding nickel-chromium-molybdenum alloy (ASTM B574, B575, B619, B622, and B628 having UNS number N06455) to itself with GTAW, GMAW, and PAW processes. They are also used with the same processes to weld nickel-chromium-molybdenum alloy to steel, to other nickel-base alloys, and for cladding steel with nickel-chromium-molybdenum weld metal.
ERNiCrMo-8	These filler metals are used for welding nickel-chromium-molybdenum alloy (ASTM B581, B582, B619, B622 and B626 having UNS number N06975) to itself with GTAW, GMAW, and PAW processes. They are also used with the same processes to weld nickel-chromium-molybdenum alloy to steel, to other nickel-base alloys, and for cladding steel with nickel-chromium-molybdenum weld metal.
ERNiCrMo-9	These filler metals are used for welding nickel-chromium-molybdenum alloy (ASTM B581, B582, B619, B622, and B626 having UNS number N06007 or N06985) to itself with GTAW, GMAW, and PAW processes. They are also used with the same processes to weld nickel-chromium-molybdenum alloy to steel, to other nickel-base alloys, and for cladding steel with nickel-chromium-molybdenum weld metal.
ERNiCrMo-10	These filler metals are used for welding nickel-chromium-molybdenum alloy (ASTM B574, B575, B619, B622 and B628 having UNS number N06022) to itself with GTAW, GMAW, and PAW processes. They are also used with the same processes to weld nickel-chromium-molybdenum alloy to steel, to other nickel-base alloys, and for cladding steel with nickel-chromium-molybdenum weld metal.
ERNiCrMo-11	These filler metals are used for welding nickel-chromium-molybdenum alloy (ASTM B581, B582, B619, B622, and B626 having UNS number N06030) to itself with GTAW, GMAW, and PAW processes. They are also used with the same processes to weld nickel-chromium-molybdenum alloy to steel, to other nickel-base alloys, and for cladding steel with nickel-chromium-molybdenum weld metal.
ERNiCrMo-13	These filler metals are used for welding nickel-chromium-molybdenum alloy (B564, B574, B575, B619, B622, and B626 having UNS number N06059) to itself with GTAW, GMAW, SAW, and PAW processes. They are also used with the same processes to weld nickel-chromium-molybdenum alloy to steel, to other nickel-base alloys, and for cladding steel with nickel-chromium-molybdenum weld metal.
ERNiCrMo-14	These filler metals are used for welding nickel-chromium-molybdenum alloy (ASTM B574, B575, B619, B622, and B628 having UNS number N06686) to itself with GTAW, GMAW, and SAW processes. They are also used with the same processes to weld nickel-chromium-molybdenum alloy to steel, to other nickel-base alloys, and for cladding steel with nickel-chromium-molybdenum weld metal.
ERNiCrMo-15	These filler metals are used for welding nickel-chromium-molybdenum alloy (ASTM B805, UNS number N07725) to itself and for cladding steel with GMAW and GTAW processes. The weld metal age-hardens on heat treatment. For specific information concerning age hardening, consult the manufacturer or the manufacturer's technical literature.
ERNiCrCoMo-1	These filler metals are used for welding nickel-chromium-cobalt-molybdenum alloy (UNS N06617) to itself with GTAW and GMAW processes.
ERNiCrWMo-1	These filler metals are used for welding nickel-chromium-tungsten-molybdenum-lanthanum alloy (ASTM B366, B435, B564, and B572 having UNS number N06230) to itself with GTAW and GMAW processes.

208T09C.EPS

Table 10 Recommended Filler Metals for Common Magnesium Alloys (1 of 2)

Base Metal	Base Metal											
	AM100A	**AZ10A**	**AZ31B AZ31C**	**AZ61A**	**AZ63A**	**AZ80A**	**AZ81A**	**AZ91C**	**AZ92A**	**EK41A**	**EZ33A**	**HK31A**
	Filler Metal[a,b]											
AM100A	AZ92A											
AZ10A	AZ92A	AZ61A AZ92A										
AZ31B AZ31C	AZ92A	AZ61A AZ92A	AZ61A AZ92A									
AZ61A	AZ92A	AZ61A AZ92A	AZ61A AZ92A	AZ61A AZ92A								
AZ63A	c	c	c	c	AZ101A AZ92A							
AZ80A	AZ92A	AZ61A AZ92A	AZ61A AZ92A	AZ61A AZ92A	c	AZ61A AZ92A						
AZ81A	AZ92A	AZ92A	AZ92A	AZ92A	c	AZ92A	AZ101A AZ92A					
AZ91C	AZ92A	AZ92A	AZ92A	AZ92A	c	AZ92A	AZ92A	AZ101A AZ92A				
AZ92A	AZ92A	AZ92A	AZ92A	AZ92A	c	AZ92A	AZ92A	AZ92A	AZ101A			
EK41A	AZ92A	AZ92A	AZ92A	AZ92A	c	AZ92A	AZ92A	AZ92A	AZ92A	EZ33A		
EK33A	AZ92A	AZ92A	AZ92A	AZ92A	c	AZ92A	AZ92A	AZ92A	AZ92A	EZ33A	EZ33A	
HK31A	AZ92A	AZ92A	AZ92A	AZ92A	c	AZ92A	AZ92A	AZ92A	AZ92A	EZ33A	EZ33A	EZ33A
HM21A	AZ92A	AZ92A	AZ92A	AZ92A	c	AZ92A	AZ92A	AZ92A	AZ92A	EZ33A	EZ33A	EZ33A
HM31A	AZ92A	AZ92A	AZ92A	AZ92A	c	AZ92A	AZ92A	AZ92A	AZ92A	EZ33A	EZ33A	EZ33A
HM32A	AZ92A	AZ92A	AZ92A	AZ92A	c	AZ92A	AZ92A	AZ92A	AZ92A	EZ33A	EZ33A	EZ33A
K1A	AZ92A	AZ92A	AZ92A	AZ92A	c	AZ92A	AZ92A	AZ92A	AZ92A	EZ33A	EZ33A	EZ33A
LA141A					c	c	c	c	c	d	d	d
M1A MG1	AZ92A	AZ61A AZ92A	AZ61A AZ92A	AZ61A AZ92A	c	AZ61A AZ92A	AZ92A	AZ92A	AZ92A	AZ92A	AZ92A	AZ92A
QE22A	d	d	AZ92A	d	c	d	d	d	d	EZ33A	EZ33A	EZ33A
ZE10A	AZ92A	AZ61A AZ92A	AZ61A AZ92A	AZ61A AZ92A	c	AZ61A AZ92A	AZ92A	AZ92A	AZ92A	EZ33A AZ92A	EZ33A AZ92A	EZ33A AZ92A
ZE41A	d	d	d	d	c	d	d	d	d	EZ33A	EZ33A	EZ33A
ZK21A	AZ92A	AZ61A AZ92A	AZ61A AZ92A	AZ61A AZ92A	c	AZ61A AZ92A	AZ92A	AZ92A	AZ92A	AZ92A	AZ92A	AZ92A
ZH62A ZK51A ZK60A ZK61A	c	c	c	c	c	c	c	c	c	c	c	c

Notes:
a. When more than one filler metal is given, they are listed in order of preference.
b. The letter prefix (ER or R), designating usability of the filler metal, has been deleted to reduce clutter in the table.
c. Welding not recommended
d. No data available

208T10A.EPS

Table 10 Recommended Filler Metals for Common Magnesium Alloys (2 of 2)

Base Metal	Base Metal										
	HM21A	**HM31A**	**HZ32A**	**K1A**	**LA141A**	**M1A MG1**	**QE22A**	**ZE10A**	**ZE41A**	**ZK21A**	**ZH62A ZK51A ZK60A ZK61A**
	Filler Metal[a,b]										
HM21A	EZ33A										
HM31A	EZ33A	EZ33A									
HZ32A	EZ33A	EZ33A									
K1A	EZ33A	EZ33A	EZ33A	EZ33A							
LA141A	EZ33A	d	d	d	EZ33A						
M1A MG1	AZ92A	AZ92A	AZ92A	AZ92A	d	AZ61A AZ92A					
QE22A	EZ33A	EZ33A	EZ33A	EZ33A	EZ33A	c	EZ33A				
ZE10A	EZ33A AZ92A	EZ33A AZ92A	EZ33A AZ92A	EZ33A AZ92A	EZ33A	AZ61A AZ92A	EZ33A AZ92A	AZ61A AZ92A			
ZE41A	EZ33A	EZ33A	EZ33A	EZ33A	d	d	EZ33A	d	EZ33A		
ZK21A	AZ92A	AZ92A	AZ92A	AZ92A	d	AZ61A AZ92A	AZ92A	AZ61A AZ92A	AZ92A	AZ61A AZ92A	
ZH62A ZK51A ZK60A ZK61A	c	c	c	c	c	c	c	c	c	c	EZ33A

Notes:

a. When more than one filler metal is given, they are listed in order of preference.

b. The letter prefix (ER or R), designating usability of the filler metal, has been deleted to reduce clutter in the table.

c. Welding not recommended

d. No data available

208T10B.EPS

Table 11 AWS Titanium Filler Metal Classifications and Uses

AWS Classification	Intended Uses
ERTi-1 ERTi-2 ERTi-3 ERTi-4	The alloys used for these filler metals are commonly referred to as commercially pure (C.P.) grades (1 through 4) of titanium with the level of impurities and mechanical properties increasing slightly from ERTi-1 to ERTi-4. C.P. Grade 2 (equivalent to ERTi-2) is the most widely used titanium alloy for industrial applications because of its good balance of strength, formability, and weldability. Typical uses are in seawater and brackish water heat exchangers, chemical process heat exchangers, pressure vessels and piping systems, pulp bleaching systems, air pollution control scrubbers, and electrochemical and chemical storage tanks. These grades also have some uses in the aerospace industry.
ERTi-5	The alloy used for these filler metals commonly referred to as 6-4 titanium and is probably the most widely used titanium alloy. Its high strength, ability to be heat treated, weldability, excellent fatigue strength, and hardness make the alloy excellent for industrial fans, pressure vessels, aircraft components, compressor blades, and automotive and jet engine parts.
ERTi-5ELI	This filler metal is a slightly purer version of ERTi-5 with ELI. With special processing, this alloy can develop high fracture toughness. Primary uses are in surgical implants, cryogenic vessels, and airframe components.
ERTi-6	The alloy used for these filler metals has good weldability, oxidation resistance, stability, and strength at elevated temperature. Typical uses of the alloy include gas turbine engine casings, aerospace structural members at elevated temperatures near engines and wing leading edges, and chemical process equipment where high elevated temperature strength is required.
ERTi-6ELI	These filler metals are a slightly purer version of ERTi-6 with ELI. They are used to fabricate pressure vessels for liquefied gases and other high pressure cryogenic vessels where better ductility and toughness with slightly lower strength are required.
ERTi-7	Welds made with these filler metals probably are the most corrosion-resistant titanium welds used in industrial applications. Mechanical and physical properties are similar to those of ERTi-2. This alloy extends the use of titanium into mildly reducing media, into much higher chloride levels, or where the environment fluctuates between oxidizing and reducing.
ERTi-9	These filler metals often are referred to as half 6-4 because the major components are roughly half that found in ERTi-5. The primary use has been in welding hydraulic tubing and fittings for aircraft. This alloy is desirable when high strength and the ability to maintain strength at elevated temperatures is required. Corrosion resistance, in most environments, appears to be similar to, or slightly less than, that of weld metal from ERTi-2 electrodes.
ERTi-9ELI	The reduced oxygen content of the ERTi-9ELI alloy results in slightly lower strength and improved toughness in comparison with weld metal from ERTi-9 filler metals.
ERTi-12	Welds made with these filler metals offer improved corrosion resistance, especially to crevice corrosion in hot brines, and higher strength levels compared to similar welds made using ERTi-2. Uses in industrial applications are similar to those of ERTi-2, but can be extended to less oxidizing conditions.
ERTi-15	Welds made with these filler metals have excellent resistance to salt water corrosion combined with good toughness and moderate strength. Typical uses are the fabrication of submersible hulls and pressure vessels using base material of a matching composition.

208T11.EPS

Table 12 Stainless Steel Flux Cored Electrode and Rod Welding Process, Shielding, and Polarity Requirements

AWS Classification[a]	External Shielding Medium[b]	Welding Polarity	
RXXXT1-5	100% Argon	DCEN	GTAW
RXXXT1-G	Not Specified	Not Specified	GTAW

Notes:

a. The letters "XXX" stand for the designation of the chemical composition. The "X" after the "T" designates the position of operation. A "0" indicates flat or horizontal operations; a "1" indicates all position operation.

b. The requirement for the use of specified external shielding medium shall not be construed to restrict the use of any other medium for which the electrodes are found suitable, for any application other than the classification tests.

208T12.EPS

Table 13 AWS Stainless Steel Alloy Flux Cored Rod Classifications and Uses

AWS Classification	Intended Uses
R308LT1-5	This flux cored rod is used primarily for root pass welding of Type 304 or 304L stainless steel piping joints when an inert gas backing purge is not possible or desirable. This rod can only be used for GTAW; however, a slag cover is produced that must be removed before additional weld layers can be deposited. Make sure to follow the manufacturer's instructions when using this rod.
R309LT1-5	This flux cored rod is used primarily for root pass welding of carbon steel pipe to austenitic stainless steel pipe when an inert gas backing purge is not possible or desirable. The high chromium and nickel content allows dilution with carbon steel while still producing a weld. This rod can only be used for GTAW; however, a slag cover is produced that must be removed before additional weld layers can be deposited. Make sure to follow the manufacturer's instructions when using this rod.
R316LT1-5	This flux cored rod is used primarily for root pass welding of Type 316 or 316L stainless steel piping joints when an inert gas backing purge is not possible or desirable. This rod can only be used for GTAW; however, a slag cover is produced that must be removed before additional weld layers can be deposited. Make sure to follow the manufacturer's instructions when using this rod.
R347T1-5	This flux cored filler rod is used primarily for the root pass welding of Type 347 stainless steel piping joints when an inert gas backing purge is not possible or desirable. This rod can only be used for GTAW; however, a slag cover is produced that must be removed before additional weld layers can be deposited. Make sure to follow the manufacturer's instructions when using this rod.

208T13.EPS

Table 14 Welding Current Requirements for Various Base Metals

Base Metal	ACHF Current	DCEN Current	DCEP Current
Aluminum	1	NR	2
Aluminum Casting	1	NR	NR
Aluminum Bronze	1	NR	2
Beryllium Copper	1	NR	2
Brass Alloys	2	1	NR
Copper Base Alloys	2	1	NR
Cast Iron	2	1	NR
Deoxidized Copper <0.09"	NR	1	NR
Dissimilar Metals	2	1	NR
Hard-Facing	1	2	NR
Hastelloy Alloys	2	1	NR
High Carbon Steel 0.015 – 0.030"	2	1	NR
High Carbon Steel 0.030" – thicker	2	1	NR
Low Carbon Steel 0.015" – 0.030"	2	1	NR
Low Carbon Steel 0.030" – 0.125"	NR	1	NR
Magnesium Casting	1	NR	2
Magnesium <1/8"	1	NR	2
Magnesium >3/16"	1	NR	NR
Nickel & Nickel Alloys	2	1	NR
Silicon Bronze	NR	1	NR
Silver	2	1	NR
Silver Cladding	1	NR	NR
Stainless Steel <0.05"	1	2	NR
Stainless Steel >0.05"	2	1	NR
Titanium	2	1	NR

1 = recommended or first choice 2 = acceptable alternate or second choice NR = not recommended < = Less than > = Greater than

208T14.EPS

Table 15 Carbon Steel – Direct Current Electrode Negative (DCEN)

Metal Thickness	Joint Type	Amperage	Electrode Diameter	Filler Rod Diameter	Shield Gas	Cubic Feet per Hour
1/16"	Butt	60–70	1/16"	1/16"	Argon	15
	Lap	70–90				
	Corner	60–70				
	Fillet	70–90				
1/8"	Butt	80–100	1/16"–3/32"	3/32"	Argon	15
	Lap	90–115				
	Corner	80–100				
	Fillet	90–115				
3/16"	Butt	115–135	3/32"	1/8"	Argon	20
	Lap	140–165				
	Corner	115–135				
	Fillet	140–170				
1/4"	Butt	160–175	1/8"	5/32"	Argon	20
	Lap	170–200				
	Corner	160–175				
	Fillet	175–210				

208T15.EPS

Table 16 Stainless Steel – Direct Current Electrode Negative (DCEN)

Metal Thickness	Joint Type	Amperage (Flat Position)	Electrode Diameter	Filler Rod Diameter	Shield Gas	Cubic Feet per Hour
1/16"	Butt	80–100	1/16"	1/16"	Argon	10
	Lap	100–120				
	Corner	80–100				
	Fillet	90–110				
1/8"	Butt	120–140	1/16"	3/32"	Argon	10
	Lap	130–150				
	Corner	120–140				
	Fillet	130–150				
3/16"	Butt	200–250	3/32" or 1/8"	1/8"	Argon	15
	Lap	225–275				
	Corner	200–250				
	Fillet	225–275				
1/4"	Butt	275–350	1/8"	5/32"	Argon	15
	Lap	300–375				
	Corner	275–350				
	Fillet	300–375				
1/2"	Butt	350–450	1/8" or 3/16"	3/16"	Argon	15
	Lap	375–475				
	Fillet	375–475				

208T16.EPS

Table 17 Aluminum – Alternating Current with High Frequency Stabilization

Metal Thickness	Joint Type	Amperage (Flat Positions)	Electrode Diameter	Filler Rod Diameter	Shield Gas	Cubic Feet per Hour
1/16"	Butt	60–85	1/16"	1/16"	Argon	15
	Lap	70–90				
	Corner	60–85				
	Fillet	75–100				
1/8"	Butt	125–150	3/32"	3/32" or 1/8"	Argon	20
	Lap	130–160				
	Corner	120–140				
	Fillet	130–160				
3/16"	Butt	180–225	1/8"	1/8"	Argon	20
	Lap	190–240				
	Corner	180–225				
	Fillet	190–240				
1/4"	Butt	240–280	3/16"	1/8" or 3/16"	Argon	25
	Lap	250–320				
	Corner	240–280				
	Fillet	250–320				
1/2"	Butt	400–450	3/16" or 1/4"	3/16" or 1/4"	Argon	30
	Lap	400–450				
	Corner	400–450				
	Fillet	420–470				

208T17.EPS

Table 18 Deoxidized Copper – Direct Current Electrode Negative (DCEN)

Metal Thickness	Joint Type	Amperage	Electrode Diameter	Filler Rod Diameter	Shield Gas	Cubic Feet per Hour
1/16"	All–S.P.	110–150	1/16"	1/16"	Argon	15
1/8"	All–S.P.	175–250	3/32"	3/32"	Argon	15
3/16"	All–S.P.	190–250	1/8"	1/8"	Helium	30
1/4"	All–S.P.	225–280	1/8"	1/8"	Helium	30
3/8"	All–M.P.	280–340	3/16"	3/16"	Helium	40
1/4"	Butt–M.P.	375–525	3/16" or 1/4"	1/4"	Helium	40

208T18.EPS

Table 19 Electrode Current Ranges

Electrode Diameter	GTAW Modes			
	AC*		Pure Tungsten, Rare Earth, or Thoriated Tungsten Electrodes	
	Pure Tungsten Electrodes	Thoriated or Rare Earth Electrodes**	DCEN	DCEP
0.020	5–15	8–20	8–20	
0.040	10–60	15–80	15–80	
1/16	50–100	70–150	70–150	10–20
3/32	100–160	140–235	150–250	15–30
1/8	150–210	225–325	250–400	25–40
5/32	200–275	300–425	400–500	40–55
3/16	250–350	400–525	500–800	55–80
1/4	325–475	500–700	800 - 1000	80–125

* Maximum values shown have been determined using an unbalanced AC power source. If a balanced AC power source is used, either reduce these values by about 30% or use the next size electrode. This is necessary because of the higher heat input to the electrode in a balanced AC setup.

** Balled electrode tip ends can best be sustained at these current levels.

208T19.EPS

Table 20 Tip Diameters

Electrode Diameter (in.)	Diameter at Tip (in.)
0.040	0.005
0.040	0.010
0.062	0.020
0.062	0.030
0.093	0.030
0.093	0.045
0.125	0.045
0.125	0.060

208T20.EPS

Additional Resources

This module is intended to be a thorough resource for task training. The following reference works are suggested for further study. These are optional materials for continued education rather than for task training.

GTAW Handbook. Current Edition. William H. Minnick. Tinley Park, IL: The Goodheart-Willcox Company, Inc.

Modern Welding, 2000. A.D. Althouse, C.H. Turnquist, W.A. Bowditch, K.E. Bowditch. Tinley Park, IL: The Goodheart-Willcox Company, Inc.

Welding Skills, Current Edition. R.T. Miller. Homewood, IL: American Technical Publishers, Inc.

Figure Credits

Miller Electric Manufacturing Company	208F01, 208F05, 208F07, 208F08, 208F09, 208F10, 208F18, 208F21
Gerald Shannon	208F15, 208SA02, 208SA03, 208F34
Rex Ball	208F19, 208SA04, 208F32
Arc-Zone.com®, Inc.	208F16, 208F17, 208SA05
Smith Equipment	208F20

Figure 14 courtesy of American Welding Society

Adapted from AWS A2 Committee on Definitions and Symbols, 2001, *Standard Welding Terms and Definitions*, AWS A3.0:2001, Miami: American Welding Society, Figure 36.

Figure 22 courtesy of American Welding Society

AWS A5 Committee on Filler Metals and Allied Materials, 2001, *Specification for Carbon Steel Electrodes and Rods for Gas Shielded Arc Welding*, AWS A5.18/A5.18M: 2001, Miami: American Welding Society, Figure A1.

Table 7 courtesy of American Welding Society

AWS Committee on Filler Metals, 1999, *Specification for Bare Aluminum and Aluminum Alloy Welding Electrodes and Rods*, AWS A5.10/A5.10M:1999, Miami: American Welding Society, Table A2.

Table 10 courtesy of American Welding Society

AWS Committee on Filler Metal, 1992, *Specification for Magnesium Alloy Welding Electrodes and Rods*, ANSI/AWS A5.19-92, Miami: American Welding Society, Table A1.

CONTREN™ LEARNING SERIES — USER UPDATES

The NCCER makes every effort to keep these textbooks up-to-date and free of technical errors. We appreciate your help in this process. If you have an idea for improving this textbook, or if you find an error, a typographical mistake, or an inaccuracy in NCCER's Contren™ textbooks, please write us, using this form or a photocopy. Be sure to include the exact module number, page number, a detailed description, and the correction, if applicable. Your input will be brought to the attention of the Technical Review Committee. Thank you for your assistance.

Instructors – If you found that additional materials were necessary in order to teach this module effectively, please let us know so that we may include them in the Equipment/Materials list in the Instructor's Guide.

Write: Curriculum Revision and Development Department
National Center for Construction Education and Research
P.O. Box 141104, Gainesville, FL 32614-1104

Fax: 352-334-0932

E-mail: curriculum@nccer.org

Craft ______________ Module Name ______________

Copyright Date ______________ Module Number ______________ Page Number(s) ______________

Description

____

____

____

____

(Optional) Correction

____

____

____

(Optional) Your Name and Address

____

____

____

Module 29209-03

GTAW – Plate

COURSE MAP

This course map shows all of the modules in the second level of the Welding curriculum. The suggested training order begins at the bottom and proceeds up. Skill levels increase as you advance on the course map. The local Training Program Sponsor may adjust the training order.

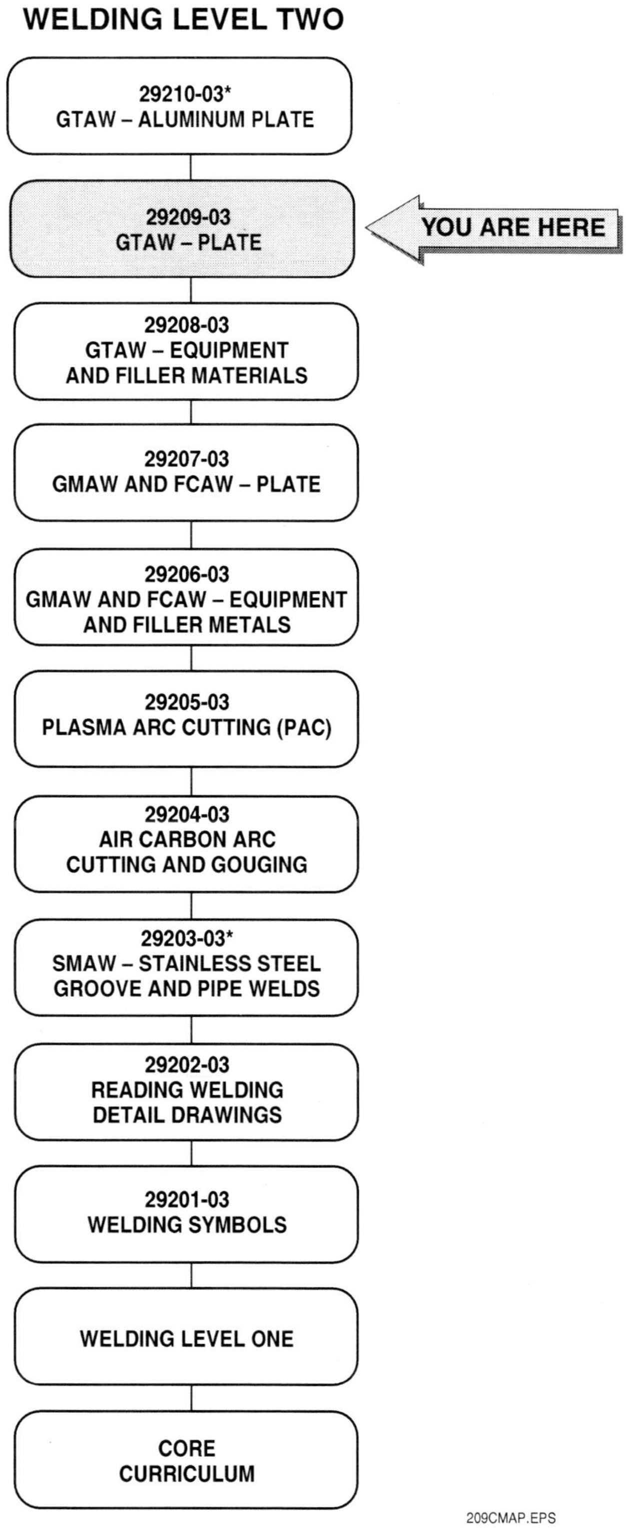

Please note that Modules 29203-03 and 29210-03 are electives for *Welding Level Two* but are requirements for *AWS Entry Level Welder.

MODULE 29209-03 CONTENTS

Figures

MODULE 29209-03

GTAW – Plate

Objectives

When you have completed this module, you will be able to do the following:

1. Build a pad in the flat position with stringer beads using GTAW and carbon steel filler metal.
2. Make multipass open V-groove welds on carbon steel plate in the 1G (flat) position using GTAW and carbon steel filler metal.
3. Make multipass open V-groove welds on carbon steel plate in the 2G (horizontal) position using GTAW and carbon steel filler metal.
4. Make multipass open V-groove welds on carbon steel plate in the 3G (vertical) position using GTAW and carbon steel filler metal.
5. Make multipass open V-groove welds on carbon steel plate in the 4G (overhead) position using GTAW and carbon steel filler metal.

Prerequisites

Before you begin this module, it is recommended that you successfully complete the following: Core Curriculum; Welding Level One; Welding Level Two, Modules 29201-03 through 29208-03.

Required Trainee Materials

1. Pencil and paper
2. Appropriate personal protective equipment

1.0.0 ◆ INTRODUCTION

Gas tungsten arc welding (GTAW) is an arc welding process that uses an arc between a tungsten electrode and the base metal to melt the base metal. The electrode, the arc, and the molten base metal are shielded from atmospheric contamination by a flow of inert gas from the torch nozzle. The filler metal is a rod of similar composition to the base metal to be welded that is usually hand-held and manually fed into the leading edge of the weld puddle. GTAW produces high-quality welds without slag or oxidation. Since there is no flux, there can be no corrosion due to flux entrapment, so no postweld cleaning is necessary. An exception is the slag left by some flux cored rods. This slag must be removed. *Figure 1* shows the GTAW process.

This module explains how to set up GTAW equipment and perform V-groove welds on carbon steel plate with carbon steel filler metal in the 1G, 2G, 3G, and 4G welding positions. The dimensions and specifications in this module are designed to be representative of codes in general and are not specific to any certain code. Always follow the proper codes for your site.

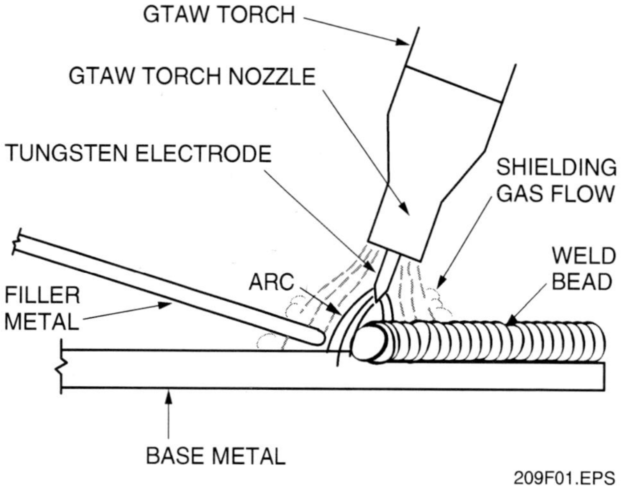

Figure 1 ◆ GTAW process.

1.1.0 Safety Practices

The following is a summary of safety procedures and practices that must be observed while cutting or welding. Keep in mind that this is just a summary. Complete safety coverage is provided in the Level One module, *Welding Safety*. If you have not completed that module, do so before continuing. Above all, be sure to wear appropriate protective clothing and equipment when welding or cutting.

1.1.1 Protective Clothing and Equipment

- Always use safety goggles with a full face shield or a helmet. The goggles, face shield, or helmet lens must have the proper light-reducing tint for the type of welding or cutting to be performed. Never directly or indirectly view an electric arc without using a properly tinted lens.
- Wear proper protective leather and/or flame retardant clothing along with welding gloves that will protect you from flying sparks and molten metal, as well as heat.
- Wear 8" or taller high-top safety shoes or boots. Make sure that the tongue and lace area of the footwear will be covered by a pant leg. If the tongue and lace area is exposed or the footwear must be protected from burn marks, wear leather spats under the pants or chaps and over the front top of the footwear.
- Wear a solid material (non-mesh) hat with a bill pointing to the rear or, if much overhead cutting or welding is required, a full leather hood with a welding face plate and the correct tinted lens. If a hard hat is required, use a hard hat that allows the attachment of rear deflector material and a face shield.
- If a full leather hood is not worn, wear a face shield and snug fitting welding goggles over safety glasses for gas welding or cutting. Either the face shield or the lenses of the welding goggles must be an approved shade 5 or 6 filter. For electric arc welding or cutting, wear safety goggles or a welding hood with the correct tinted lens (shade 9 to 14).
- If a full leather hood is not worn, wear earplugs to protect your ear canals from sparks.

1.1.2 Fire/Explosion Prevention

- Never carry matches or gas-filled lighters in your pockets. Sparks can cause the matches to ignite or the lighter to explode, causing serious injury.
- Never perform any type of heating, cutting, or welding until a hot-work permit is obtained and an approved fire watch is established. Most work-site fires caused by these types of operations are started by cutting torches.
- Never use oxygen to blow off clothing. The oxygen can remain trapped in the fabric for a time. If a spark hits the clothing during this time, the clothing can burn rapidly and violently out of control.
- Make sure that any flammable material in the work area is moved or shielded by a fire-resistant covering. Approved fire extinguishers must be available before attempting any heating, welding, or cutting operations.
- Always comply with any site requirement for a hot-work permit and/or a fire watch.
- Never release a large amount of oxygen or use oxygen as compressed air. Its presence around flammable materials or sparks can cause rapid and uncontrolled combustion. Keep oxygen away from oil, grease, and other petroleum products.
- Never release a large amount of fuel gas, especially acetylene. Methane and propane tend to concentrate in and along low areas and can ignite at a considerable distance from the release point. Acetylene is lighter than air but is even more dangerous than methane. When mixed with air or oxygen, it will explode at much lower concentrations than any other fuel.
- To prevent fires, maintain a neat and clean work area and make sure that any metal scrap or slag is cold before disposal.
- Before cutting or welding containers such as tanks or barrels, check to see if they have contained any explosive, hazardous, or flammable materials including petroleum products, citrus products, or chemicals that decompose into toxic fumes when heated. As a standard practice, always clean and then fill any tanks or barrels with water or purge them with a flow of inert gas to displace any oxygen.

1.1.3 Work Area Ventilation

- Make sure confined space procedures are followed before conducting any welding or cutting in the confined space.
- Never use oxygen in confined spaces for ventilation purposes.

- Always perform cutting or welding operations in a well-ventilated area. Cutting or welding operations involving zinc or cadmium materials or coatings result in toxic fumes. For long-term cutting or welding of such materials, always wear an approved full-face, supplied-air respirator (SAR) that uses breathing air supplied outside of the work area. For occasional, very short-term exposure, a high-efficiency particulate arresting (HEPA)-rated or metal-fume filter may be used on a standard respirator.
- Make sure confined spaces are ventilated properly for cutting or welding purposes.

2.0.0 ◆ GTAW WELDING EQUIPMENT SETUP

Before welding can take place, the work area has to be made ready, the welding equipment set up, and the metal to be welded prepared. The following sections explain how to prepare the area and set up the equipment to perform GTAW welding of carbon steel plate.

2.1.0 Setting Up the Welding Area

Practice welding on a welding table, bench, or stand. The welding surface must be steel, and provisions must be made for mounting practice welding coupons out of position. Follow these steps to set up the area for welding:

Step 1 Make sure the area is properly ventilated. Make use of doors, windows, and fans.

Step 2 Check the area for fire hazards. Remove any flammable materials before proceeding.

Step 3 Check the location of the nearest fire extinguisher. Do not proceed unless the extinguisher is charged and you know how to use it.

Step 4 Position a welding table within convenient distance to the GTAW welding machine.

Step 5 Set up flash shields around the welding area.

2.1.0 Preparing Weld Practice Coupons

If possible, the weld coupons should be ⅜" thick carbon steel. If this size is not readily available, ¼" to ¾" thick steel can be used for practice welds. Open-root V-groove welds generally have a bevel angle of 30° for carbon steel or 37½° for stainless steel plate.

NOTE

Carbon steel ⅜" thick will be required for the performance qualification tests.

Clean the steel plate prior to welding by using a wire brush or grinder to completely remove all mill scale or corrosion. This is because the direct current electrode negative (DCEN) used in GTAW has little or no cleaning action. Prepare weld coupons to practice the following welds:

- *Stringer and weave bead welds* — The coupons can be any size or shape that is easily handled.
- *Open-root V-groove welds* — As shown in *Figure 2*, cut the metal into 3" × 5" rectangles with one (or both) of the 5" lengths beveled at 30° or 37½°. Grind a 0" to ⅛" root face on one bevel as directed by your instructor.

NOTE

Check with your instructor to determine which bevel angle should be used.

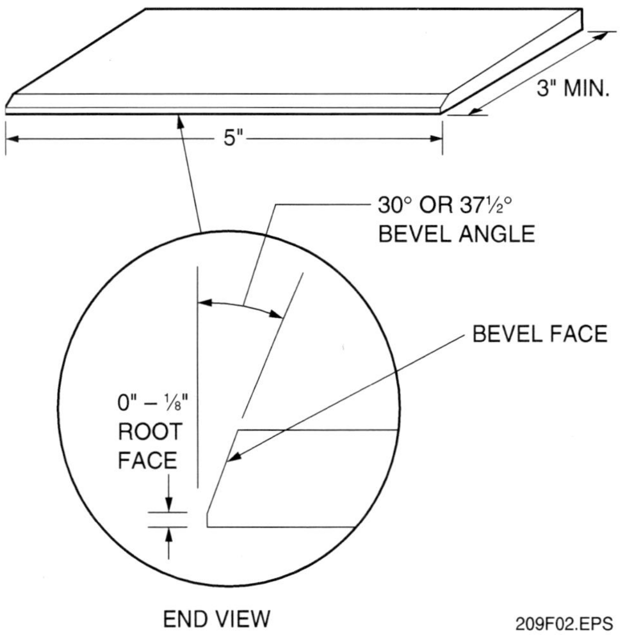

Figure 2 ◆ Metal cut for open-root V-groove weld coupons.

> **NOTE**
> The welding codes allow the root face and opening on open-root welds to be 0" to ⅛". Adjust the root face and root opening as needed when you start your welding practices.

Follow these steps to prepare each V-groove weld coupon:

Step 1 Check the bevel face. There should be no dross and a 0" to ⅛" root face. The bevel angle should be 30° or 37½°. An example of this is shown in *Figure 3*.

Step 2 Center the beveled strips with a root opening determined by your instructor and tack weld them in place. Place the tack welds on the back side of the joint. Use three to four ¼" tack welds.

2.3.0 Welding Equipment

Identify a proper welding machine for GTAW use, and follow these steps to set it up for welding:

Step 1 Verify that the welding machine can be used for GTAW with or without internal gas shielding control. If desired, identify an optional cooling unit.

Step 2 Identify an air-cooled or water-cooled GTAW torch. Make sure that it is compatible with the welding machine and any cooling unit.

Step 3 Check to be sure that the welding machine is properly grounded through the primary current receptacle.

Conserve Weld Coupons

Steel for practice welding is expensive and difficult to obtain. Every effort should be made to conserve the material that is available. Reuse welding coupons until all the surfaces have been used for welding. Then, cut the weld coupon apart and reuse the pieces. Use material that cannot be cut into weld coupons to practice running beads.

Step 4 Verify the location of the primary disconnect.

Step 5 Configure the welding machine and torch for GTAW welding (*Figure 4*) as directed by your instructor. Configure the torch polarity and equip the torch with the correct diameter and properly prepared type of tungsten electrode for the filler metal used for the application.

Step 6 Connect the proper shielding gas for the application as directed by your instructor, filler metal manufacturer, welding procedure specification (WPS), or site quality standards.

Step 7 Connect the clamp of the workpiece lead to the workpiece.

Step 8 Turn on the welding machine, and purge the torch as directed by the manufacturer's instructions.

Step 9 Set the initial welding current for the desired GTAW application.

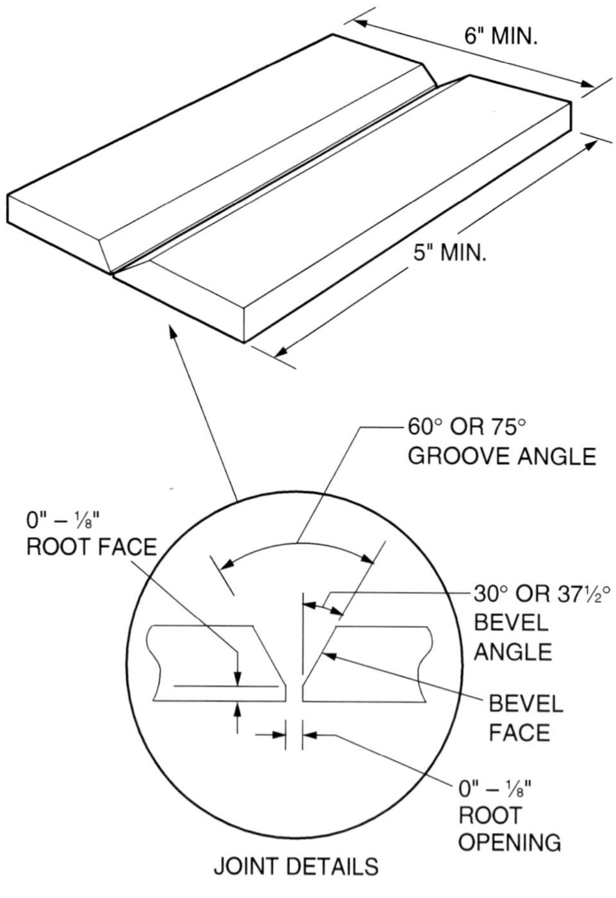

Figure 3 ◆ Open-root V-groove weld coupon.

3.0.0 ◆ GAS TUNGSTEN ARC WELDING TECHNIQUES

GTAW weld bead characteristics and quality are affected by several factors resulting from the way the individual performing the welding handles the torch. These factors include:

- Torch travel speed and arc length
- Torch angles
- Torch and filler metal handling techniques

3.1.0 Torch Travel Speed and Arc Length

Torch travel speed and arc length affect the GTAW weld puddle and the penetration of the weld. A slow travel speed allows more heat to concentrate and forms a larger, more deeply penetrating puddle. Faster travel speeds prevent heat buildup and form smaller and shallower puddles. Arc length is the major control for bead width. As the torch is raised, arc length increases, voltage increases, and the bead width increases.

3.2.0 Torch Angles

There are two basic torch angles that must be controlled when performing GTAW. These are the work angle and the travel angle. The definition of these angles is the same as for all other methods of arc welding.

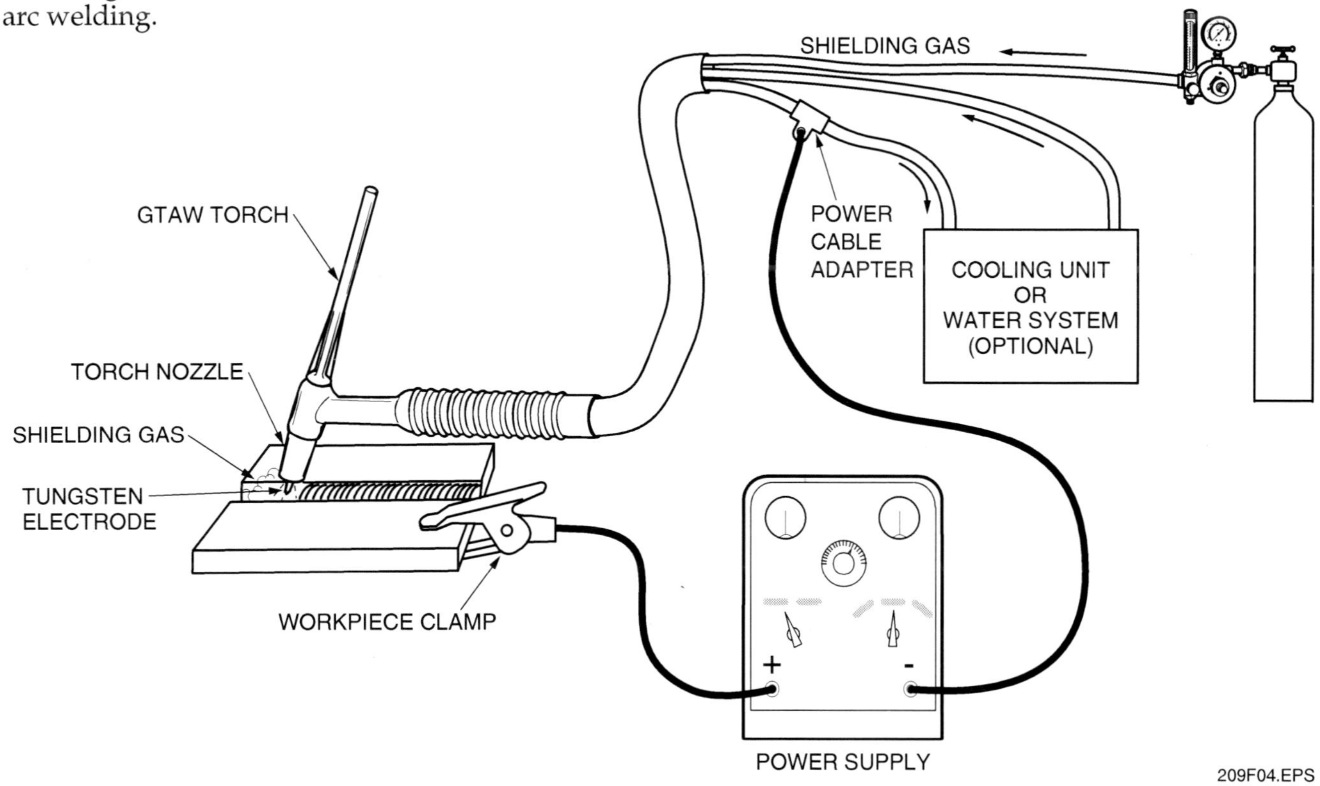

Figure 4 ◆ Configuration diagram of typical GTAW welding equipment.

3.2.1 Work Angle

The torch work angle (*Figure 5*) is defined as an angle less than 90° between a line perpendicular to the major workpiece surface at the point of electrode contact and a plane determined by the electrode axis and the weld axis. For a T-joint or corner joint, the line is perpendicular to the nonbutting member. For pipe, the plane is determined by the electrode axis and a line tangent to the pipe surface at the same point.

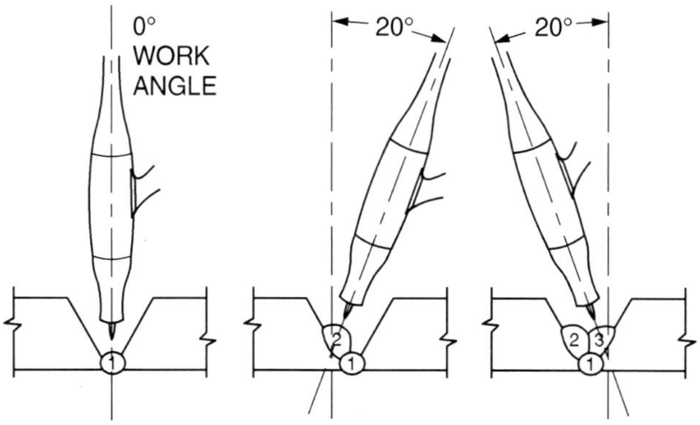

Figure 5 ◆ Typical torch work angles.

3.2.2 Travel Angle

The torch travel angle (*Figure 6*) is defined as an angle less than 90° between the electrode axis and a line perpendicular to the weld axis at the point of electrode contact in a plane determined by the electrode axis and the weld axis. For pipe, the plane is determined by the electrode axis and a line tangent to the pipe surface at the same point. A push angle is used for GTAW. A push angle is the angle created when the torch is tilted back so that the electrode tip precedes the torch in the direction of the weld. In this position, the electrode tip and shielding gas are being directed ahead of the weld bead. Push angles of 15° to 20° are normally used for GTAW.

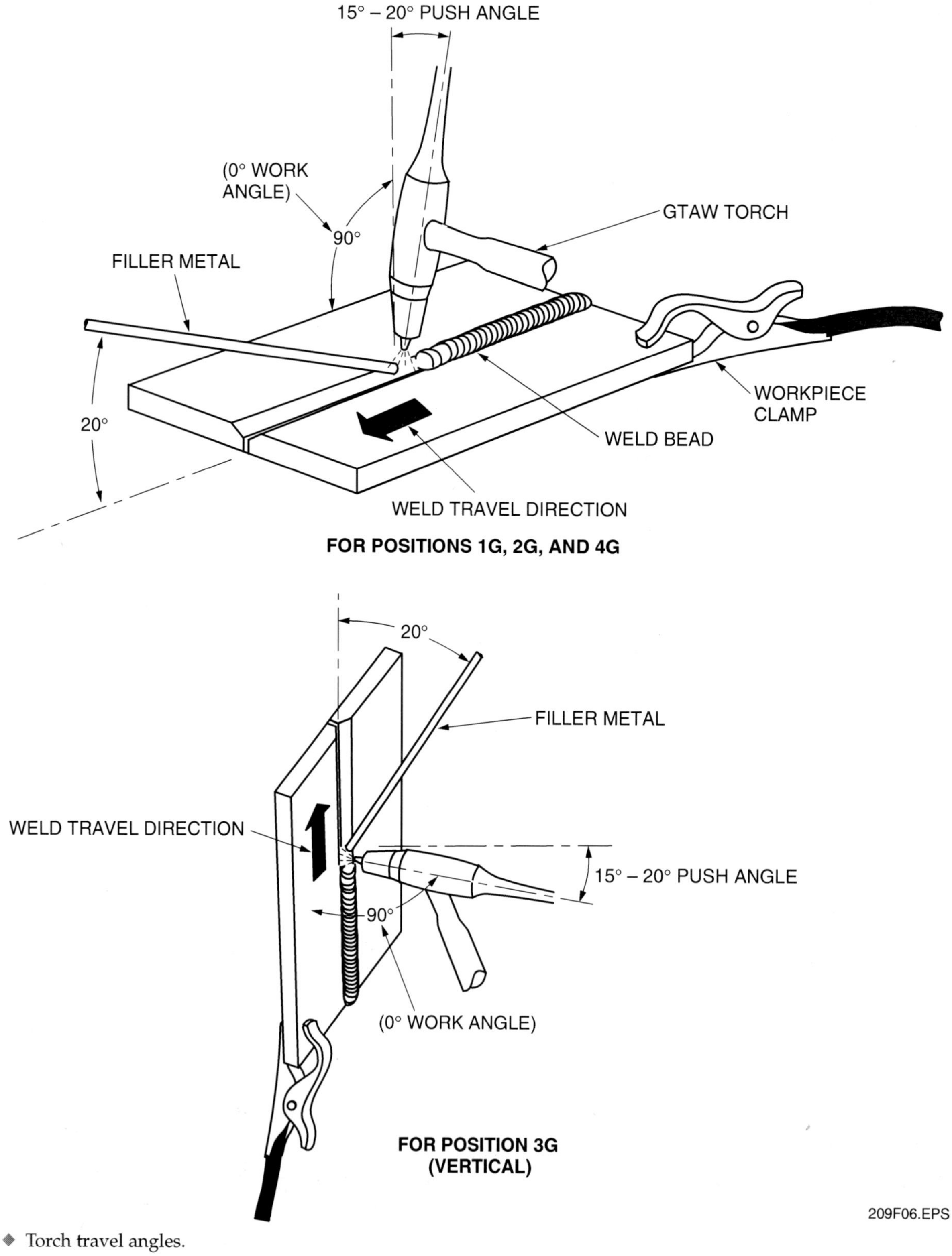

Figure 6 ◆ Torch travel angles.

Hot Tip

Excessive Push Angle

Too great a push angle will tend to draw air from under the back edge of the torch nozzle where it will mix with the shielding gas stream and contaminate the weld.

3.3.0 Torch and Filler Metal Handling Techniques

The two basic handling techniques used to perform GTAW are known as freehand and walking-the-cup. Try both techniques and use the one that gives the best results. Both techniques are explained in the following sections.

3.3.1 Freehand Technique

In the freehand technique, the torch electrode tip is held just above the weld puddle or base metal. The torch is supported by the welder's hand, which is usually steadied by resting some part of it on or against the base metal to maintain the proper arc length (electrode tip to puddle distance). If required, the welder can move the torch tip in a small circular motion within the molten puddle to maintain the puddle size and advance the puddle. Filler metal is added as it is needed.

The GTAW filler metal is held in the hand not holding the torch. For all positions, the filler rod is held at an angle of about 20° above the base metal surface and in line with the weld. The tip of the filler rod is always kept within the shielding gas envelope to protect the filler rod from atmospheric contamination and to keep it preheated.

Dab the filler metal into the leading edge of the weld puddle, using extreme care not to touch the tungsten electrode with the end of the filler metal. If the tungsten electrode touches the filler metal or weld puddle, it will become contaminated with filler metal. The electrode must then be removed and cleaned by grinding (or chemical cleaning) before proceeding. A technique for preventing contamination of the electrode is to move the electrode to the back edge of the weld puddle as the filler metal is dabbed into the leading edge of the weld puddle.

Figure 7 shows the use of the freehand technique for running stringer beads.

3.3.2 Walking-the-Cup Technique

In the walking-the-cup technique, the welder rests the edge of the torch nozzle (cup) against the base metal or groove edges to steady the torch and maintain a constant arc length (electrode tip to weld pool distance). The torch is rocked from side to side on the edge of the cup as it is advanced to maintain the puddle size and heat both sides of the groove.

Add the filler metal in the same manner as with the freehand technique, using care not to contaminate the electrode.

Figure 8 shows the use of the walking-the-cup technique for running stringer beads.

4.0.0 ◆ BEAD TYPES

The basic bead types that can be made with GTAW are stringer beads and weave beads.

4.1.0 Stringer Beads

Stringer beads are made with little or no side-to-side (oscillation) movement of the torch. The bead is generally no more than three times the diameter of the electrode being used. *Figure 9* shows a stringer bead.

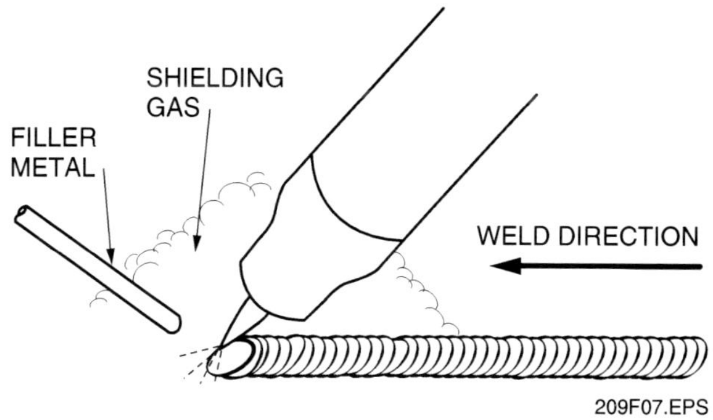

Figure 7 ◆ Freehand technique.

Hot Tip

Filler Metal Feeding Precautions

Do not insert the end of the filler metal rod into the molten puddle under the electrode and then attempt to melt it off. This can cause hard spots and weld defects.

209F08.EPS

Figure 8 ◆ Walking-the-cup technique.

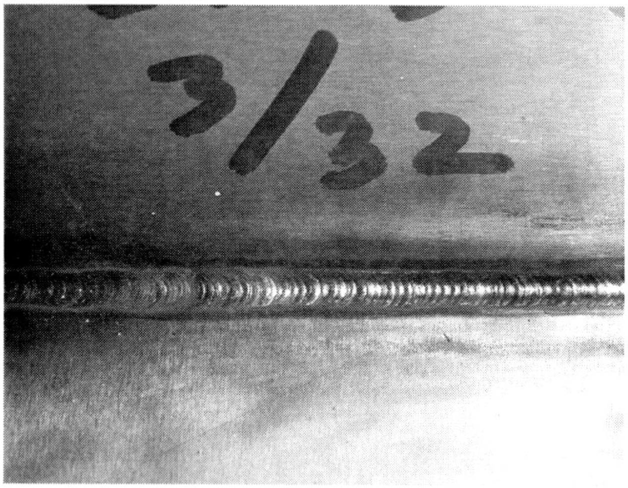

209F09.EPS

Figure 9 ◆ Stringer bead.

4.2.0 Weave Beads

Weave beads are made by working the weld puddle back and forth across the axis of the weld. This produces a weld bead much wider than the puddle width. The width of a weave bead is determined by the amount of oscillation.

Weave beads put more heat into the base metal. For this reason, they are not recommended for welding metals sensitive to heat input, such as most types of stainless steels. Weave beads will not be practiced or used in this module.

CAUTION

Check the WPS or site quality standards to determine if stringer or weave beads should be used.

5.0.0 ◆ PRACTICING STRINGER BEADS

Practice running stringer beads in the flat position using both the freehand and walking-the-cup techniques. Flat-position stringer beads should be run with a 15° to 20° push angle. Practice stringer beads without filler metal until you are comfortable with the torch and get consecutive uniform rippled beads.

To run stringer beads with filler metal, hold the filler metal in front of the weld bead, with the end of the filler metal within the torch gas envelope. The filler rod should be held at a 20° angle with the plate surface. *Figure 10* shows the flat-position torch and filler metal angles.

Follow these steps to run flat-position stringer beads:

Step 1 Position the torch at a 15° to 20° push angle, with the electrode tip directly over the point where the weld is to begin.

Step 2 Press the remote control to energize the electrode and start the shielding gas flow.

NOTE

If the equipment being used does not have remote control, the electrode will be energized when the power source is turned on. Start the shielding gas flow by opening the manual shielding gas valve on the torch or in the line.

Step 3 Initiate the arc by bringing the electrode close to the base metal (with high frequency) or by lightly touching the electrode to the base metal (no high frequency).

Step 4 Oscillate the arc slightly until the weld puddle begins to form.

Step 5 While using either the freehand or the walking-the-cup technique, slowly advance the torch and add filler metal as needed.

Step 6 Continue to weld until a bead about 2" to 3" long is formed, and then stop the arc without withdrawing the torch.

Hot Tip

Stopping the Arc without Remote Control

If the equipment being used does not have remote control, quickly lift the torch to break the arc, and then immediately return it to shield the weld and the end of the filler metal.

Cooling Practice Coupons

If the practice coupon gets too hot between passes, cool it in water. Cooling with water is only done on practice coupons. Never cool test coupons or any other weld with water. Cooling with water can cause weld cracks and affect the mechanical properties of the base metal.

Step 7 Continue to hold the torch in place so that the postflow gas protects the hot weld and filler metal until they are sufficiently cool to prevent oxidation damage.

CAUTION

Do not remove the torch or filler metal from the torch-shielding gas flow until the puddle has solidified and cooled to a temperature at which it will not be affected by the atmosphere. The gas postflow that continues after the welding stops protects the molten metal. If the torch and shielding gas are removed before the weld has solidified, crater porosity or cracks can occur. Also, the filler metal will be contaminated, and the contaminated end will have to be discarded.

Step 8 Remove the torch, and close the shielding gas valve if the equipment does not have remote control.

Step 9 Inspect the bead for:

- Straightness of the bead
- Uniform rippled appearance of the bead face
- Smooth, flat transition with complete fusion at the toes of the weld
- No porosity
- No excessive undercut

Step 10 Continue practicing stringer beads until acceptable beads are produced consistently.

WARNING!

Use pliers to handle the hot practice coupons. Wear gloves when placing the practice coupons in water. Steam will rise off the coupons and can burn or scald unprotected hands.

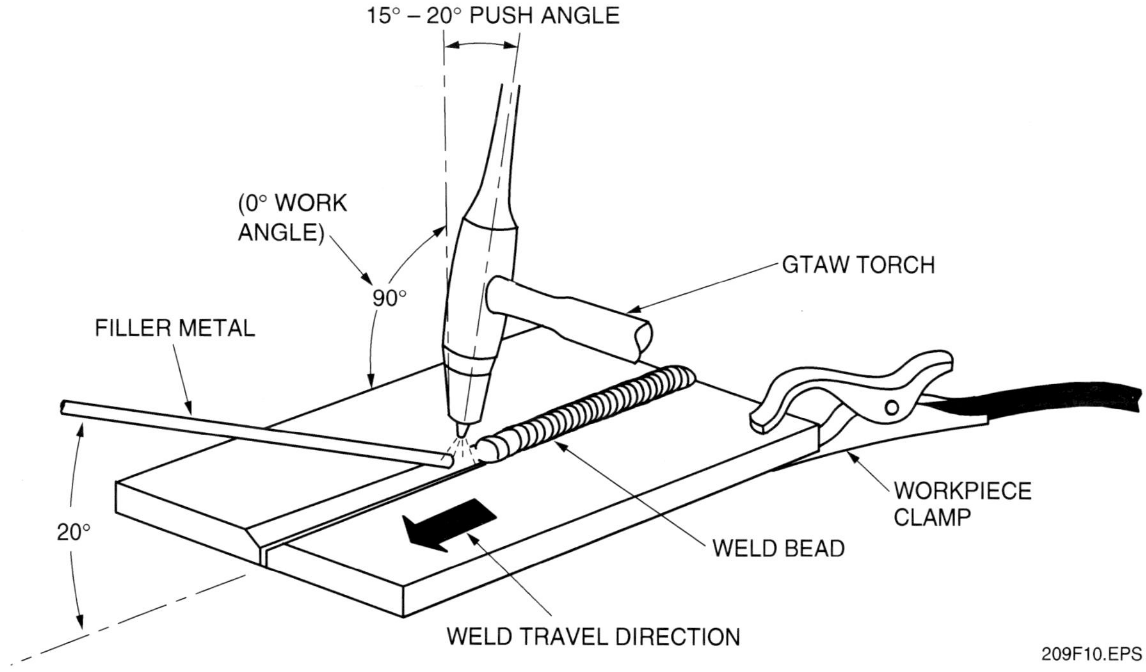

Figure 10 ◆ Flat-position torch and filler metal angles.

6.0.0 ◆ WELD RESTARTS

A restart is the junction where a new weld connects to continue the bead of a previous weld. Restarts are important because an improperly made restart will create a weld defect. A restart must be made so that it blends smoothly with the previous weld and does not stand out. If possible, avoid restarts by running a bead the full length of the weld joint to be made.

Follow these steps to make a restart:

Step 1 Use a wire brush to clean the restart area.

Step 2 Hold the torch at the proper angle and arc distance, and then restart the arc directly over the center of the crater. (The welding codes do not allow arc strikes outside the area to be welded.)

Step 3 Move the electrode tip in a small circular motion over the crater, and add filler metal when the molten puddle is the same size as the crater.

Step 4 As soon as the puddle fills the crater, advance the puddle slightly and continue to add filler metal as needed.

Step 5 Inspect the restart. A properly made restart will blend into the bead, making it hard to detect.

If the restart has underfill or undercut, not enough time was spent in the crater to completely melt it, or not enough filler metal was added. If the restart is higher than the rest of the bead, too much filler metal was added.

Continue to practice restarts until they are correct. Use the same techniques for making restarts whenever performing GTAW.

7.0.0 ◆ WELD TERMINATIONS

A weld termination is made at the end of a weld. Most welding codes require that when terminating a weld, the crater must be filled to the full cross section of the weld to prevent crater cracking. This can be difficult because most terminations are at the edge of a plate where welding heat tends to build up. This makes filling the crater more difficult. Filling the crater is much easier if remote control equipment with a potentiometer is used. The potentiometer allows the welding current to be reduced as the end of the weld is approached.

Follow these steps to make a termination using remote control equipment and a potentiometer:

Step 1 As you approach the end of the weld, slowly back off (reduce) the welding current. The amount of current reduction is determined by the weld puddle width. If the puddle starts to become wider, there is too much heat; if the puddle starts to become narrower, there is not enough heat.

Step 2 Continue to back off the welding current while adding filler metal until the crater is filled.

Step 3 Stop the arc and hold the torch in place until the gas postflow cools the weld metal and filler metal.

Follow these steps to make a termination without a potentiometer:

Step 1 As you approach the end of the weld, add filler metal at a faster pace. The filler metal will absorb excess heat.

Step 2 Continue to add filler metal until the crater is filled, and then stop the arc.

Step 3 Hold the torch in place until the gas postflow cools the weld metal and filler metal sufficiently to prevent oxidation damage.

CAUTION

Do not remove the torch or filler metal from the torch-shielding gas flow until the puddle has solidified and cooled to a temperature at which it will not be affected by the atmosphere. The gas postflow that continues after the welding stops protects the molten metal and the tungsten electrode. If the torch and shielding gas are removed before the weld has solidified, crater porosity or cracks can occur. In addition, the tungsten electrode can oxidize (turn a bluish-black color), which will require re-pointing the electrode. Also, the filler metal will be contaminated, and the contaminated end will have to be discarded.

Inspect the termination for:

- Crater filled to the full cross section of the weld (no underfill)
- No crater cracks
- No crater porosity
- No excessive undercut

Thin Materials

On thin materials, stop about 1" or 2" short of the edge without terminating the weld. Then, backweld from the edge to the crater and fill the crater.

8.0.0 ◆ OVERLAPPING BEADS

Overlapping beads are made by depositing overlapping stringer (or weave) beads parallel to one another. Each successive bead overlaps the toe of the previous bead to form an approximately flat surface. This is called padding. Overlapping beads are used to build up surfaces and to make multipass welds.

Properly overlapped beads, when viewed from their ends, will form a nearly flat surface. *Figure 11* shows proper and improper overlapping beads.

Follow these steps to make GTAW overlapping stringer beads using 3/32" or 1/8" carbon steel filler metal:

Step 1 Mark out a square on a piece of steel.

Step 2 Weld a stringer bead along one edge.

Step 3 Clean the weld.

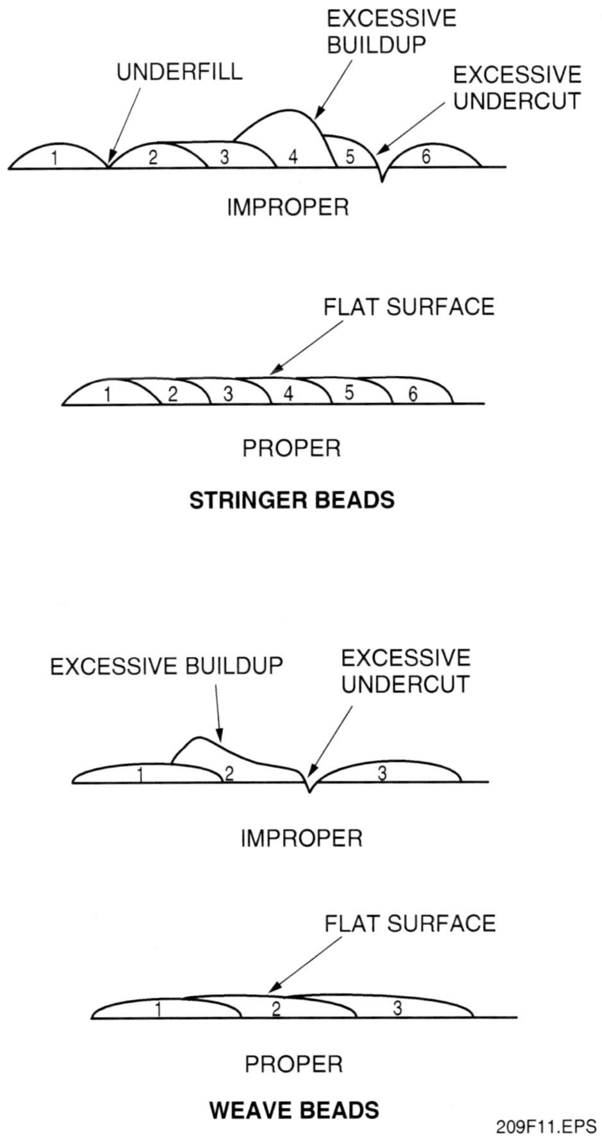

Figure 11 ◆ Proper and improper overlapping beads.

Step 4 Run a second bead along the previous bead. Be sure to overlap the previous bead to obtain a good tie-in and produce a flat profile.

Step 5 Continue running overlapping stringer beads until the face of the metal square is covered.

Step 6 Continue building layers of stringer beads (pads), one on top of the other, until the technique is perfected.

9.0.0 ◆ GROOVE WELDS

The open V-groove weld is a common groove weld normally made on plate and pipe. Practicing the open V-groove weld on plate will prepare the welder to make the more difficult pipe welds.

9.1.0 Open V-Groove Root Pass Techniques

The most difficult part of welding a V-groove butt joint is the root pass. The root must have complete penetration, but not an excessive amount of penetration or burn-through. The penetration is controlled by using the correct amperage and torch travel speed for the root opening and root face being used. Root reinforcement should be flush to 1/8", depending on the site specifications.

There are two techniques for running the root pass. These are:

- **On the wire**
- Keyhole

Regardless of the technique used, always start the root pass from a tack weld.

9.1.1 On-the-Wire Root Pass Technique

To perform the on-the-wire root pass technique, secure or hold the filler metal in the bottom of the root opening of the V-groove parallel with the plates. Fuse it in place by moving the torch along the groove using either the freehand or walking-the-cup torch technique. Advance the torch slowly enough to thoroughly fuse the filler metal with the plate edges, but not so slowly that it can burn through. This technique is used with a filler metal that is the same size as the width of the root opening and is similar to pipe welding using a consumable insert. With this technique, the filler metal should not be moved into or out of the weld puddle. This will ensure a uniform weld bead thickness. *Figure 12* shows the on-the-wire root pass technique.

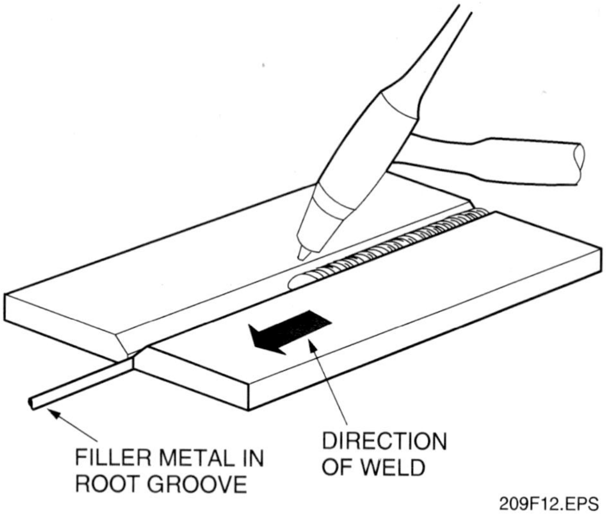

Figure 12 ◆ On-the-wire root pass technique.

Backing and/or Backing Gas

Normally, backing and/or backing gas is not required for GTAW of carbon steel and some low-alloy steel. However, it is required for stainless and low-alloy steel and most non-ferrous metals.

9.1.2 Keyhole Technique

To perform the keyhole root pass technique (*Figure 13*), hold the filler metal at an angle to the plate surfaces and feed it into the groove root from the weld side. This technique uses a filler metal rod diameter slightly larger than the root opening so that the rod can rest on the groove. When making a root pass, start on a tack weld. When a keyhole weld puddle has been established by melting the root faces of both plates together, move the arc to the back of the puddle and dab the end of the filler metal into the leading edge of the weld puddle. Then, move the filler metal away from the weld puddle. Advance the torch, using either the walking-the-cup or the freehand technique. Dab the filler metal back into the leading edge of the molten puddle and forward out of the molten puddle to add filler metal as needed. However, be sure the filler metal is always kept in the shielding gas stream to keep the filler metal preheated and protected from atmospheric contamination.

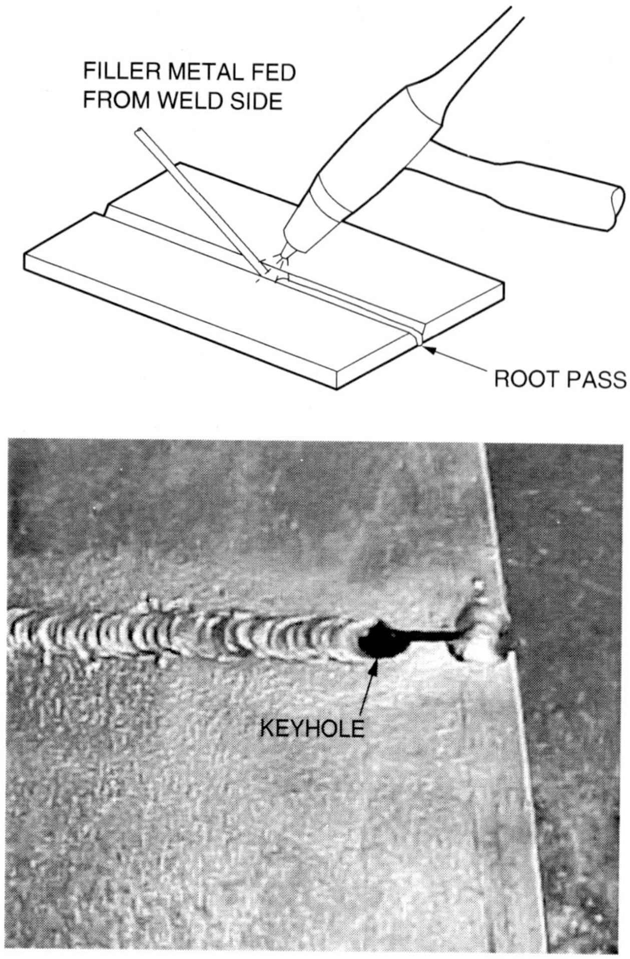

Figure 13 ◆ Root pass with oversized keyhole.

CAUTION

When adding filler metal, always add it to the molten puddle created by the base metal. This prevents lack of fusion, which is caused when filler metal is melted onto the solid base metal in an attempt to fuse the two together.

After the root pass is completed, clean and inspect it for:

- Uniform and smooth face and root
- Excessive root reinforcement
- Excessive face buildup
- Excessive undercut
- Porosity
- Cracks

9.2.0 Groove Weld Positions

Open groove welds can be made in all positions. The weld position is determined by the axis of the weld. Groove weld positions are flat (1G), horizontal (2G), vertical (3G), and overhead (4G). *Figure 14* shows the groove weld positions.

9.3.0 Acceptable and Unacceptable Groove Weld Profiles

Groove welds should be made with slight reinforcement (not exceeding ⅛") and a gradual transition to the base metal at each toe. Groove welds must not have excess reinforcement, underfill, excessive undercut, or overlap. If a groove weld has any of these defects, it must be repaired.

CAUTION

Refer to your site's WPS for specific requirements on groove welds. The information in this module is provided as a general guideline only. The WPS or site quality specifications must be followed for all welds. Check with your supervisor if you are unsure of the specifications for your application.

Figure 15 shows acceptable and unacceptable groove weld profiles.

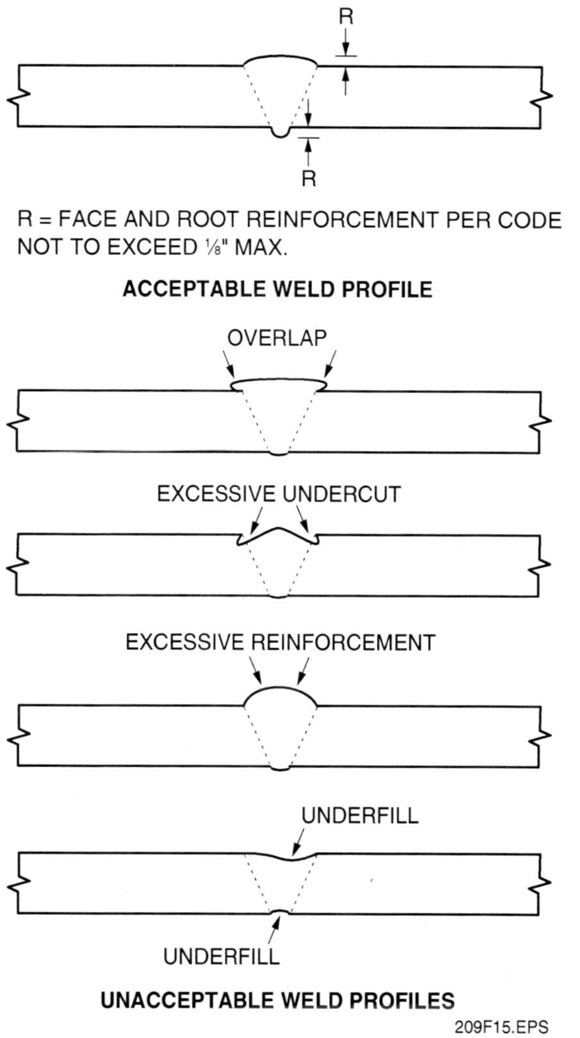

Figure 15 ◆ Acceptable and unacceptable groove weld profiles.

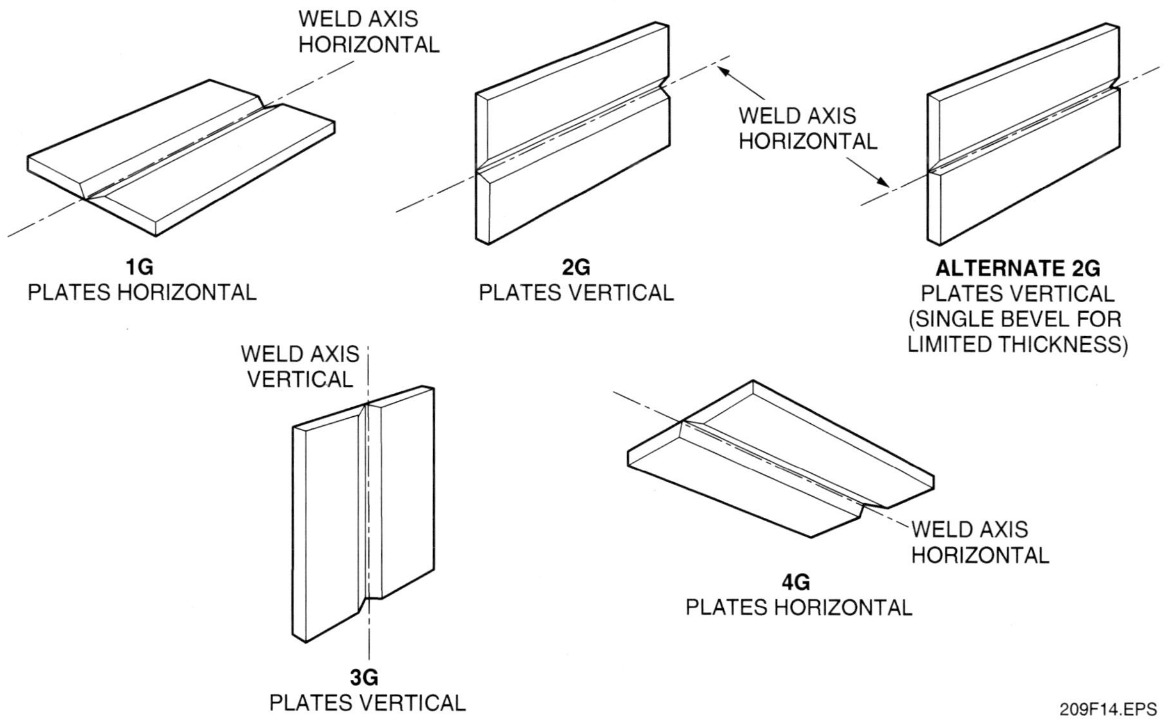

Figure 14 ◆ Groove weld positions.

10.0.0 ◆ OPEN V-GROOVE WELDS

Practice making GTAW open V-groove welds in the 1G, 2G, 3G, and 4G positions, using appropriate diameter carbon steel filler metal for the root pass, filler, and cover passes with stringer beads. Pay particular attention at the termination of the weld to fill the crater. Cooling with water can cause weld cracks and affect the mechanical properties of the base metal. Practice coupons can be cooled with water, but never cool test coupons with water.

> **WARNING!**
> Use pliers to handle the hot practice coupons after welding. Wear gloves when placing the practice coupons in water to cool them. Steam will rise off the coupons and can burn or scald unprotected hands.

10.1.0 Practicing Flat (1G) Position Open V-Groove Welds

Follow these steps to make open V-groove welds in the flat (1G) position (*Figure 16*):

Step 1 Tack weld the practice coupon together, following the example given in the section on preparing welding coupons.

Step 2 Clamp or tack weld the welding coupon in a flat position above the welding table surface.

Step 3 Run the root pass using an appropriate diameter carbon steel filler metal, and either the on-the-wire or the keyhole technique. Use a 15° to 20° push angle, a 20° filler metal angle to the plate surface, and the work angle for pass 1 (*Figure 16*).

Step 4 Clean the root pass.

Step 5 Run the remaining passes using either 3/32" or 1/8" carbon steel filler metal, a 15° to 20° push angle, and the work angles shown in *Figure 16*. Hold the filler metal about 20° above the plate surface. Clean the weld between each pass.

10.2.0 Practicing Horizontal Beads

Before welding a joint in the horizontal (2G) position, practice running horizontal stringer beads by building a pad in the horizontal position. Use 3/32" or 1/8" carbon steel filler metal to build the pad. Use a push angle of 15° to 20°, and hold the filler metal in front of the weld at a 20° angle with the plate face.

Follow these steps to practice overlapping horizontal welding beads in the horizontal (2G) position:

Step 1 Clamp or tack weld the welding coupon in the horizontal position.

Step 2 Run the first pass along the bottom edge of the coupon, using either 3/32" or 1/8" carbon steel filler metal.

Step 3 Clean the weld.

Step 4 Run the second bead along the top edge, overlapping the first bead. Adjust the work angle to obtain a good tie-in.

Step 5 Continue running beads to complete the pad. Be sure to clean the weld between each pass.

Figure 17 shows how to build a pad in the horizontal position.

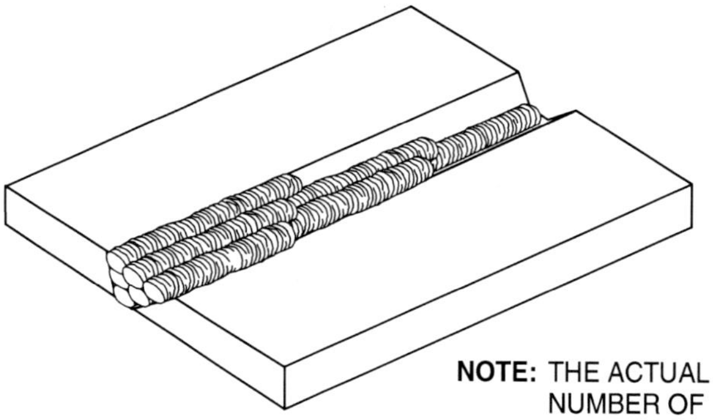

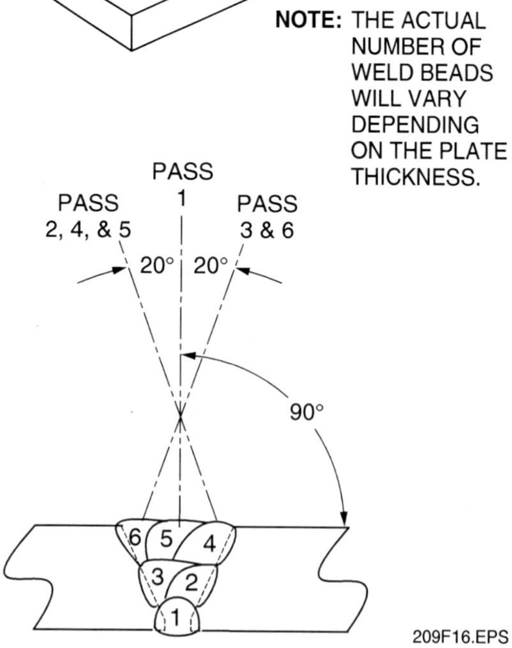

Figure 16 ◆ Multipass 1G weld sequences and work angles.

10.3.0 Practicing Horizontal (2G) Position Open V-Groove Welds

Follow these steps to practice welding open V-groove welds in the horizontal (2G) position (*Figure 18*):

Step 1 Tack weld the practice coupon together as explained earlier. Use the standard or alternate weld coupon as directed by your instructor.

Step 2 Clamp or tack weld the coupon in the horizontal position.

Step 3 Run the root pass using either the on-the-wire or the keyhole welding technique, and using an appropriate diameter carbon steel filler metal. Use a 15° to 20° push angle, a 20° filler metal angle with the plate surface, and the appropriate work angle for pass 1 (*Figure 18*).

Step 4 Clean the weld.

Step 5 Run the remaining passes using either 3/32" or 1/8" carbon steel filler metal, a 15° to 20° push angle, and the work angles shown in *Figure 18*. Clean the weld between each pass.

10.4.0 Practicing Vertical Beads

Before welding a joint in the vertical position, practice running vertical stringer beads by building a pad in the vertical position. Use 3/32" or 1/8" carbon steel filler metal to build the pad. Weld vertically with a push angle of 15° to 20° and a filler metal angle of about 20° with the plate surface.

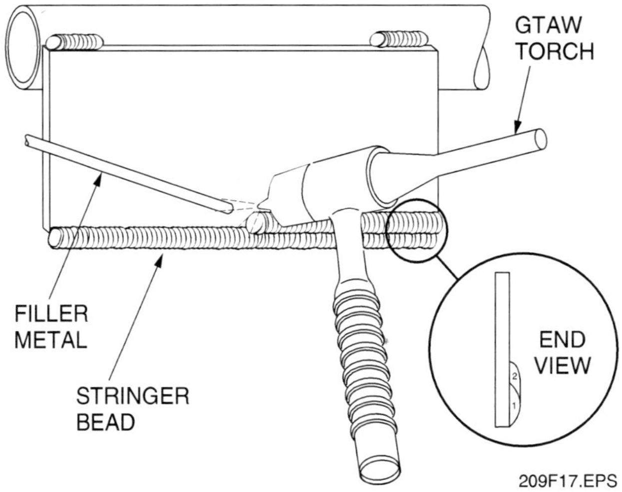

Figure 17 ◆ Building a pad in the horizontal position.

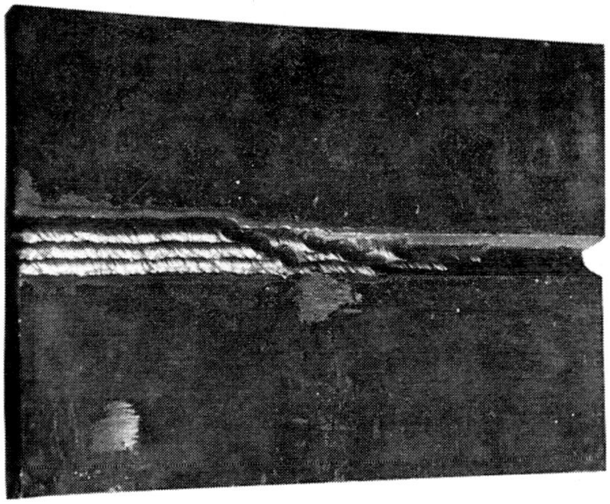

NOTE: THE ACTUAL NUMBER OF WELD BEADS WILL VARY DEPENDING ON THE PLATE THICKNESS.

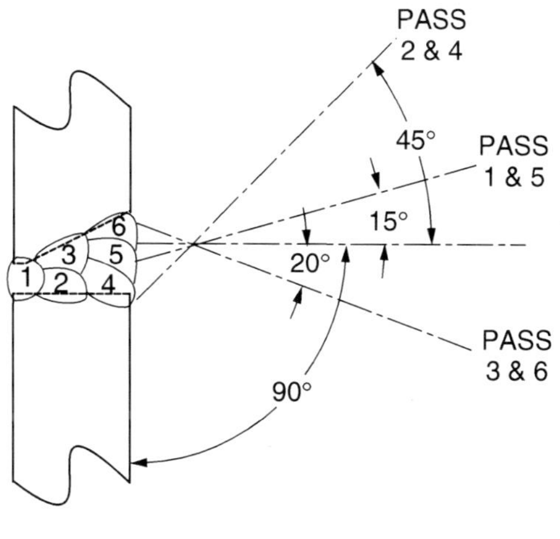

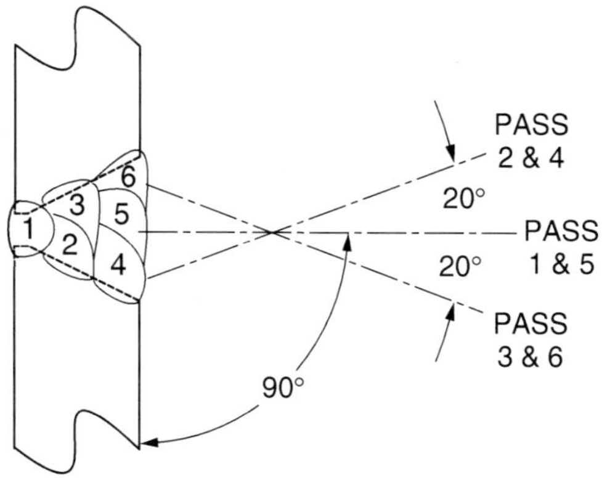

Figure 18 ◆ Multipass 2G weld sequences and work angles.

Follow these steps to practice overlapping vertical beads in the vertical position:

Step 1 Clamp or tack weld the practice coupon in the vertical position.

Step 2 Run the first bead uphill along a vertical edge of the coupon using either 3⁄32" or 1⁄8" carbon steel filler metal.

Step 3 Clean the weld.

Step 4 Run the second bead, overlapping the edge of the first bead. Adjust the work angle to obtain a good tie-in.

Step 5 Continue running beads to complete the pad. Clean the weld between each pass.

Figure 19 shows how to build a pad in the vertical position.

10.5.0 Practicing Vertical (3G) Position Open V-Groove Welds

Follow these steps to practice open V-groove welds in the vertical (3G) position (*Figure 20*):

Step 1 Tack weld the practice coupon together as explained earlier.

Step 2 Clamp or tack weld the coupon in the vertical position.

Step 3 Run the root pass uphill with either the on-the-wire or the keyhole welding technique, and using an appropriate diameter carbon steel filler metal. Use a push angle of 15° to 20°, a filler metal angle of 20° with the plate surface, and the pass 1 work angle (*Figure 20*).

Step 4 Clean the weld.

Step 5 Run the remaining passes uphill using either 3⁄32" or 1⁄8" carbon steel filler metal, a 15° to 20° push angle, and the work angles shown in *Figure 20*. Clean the weld between each pass.

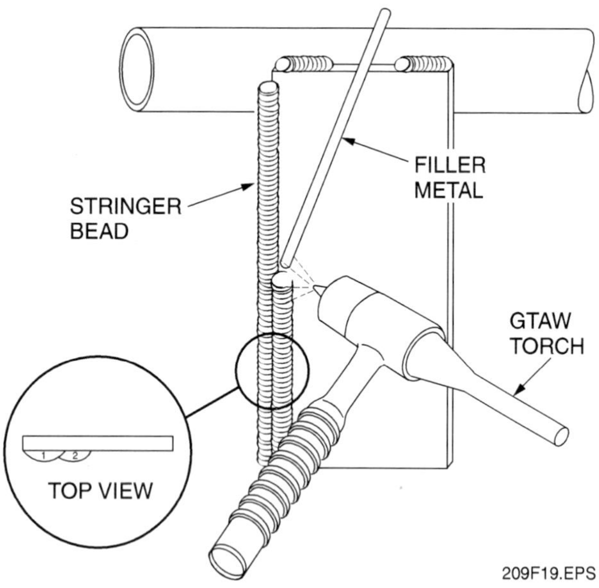

Figure 19 ◆ Building a pad in the vertical position.

NOTE: THE ACTUAL NUMBER OF WELD BEADS WILL VARY WITH THE METAL THICKNESS.

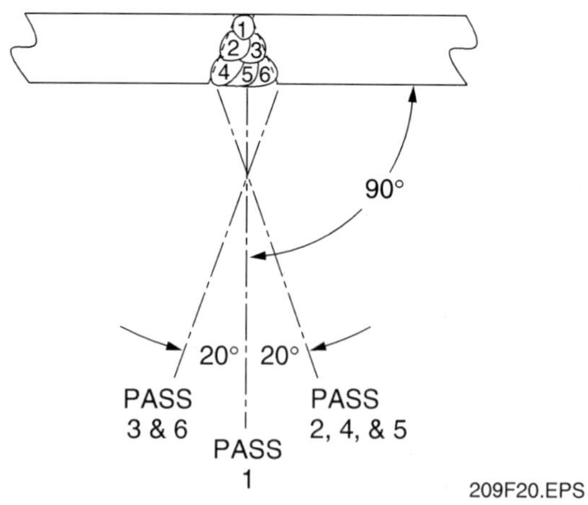

Figure 20 ◆ Multipass 3G weld sequences and work angles.

10.6.0 Practicing Overhead Beads

Before welding a joint in the overhead position, practice running overhead stringer beads by building a pad in the overhead position. Use ³⁄₃₂" or ⅛" carbon steel filler metal to build the pad. Weld overhead stringer beads with a push angle of 15° to 20°. Hold the filler metal ahead of the weld at about 20° with the plate surface. Follow these steps to practice overlapping vertical beads in the vertical position:

Step 1 Clamp or tack weld the coupon in the overhead position.

Step 2 Run the first bead along one edge of the coupon using either ³⁄₃₂" or ⅛" carbon steel filler metal.

Step 3 Clean the weld.

Step 4 Run the second bead, overlapping the edge of the first bead. Adjust the work angle to obtain a good tie-in.

Step 5 Continue running beads to complete the pad. Be sure to clean the weld between each pass.

Figure 21 shows how to build a pad in the overhead position.

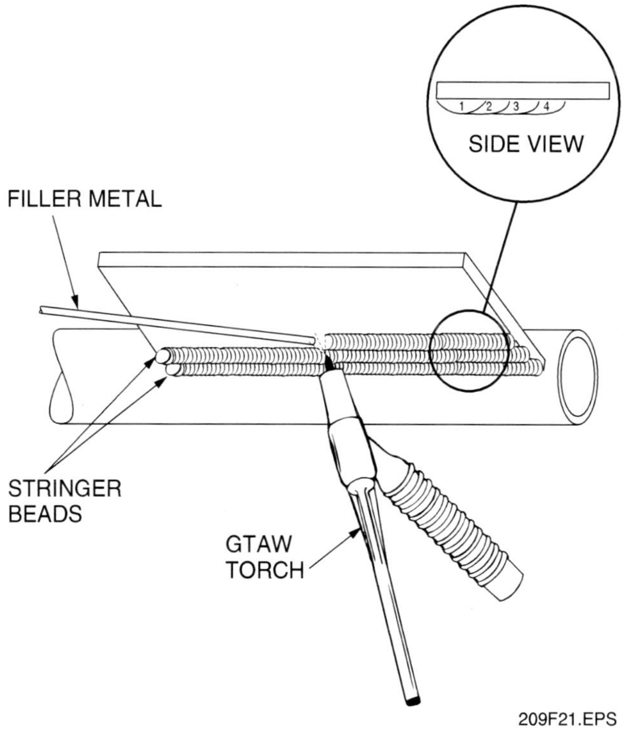

Figure 21 ◆ Building a pad in the overhead position.

10.7.0 Practicing Overhead (4G) Position Open V-Groove Welds

Follow these steps to practice open V-groove welds in the overhead (4G) position (*Figure 22*):

Step 1 Tack weld the practice coupon together as explained earlier.

Step 2 Clamp or tack weld the coupon in the overhead position.

Step 3 Run the root pass with either the on-the-wire or the keyhole welding technique, using an appropriate diameter carbon steel filler metal. Use a 15° to 20° push angle, a 20° filler metal angle to the plate surface, and the pass 1 work angle (*Figure 22*).

Step 4 Clean the weld.

Step 5 Run the remaining passes using either ³⁄₃₂" or ⅛" carbon steel filler metal, a 15° to 20° push angle, and the work angles shown in *Figure 22*. Clean the weld between each pass.

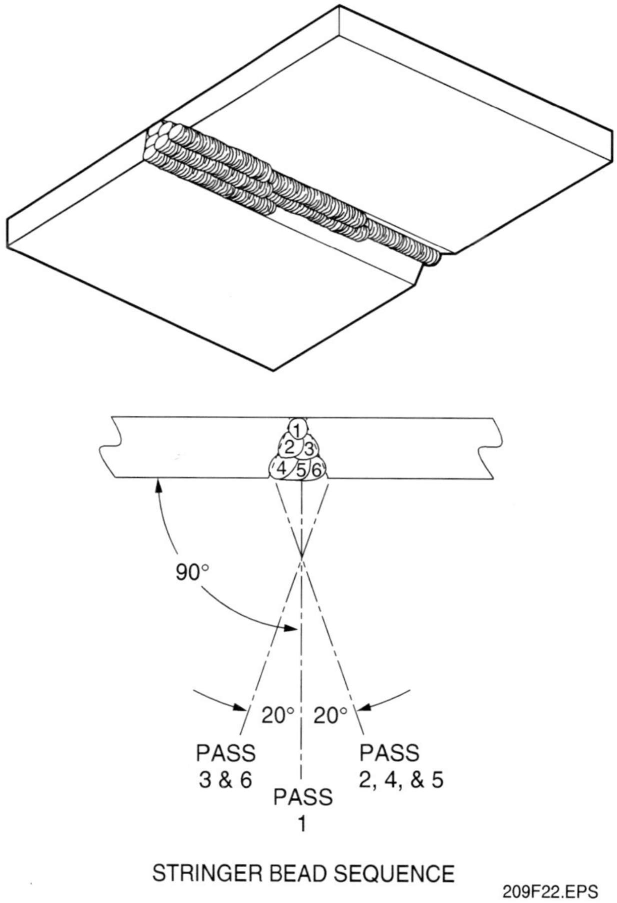

Figure 22 ◆ Multipass 4G weld sequences and work angles.

Summary

Setting up GTAW equipment, preparing the welding work area, running stringer beads, and making acceptable groove welds in all positions are essential skills a welder must have to perform basic welding jobs or to progress to more difficult welding procedures. Practice these welds until you can consistently produce acceptable welds as defined in the criteria for acceptance.

Review Questions

1. Open-root V-groove welds generally use a bevel angle of _____ for carbon steel.
   a. 20° or 25°
   b. 30° or 35°
   c. 45° or 50°
   d. 60° or 65°

2. Weld travel speed affects the size of the weld puddle and the weld _____.
   a. temperature
   b. convexity
   c. appearance
   d. penetration

3. The two basic GTAW handling techniques are known as freehand and _____.
   a. side-to-side
   b. on-the-wire
   c. walking-the-cup
   d. keyhole

4. In the freehand technique, the torch electrode tip is manually held _____.
   a. against the base metal
   b. against the groove edges
   c. just below the weld puddle or base metal
   d. just above the weld puddle or base metal

5. A weave bead is much _____ than the puddle width used to produce it.
   a. deeper
   b. higher
   c. wider
   d. hotter

6. Flat-position stringer beads should be run with a _____ push angle.
   a. 15° to 20°
   b. 20° to 30°
   c. 30° to 40°
   d. 60° to 70°

7. The junction where a new weld connects to a previous weld is called a _____.
   a. bead face
   b. termination
   c. restart
   d. blend

8. Overlapping beads to form a flat surface is called _____.
   a. padding
   b. stringer beads
   c. weave beads
   d. fillet welding

9. Cooling coupons in water can cause _____.
   a. grooves
   b. convexity
   c. annealing
   d. weld cracks

10. When running stringer beads in the horizontal position, always start the first bead _____ of the workpiece.
    a. on the top
    b. on the bottom
    c. at the left
    d. at the right

GLOSSARY

Trade Term Introduced in This Module

On-the-wire: A root pass technique in which the filler metal is held in the bottom of the groove parallel to the plates and then melted and fused in place by moving the torch along the groove.

APPENDIX

Performance Accreditation Tasks

The Performance Accreditation Tasks (PATCs) correspond to and support the learning objectives in *AWS EG2.0-95, Guide for the Training and Qualification of Welding Personnel: Entry-Level Welder*.

PATCs provide specific acceptable criteria for performance and help to ensure a true competency-based welding program for students.

The following tasks are designed to evaluate your ability to run open-root V-groove welds on plate in all positions using GTAW equipment. Perform each task when you are instructed to do so by your instructor. As you complete each task, take it to your instructor for evaluation. Do not proceed to the next task until told to do so by your instructor.

OPEN-ROOT V-GROOVE WELDS IN THE FLAT POSITION

Using the appropriate carbon steel filler rods, make an open-root V-groove weld on plate in the flat (1G) position, as shown.

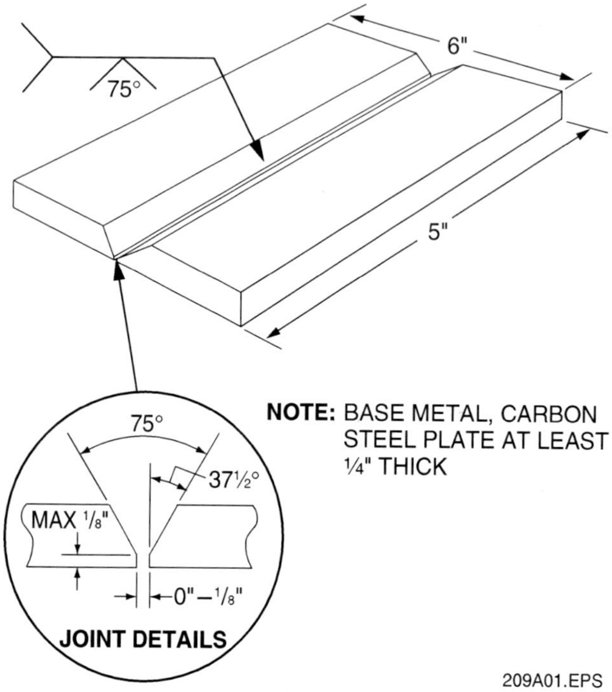

Criteria for Acceptance

- Uniform rippled appearance on the bead face ______
- Craters and restarts filled to the full cross section of the weld ______
- Uniform weld size ±1⁄16" ______
- Acceptable weld profile in accordance with the *ASME Boiler and Pressure Vessel Code* ______
- Smooth transition with complete fusion at the toes of the weld ______
- Complete uniform root reinforcement at least flush with the base metal to a maximum buildup of 3⁄32" ______
- No porosity ______
- No overlap ______
- No excessive undercut ______
- No inclusions ______
- No cracks ______
- No pinholes (fisheyes) ______

OPEN-ROOT V-GROOVE WELDS IN THE HORIZONTAL POSITION

Using the appropriate carbon steel filler rods, make an open-root V-groove weld on plate in the horizontal (2G) position, as shown.

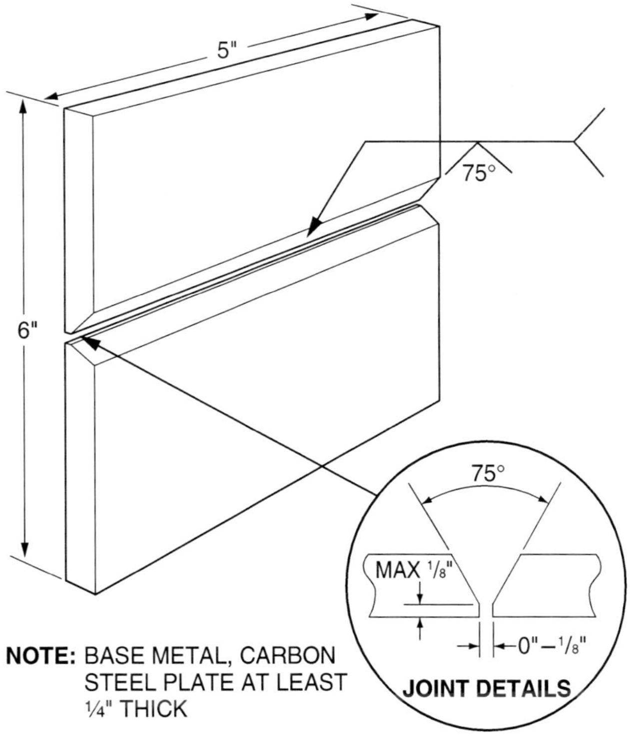

209A02.EPS

Criteria for Acceptance

- Uniform rippled appearance on the bead face ____________
- Craters and restarts filled to the full cross section of the weld ____________
- Uniform weld size ±1⁄16" ____________
- Acceptable weld profile in accordance with the *ASME Boiler and Pressure Vessel Code* ____________
- Smooth transition with complete fusion at the toes of the weld ____________
- Complete uniform root reinforcement at least flush with the base metal to a maximum buildup of 3⁄32" ____________
- No porosity ____________
- No overlap ____________
- No excessive undercut ____________
- No inclusions ____________
- No cracks ____________
- No pinholes (fisheyes) ____________

OPEN-ROOT V-GROOVE WELDS IN THE VERTICAL POSITION

Using the appropriate carbon steel filler rods, make an open-root V-groove weld on plate in the vertical (3G) position, as shown.

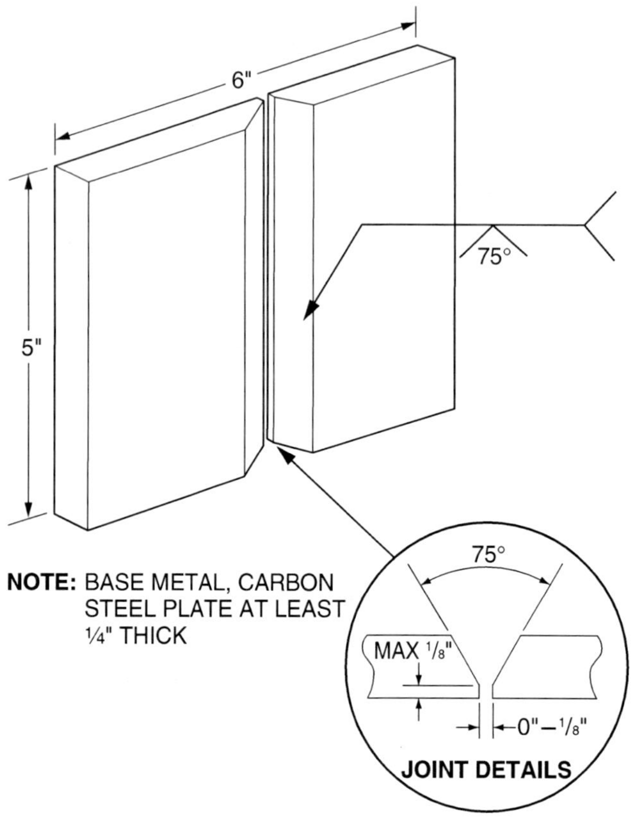

209A03.EPS

Criteria for Acceptance

- Uniform rippled appearance on the bead face ____________
- Craters and restarts filled to the full cross section of the weld ____________
- Uniform weld size ±1/16" ____________
- Acceptable weld profile in accordance with the *ASME Boiler and Pressure Vessel Code* ____________
- Smooth transition with complete fusion at the toes of the weld ____________
- Complete uniform root reinforcement at least flush with the base metal to a maximum buildup of 3/32" ____________
- No porosity ____________
- No overlap ____________
- No excessive undercut ____________
- No inclusions ____________
- No cracks ____________
- No pinholes (fisheyes) ____________

OPEN-ROOT V-GROOVE WELDS IN THE OVERHEAD POSITION

Using the appropriate carbon steel filler rods, make an open-root V-groove weld on plate in the overhead (4G) position, as shown.

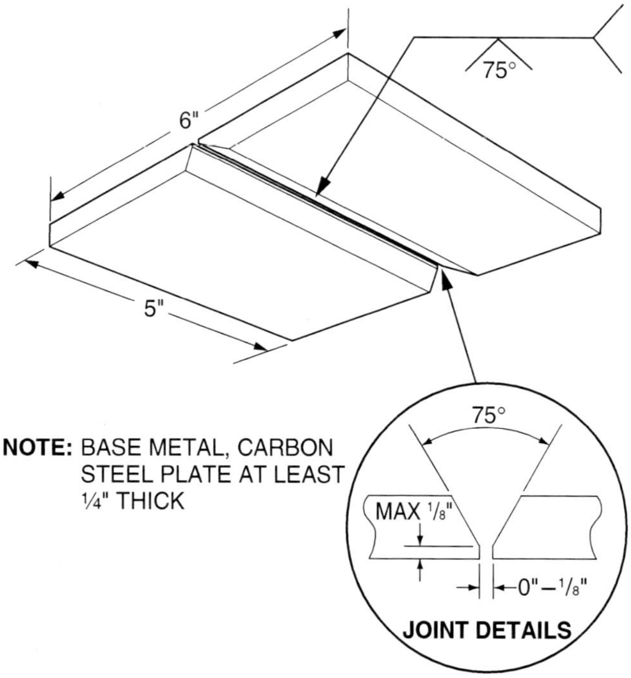

209A04.EPS

Criteria for Acceptance

- Uniform rippled appearance on the bead face ________
- Craters and restarts filled to the full cross section of the weld ________
- Uniform weld size ±1⁄16" ________
- Acceptable weld profile in accordance with the *ASME Boiler and Pressure Vessel Code* ________
- Smooth transition with complete fusion at the toes of the weld ________
- Complete uniform root reinforcement at least flush with the base metal to a maximum buildup of 3⁄32" ________
- No porosity ________
- No overlap ________
- No excessive undercut ________
- No inclusions ________
- No cracks ________
- No pinholes (fisheyes) ________

Additional Resources

This module is intended to be a thorough resource for task training. The following reference works are suggested for further study. These are optional materials for continued education rather than for task training.

Gas Tungsten Arc Welding Handbook. William H. Minnick. Tinley Park, IL: The Goodheart-Willcox Co., Inc.

Modern Welding, 2000. W.A. Bowditch, K.E. Bowditch, A.D. Althouse, C.H. Turnquist. Tinley Park, IL: The Goodheart-Willcox Company, Inc.

OSHA Standard 1926.351, Arc Welding and Cutting.

Welding Skills. R. T. Miller. Homewood, IL: American Technical Publishers, Inc.

Figure Credits

Tom Atkinson	209F08
John Yochum	209F09
Charles Rogers	209F18 (photo), 209F20 (photo)

CONTREN™ LEARNING SERIES — USER UPDATES

The NCCER makes every effort to keep these textbooks up-to-date and free of technical errors. We appreciate your help in this process. If you have an idea for improving this textbook, or if you find an error, a typographical mistake, or an inaccuracy in NCCER's Contren™ textbooks, please write us, using this form or a photocopy. Be sure to include the exact module number, page number, a detailed description, and the correction, if applicable. Your input will be brought to the attention of the Technical Review Committee. Thank you for your assistance.

Instructors – If you found that additional materials were necessary in order to teach this module effectively, please let us know so that we may include them in the Equipment/Materials list in the Instructor's Guide.

Write: Curriculum Revision and Development Department
National Center for Construction Education and Research
P.O. Box 141104, Gainesville, FL 32614-1104

Fax: 352-334-0932

E-mail: curriculum@nccer.org

Craft ______ Module Name ______

Copyright Date ______ Module Number ______ Page Number(s) ______

Description

______

______

______

______

(Optional) Correction

______

______

______

(Optional) Your Name and Address

______

______

______

Module 29210-03

GTAW – Aluminum Plate

COURSE MAP

This course map shows all of the modules in the second level of the Welding curriculum. The suggested training order begins at the bottom and proceeds up. Skill levels increase as you advance on the course map. The local Training Program Sponsor may adjust the training order.

WELDING LEVEL TWO

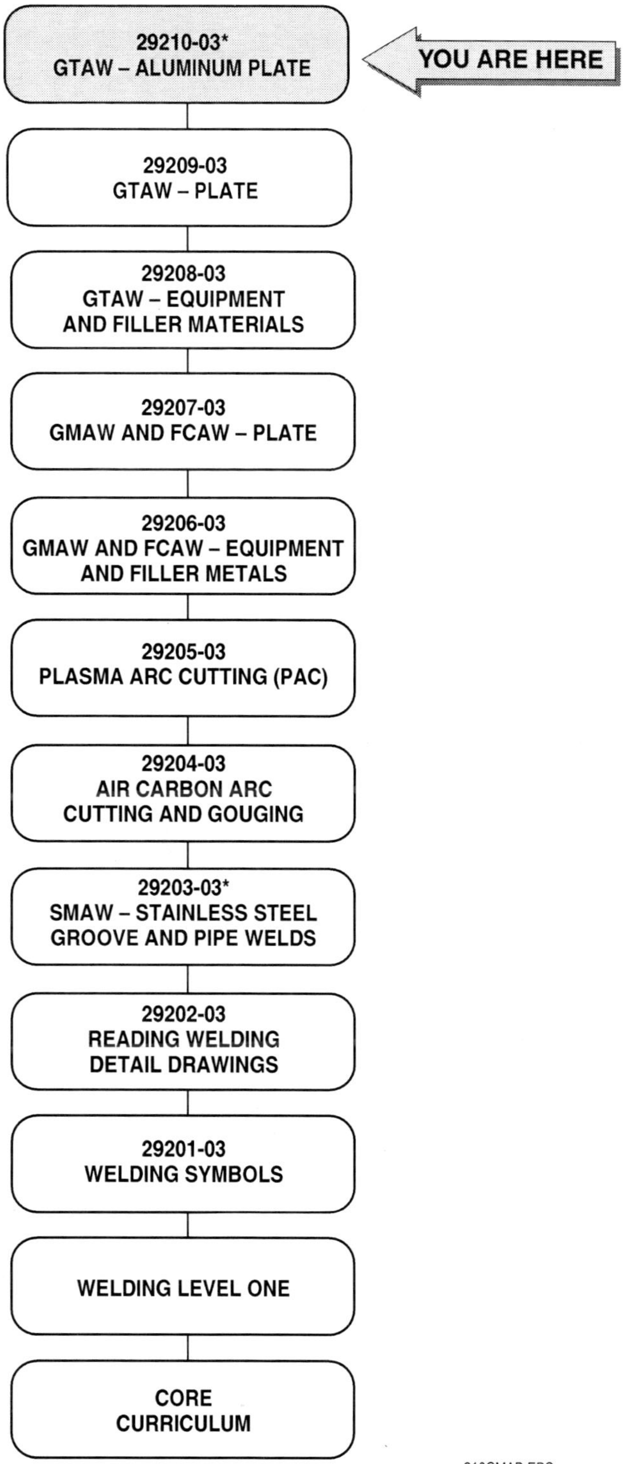

210CMAP.EPS

Please note that Modules 29203-03 and 29210-03 are electives for *Welding Level Two* but are requirements for *AWS Entry Level Welder.

MODULE 29210-03 CONTENTS

Figures

Tables

MODULE 29210-03

GTAW – Aluminum Plate

Objectives

When you have completed this module, you will be able to do the following:

1. Identify and explain aluminum metallurgy.
2. Explain and identify characteristics of aluminum.
3. Explain GTAW and set up equipment to weld aluminum plate.
4. Explain and practice GTAW techniques for plate, including padding in the flat position with stringer beads, using aluminum filler metal.
5. Make fillet welds on aluminum plate in the following positions:
   - 1F (flat)
   - 2F (horizontal)
   - 3F (vertical)
   - 4F (overhead)
6. Make multipass open V-groove welds with backing on aluminum plate in the following positions:
   - 1G (flat)
   - 2G (horizontal)
   - 3G (vertical)
   - 4G (overhead)

Prerequisites

Before you begin this module, it is recommended that you successfully complete the following: Core Curriculum; Welding Level One; Welding Level Two, Modules 29201-03 through 29209-03.

Required Trainee Materials

1. Pencil and paper
2. Appropriate personal protective equipment

1.0.0 ◆ INTRODUCTION

Gas tungsten arc welding (GTAW) is an arc welding process used for making high-quality welds on aluminum. In the past, this method of welding was called tungsten inert gas (TIG) welding and, occasionally, is referred to today as TIG welding. Tungsten refers to the tungsten electrode that delivers the arc. *Figure 1* shows the GTAW process.

This module describes aluminum metallurgy and the characteristics of aluminum welding. It also explains how to set up GTAW equipment and prepare aluminum plate for fillet welds in the 1F, 2F, 3F, and 4F positions as well as open V-groove welds in the 1G, 2G, 3G, and 4G welding positions.

1.1.0 Safety Practices

The following is a summary of safety procedures and practices that must be observed while cutting or welding. Keep in mind that this is just a summary. Complete safety coverage is provided in the Level One module, *Welding Safety*. If you have not completed that module, do so before continuing. Above all, be sure to wear appropriate protective clothing and equipment when welding or cutting.

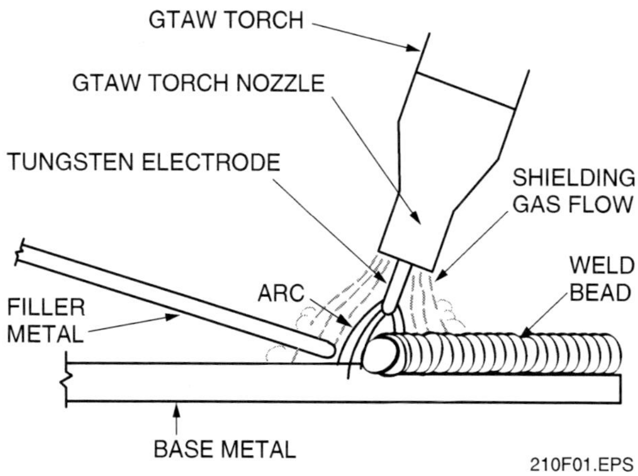

Figure 1 ◆ GTAW process.

Hot Tip: Dimensions and Specifications

The dimensions and specifications in this module are designed to be representative of codes in general and are not specific to any certain code. Always follow the proper codes for your site.

1.1.1 Protective Clothing and Equipment

- Always use safety goggles with a full face shield or a helmet. The goggles, face shield, or helmet lens must have the proper light-reducing tint for the type of welding or cutting to be performed. Never directly or indirectly view an electric arc without using a properly tinted lens.
- Wear proper protective leather and/or flame retardant clothing along with welding gloves that will protect you from flying sparks and molten metal, as well as heat.
- Wear 8" or taller high-top safety shoes or boots. Make sure that the tongue and lace area of the footwear will be covered by a pant leg. If the tongue and lace area is exposed or the footwear must be protected from burn marks, wear leather spats under the pants or chaps and over the front top of the footwear.
- Wear a solid material (non-mesh) hat with a bill pointing to the rear or, if much overhead cutting or welding is required, a full leather hood with a welding face plate and the correct tinted lens. If a hard hat is required, use a hard hat that allows the attachment of rear deflector material and a face shield.
- If a full leather hood is not worn, wear a face shield and snugly fitting welding goggles over safety glasses for gas welding or cutting. Either the face shield or the lenses of the welding goggles must be an approved shade 5 or 6 filter. For electric arc welding or cutting, wear safety goggles or a welding hood with the correct tinted lens (shade 10 to 14).
- If a full leather hood is not worn, wear earplugs to protect your ear canals from sparks.

1.1.2 Fire/Explosion Prevention

- Never carry matches or gas-filled lighters in your pockets. Sparks can cause the matches to ignite or the lighter to explode causing serious injury.
- Never perform any type of heating, cutting, or welding until a hot-work permit is obtained and an approved fire watch is established. Most work-site fires caused by these types of operations are started by cutting torches.
- Never use oxygen to blow off clothing. The oxygen can remain trapped in the fabric for a time. If a spark hits the clothing during this time, it can burn rapidly and violently out of control.
- Make sure that any flammable material in the work area is moved or shielded by a fire-resistant covering. Approved fire extinguishers must be available before attempting any heating, welding, or cutting operations.
- Always comply with any site requirement for a hot-work permit and/or fire watch.
- Never release a large amount of oxygen or use oxygen as compressed air. Its presence around flammable materials or sparks can cause rapid and uncontrolled combustion. Keep oxygen away from oil, grease, and other petroleum products.
- Never release a large amount of fuel gas, especially acetylene. Methane and propane tend to concentrate in and along low areas and can ignite at a considerable distance from the release point. Acetylene is lighter than air but is even more dangerous than methane. When mixed with air or oxygen, it will explode at much lower concentrations than any other fuel.
- To prevent fires, maintain a neat and clean work area and make sure that any metal scrap or slag is cold before disposal.
- Before cutting or welding containers such as tanks or barrels, check to see if they have contained any explosive, hazardous, or flammable materials including petroleum products, citrus products, or chemicals that decompose into toxic fumes when heated. As a standard practice, always clean and then fill any tanks or barrels with water or purge them with a flow of inert gas to displace any oxygen.

1.1.3 Work Area Ventilation

- Make sure confined space procedures are followed before conducting any welding or cutting in the confined space.

- Never use oxygen in confined spaces for ventilation purposes.
- Always perform cutting or welding operations in a well-ventilated area. Cutting or welding operations involving zinc or cadmium materials or coatings result in toxic fumes. For long-term cutting or welding of such materials, always wear an approved full face, supplied-air respirator (SAR) that uses breathing air supplied outside of the work area. For occasional, very short-term exposure, a high-efficiency particulate arresting (HEPA)-rated or metal-fume filter may be used on a standard respirator.
- Make sure confined spaces are ventilated properly for cutting or welding purposes.

2.0.0 ◆ ALUMINUM METALLURGY

Aluminum and aluminum alloys are frequently used in fabrication because of their low weight and good corrosion resistance. Aluminum is one-third the weight of steel and has six times the thermal conductivity. Pure aluminum is fairly soft with low strength and is not suitable for load carrying. However, the recent introduction of lithium alloys, with their lower density and higher strength, has increased the use of aluminum in structural applications.

Aluminum is not toxic and is nonsparking and nonmagnetic. Molten aluminum has a silvery sheen under the glow of the arc. When heated, aluminum does not change color as steel does.

Aluminum is a highly reactive element, which explains why it is never found in nature in its free state, even though it is the most common metal on earth. Aluminum has a strong chemical inclination toward oxygen, causing it to oxidize as soon as it is exposed to air. Aluminum oxide is extremely hard and has a high melting point. It is also porous, allowing it to hold moisture and contaminants, which can result in weld porosity. This is especially true of alloys containing magnesium.

Aluminum can be strengthened by the addition of small amounts of alloy metals to the base metal. Alloys are also used to enhance various other properties of aluminum. Alloys are grouped into series by their alloying elements. The alloys in each series are further defined by whether they are treatable by heat. Heat-treatable and nonheat-treatable aluminum alloys react differently to welding. **Strain hardening** is another method of increasing the strength of a metal by **cold working** it; alloys that respond to cold working are said to be strain-hardenable. *Table 1* shows aluminum alloy characteristics.

Aluminum metallurgy can be divided into the following categories:

- Nonheat-treatable alloys
- Heat-treatable alloys
- Filler metal alloys
- Cast aluminum
- Wrought aluminum

Table 1 Aluminum Alloy Characteristics

Series	Classification	Major Alloying Elements	Characteristics
1xxx 1050A	Nonheat-treatable	High purity with small amounts of iron and silicon	Relatively soft, low-strength, nonload carrying; excellent corrosion resistance
2xxx 2014A	Heat-treatable	Copper	High strength; prone to liquation under welding heat
3xxx 3103	Nonheat-treatable	Manganese	Stronger and more ductile than 1xxx grades; excellent corrosion resistance
4xxx 4043A	Filler wires	Silicon	Excellent corrosion resistance
5xxx 5083	Nonheat-treatable	Magnesium	Good ductility; excellent corrosion resistance; good for situations involving impact and shock-loading; not recommended for service with prolonged exposure to high temperatures
6xxx 6082	Heat-treatable	Silicon and magnesium	Good as-welded strength; easily extruded; not recommended for exposure to electrolytes, such as sea water
7xxx 7020	Heat-treatable	Zinc	Good weldability and high as-welded strength
8xxx 8090	Nonheat-treatable and heat-treatable	Other	Not normally welded

210T01.EPS

2.1.0 Nonheat-Treatable Alloys

The strength of nonheat-treatable aluminum alloys depends on elements such as silicon (Si), manganese (Mn), iron (Fe), and magnesium (Mg). These alloying elements are mainly found in the 1xxx, 3xxx, and 5xxx series of alloys. When welded, these alloys may lose the effects of **work hardening**, which results in softening of the **heat-affected zone (HAZ)** next to the weld.

Additional strengthening comes from work hardening (strain hardening) in the forming process, such as by cold-rolling. Welds in nonheat-treatable alloys generally have excellent ductility and can accommodate severe distortion before failing. Welds in heat-treatable alloys generally have lower ductility than welds in nonheat-treatable alloys. The basic temper designations for nonheat-treatable or heat-treatable aluminum alloys, such as castings, are as follows:

- *O* – **Annealed**, recrystallized
- *F* – As fabricated
- *H1(x)* – Strain hardened only, where (x) = 2, 4, 6, 8, or 9 for ¼, ½, ¾, full hardness, and extra hardness
- *H2(x)* – Strain hardened and then partially annealed, where (x) = 2, 4, 6, 8, or 9 for ¼, ½, ¾, full hardness, and extra hardness
- *H3(x)* – Strain hardened and then stabilized, where (x) = 2, 4, 6, 8, or 9 for ¼, ½, ¾, full hardness, and extra hardness
- *W(HR)* – Solution heat treated, where (HR) = fractional and full hours of aging at room temperature after heat treatment
- *T(N)* – Thermally treated (other than O, F, or H), where (N) = 1 through 10, which indicates a specific treatment, sometimes followed by another digit representing a variation of the treatment

2.2.0 Heat-Treatable Alloys

Heat-treatable aluminum alloys develop their properties by solution heat-treating and quenching, followed by either natural or artificial aging. Heat-treatable alloys contain copper, magnesium, zinc, or silicon. These elements become more soluble when heated. Welding heat reduces the strength of heat-treatable alloys, but, because longer times are required to modify their structure, factors such as welding speed and cooling rate have an influence on their final properties. Because of the time factor, the HAZ in heat-treatable alloys is usually only partially annealed. Fusion welding redistributes the hardening components in the HAZ, which locally reduces strength. The 2xxx, 6xxx, 7xxx, and some of the 8xxx series contain the principal heat-treatable alloying elements. Heat-treatable temper designations are O, F, W, and T1 through T10.

Postweld heat treatment increases the strength of the HAZ, but it also reduces the amount of ductility. The strength improvement depends somewhat on the filler metal used. Preheating is seldom recommended for welding heat-treatable alloys.

Joint thickness can also have an effect on weld properties in heat-treatable alloys. Thinner metal can be welded with a minimum of heat input and with a high cooling rate, resulting in higher as-welded strengths. For thicker metal, a number of string beads are preferable to a few large beads.

2.3.0 Filler Metal Alloys

Filler metal composition is determined by the weldability of the base metal, the minimum properties of the weld metal, the corrosion resistance, and the **anodic coating** requirements. For nonheat-treatable alloys, matching fillers are often used. For low-alloy materials and heat-treatable alloys, nonmatching fillers are used to prevent solidification cracking.

2.4.0 Cast Aluminum

Cast aluminum has high silicon content for increased fluidity and does not anodize as well as wrought aluminum. Cast aluminum is identifiable by three-digit grade numbers, such as 206, 319, and 356. The first number indicates the alloying element. The second and third numbers identify the specific alloy by formula. The number after the decimal indicates whether the alloy composition is for the final casting (.0) or for the **ingot** (.1 or .2, depending on purity). Modifications to cast aluminum are indicated by a letter in front of the number. An example of a typical silicon-magnesium alloy for final casting is A356.0, and the same alloy in ingot form would be A356.1 or A356.2. *Table 2* lists the numbers representing the alloying elements used with cast aluminum.

Table 2 Alloy Numbers for Cast Aluminum

Alloy Series	Principal Alloying Elements
1xx.x	Essentially pure aluminum
2xx.x	Copper
3xx.x	Silicon plus copper and/or magnesium
4xx.x	Silicon
5xx.x	Magnesium
6xx.x	Unused series
7xx.x	Zinc
8xx.x	Tin
9xx.x	Other elements

210T02.EPS

2.5.0 Wrought Aluminum

Wrought aluminum alloys are designated by four numbers. The first number indicates the principal alloying element or elements. The second number, if not a zero, indicates some modification to the alloy. The third and fourth numbers identify the specific alloy by formula. For example, alloy 5054 is an aluminum-magnesium alloy without modifications and has the assigned alloy number of 54. Alloy 5154 is the same alloy after modification. The only exception is the 1xxx series, which is for pure aluminum. The last two numbers represent the degree of purity above 99 percent: for example, 1060 has a purity of 99.60 percent. *Table 3* shows the elements used with the various wrought aluminum alloys.

Most of the wrought grades in the 1xxx, 3xxx, 5xxx, 6xxx, and medium-strength 7xxx (for example, 7020) series can be fusion-welded with GTAW and GMAW. The 5xxx series alloys have excellent weldability.

Wrought aluminum is available in five basic categories:

- Plate or sheet
- Rod, bar, or wire
- Tubing
- Extrusions
- Forgings

Table 3 Alloy Numbers for Wrought Aluminum

Alloy Series	Principal Alloying Elements
1xxx	Aluminum (99.00% minimum and greater)
2xxx	Copper
3xxx	Manganese
4xxx	Silicon
5xxx	Magnesium
6xxx	Magnesium and Silicon
7xxx	Zinc
8xxx	Other elements
9xxx	Unused series

210T03.EPS

3.0.0 ◆ CHARACTERISTICS OF ALUMINUM WELDING

Normally, GTAW and GMAW do not require preheating of aluminum. However, the high thermal conductivity of aluminum (as compared to steel) requires a high rate of heat input for fusion welding. In some cases, extremely thick sections may require some preheating (less than 250°F). To achieve the necessary welding heat for aluminum, higher current with more precise control of the welding variables are required for GTAW and GMAW than when welding steel.

Although GTAW can use either alternating current (AC) or direct current (DC) power sources, AC is recommended because the alternating polarity of the welding pulses provides both the cleaning action and the heat needed to weld aluminum. In some instances, DCEN with **helium** shielding gas is used in automatic welding of thick aluminum materials when high heat and deeper penetration are required. However, complete oxide removal must be accomplished just prior to welding because no cleaning action occurs with the negative electrode. For this reason, AC is used almost exclusively when welding aluminum.

Welding current is principally related to the metal thickness, because the arc must be hot enough to give the required penetration. Oxide contamination in the base metal, or too little current, will cause excessive delay between the time the arc is struck and the time the weld pool becomes large enough for the filler metal to be added. If the weld becomes too hot, the welding speed must be increased beyond that which is comfortable for the welder. If the speed is too slow, a characteristic of non-ferrous metals called hot shortness can occur. Hot shortness is a condition in which a metal loses its strength prior to melting. On thin materials, this causes the area around a molten puddle to crumble and fall away without warning. The result is a large hole in the base metal. To prevent this from occurring on thin materials and on root passes with complete penetration, backing strips or rings made of stainless steel or a non-metallic material are sometimes used. The backing strips or rings are often made with a shallow groove in the surface to ensure that penetrating welds have the required size and shape when cool. On advanced welding machines used for GTAW welding, careful control of AC balance, frequency, and the use of pulsing control can help eliminate hot shortness problems.

Hot Tip

Thickness versus Required Current

A guideline for approximating the amperage for welding aluminum is that for each 0.001" of metal thickness under the weld, a welding current of about one amp is required.

3.1.0 Surface Preparation

Aluminum combines with oxygen from the atmosphere, even at room temperatures, and forms a surface film of very hard oxide. While pure aluminum melts at 1,220°F, its oxide melts at 3,600°F to 3,900°F. Aluminum oxide is second only to a diamond in hardness and can destroy a milling cutter if the oxides are trapped in the welds. Because of this, aluminum oxide is commonly used for the grit in grinding wheels. Aluminum oxide also tends to absorb moisture, which, under the extreme heat of welding, breaks down to free hydrogen. This free hydrogen often leads to porosity in the weld. Always remove oxidation from the base metal prior to welding.

CAUTION

Grinding wheels must be made for aluminum and should not be used on any other type of metal or on surfaces contaminated with grease or oil. Using other types of grinding wheels may cause them to shatter.

3.1.1 Aluminum Storage

Aluminum can also pick up hydrogen gas from condensation, which occurs during rapid rises in temperature. Because materials are often stored in unheated areas, special care should be taken to raise the temperature of aluminum slowly. It is also recommended that aluminum be stored in the vertical position to allow any condensation to run off.

3.1.2 Aluminum Cleaning

Oil and grease are contaminants often found on aluminum. Since wire brushing will only smear the contaminants into the surface or will damage the aluminum itself, the oil and grease must be chemically removed before wire brushing.

A grinder equipped with a buffer wheel can be used to remove most contaminants. Both water-based aluminum deoxidizers and solvent cleaners, such as acetone, mineral spirits, or denatured alcohol, can also be used to remove grease. When using water-based cleaners, any water must be removed before welding. Most hydrocarbon solvents are highly volatile and evaporate quickly, but the water-based cleaners must be thoroughly wiped away or heat-dried.

Preweld cleaning solutions are available that can remove both the aluminum oxide and any contaminants. Their disadvantage is that deoxidizers for aluminum are very aggressive chemicals and can be dangerous to the people using them.

WARNING!

To prevent injury when using cleaning solvents, always comply with the guidelines of the material safety data sheet (MSDS) for the solvent(s) being used.

3.1.3 Chemical Aluminum Preparation Precautions

Preweld cleaning products usually contain dilute acids, and it is very important to follow the manufacturer's instructions and take the recommended safety precautions. When using etchants or deoxidizers, the metal must be fully rinsed and completely dried to avoid creating a surface film of hydrated oxide.

3.2.0 Weld Problems

Before fusion welding can take place, the oxide must be removed. The melting point of the oxide is about three times that of aluminum. Because of this, the welding heat needed to break down the oxide causes the aluminum to melt long before the oxide does, and **coalescence** cannot take place.

Hot Tip

Clean, Clean, and Clean Again

Cleanliness of the welding surfaces cannot be overemphasized. Oxides, greases, and oil films contain oxygen and hydrogen that can contaminate and cause unsound welds. Aluminum should be welded as soon as possible after cleaning or within one-half hour of cleaning to prevent having to clean it again. If oxides are not removed to the maximum extent possible, the weld is subject to poor fusion and inclusions due to the reduced wetting action of the molten weld pool with any remaining oxide-contaminated weld joint surfaces. Aluminum should be hand-brushed between passes with a stainless steel brush that has not been used on other materials or for any other purpose.

The following are the most common problems in aluminum welding:

- Porosity
- Cracking
- Poor weld bead profile

3.2.1 Porosity

Porosity often occurs with GTAW, especially when welding aluminum, because of its high susceptibility to porosity. The main cause of porosity is the absorption of hydrogen into the weld pool; the hydrogen forms individual pores in the weld metal as the weld pool solidifies. The most common sources of hydrogen are hydrocarbons, atmosphere, and moisture from contaminants on both the parent material and the filler wire surfaces. In some cases, a torch gas flow that is set too high can cause turbulence that draws air in the gas and causes porosity in the weld. *Figure 2* shows porosity in a GTAW sample caused by hydrocarbon contamination.

Thoroughly clean the surface to reduce the risks of porosity. Solvent degreasing, chemical etching, and mechanical cleaning, such as grinder buffing, are proper cleaning techniques. *Table 4* shows the causes and cures for porosity and cracking in GTAW.

3.2.2 Weld Cracking in General

Cracking occurs in aluminum alloys because of the high stresses that are generated across the weld by the high thermal expansion (twice that of steel).

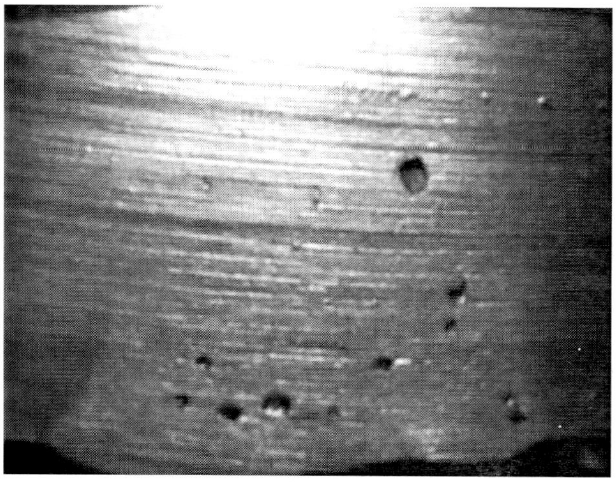

210F02.EPS

Figure 2 ◆ Porosity in a GTAW sample.

Table 4 Causes and Cures for Porosity and Cracking in GTAW

Cause	Contributing Factors	Corrective Measures
Hydrogen	Hydrogen is the result of dirt-containing oils or other hydrocarbons; moisture in atmosphere or on metal; a hydrated oxide film on metal; or moisture in gas or gas lines. Base metal may be the source of entrapped hydrogen (the thicker the metal, the greater the possibility for hydrogen). Spatter contributes to hydrogen formation.	Degrease and mechanically or chemically remove oxide from weld area. Avoid humidity; use dry metal or wipe dry. Reduce moisture content of gas. Check gas and water lines for leaks. Increase gas flow to compensate for increased hydrogen in thicker sections. To minimize spatter, adjust welding conditions.
Impurities	Cleaning or other compounds, especially those containing calcium.	Use recommended cleaning compounds; keep work free of contaminants.
Incomplete root penetration	Insufficient welding current or preheat.	Use a higher welding current, or redesign the joint geometry. Preheat.
Temperature	Running too cool tends to increase porosity that is caused by premature solidification of molten metal.	Maintain the proper current, arc-length, and torch-travel speed relationship.
Welding speed	Too great a welding speed may increase porosity.	Decrease the welding speed and establish and maintain proper arc-length and current relationship.
Solidification time	Quick freezing of the weld pool entraps any gases present, causing porosity.	Establish the correct welding current and speed. If the work is appreciably below room temperature, use supplemental heating.
Chemical composition of weld material	Pure aluminum weld metal is more susceptible to porosity than is an aluminum alloy.	If porosity is excessive, try an alloy filler material.
Cracking	Such causes of porosity as temperature, welding time, and solidification may also be contributing causes of cracking. Other causes may be discontinuous welds, welds that intersect, repair welds, cold-working either before or after welding, and weld-metal composition. In general, crack-sensitive alloys include those containing 0.4 to 0.6% silicon, 1.5 to 3.0% magnesium, or 1.0% copper.	Lower current and faster speeds often prevent cracking. However, a change to a filler alloy that brings the weld-metal composition out of the cracking range is recommended where possible.

210T04.EPS

Thermal expansion is the fractional increase in size due to a rise in temperature. Cracking also occurs because of the amount of contraction when the weld solidifies—usually 5% more than in similar welds on steel. High-strength aluminum alloys (for example, 7010 and 7050) and most of the 2xxx series are not recommended for fusion welding, because they are prone to both solidification and **liquation** cracking.

3.2.3 Solidification Cracking

Solidification cracks (also called hot cracks) form in the center of the weld and run along the centerline of the weld. Solidification cracks also occur in the weld crater at the end of the weld. There are two main causes of solidification cracks:

- Incorrect combination of filler wire and base metal
- Welding under high restraint conditions

Use of a non-matching, crack-resistant filler (usually from the 4xxx and 5xxx series alloys) can reduce the risk of cracking. Unfortunately, use of these fillers can result in a weld of lower strength and the weld may not respond to heat treatment. The weld bead must be thick enough to withstand the stresses caused by contraction as the weld solidifies. Correct edge preparation, accurate joint setup, more compatible filler alloy, and correct weld sequence can reduce the degree of restraint on the weld. Higher weld speeds can also reduce solidification cracking.

Preheating reduces stress on the solidifying weld metal by reducing the temperature change across the weld zone and by permitting faster joining. However, to reduce the risk of stress cracking, preheating should be used only when the joint is unrestrained. Preheat should be limited to under 250°F to prevent destroying the mechanical properties of the base metal. *Figure 3* shows solidification cracking.

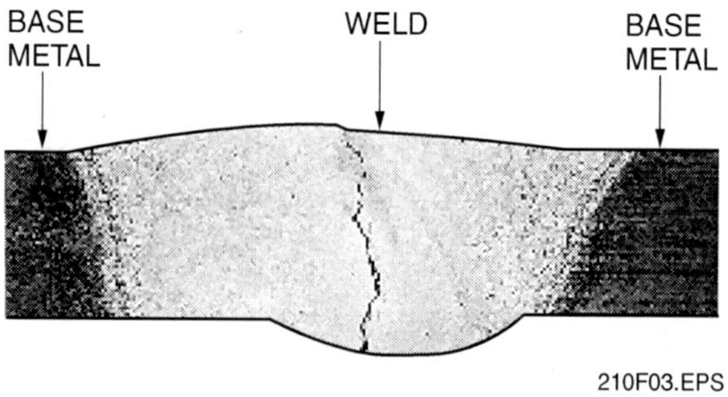

Figure 3 ◆ Solidification cracking.

3.2.4 Liquation Cracking

Liquation cracking occurs in the HAZ at the grain boundaries when low-melting-point films are formed. These films are not strong enough to withstand the contraction stresses created when the weld metal solidifies and cools. Heat-treatable alloys such as the 6xxx, 7xxx, and 8xxx series are more likely than other alloys to crack in this way.

Use of a filler metal with a lower melting temperature than that of the base metal can reduce the risk of cracking: for example, using a 4xxx filler metal to weld a 6xxx-series-alloy base metal. However, 4xxx filler metal should not be used to weld high-magnesium alloys such as 5083, because excessive magnesium silicide may form at the fusion boundary, resulting in the loss of ductility and an increased risk of cracking. *Figure 4* shows liquation cracking.

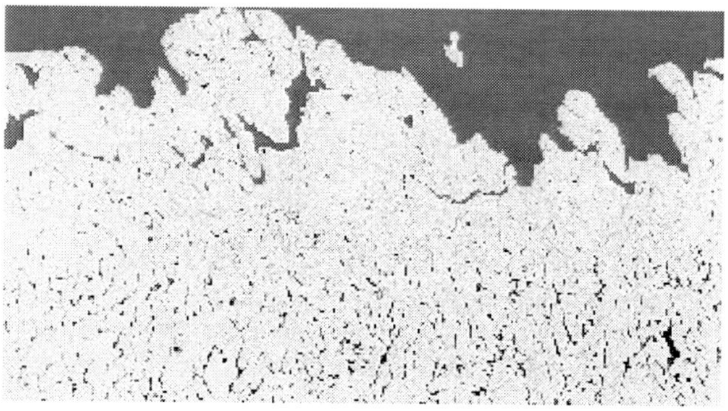

Figure 4 ◆ Liquation cracking.

3.2.5 Poor Weld Bead Profile

Incorrect welding parameter settings or poor welder technique can introduce weld profile imperfections such as lack of fusion, lack of penetration, and undercut. The high thermal conductivity of aluminum and the rapidly solidifying weld pool make aluminum alloys particularly prone to profile imperfections.

4.0.0 ◆ WELDING EQUIPMENT SETUP

Before welding can take place, the work area must be prepared, the welding equipment must be set up, and the metals to be welded must be cleaned and prepared. The following sections explain how to prepare the area and set up the GTAW equipment to weld aluminum plate.

4.1.0 Preparing the Welding Area

To practice welding, a welding table, bench, or stand is needed. The welding surface must be steel, and provisions must be made for mounting practice welding coupons for out-of-position welding.

Follow these steps to set up the area for welding:

Step 1 Check to be sure the area is properly ventilated. Use doors, windows, and fans as necessary.

Step 2 Check the area for fire hazards. Remove any flammable materials before proceeding.

Step 3 Check the location of the nearest fire extinguisher.

Step 4 Position a welding table near the welding machine.

Step 5 Set up flash shields around the welding area.

4.2.0 Selecting Aluminum Filler Metals

The aluminum filler metal typically used for the GTAW process is an uncoated solid-metal filler rod. The metallurgical content of the filler metal used with the GTAW process is chosen to be compatible with the base metal being welded. Filler metals are available for almost all types of aluminum.

4.2.1 Manufacturers' Classifications

All filler rods manufactured in accordance with AWS specifications must have the AWS classification number printed on the shipping container. In addition, filler rod manufacturers generally print their own unique classification name (or number) on the shipping container. Manufacturers may also make more than one filler rod with different characteristics within the same AWS classification.

When there is more than one filler rod in an AWS classification and you are unsure of which one to use, check with your supervisor or quality assurance/quality control personnel.

4.2.2 Filler Rod Sizes

Aluminum filler rods come in standard lengths of 18 and 36 inches. Common sizes (diameters) are 1/16", 3/32", 1/8", 5/32", 3/16", and 1/4". They are commonly packaged in 5-, 10-, 25-, and 50-pound packages.

4.2.3 Filler Metal Selection Considerations

Because there are so many types of filler metals available, the welder must be able to distinguish between them and select the correct filler metal for the job at hand. There are two basic factors to be considered when selecting aluminum filler metals:

- Welding Procedure Specification (WPS)
- Base metal composition

If there is a WPS or site quality standard, the filler rod type will be specified. It will be given as an AWS classification and/or manufacturer's standard and will sometimes include a manufacturer's name. If there is a WPS, the filler metal specified must be used.

> **CAUTION**
>
> Filler metal specifications in the WPS or site quality standards must be followed. Failure to use the specified filler metal will result in a weld quality issue.

As a general rule, aluminum filler rods for welding aluminum with GTAW are selected to be compatible with the chemical composition of the base metal. Because there are so many aluminum base metals and filler metal types, it is important to refer to a filler metal manufacturer's table or chart.

Table 5 shows a filler metal guide for general-purpose welding of aluminum.

In many cases, only a few filler alloys are used to weld most aluminum alloys. Filler alloys such as 5356 or 4043 can be used for most aluminum fusion-welding applications. Always check the WPS or job standards to determine the filler metal needed for the job.

Manufacturers' Data for Filler Metal

Most filler metal manufacturers publish paperback welding guides that include information about their filler materials. These booklets can be obtained through your distributor free or for a small charge.

Filler Rod Size Selection

A guideline for GTAW filler rod size selection is that the filler rod diameter should be close to the diameter of the tungsten electrode.

Table 5 Filler Metal Guide for General-Purpose Welding of Aluminum

BASE METAL	319, 333 354, 355 C355	13, 43, 344 356, A356 A357, 359	214, A214 B214, F214	7039 A612, C612 D612, 7005 k	6070	6061, 6063 6101, 6151 6201, 6951	5456	5454	5154 5254 a	5086	5083	5052 5652 a	5005 5050	3004 Alc. 3004	2219	2014 2024	1100 3003 Alc. 3003	1060 EC
1060, EC	ER4145 c,i	ER4043 i,f	ER4043 e,i	ER4043 i	ER4043 i	ER4043 i	ER5356 c	ER4043 e,i	ER4043 e,i	ER5356 c	ER5356 c	ER4043 i	ER1100 c	ER4043	ER4145	ER4145	ER1100 c	ER1260 c,j
1100, 3003 Alclad 3003	ER4145 c,i	ER4043 i,f	ER4043 e,i	ER4043 i	ER4043 i	ER4043 i	ER5356 c	ER4043 e,i	ER4043 e,i	ER5356 c	ER5356 c	ER4043 e,i	ER4043 e	ER4043 e	ER4145	ER4145	ER1100 c	
2014, 2024	ER4145 g	ER4145			ER4145	ER4145									ER4145 g	ER4145 g		
2219	ER4145 g,c,i	ER4145 c,i	ER4043 i	ER4043 i	ER4043 f,i	ER4043 f,i	ER4043	ER4043 i	ER4043 i	ER4043	ER4043	ER4043 i	ER4043	ER4043	ER2319 c,f,i			
3004 Alclad 3004	ER4043 i	ER4043 i	ER5654 b	ER5356 e	ER4043 e	ER4043 b	ER5356 e	ER5654 b	ER5654 b	ER5356 e	ER5356 e	ER4043 e,i	ER4043 e	ER4043 e				
5005, 5050	ER4043 i	ER4043 i	ER5654 b	ER5356 e	ER4043 e	ER4043 b	ER5356 e	ER5654 b	ER5654 b	ER5356 e	ER5356 e	ER4043 e,i	ER4043 d,e					
5052, 5652 a	ER4043 i	ER4043 b,i	ER5654 b	ER5356 e,h	ER5356 b,c	ER5356 b,c	ER5356 b	ER5654 b	ER5654 b	ER5356 e	ER5356 e	ER5654 a,b,e						
5083		ER5356 c,e,i	ER5356 e	ER5183 e,h	ER5356 e	ER5356 e	ER5183 e	ER5356 e	ER5356 e	ER5356 e	ER5183 e							
5086		ER5356 c,e,i	ER5356 e	ER5356 e,h	ER5356 e	ER5356 e	ER5356 e	ER5356 b	ER5356 b	ER5356 e								
5154, 5254 a		ER4043 b,i	ER5654 b	ER5356 b,h	ER5356 b,c	ER5356 b,c	ER5356 b	ER5654 b	ER5654 a,b									
5454	ER4043 i	ER4043 b,i	ER5654 b	ER5356 b,h	ER5356 b,c	ER5356 b,c	ER5356 b	ER5554 c,e										
5456		ER5356 c,e,i	ER5356 e	ER5556 e,h	ER5356 e	ER5356 e	ER5556 e											
6061, 6063, 6101 6201, 6151, 6951	ER4145 c,i	ER4043 b,i	ER5356 b,c	ER5356b,c,h,i	ER4043 b,i	ER4043 b,i												
6070	ER4145 c,i	ER4043 e,i	ER5356 c,e	ER5356c,e,h,i	ER4043 e,i													
7039 A612, C612 D612, 7005 k	ER4043 i	ER4043 b,h,i	ER5356 b,h	ER5039 e														
214, A214 B214, F214		ER4043 b,i	ER5654 b,d															
13, 43, 344 356, A356 A357, 359	ER4145 c,i	ER4043 d,i																
319, 333 354, 355, C355	ER4145 d,c,i																	

NOTES:

1. Service conditions such as immersion in fresh or salt water, exposure to specific chemicals, or a sustained high temperature (over 150°F) may limit the choice of filler metals.
2. Recommendations in this table apply to gas shielded arc welding processes. For gas welding, only R1100, R1260, and R4043 filler metals are ordinarily used.
3. Filler metals designated with ER prefix are listed in AWS specification A5.10.
   a. Base metal alloys 5652 and 5254 are used for hydrogen peroxide service. ER5654 filler metal is used for welding both alloys for low-temperature service (150°F and below).
   b. ER5183, ER5356, ER5554, ER5556, and ER5654 may be used. In some cases they provide (1) improved color match after anodizing treatment, (2) highest weld ductility, and (3) higher weld strength. ER5554 is suitable for elevated temperature service.
   c. ER4043 may be used for some applications.
   d. Filler metal with the same analysis as the base metal is sometimes used.
   e. ER5183, ER5356, or ER5556 may be used.
   f. ER4145 may be used for some applications.
   g. ER2319 may be used for some applications.
   h. ER5039 may be used for some applications.
   i. ER4047 may be used for some applications.
   j. ER1100 may be used for some applications.
   k. This refers to 7005 extrusions only.
4. Where no filler metal is listed, the base metal combination is not recommended for welding.

210T05.EPS

High-quality aluminum welds can only be obtained if the filler metal used is both clean and of a high quality. Unclean filler metal transfers contaminants into the weld pool. Filler metal needs to be used as soon as possible after being removed from the shipping or storage packaging or stored in air-tight storage containers to prevent continued oxidation of the rod.

For the welding exercises in this module, 3⁄32", 1⁄8", 5⁄32", and/or 3⁄16" aluminum filler material classifications as shown in *Table 5,* or their equivalents, should be used to weld the aluminum plate coupons. Remove only a small number of filler rods at a time. Keep the rest in their package to keep them clean. Before using a filler metal, check it for contamination, such as corrosion, dirt, oil, grease, or burned ends. All these forms of contamination will cause weld defects. Use denatured alcohol or acetone and a clean, oil-free rag to clean the filler metal. Use emery cloth or stainless steel wool to remove oxides. Snip off any burned ends after the filler metal has been cleaned. If the filler metal cannot be cleaned, do not use it. The filler metal should be cleaned using the same method used for cleaning the base metal.

4.3.0 Preparing Welding Coupons

Because aluminum tends to be more easily contaminated, coupon preparations become even more important. Aluminum coupons for V-groove welds must be cut to the desired lengths, cleaned, and beveled to the desired angles before they are positioned for welding.

4.3.1 Cutting and Cleaning Aluminum Plate Coupons

If possible, welding coupons should be cut from 3⁄8" thick aluminum plate. If this size is not readily available, aluminum plate 1⁄4" to 3⁄4" thick can be used. Use an approved chemical such as acetone to remove oil and grease from the coupons. Use a grinder with a stainless steel buffer wheel as necessary to remove all contaminants and oxidation. Use only light pressure.

WARNING!
To prevent injury when using cleaning solvents, always comply with the guidelines of the MSDS for the solvent(s) being used.

4.3.2 Preparing Practice Weld Coupons

The coupons must be shaped to allow the following welds:

- *Stringer or weave beads* – The plate coupons can be any size or shape that can be easily handled.
- *Running beads* – The coupons can be any size or shape that can be easily handled.
- *Overlapping beads* – The coupons can be any size or shape that can be easily handled.
- *Fillet welds* – Cut the metal into 4" × 5" rectangles for the base and 3" × 5" rectangles for the web. *Figure 5* shows the weld coupons for fillet welding.
- *V-groove welds with backing (joint H of Figure 6)* – As shown in *Figure 7,* cut the metal into 3" × 5" rectangles with one (or both) of the 5" lengths beveled at 30°. Grind a 3⁄16" root face on the bevel as directed by your instructor. Cut backing strips at least 1½" wide and 6" long from the same base metal as the beveled pieces. Follow these steps to prepare each V-groove weld coupon with backing.

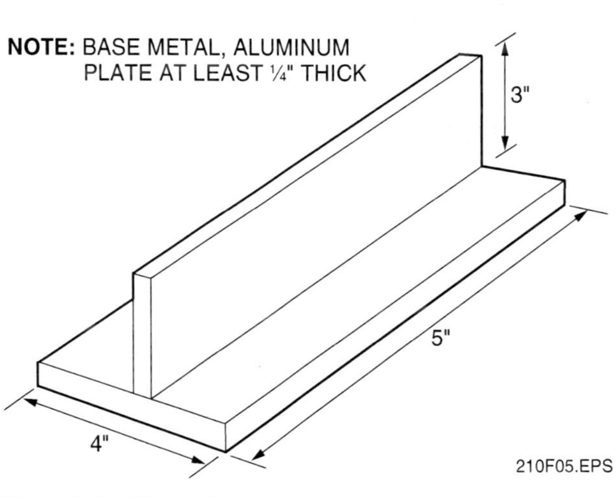

Figure 5 ◆ Fillet weld coupons.

Hot Tip

Conserve Aluminum Practice Coupons

Aluminum for practice welding is expensive and difficult to obtain. Every effort should be made to conserve the material that is available. Reuse welding coupons until all surfaces have been welded. Weld on both sides of the joint and then cut the welding coupon apart and reuse the pieces. Use material that cannot be cut into welding coupons to practice running beads.

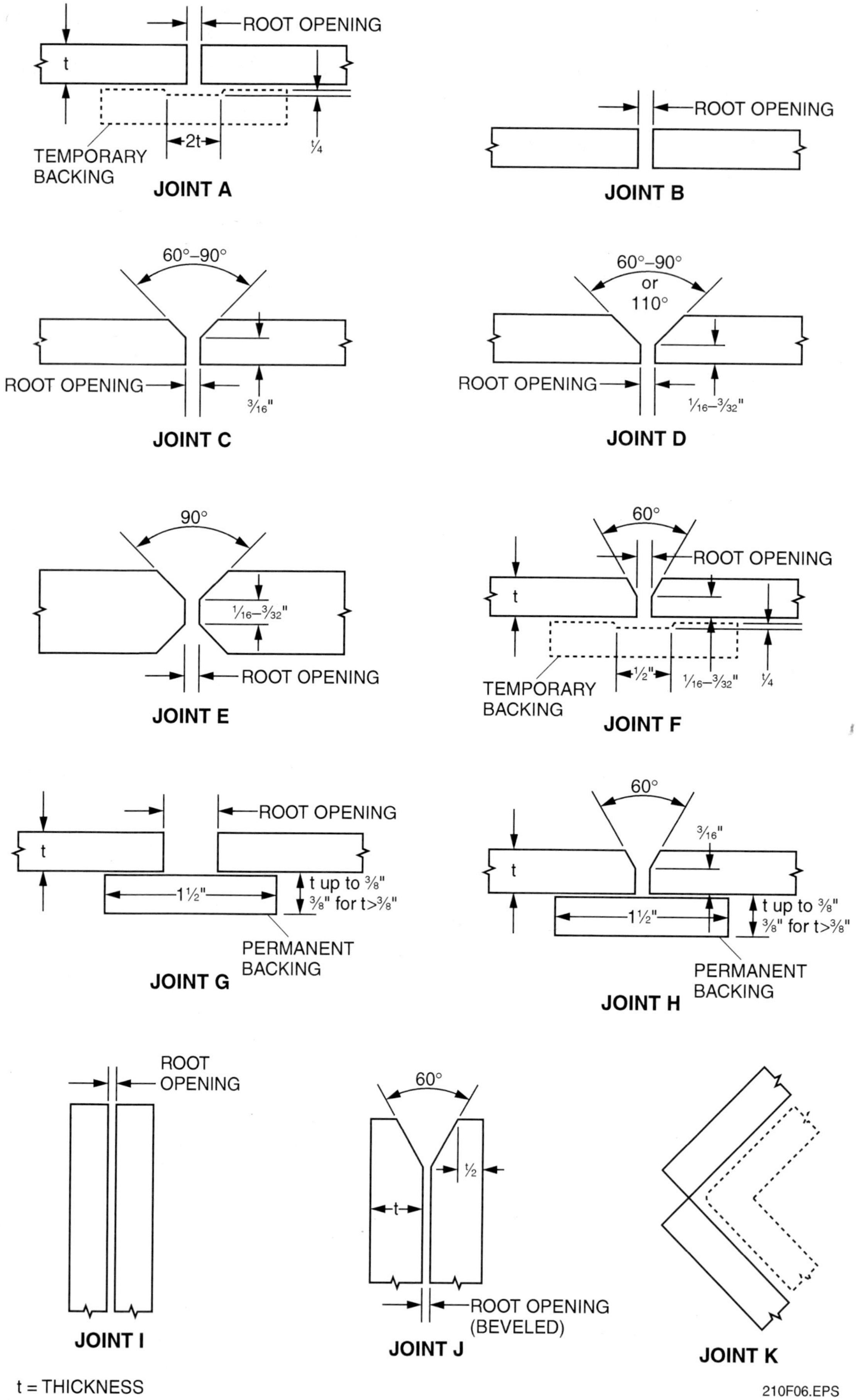

Figure 6 ◆ Joint preparations and geometries.

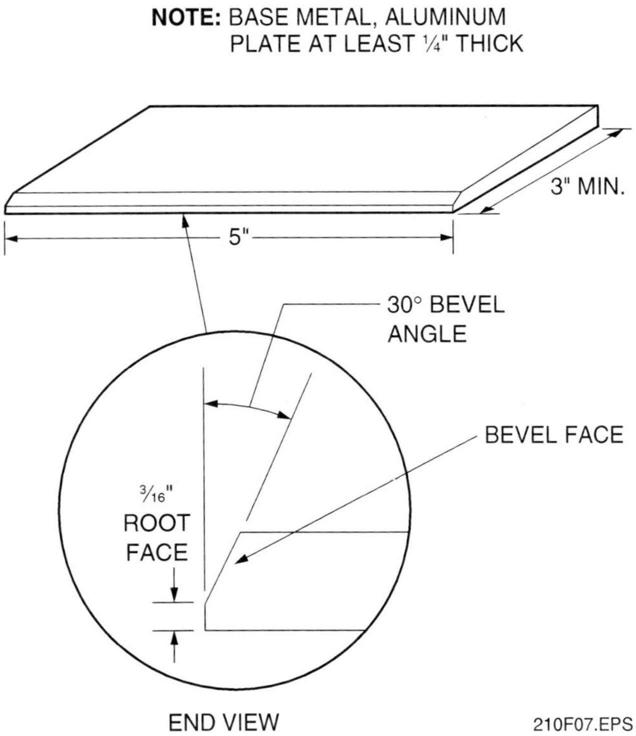

Figure 7 ◆ Preparing the weld coupon.

> **NOTE**
> While this module uses joint H shown in *Figure 6* for practice purposes, the other configurations could also be used.

> **NOTE**
> If the bevel has been cut thermally, machine the bevel back ⅛" before machining the root face.

Step 1 Check the bevel. The bevel angle should be approximately 30° and the root face should be ³⁄₁₆".

Step 2 Center the beveled strips on the backing strip with a ⅜" root opening (*Table 6*) and tack-weld them in place. If desired, use a piece of metal or rod with the required thickness as a temporary spacer. Place the tack welds on the back side of the joint in the lap formed by the backing strip and the beveled plate. Use three to four ½" tack welds on each beveled plate. Be sure the backing strip is tight against the beveled plates. *Figure 8* shows the V-groove with metal backing weld coupon.

> **NOTE**
> Make sure that the surface of the backing strip under the root opening has been cleaned of all oxide coating before tacking the strip in place.

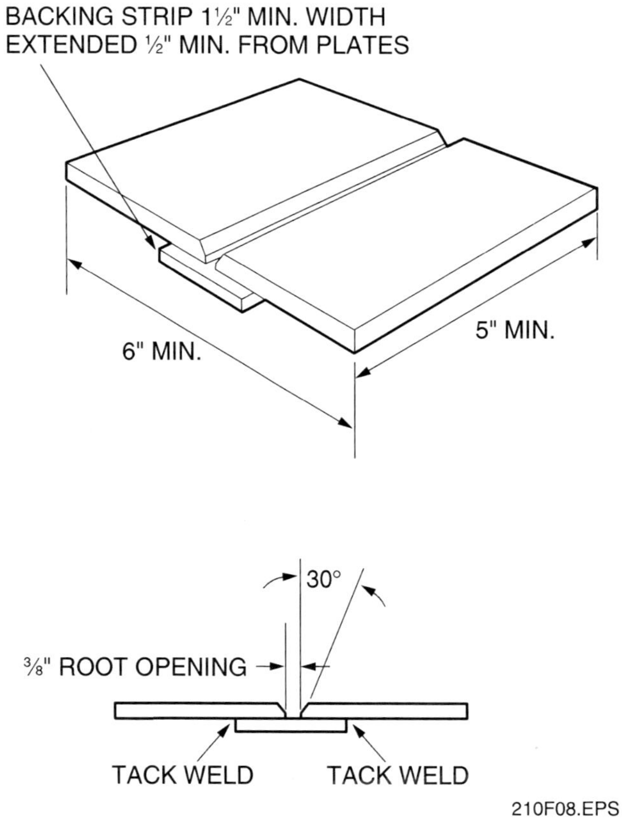

Figure 8 ◆ V-groove with metal backing weld coupon.

When welding in the horizontal position, an alternate joint preparation can be used. The alternate joint has one plate beveled at approximately 45° and the other plate at approximately 15°. The plates are positioned with the 45° beveled plate above the 15° plate and with a ⅜" root opening.

> **Hot Tip — Backing Strip**
>
> When the backing strip is ½" to 1" longer at each end of the weld groove, it allows the welder to start and stop the bead outside the weld groove.

Figure 9 shows the alternate horizontal weld coupons (degrees shown are approximate).

> **NOTE**
>
> Check with your instructor on whether to use the standard V-groove preparation or the alternate preparation for horizontal weld coupons.

4.3.3 Positioning Aluminum Plate Coupons for Welding

Different welding techniques are needed for different welding situations. This section explains both the different positions in which the aluminum plate coupons can be placed and the welding requirements for each respective position. GTAW welding of aluminum plate requires specific preparation and shapes. Refer to *Figure 6* for different joint preparations and geometries.

The specifications of each weld are determined by a variety of factors. The following tables provide the specifications for groove welds, fillet and lap welds, along with edge and corner welds. *Table 6* shows the specifications for making GTAW groove welds on aluminum plate without backing. *Table 7* shows the specifications for making GTAW fillet and lap welds on aluminum plate. *Table 8* shows the specifications for making GTAW edge and corner welds on aluminum plate.

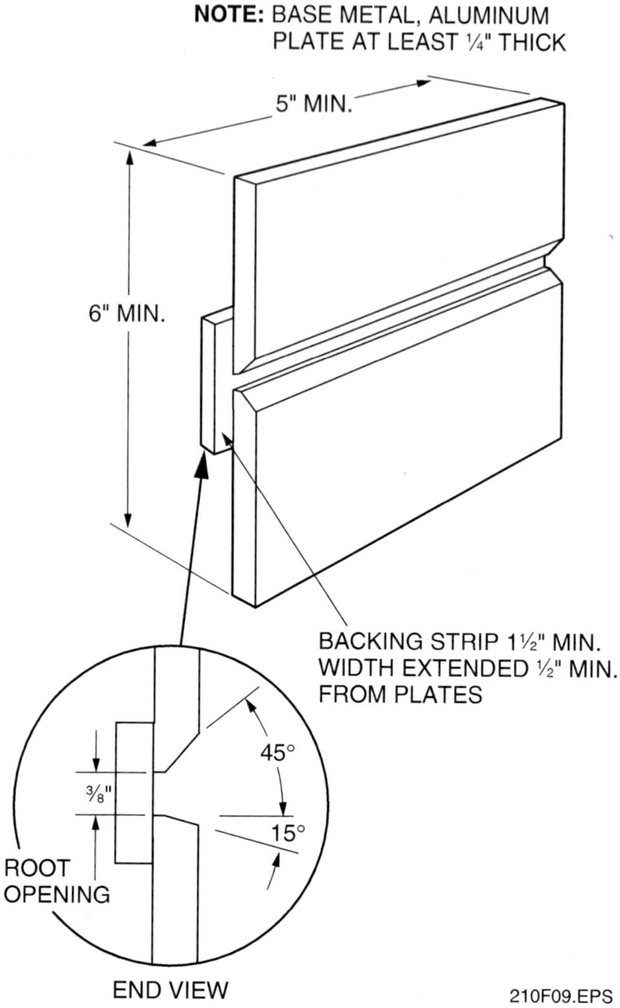

Figure 9 ◆ Alternate horizontal weld coupons.

> **Hot Tip — Always Use the Correct Gas**
>
> Accidental use of a GMAW gas mix can cause rapid destruction of the tungsten and contamination of the weld.

4.4.0 Selecting Shielding Gas

Shielding gas is selected according to the material being welded. Argon is the most commonly used shielding gas. It can be used for welding a wide range of materials, including aluminum, titanium, steel, and stainless steel.

Helium and **helium/argon mixtures** raise the temperature of the arc. This promotes higher welding speeds and deeper weld penetration. Disadvantages of using helium or a helium/argon mixture are the high cost of this gas and the difficulty in starting the arc.

Impurities or mixtures with other gases such as carbon dioxide will contaminate the weld and change the color of the aluminum, usually to black.

Table 6 Specifications for GTAW Groove Welds on Aluminum Plate without Backing

ALUMINUM THICKNESS (inch)	WELD POSITION[1]	EDGE PREP[2] (joint)	ROOT OPENING[3] (inch)	PREHEAT[4] (°F)	WELD PASSES	FILLER DIAMETER (inch)	TUNGSTEN ELECTRODE DIAMETER (inch)	GAS CUP INSIDE DIAMETER (inch)	ARGON (cfh)[5]	AC (amps)[5]	ARC TRAVEL SPEED (ipm)[5]	APPROX. FILLER ROD CONSUMP. (lb./100 ft.)[5]
1/16	F, V, H	A or B	1/16	None	1	3/32	1/16–3/32	3/8	20	70–100	8–10	0.5
	O	A or B	1/16	None	1	3/32	1/16	3/8	25	60–75	8–10	0.5
3/32	F	A or B	3/32	None	1	1/8	3/32–1/8	3/8	20	95–115	8–10	1
	V, H	A or B	3/32	None	1	3/32–1/8	3/32	3/8	20	85–110	8–10	1
	O	A or B	3/32	None	1	3/32–1/8	3/32–1/8	3/8	25	90–110	8–10	1
1/8	F	A or B	1/8	None	1–2	1/8–5/32	1/8	7/16	20	125–150	10–12	2
	V, H	A or B	3/32	None	1–2	1/8	1/8	7/16	20	110–140	10	2
	O	A or B	3/32	None	1–2	1/8–5/32	1/8	7/16	25	115–140	10–12	2
3/16	F	D-60°	1/8	None	2	5/32–3/16	5/32–3/16	7/16–1/2	25	170–190	10–12	4.5
	V	D-60°	3/32	None	2	5/32	5/32	7/16	25	160–175	10–12	4.5
	H	D-90°	3/32	None	2	5/32	5/32	7/16	25	155–170	10–12	5
	O	D-110°	3/32	None	2	5/32	5/32	7/16	30	165–180	10–12	6
1/4	F	D-60°	1/8	None	2	3/16	3/16–1/4	1/2	30	220–275	8–10	8
	V	D-60°	3/32	None	2	3/16	3/16	1/2	30	200–240	8–10	8
	H	D-90°	3/32	None	2–3	5/32–3/16	5/32–3/16	1/2	30	190–225	8–10	9
	O	D-110°	3/32	None	2	3/16	3/16	1/2	35	210–250	8–10	10
3/8	F	D-60°	1/8	Optional	2	3/16–1/4	1/4	5/8	35	315–375	8–10	15.5
	F	E	3/32	up to	2	3/16–1/4	1/4	5/8	35	340–380	8–10	14
	V	D-60°	3/32	350°F	3	3/16	3/16–1/4	5/8	35	260–300	8–10	19
	V, H, O	E	3/32	maximum	2	3/16	3/16–1/4	5/8	35	240–300	8–10	17
	H	D-90°	3/32		3	3/16	3/16–1/4	5/8	35	240–300	8–10	22
	O	D-110°	3/32		3	3/16	3/16–1/4	5/8	40	260–300	8–10	32

1 F = flat; V = vertical; H = horizontal; O = overhead
2 Refer to Figure 6 for edge preparation and joint specifications.
3 Always check the WPS or site-specific specifications.
4 Preheating at excessive temperatures or for extended periods of time will reduce weld strength. This is particularly true for alloys in heat-treated tempers.
5 chf = cubic feet per hour; amps = amperes; ipm = inches per minutes; lb./100 ft. = pounds per hundred feet

210T06.EPS

Table 7 Specifications for Making GTAW Fillet and Lap Welds on Aluminum Plate

ALUMINUM THICKNESS (inch)	WELD POSITION[1]	PREHEAT[2] (°F)	WELD PASSES	FILLER DIAMETER (inch)	PURE TUNGSTEN ELECTRODE DIAMETER (inch)	GAS CUP INSIDE DIAMETER (inch)	ARGON FLOW (cfh)	AC (amps)	ARC TRAVEL SPEED (ipm)	APPROX. FILLER ROD CONSUMP. (lb./100 ft.)
1/16	F, V, H	None	1	3/32	1/16–3/32	3/8	16	70–100	8–10	0.5
	O	None	1	3/32	1/16–3/32	3/8	20	65–90	8–10	0.5
3/32	F	None	1	3/32–1/8	1/8–5/32	3/8	18	110–145	8–10	1
	V, H	None	1	3/32	3/32–1/8	3/8	18	90–125	8–10	1
	O	None	1	3/32	3/32–1/8	3/8	20	110–135	8–10	1
1/8	F	None	1	1/8	1/8–5/32	7/16	20	135–175	10–12	2
	V, H	None	1	1/8	3/32–1/8	3/8	20	115–145	8–10	2.5
	O	None	1	1/8	3/32–1/8	7/16	25	125–155	8–10	2
3/16	F	None	1	5/32	5/32–3/16	1/2	25	190–245	8–10	4.5
	V, H	None	1	5/32	5/32–3/16	1/2	25	175–210	8–10	5.5
	O	None	1	5/32	5/32–3/16	1/2	30	185–225	8–10	4.5
1/4	F	None	1	3/16	3/16–1/4	1/2	30	240–295	8–10	7
	V, H	None	1	3/16	3/16	1/2	30	220–265	8–10	9
	O	None	1	3/16	3/16	1/2	35	230–275	8–10	7
3/8	F	Optional	2	3/16	1/4	5/8	35	325–375	8–10	17
	V	up to	2	3/16	3/16–1/4	5/8	35	280–315	8–10	20
	H	350°F	3	3/16	3/16–1/4	5/8	35	270–300	8–10	20
	O	maximum	3	3/16	3/16–1/4	5/8	40	290–335	8–10	17

1 F = flat; V = vertical; H = horizontal; O = overhead
2 Preheating at excessive temperatures or for extended periods of time will reduce weld strength. This is particularly true for alloys in heat-treated tempers.

210T07.EPS

Table 8 Specifications for Making GTAW Edge and Corner Welds on Aluminum Plate

ALUMINUM THICKNESS (inch)	EDGE PREP	WELD PASSES	FILLER ROD DIAMETER (inch)	TUNGSTEN ELECTRODE DIAMETER (inch)	GAS CUP INSIDE DIAMETER (inch)	ARGON FLOW (cfh)	AC[1,2] (amps)	ARC TRAVEL SPEED (ipm)	APPROX. FILLER ROD CONSUMP. (lb./100 ft.)
1/16	I,K	1	3/32	1/16	3/8	20	60–85	10–16	0.5
3/32	I,K	1	1/8	3/32	3/8	20	90–120	10–16	1
1/8	I,K	1	1/8–5/32	1/8	3/8	20	115–150	10–16	2
3/16	J,K	1	5/32	5/32	7/16	25	160–220	10–16	4.5
1/4	J,K	2	3/16	3/16	1/2	30	200–250	8–12	7

1 Higher currents and welding speeds can be employed if a temporary backing is used for corner joints.
2 Use the low side of the current range for horizontal and vertical welds.

210T08.EPS

4.5.0 Welding Equipment

Identify a proper welding machine for GTAW use and follow these steps to set it up for welding.

Step 1 Verify that the welding machine can be used for GTAW with or without internal gas shielding control. If desired, identify an optional cooling unit.

Step 2 Identify an air-cooled or water-cooled GTAW torch. Make sure that it is compatible with the welding machine and any optional cooling unit.

Step 3 Check to be sure that the welding machine is properly grounded through the primary current receptacle.

Step 4 Verify the location of the primary disconnect.

Step 5 Configure the welding machine for aluminum GTAW welding (*Figure 10*) as directed by your instructor. Configure the torch polarity to AC and equip the torch with a properly prepared 3/32" or 1/8" EWP or EWZR tungsten electrode.

Step 6 Connect the proper argon or argon/helium shielding gas for the application as directed by your instructor, filler metal manufacturer, WPS, or site quality standards.

Step 7 Connect the clamp of the workpiece lead to the workpiece.

Step 8 Turn on the welding machine and purge the torch as directed by the manufacturer's instructions.

Step 9 Set the initial welding current for the desired GTAW application.

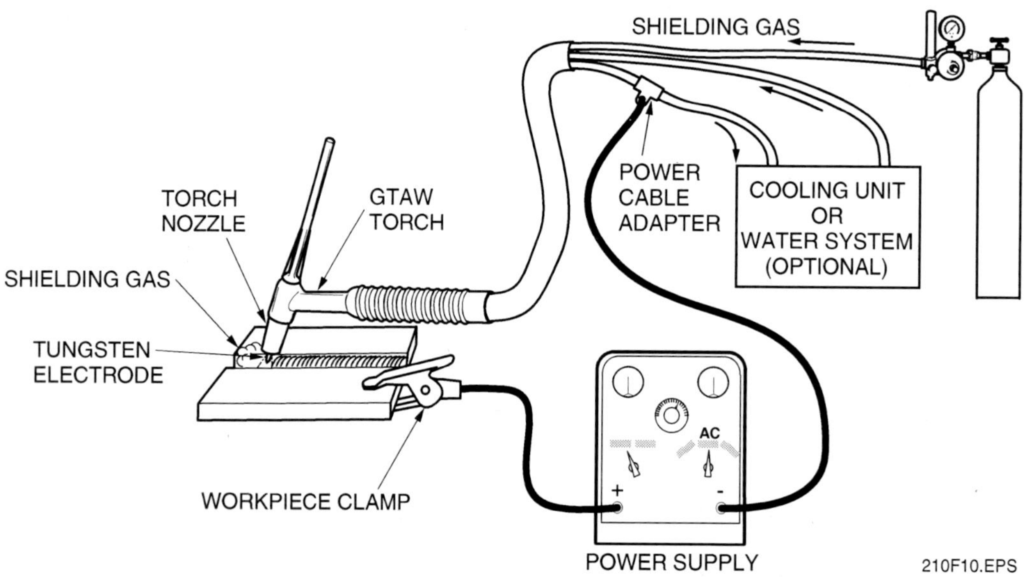

Figure 10 ◆ Configuration diagram of typical GTAW welding equipment.

Hot Tip: *Advanced Inverter Power Sources*

An advanced inverter power source, like the one shown here, can make welding aluminum much easier. With features such as HF start only, current pulsing, AC frequency adjustment, and weld current sequencing (start/stop ramping), these machines provide much more control over some of the variables that affect the quality of a weld. These types of machines also allow the use of tapered and blunted 2% thoriated or ceriated tungsten electrodes for aluminum welding in an AC mode as well as a DC mode. With a tapered electrode, an operator can place a ⅛" bead on a fillet weld made from ⅛" plate. Using a normal balled electrode makes a much larger bead that has to be ground down to final size.

210SA01.EPS

5.0.0 ◆ GTAW TECHNIQUES

The weld bead characteristics and quality of GTAW are affected by several factors resulting from the way in which the welder handles the torch. These factors include:

- Torch travel speed and arc length
- Torch angles
- Torch and filler-metal handling techniques

5.1.0 Torch Travel Speed and Arc Length

Torch travel speed and arc length affect the GTAW weld puddle and the penetration of the weld. A slow travel speed allows more heat to concentrate and forms a larger, more deeply penetrating puddle. Faster travel speeds prevent heat buildup and form smaller and shallower puddles. Arc length is the major control for bead width. As the torch is raised, arc length increases, voltage increases, and the bead width increases.

5.2.0 Torch Angles

There are two basic torch angles that must be controlled when performing GTAW: the work angle and the travel angle. The definition of these angles is the same as for all other methods of welding.

5.2.1 Work Angle

The torch work angle (*Figure 11*) is defined as an angle less than 90° between a line perpendicular to the major workpiece surface at the point of electrode contact and a plane determined by the electrode axis and the weld axis. For a T-joint or corner joint, the line is perpendicular to the nonbutting member. For pipe, the plane is determined by the electrode axis and a line tangent to the pipe surface at the same point.

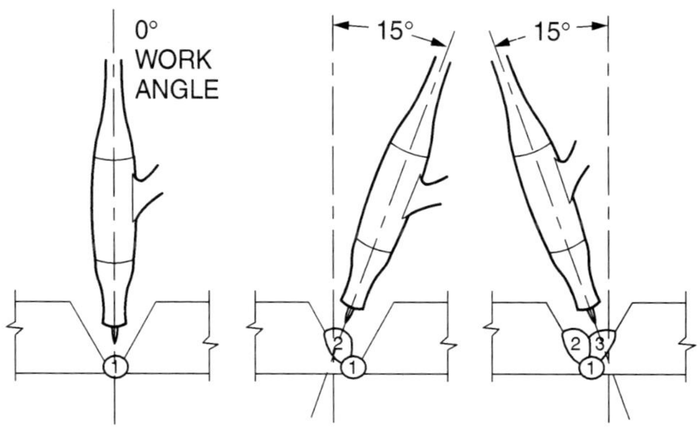

210F11.EPS

Figure 11 ◆ Typical torch work angles.

5.2.2 Travel Angle

The torch travel angle (*Figure 12*) is defined as an angle less than 90° between the electrode axis and a line perpendicular to the weld axis at the point of electrode contact in a plane determined by the electrode axis and the weld axis. For pipe, the plane is determined by the electrode axis and a line tangent

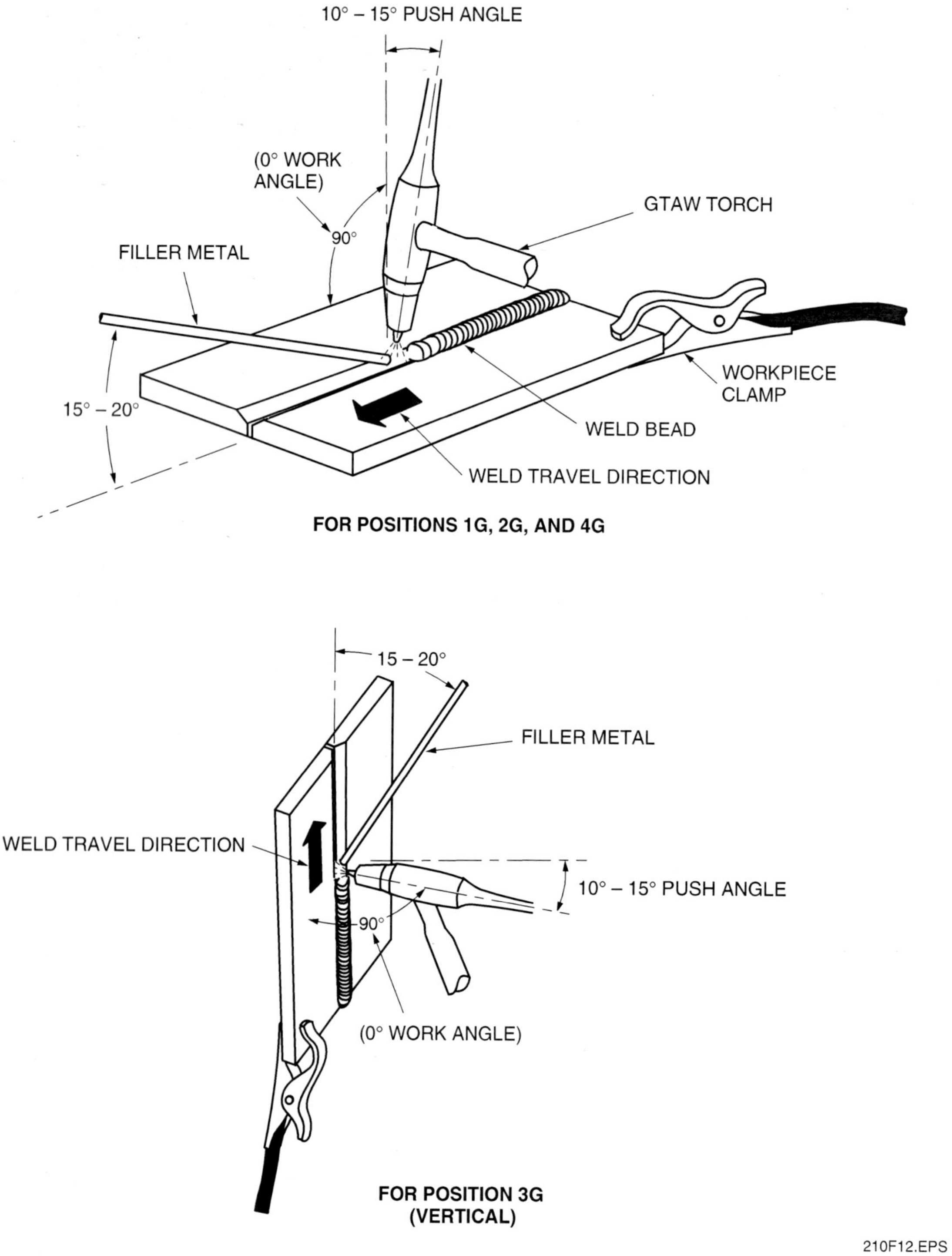

Figure 12 ◆ Torch travel angles.

to the pipe surface at the same point. A push angle is used for GTAW. A push angle is the angle created when the torch is tilted back so that the electrode tip precedes the torch in the direction of the weld. In this position, the electrode tip and shielding gas are being directed ahead of the weld bead to provide a cleaning action of the positive portion of the AC voltage cycle. Push angles of 10° to 15° are normally used for GTAW of aluminum plate.

5.3.0 Torch and Filler Metal-Handling Techniques

The two basic handling techniques used to perform GTAW are known as freehand and walking-the-cup. Try both techniques and use the one that gives the best results. Both techniques are explained in the following sections.

5.3.1 Freehand Technique

In the freehand technique, the preferred method for aluminum, the torch electrode tip is held just above the weld puddle or base metal. The torch is supported by the welder's hand like a pencil. The hand is usually steadied by resting some part of it on or against the base metal to maintain the proper arc length (electrode tip to puddle distance). If required, the welder can move the torch tip in a small circular motion within the molten puddle to maintain the puddle size and advance the puddle. Filler metal is added as it is needed.

The GTAW filler metal is held in the hand that is not occupied by the torch. The filler rod is held at an angle of about 15° to 20° above the base metal surface and in line with the weld. The tip of the filler rod is always kept within the shielding gas envelope to protect the filler rod from atmospheric contamination and to keep it preheated.

Jab the filler metal into the weld puddle, using extreme care not to touch the tungsten electrode with the end of the filler metal. If the tungsten electrode touches the filler metal or weld puddle, it will become contaminated with filler metal. The electrode must then be removed, the contaminated portion broken off, and the end reformed before welding can be continued. A technique for preventing contamination of the electrode by the filler metal is to move the electrode to the back edge of the weld puddle as the filler metal is jabbed into the weld puddle. *Figure 13* shows the freehand technique.

5.3.2 Walking-the-Cup Technique

In the walking-the-cup technique, the welder rests the edge of the torch nozzle (cup) against the base metal or groove edges to steady the torch and maintain a constant arc length (distance from the electrode tip to the weld pool). The torch is rocked from side to side on the edge of the cup as it is advanced, to maintain the puddle size and heat both sides of the groove. Add the filler metal in the same manner as with the freehand technique, using care not to contaminate the electrode.

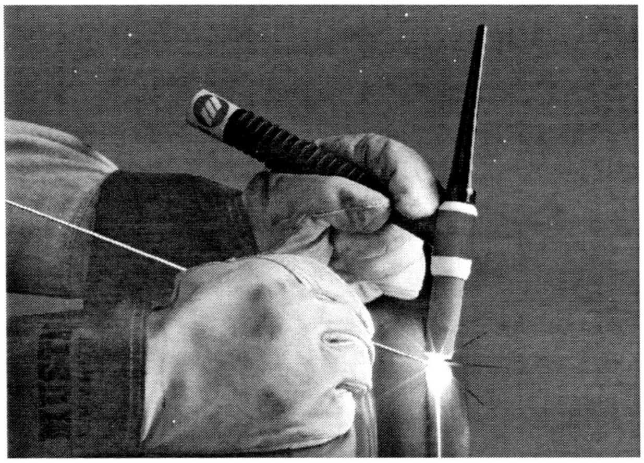

210F13.EPS

Figure 13 ◆ Freehand technique.

6.0.0 ◆ BEAD TYPES

The two basic bead types used with GTAW are weave beads and stringer beads. Stringer and weave beads are used to make welds. Overlapping stringer and weave beads are used to build pads.

6.1.0 Practicing Weave Beads

Weave beads are made by working the weld puddle back and forth across the axis of the weld. This produces a weld bead much wider than the puddle width. The width of a weave bead is determined by the amount of cross motion. Weave beads put more heat into the base metal. *Figure 14* shows a weave bead.

NOTE
Always check the WPS or site quality standards to determine whether stringer or weave beads should be used.

Hot Tip

Excessive Push Angle

Too great a push angle will tend to draw air from under the back edge of the torch nozzle, where it will mix with the shielding gas stream and contaminate the weld.

Walking-the-Cup Technique on Aluminum

The walking-the-cup technique is extremely difficult when welding aluminum because the cup tends to stick to the freshly welded metal.

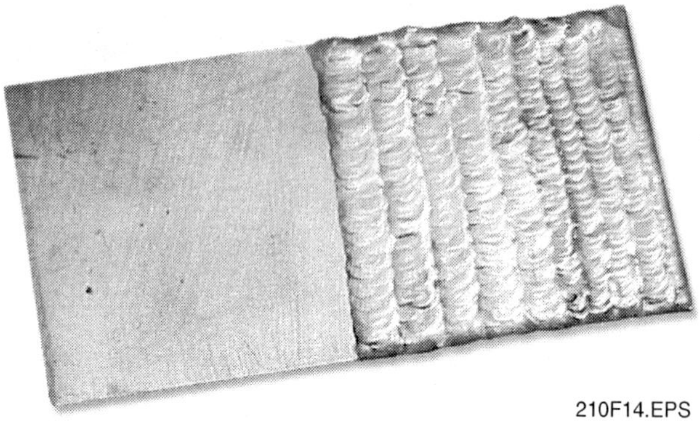

210F14.EPS

Figure 14 ◆ Weave bead.

When making a weave bead, care must be used at the toes to ensure proper tie-in to the base metal. To ensure proper tie-in at the toes, slow down or pause slightly at the edges. The pause at the edges will also flatten out the weld, giving it the proper profile.

Practice running weave beads in the flat position. Experiment with different weave motions and push angles.

Follow these steps to run weave beads:

Step 1 Hold the torch at the desired angle with the electrode tip directly over the point where the weld is to begin.

Step 2 Lower the torch and filler rod to the weld.

Step 3 Hold the arc in place until the weld puddle begins to form. Add filler metal.

Step 4 Slowly advance the arc in a weaving motion while maintaining the torch angle and adding filler metal. *Figure 15* shows the weave motion.

Step 5 Continue to weld until a bead about 2" to 3" long is formed.

Step 6 Stop, and then fill the crater while slowly reducing amperage with a current control to allow the weld puddle to solidify. Turn off the current and hold the filler rod and torch in place until the tungsten electrode and weld cool.

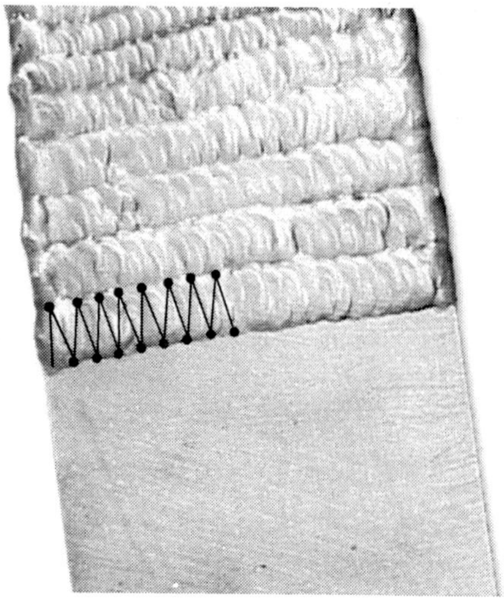

210F15.EPS

Figure 15 ◆ Weave motion.

Step 7 Inspect the bead for the following:
- Straightness of the bead
- Uniform appearance of the bead face
- Smooth, flat transition with complete fusion at the toes of the weld
- No porosity
- No excessive undercut
- Crater filled
- No cracks

Step 8 Continue practicing weave beads until you can make acceptable welds every time.

> **WARNING!**
> Use pliers to handle the hot practice coupons. Wear gloves when placing the practice coupons in water. Steam will rise off the coupons and can burn or scald unprotected hands.

Hot Tip

Weave Bead Width

The width of weave beads is often specified in the welding code or WPS being used at a site. Do not exceed the widths specified for the site.

Cooling Practice Coupons

If the practice coupon gets too hot between passes, cool it in water. Cooling with water is only done on practice coupons. Never cool test coupons or any other weld with water. Cooling with water can cause weld cracks and affect the mechanical properties of the base metal.

6.2.0 Weld Restarts

A restart is the junction where a new weld connects to the bead of a previous weld to continue the weld. Restarts are important because an improperly made restart will create a weld defect. A restart must be made so that it blends smoothly with the previous weld and does not stand out. If possible, avoid restarts by running a bead the full length of the weld joint to be made.

Follow these steps to make a restart:

Step 1 Clean the area of the restart, using approved cleaning method(s).

Step 2 Hold the torch at the proper angle and arc distance, and then restart the arc directly over the center of the crater. The welding codes do not allow arc strikes outside the area that is to be welded.

Step 3 Move the electrode tip in a small circular motion over the crater, and add filler metal when the molten puddle is the same size as the crater.

Step 4 As soon as the puddle fills the crater, advance the puddle slightly, and continue to add filler metal as needed.

Step 5 Inspect the restart. A properly made restart will blend into the bead, making it hard to detect.

Step 6 Continue to practice restarts until they are correct. Use the same techniques for making restarts whenever performing GTAW.

6.3.0 Weld Terminations

A weld termination is made at the end of a weld. Most welding codes require that when terminating a weld, the crater must be filled to the full cross section of the weld to prevent crater cracking. This can be difficult since most terminations are at the edge of a plate where welding heat tends to build up. This makes filling the crater more difficult. Filling the crater is much easier if remote control equipment with a potentiometer is used. The potentiometer allows the welding current to be reduced as the end of the weld is approached.

Follow these steps to make a termination using remote control equipment and a potentiometer:

Step 1 As you approach the end of the weld, slowly back off (reduce) the welding current. The amount to reduce the welding current is determined by the weld puddle width. If the puddle starts to become wider, there is too much heat; if the puddle starts to become narrower, there is not enough heat.

Step 2 Continue to back off the welding current while adding filler metal until the crater is filled.

Step 3 Stop the arc and hold the torch in place until the gas postflow cools the weld metal, filler metal, and tungsten electrode.

Follow these steps to make a weld termination without a potentiometer:

Step 1 As you approach the end of the weld, add filler metal at a faster pace. The filler metal will absorb excess heat.

Step 2 Continue to add filler metal until the crater is filled, and then stop the arc.

Step 3 Hold the torch in place until the gas postflow cools the weld metal, filler metal, and tungsten electrode.

Improper Restart Problems

If the restart has undercut, not enough time was spent in the crater to completely melt it, or not enough filler metal was added. If the restart is higher than the rest of the bead, too much filler metal was added.

Advanced GTAW Machines with Current Ramping

Advanced GTAW machines with programmable current ramping for the start and finish of a weld simplify repetitive weld starts and terminations.

Thin Materials

On thin materials, stop about 1" or 2" short of the edge without terminating the weld. Then, back-weld from the edge to the crater and fill the crater.

> **CAUTION**
>
> Do not remove the torch or filler metal from the torch shielding gas flow until the puddle has solidified and cooled to a temperature at which it will not be affected by the atmosphere. The gas postflow that continues after the welding stops protects the molten metal and the tungsten electrode. If the torch and shielding gas are removed before the weld has solidified, crater porosity or cracks can occur. Also, the filler metal will be contaminated, and the contaminated end will have to be discarded. In addition, the tungsten electrode can oxidize (bluish-black color), which will require reballing or re-pointing the electrode.

Step 4 Inspect the weld termination for the following:

- Crater filled to the full cross section of the weld
- No crater cracks
- No crater porosity
- No excessive undercut

6.4.0 Overlapping Beads

Overlapping beads are made by depositing connective weld beads parallel to one another. Each successive bead overlaps the toe of the previous bead to form an approximately flat surface. This is also called padding. Overlapping beads are used to build up surfaces (padding) and to make multipass welds.

Properly overlapped beads, when viewed from their ends, will form a nearly flat surface. *Figure 16* shows proper and improper overlapping beads.

Follow these steps to make GTAW overlapping stringer beads, using $\frac{3}{32}$" or $\frac{1}{8}$" aluminum filler metal:

Step 1 Mark out a square on a piece of aluminum.

Step 2 Deoxidize the base metal.

Step 3 Weld a stringer bead along one edge.

> **CAUTION**
>
> Between weld beads, always clean the weld with a stainless steel wire brush.

Step 4 Run a second bead along the previous bead. Be sure to overlap the previous bead to obtain a good tie-in and produce a flat profile.

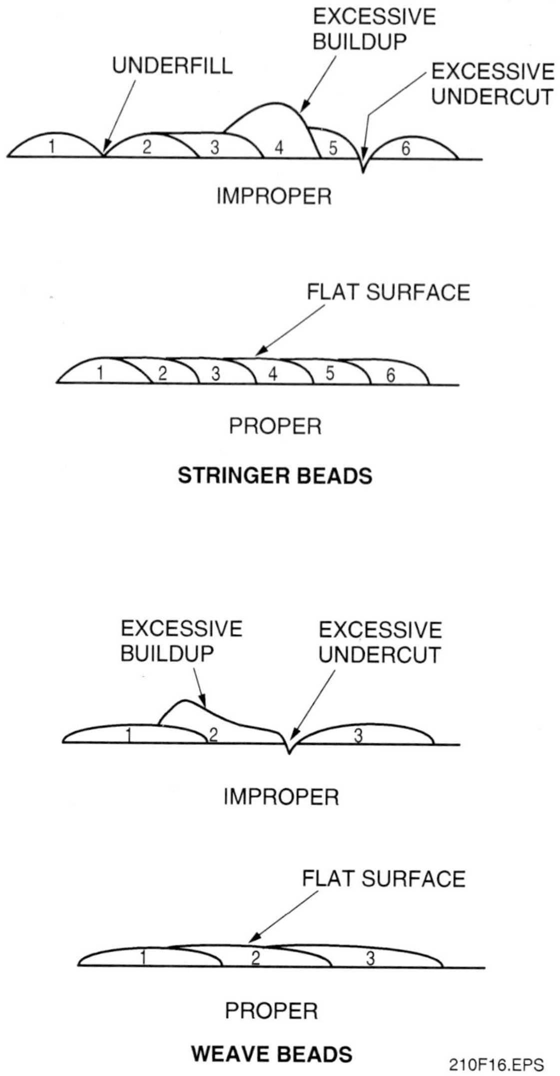

Figure 16 ◆ Proper and improper overlapping beads.

Step 5 Continue running overlapping stringer beads until the face of the metal square is covered.

Step 6 Continue building layers of stringer beads (pads), one on top of the other, until the technique is perfected.

6.5.0 Building a Pad with Stringer or Weave Beads

Continue to weld overlapping beads, using $\frac{3}{32}$" or $\frac{1}{8}$" filler rods. Build a pad with stringer beads, and then build a pad with weave beads, as indicated in *Figure 17*.

7.0.0 ◆ FILLET WELDS

The most common fillet welds are made in lap and T-joints. The weld position is determined by the axis of the weld. The positions for fillet welding are 1F (flat), 2F (horizontal), 3F (vertical), and 4F

Figure 17 ◈ Pad built with stringer beads.

(overhead). In the 1F and 2F positions, the weld axis can be inclined up to 15°. Any weld axis inclination for the other positions varies with the rotational position of the weld face as specified in AWS Standards. *Figure 18* shows the fillet welding positions.

Fillet welds can be concave or convex, depending on the WPS or site quality standards. The welding codes require a fillet weld to have a uniform concave or convex face, although a slightly non-uniform face is acceptable. The convexity of a fillet weld or individual surface bead must not exceed that permitted by the applicable code or standard. A fillet weld with profile defects is unacceptable and must be repaired. *Figure 19* shows acceptable and unacceptable fillet weld profiles.

7.1.0 Practicing Flat (1F) Position Fillet Welds

Practice 1F-position (flat) fillet welds by making multipass (six-pass) convex fillet welds in a T-joint with an appropriate aluminum filler metal. When making flat fillet welds, pay close attention to the torch angle and travel speed. For the first bead, the torch angle is vertical (45 degrees to both plate surfaces). The angle is adjusted for all subsequent beads. Clean each completed bead with a stainless steel wire brush before starting the next bead.

Figure 20 shows the torch angles for 1F-position fillet welds.

Follow these steps to make a flat fillet weld:.

> **NOTE**
> In the following steps, clean all weld beads before beginning the next bead.

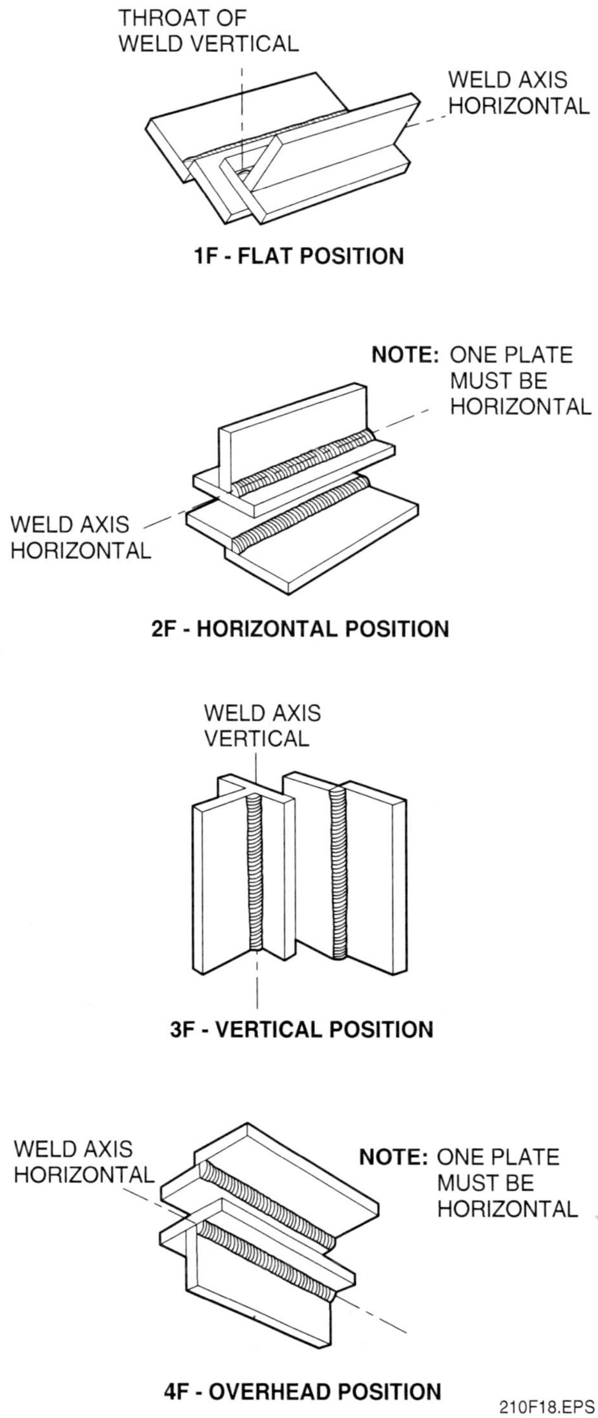

Figure 18 ◈ Fillet welding positions.

Step 1 Tack two plates together to form a T-joint for the fillet weld coupon. Clean the tack welds. *Figure 21* shows the fillet weld coupon.

Step 2 Clamp or tack-weld the coupon in the 1F position (flat).

Step 3 Run the first bead along the root of the joint, using a work angle of 45 degrees with a 10° to 15° push angle.

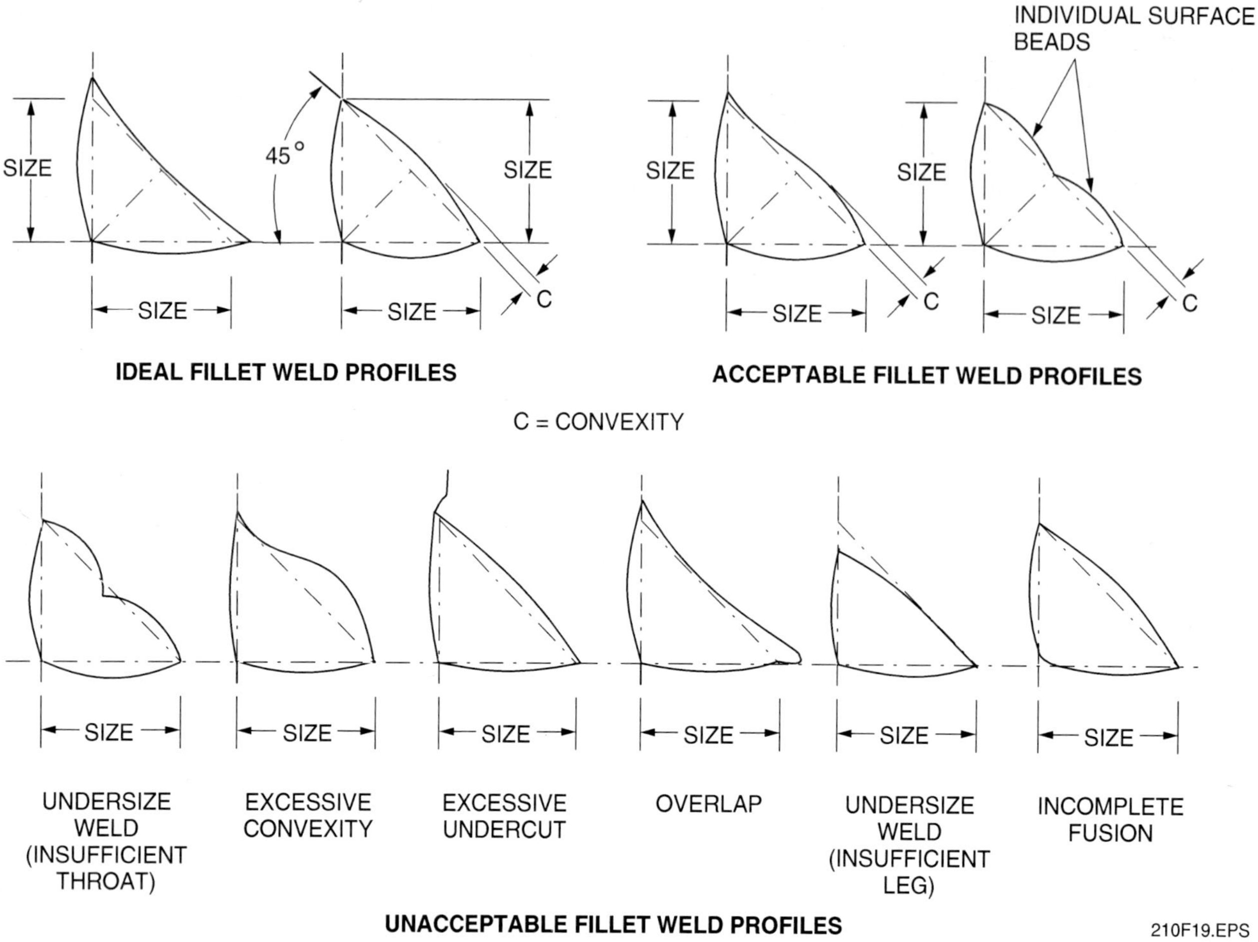

Figure 19 ◆ Acceptable and unacceptable fillet weld profiles.

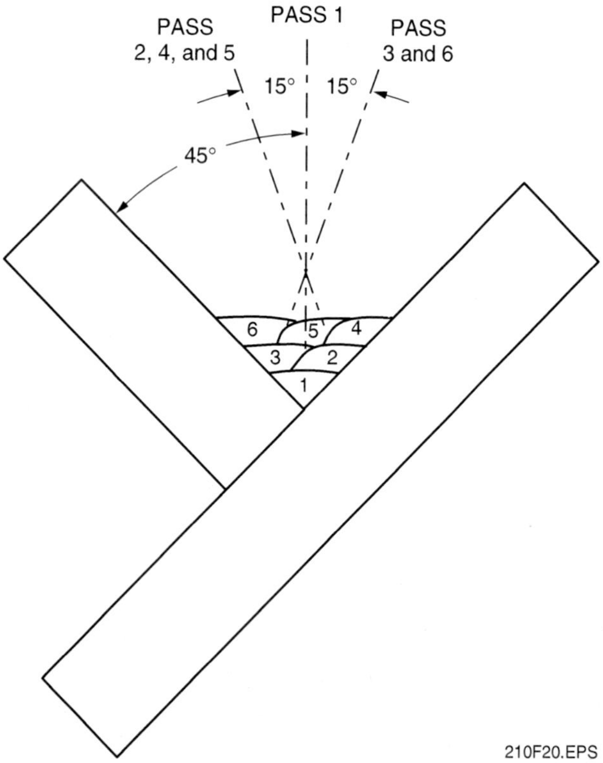

Figure 20 ◆ Multipass 1F weld sequences and work angles.

NOTE: BASE METAL, ALUMINUM PLATE AT LEAST ¼" THICK

Figure 21 ◆ Fillet weld coupon.

Step 4 Run the second bead along a toe of the first weld, overlapping about 75% of the first bead. Alter the work angle as shown in *Figure 20* and use a 10° to 15° push angle with a slight oscillation.

Step 5 Run the third bead along the other toe of the first weld, filling the groove created when the second bead was run. Use the work angle shown in *Figure 20* and a 10° to 15° push angle with a slight oscillation.

Tacking and Aligning Workpieces

When tacking workpieces together, both sides of the workpieces are usually tacked with welds that are about ½" long to position the workpieces and minimize distortion when the final welds are made. After the first tack weld, use a hammer or other tool to align the workpieces side to side and end to end: then tack the opposite side. Tack the far ends of the workpieces in the same manner. Intermediate tack welds can be made every 5" to 6" as necessary to minimize length-wise distortions.

T-Joint Heat Dissipation

In T-joints, the welding heat dissipates more rapidly in the thicker or non-butting member. On various bead passes, the arc may have to be concentrated slightly more on the thicker or the nonbutting member to compensate for the heat loss.

Step 6 Run the fourth bead along the outside toe of the second weld, overlapping about half the second bead. Use the work angle shown in *Figure 20* and a 10° to 15° push angle with a slight oscillation.

Step 7 Run the fifth bead along the inside toe of the fourth weld, overlapping about half the fourth bead. Use the work angle shown in *Figure 20* and a 10° to 15° push angle with a slight oscillation.

Step 8 Run the sixth bead along the toe of the fifth weld, filling the groove created when the fifth bead was run. Use the work angle shown in *Figure 20* and a 10° to 15° push angle with a slight oscillation.

Step 9 Inspect the weld. The weld is acceptable if it has the following:

- Uniform appearance on the bead face
- Craters and restarts filled to the full cross section of the weld
- Uniform weld size, plus or minus ¹⁄₁₆"
- Acceptable weld profile in accordance with the applicable code or standard
- Smooth transition with complete fusion at the toes of the weld
- No porosity
- No excessive undercut
- No overlap
- No inclusions
- No cracks

7.2.0 Practicing Horizontal (2F) Position Fillet Welds

Practice horizontal (2F) fillet welding by placing multiple-pass fillet welds in a T-joint using an appropriate filler metal as directed by your instructor. When making horizontal fillet welds, pay close attention to the torch angles. For the first bead, the electrode work angle is 45°. The work angle is adjusted for all other welds.

Follow these steps to make a horizontal fillet weld:

Step 1 Tack two plates together to form a T-joint for the fillet weld coupon. Clean the tack welds.

Step 2 Clamp or tack-weld the coupon in the horizontal position.

Step 3 Run the first bead along the root of the joint using a work angle of approximately 45° with a 10° to 15° push angle (*Figure 22*).

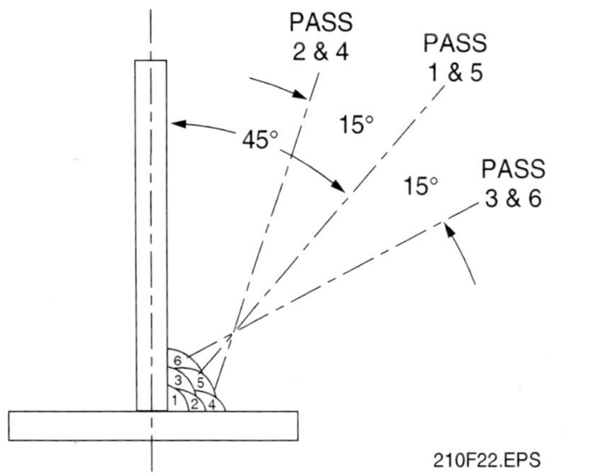

Figure 22 ◆ Multipass 2F weld sequences and work angles.

Raising the Working Height of a GTAW Torch

In some cases, raising the torch height (within limits) can be used to widen second and subsequent weld beads without oscillating the torch.

Step 4 Clean the weld.

Step 5 Run the remaining passes using a 10° to 15° push angle and a slight oscillation at the appropriate work angles (*Figure 22*). Overlap each previous pass. Clean the weld after each pass.

Step 6 Have the instructor inspect the weld. The weld is acceptable if it has:

- Uniform appearance on the bead face
- Craters and restarts filled to the full cross section of the weld
- Uniform weld size $\pm\frac{1}{16}$"
- Acceptable weld profile in accordance with the applicable code or standard
- Smooth transition with complete fusion at the toes of the weld
- No porosity
- No excessive undercut
- No overlap
- No inclusions
- No cracks

7.3.0 Practicing Vertical (3F) Position Fillet Welds

Practice vertical (3F) fillet welding by placing multiple-pass fillet welds in a T-joint using an appropriate filler metal as directed by your instructor. Normally, vertical welds are accomplished by welding uphill from the bottom to the top using a torch push angle (up-angle). Because of the uphill welding and push angle, this type of weld is sometimes called vertical-up fillet welding. When vertical welding, either stringer or weave beads can be used. On the job, the site WPS or site quality standard will specify which technique to use.

> **NOTE**
> Check with your instructor to see if you should run stringer beads or weave beads, or practice both techniques.

> **Hot Tip**
>
> *Vertical Fillet Welds*
>
> When making vertical fillet welds, pay close attention to the torch angles. For the first bead, the work angle is approximately 45°. The work angle is adjusted for all other welds.

7.3.1 Vertical Weave Bead Fillet Welds

Follow these steps to make an uphill fillet weld:

Step 1 Tack two plates together to form a T-joint for the fillet weld coupon.

Step 2 Clamp or tack-weld the coupon in the vertical position.

Step 3 Run the first bead along the root of the joint, starting at the bottom and using a work angle of approximately 45° with a 10° to 15° push angle. Pause in the weld puddle to fill the crater.

Step 4 Clean the weld.

Step 5 Run the remaining passes using a 10° to 15° push angle and a side-to-side weave technique with a 45° work angle (*Figure 23*). Use a slow motion across the face of the weld, pausing at each toe for penetration and to fill the crater. Clean the weld after each pass.

Step 6 Have the instructor inspect the weld. The weld is acceptable if it has:

- Uniform appearance on the bead face
- Craters and restarts filled to the full cross section of the weld
- Uniform weld size $\pm\frac{1}{16}$"
- Acceptable weld profile in accordance with the applicable code or standard
- Smooth transition with complete fusion at the toes of the weld
- No porosity
- No excessive undercut
- No overlap
- No inclusions
- No cracks

7.3.2 Vertical Stringer Bead Fillet Welds

Repeat vertical fillet (3F) welding using stringer beads. Use a slight oscillation and side-to-side motion, pausing slightly at each toe to prevent undercut. For stringer beads, use a 10° to 15° push angle and the work angles shown in *Figure 23* as required.

7.4.0 Practicing Overhead (4F) Position Fillet Welds

Practice overhead (4F) fillet welding by welding multiple-pass fillet welds in a T-joint using an appropriate filler metal as directed by your instructor. When making overhead fillet welds, pay close attention to the torch angles. For the first bead, the work angle is approximately 45°. The work angle is adjusted for all other welds.

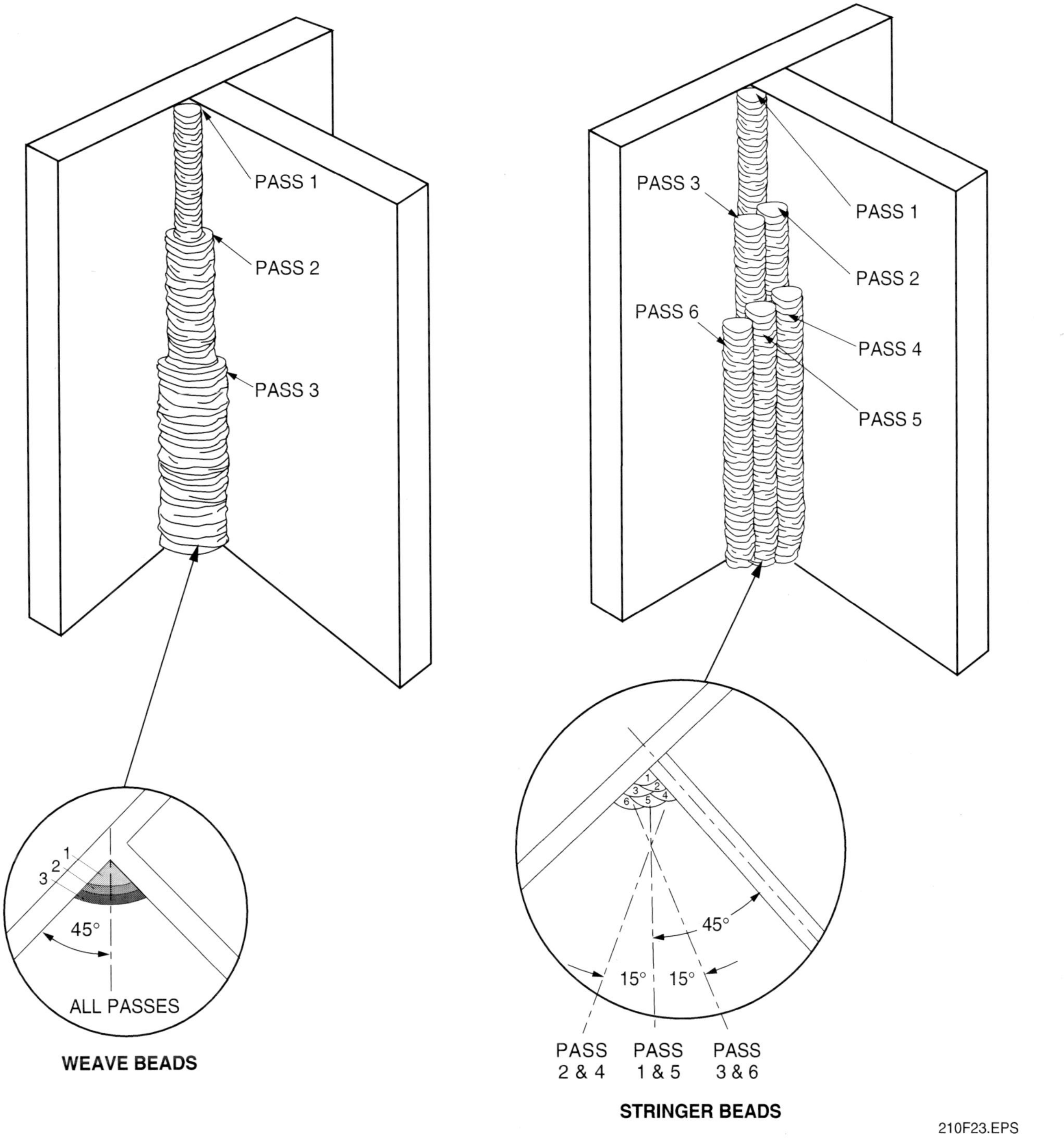

Figure 23 ◆ Multiple pass 3F weld sequences and work angles for stringer and weave beads.

Follow these steps to make an overhead fillet weld:

Step 1 Tack two plates together to form a T-joint for the fillet weld coupon.

Step 2 Clamp or tack-weld the coupon so it is in the overhead position.

Step 3 Run the first bead along the root of the joint using a work angle of approximately 45° with a 10° to 15° push angle.

Step 4 Clean the weld.

Step 5 Using a slight oscillation, run the remaining passes using a 10° to 15° push angle and the work angles shown in *Figure 24*. Overlap each previous pass. Clean the weld after each pass.

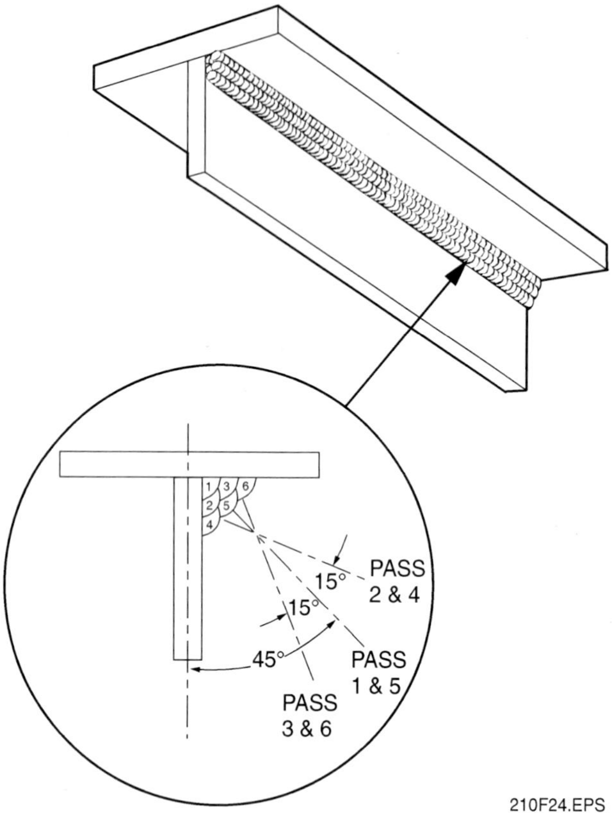

Figure 24 ◆ Multiple pass 4F weld sequences and work angles.

8.0.0 ◆ V-GROOVE WELDS

The V-groove weld is a common groove weld normally made on plate and pipe. Practicing the V-groove welds on plate will prepare the welder to make the more difficult pipe welds.

8.1.0 V-Groove Root with Backing Pass Techniques

In this module, the backing method of welding an aluminum V-groove joint with GTAW is used because it is easier to master.

After the root pass is completed, clean and inspect it. Use the appropriate cleaning method(s) to clean the root pass of the oxidation that tends to form along the joint. Inspect the root pass for the following:

- Uniform smooth face
- Excessive buildup
- Excessive undercut
- Porosity
- Non-fusion

8.2.0 V-Groove Weld Positions

V-groove welds with backing can be made in all positions. The weld position is determined by the axis of the weld. Groove weld positions are flat (1G), horizontal (2G), vertical (3G), and overhead (4G). *Figure 25* shows the groove weld positions.

8.3.0 Acceptable and Unacceptable V-Groove Welds with Backing

V-groove welds with backing should be made with slight reinforcement (not exceeding ⅛") and a gradual transition to the base metal at each toe. Groove welds must not have excess reinforcement, underfill, excessive undercut, or overlap. If a groove weld has any of these defects, it must be repaired. *Figure 26* shows acceptable and unacceptable V-groove weld profiles.

Hot Tip

Specific V-Groove Requirements

Refer to your site's WPS for specific requirements on groove welds. The information in this module is provided as a general guideline only. The WPS or site quality standards must be followed for all welds. Check with your supervisor if you are unsure of the specifications for your application.

8.4.0 V-Groove Welds with Backing

Practice making GTAW V-groove welds with backing in the 1G, 2G, 3G, and 4G positions, using 3⁄32" or ⅛" aluminum filler metal for the root and other passes. Pay particular attention to fill the crater at the termination of the weld. Clean each completed bead with a stainless steel wire brush before starting the next bead.

WARNING!

Use pliers to handle hot practice coupons. Wear gloves when handling the practice coupon.

Hot Tip

Cooling Practice Coupons

Cooling coupons with water can cause weld cracks and affect the mechanical properties of the base metal. Practice coupons can be cooled with water, but never cool test coupons with water.

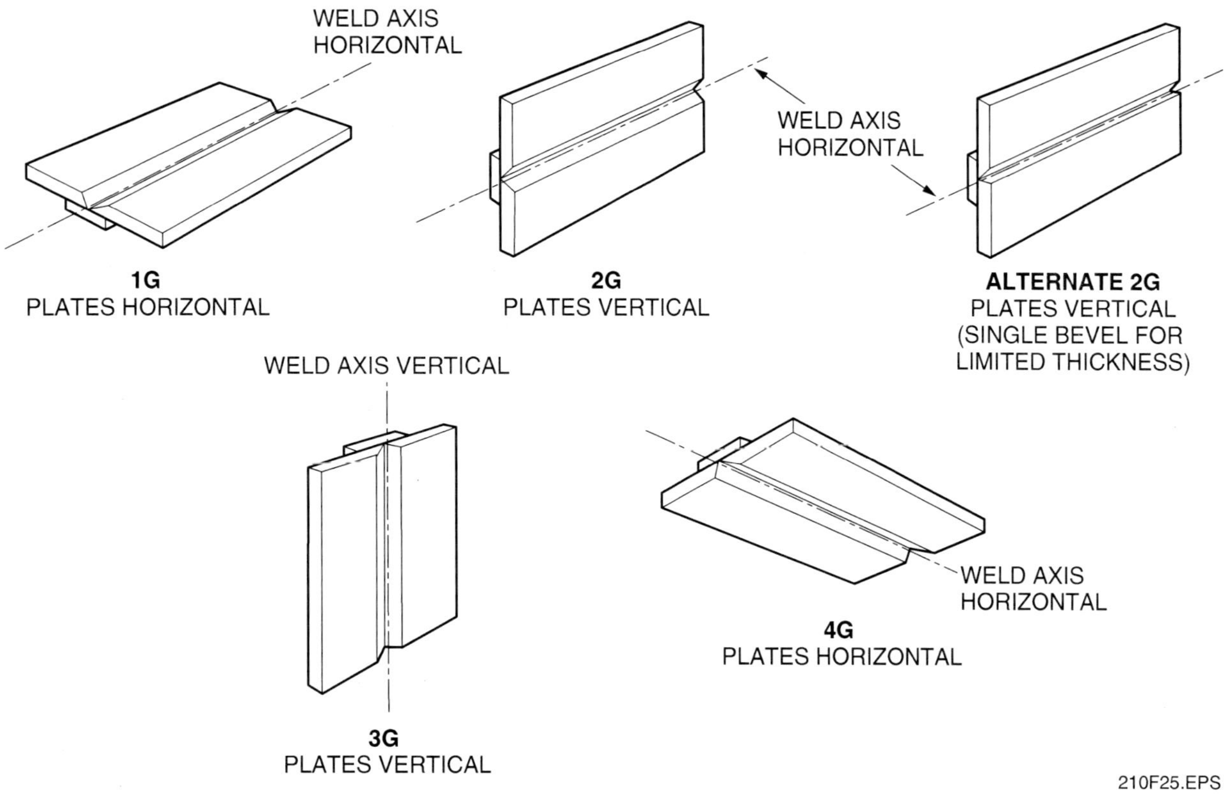

Figure 25 ◆ V-groove weld positions.

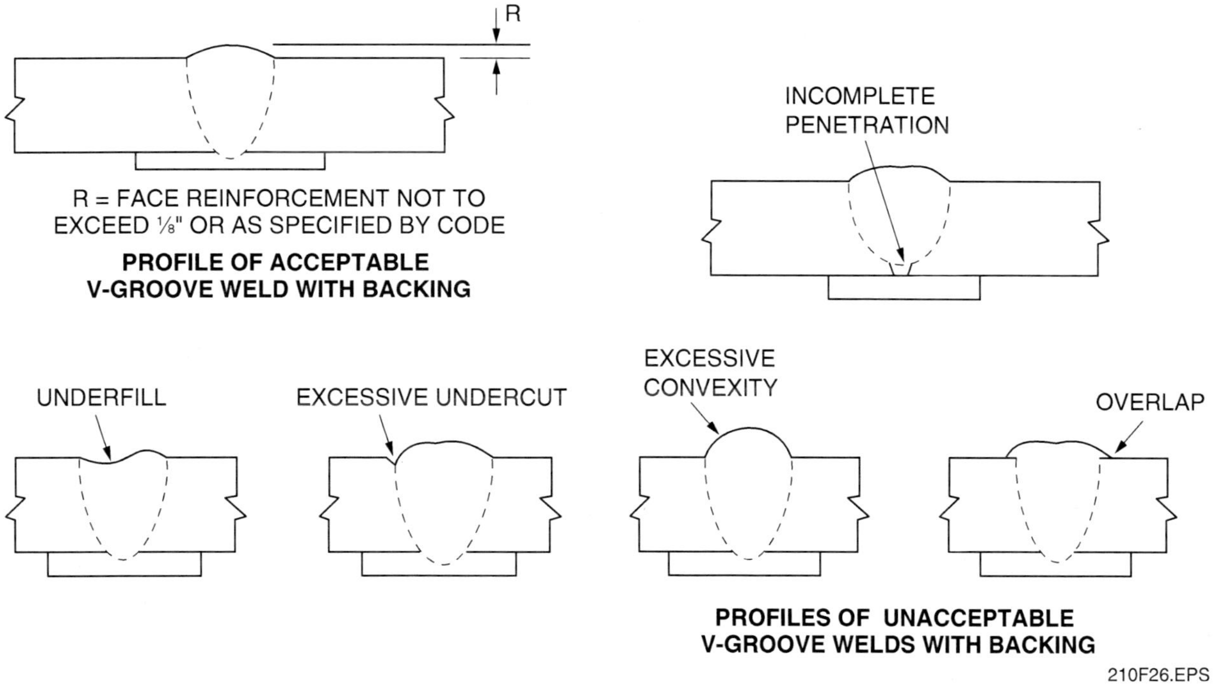

Figure 26 ◆ Acceptable and unacceptable V-groove weld profiles.

8.4.1 Practicing Flat (1G) Position V-Groove Welds with Backing

Follow these steps to make V-groove welds with backing in the flat (1G) position (*Figure 27*):

Step 1 Tack weld the practice coupon together following the example given in the section on preparing welding coupons.

Step 2 Position the weld coupon in the flat position on the welding table.

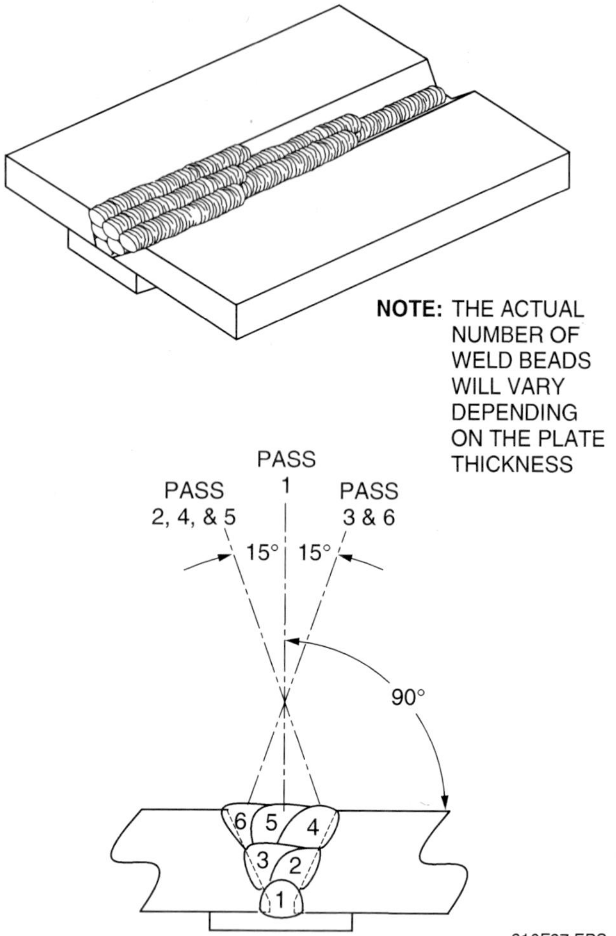

Figure 27 ◆ Multipass 1G weld sequences and work angles.

Step 3 Run the root pass using an appropriate aluminum filler metal. Use a 10° to 15° push angle and a 0° work angle for the root pass. Feed the filler metal at a 15° to 20° angle to the plate.

Step 4 Clean the root pass.

Step 5 Run the remaining passes using a 10° to 15° push angle and the work angles shown in *Figure 27*. Hold the filler metal about 15° to 20° above the plate face. Clean the weld between each pass.

8.4.2 Practicing Horizontal (2G) Position V-Groove Welds with Backing

Follow these steps to practice V-groove welds with backing in the horizontal (2G) position (*Figure 28*):

Step 1 Tack weld the practice coupon together as explained earlier. Use the standard or alternate weld coupon as directed by your instructor.

Step 2 Clamp or tack weld the coupon in the horizontal position.

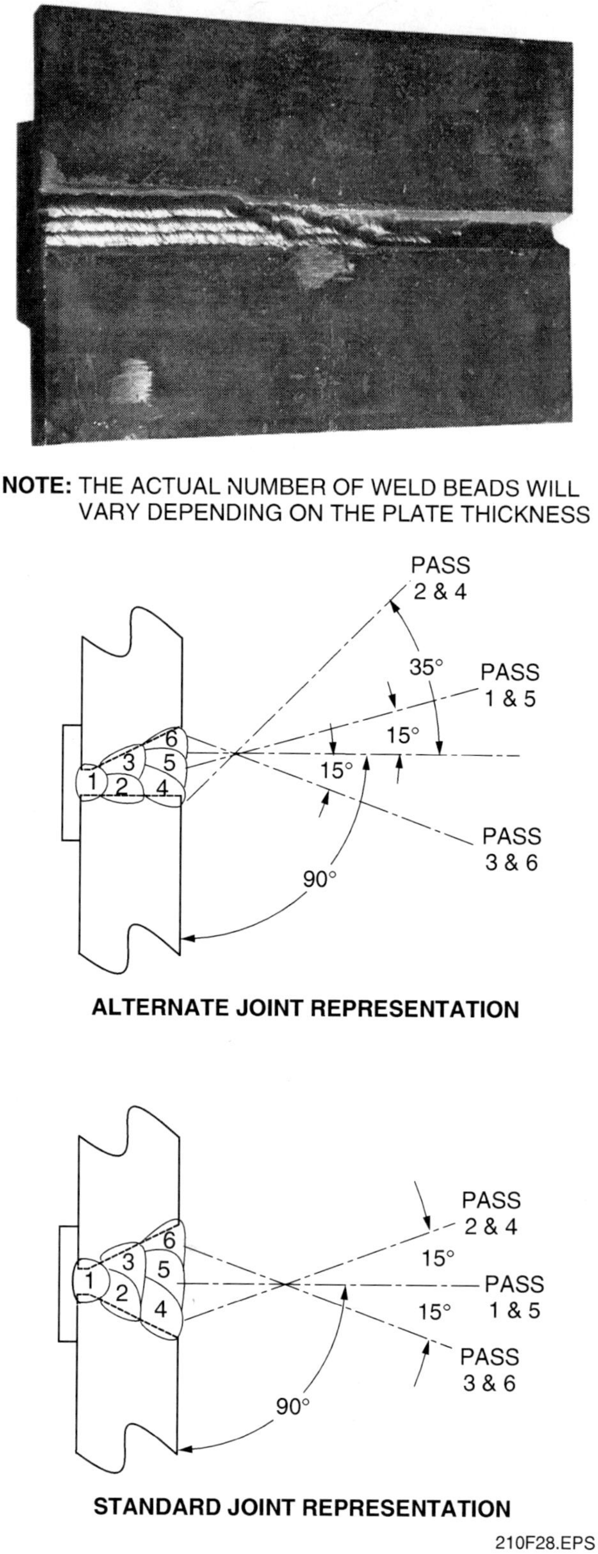

Figure 28 ◆ Multipass 2G weld sequences and work angles.

Step 3 Run the root pass using an appropriate aluminum filler metal. Use a 10° to 15° torch push angle and a 15° to 20° filler metal angle with the plate surface. Use the appropriate work angle for pass 1 shown in *Figure 28*.

Step 4 Clean the weld.

Step 5 Run the remaining passes using a 15° to 20° torch push angle and the work angles shown in *Figure 28*. Clean the weld between each pass.

8.4.3 Practicing Vertical (3G) Position V-Groove Welds with Backing

Follow these steps to practice V-groove welds with backing in the vertical (3G) position (*Figure 29*):

Step 1 Tack weld the practice coupon together as explained earlier.

Step 2 Clamp or tack weld the coupon in the vertical position.

Step 3 Run the root pass uphill using an appropriate aluminum filler metal. Use a torch push angle of 10° to 15° and a filler metal angle of 15° to 20° with the plate surface. Use a 0° work angle for pass 1 as shown in *Figure 29*.

NOTE: THE ACTUAL NUMBER OF WELD BEADS WILL VARY DEPENDING ON THE METAL THICKNESS.

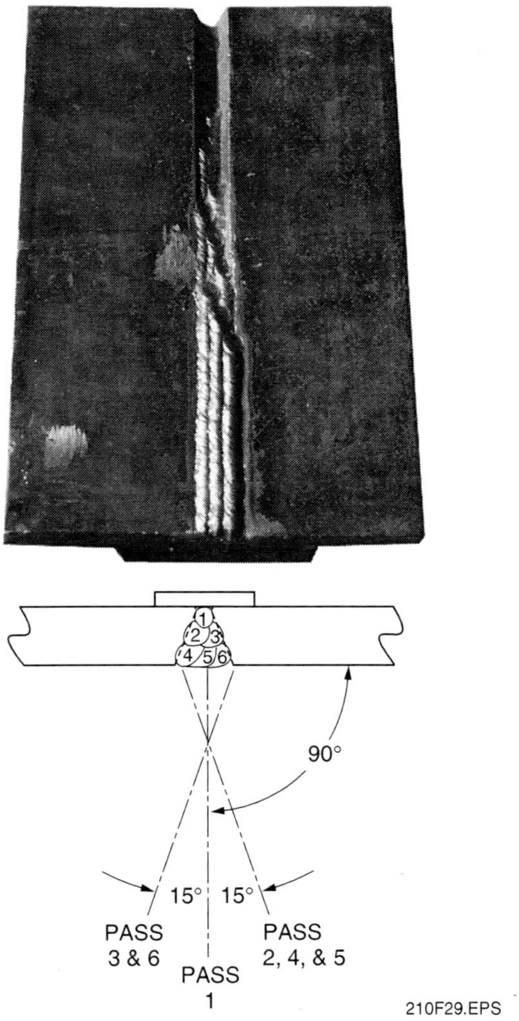

Figure 29 ◆ Multipass 3G weld sequences and work angles.

Step 4 Clean the weld.

Step 5 Run the remaining passes uphill to complete the weld. Use a 10° to 15° torch push angle and the work angles shown in *Figure 29*. Clean the weld between each pass.

8.4.4 Practicing Overhead (4G) Position V-Groove Welds with Backing

Follow these steps to practice V-groove welds with backing in the overhead (4G) position (*Figure 30*):

Step 1 Tack weld the practice coupon together as explained earlier.

Step 2 Clamp or tack weld the coupon in the overhead position.

Step 3 Run the root pass using an appropriate aluminum filler metal. Use a 10° to 15° push angle and a 15° to 20° filler metal angle with the plate surface. Use a 0° work angle for pass 1 as shown in *Figure 30*.

Step 4 Clean the weld.

Step 5 Run the remaining passes using a 10° to 15° torch push angle and the work angles shown in *Figure 30*. Clean the weld between each pass.

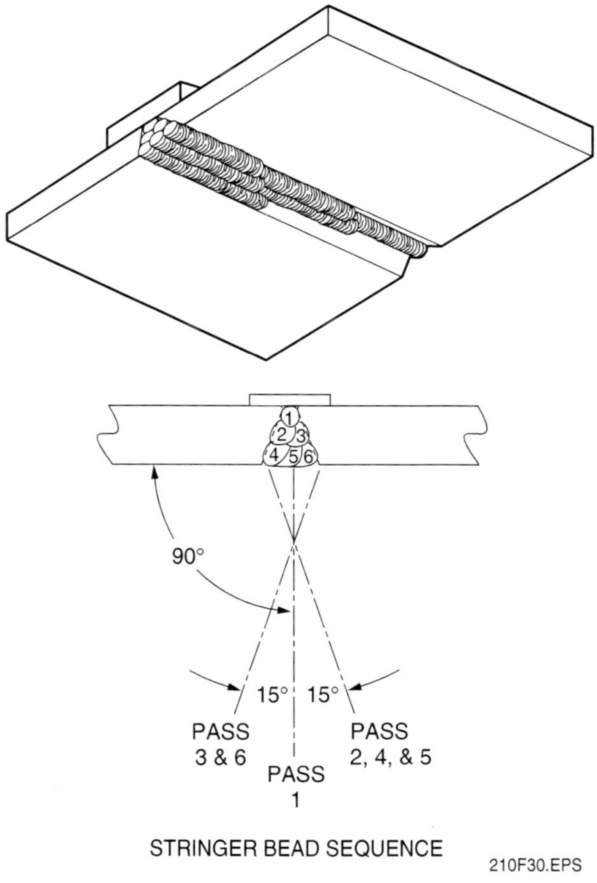

Figure 30 ◆ Multipass 4G weld sequences and work angles.

Summary

Welders must understand aluminum metallurgy and be familiar with its characteristics before they can successfully perform GTAW of aluminum. This module explains aluminum characteristics, the special storage and cleaning requirements of aluminum, the problems with welding it, and the selection of aluminum filler metals. The welder then learns procedures for setting up GTAW equipment for aluminum and how to make weave and overlapping beads. Next, the welder learns methods of terminating and restarting welds and techniques for making fillet and groove welds on aluminum plate.

Review Questions

1. Because of its strong inclination toward oxygen, aluminum oxidizes as soon as it is exposed to _____.
   a. water
   b. air
   c. carbon
   d. hydrogen

2. When welded, nonheat-treatable alloys may lose the effects of work hardening, which results in softening of the _____.
   a. filler metal
   b. heat-affected zone (HAZ)
   c. weld
   d. backing

3. Fusion welding of heat-treatable alloys redistributes the hardening constituents in the HAZ, which locally _____.
   a. increases strength
   b. reduces strength
   c. improves ductility
   d. increases corrosion resistance

4. For nonheat-treatable alloys, _____ fillers are often used.
   a. more ductile
   b. nonmatching
   c. matching
   d. magnesium

5. Cast aluminum has a high silicon content for increased _____.
   a. strength
   b. oxidation resistance
   c. corrosion resistance
   d. fluidity

6. Alternating current (AC) is recommended for GTAW because it provides both the heat and the _____ needed to weld aluminum.
   a. work-hardening action
   b. cleaning action
   c. low voltage
   d. low current

7. Aluminum's high thermal conductivity requires _____ for welding than steel.
   a. lower current
   b. higher current
   c. less precision
   d. less voltage

8. Special care should be taken to raise the temperature of cold aluminum slowly because rapid rises in temperature cause _____.
   a. distortion
   b. loss of strength
   c. loss of ductility
   d. condensation

9. When using etchants or deoxidizers, the metal must be fully rinsed and completely dried to avoid creating a surface film of _____.
   a. hydrated oxide
   b. dehydrated oxide
   c. hydrated acetone
   d. aluminum deoxide

10. The risk of liquation cracking can be reduced by use of a filler metal _____.
    a. with a lower melting temperature than the base metal
    b. with a higher melting temperature than the base metal
    c. of the same composition as the base metal
    d. made of a heat-treatable alloy

11. The aluminum filler metal typically used for the GTAW process is _____ filler rod.
    a. dual shield continuous
    b. composite core
    c. solid-metal
    d. carbon-coated steel

12. As a general rule, aluminum filler rods for welding aluminum with GTAW are selected to _____.
    a. add strength to the soft base metal
    b. increase the corrosion resistance
    c. be compatible with the base metal
    d. reduce the thermal expansion of the base metal

13. The shielding gas most commonly used in GTAW is _____.

a. helium
b. carbon dioxide
c. hydrogen
d. argon

14. Push angles of _____ are normally used for GTAW of aluminum plate.

a. 0° to 10°
b. 10° to 15°
c. 20° to 30°
d. 55° to 65°

15. When using the freehand GTAW technique, the tungsten electrode should _____ the weld puddle.

a. remain in contact with
b. be held just above
c. barely touch
d. rock from side to side in

16. To ensure proper tie-in at the weave bead toes, _____ slightly at the edges.

a. speed up
b. tilt the torch
c. slow down or pause
d. lift the torch

17. To make a good weld termination, as you approach the end of the weld, _____ the welding current.

a. immediately turn off
b. slowly increase
c. slowly reduce
d. rapidly increase

18. The weld position in a fillet weld is determined by the _____.

a. size of the electrode
b. type of joint
c. welder
d. axis of the weld

19. In the 2F position, the first bead of the joint should be made with a work angle of approximately _____ degrees.

a. 20
b. 45
c. 60
d. 70

20. V-groove welds with backing should be made with _____.

a. the on-the-wire technique
b. a small amount of overlap
c. reinforcement to 3/8"
d. slight reinforcement (not exceeding 1/8")

GLOSSARY

Trade Terms Introduced in This Module

Anneal: To expose a metal to a thermal treatment process called annealing that returns a metal to its softest, most ductile, and lowest-strength condition.

Anodic coating: A buildup of aluminum oxides that is many times thicker than a coating of naturally occurring oxides.

Coalescence: The growing together or growth into one body of the materials being joined.

Cold working: Plastic deformation of a metal below the annealing temperature that causes permanent strain hardening.

Heat-affected zone (HAZ): The area of the base metal that is not melted but whose mechanical properties have been altered by the heat of the welding.

Helium: A fairly expensive, colorless, odorless inert gas used as shielding gas to produce a very hot arc and deeper penetration.

Helium/argon mixtures: A shielding gas mixture in which helium is added to argon to raise the temperature of the arc, which promotes higher welding speeds and deeper weld penetration.

Ingot: A mass of metal shaped for convenience in storage and transport.

Liquation: The process of separation by melting, as in an alloy when one element melts while the others remain solid.

Strain hardening: A method of increasing the strength of a metal by cold-working it.

Work hardening: The increase in strength that sometimes accompanies plastic deformation of a metal.

APPENDIX

Performance Accreditation Tasks

The Performance Accreditation Tasks (PATCs) correspond to and support the learning objectives in *AWS EG2.0-95, Guide for the Training and Qualification of Welding Personnel: Entry-Level Welder*.

PATCs provide specific acceptable criteria for performance and help to ensure a true competency-based welding program for students.

The following tasks are designed to evaluate your ability to run beads and fillet welds with GTAW equipment. Perform each task when you are instructed to do so by your instructor. As you complete each task, show it to your instructor for evaluation. Do not proceed to the next task until instructed to do so by your instructor.

MAKE A FILLET WELD IN THE (1F) FLAT POSITION

As directed by the instructor, use an appropriate aluminum filler metal to make a six-pass fillet weld using stringer beads on aluminum plate, as shown.

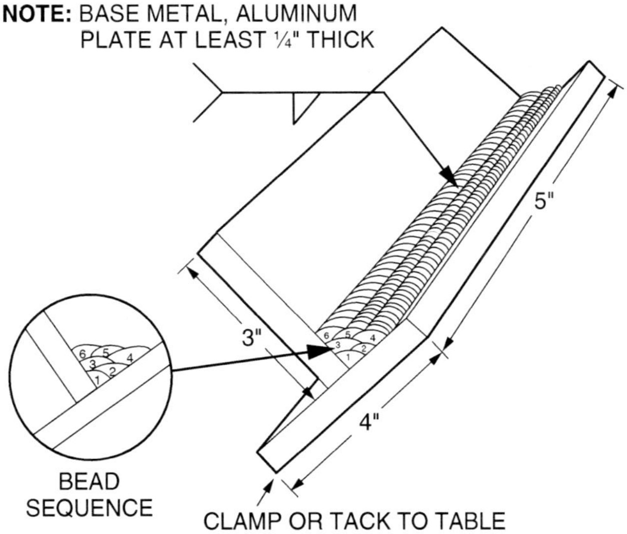

Criteria for Acceptance

- Uniform rippled appearance on the bead face __________
- Craters and restarts filled to the full cross section of the weld __________
- Uniform weld size ±1/16" __________
- Smooth flat transition with complete fusion at the toes of the welds __________
- Acceptable weld profile in accordance with the applicable code or standard __________
- No porosity __________
- No excessive undercut __________
- No overlap __________
- No inclusions __________

MAKE A FILLET WELD IN THE (2F) HORIZONTAL POSITION

As directed by the instructor, use an appropriate aluminum filler metal to make a six-pass fillet weld using stringer beads on aluminum plate, as shown.

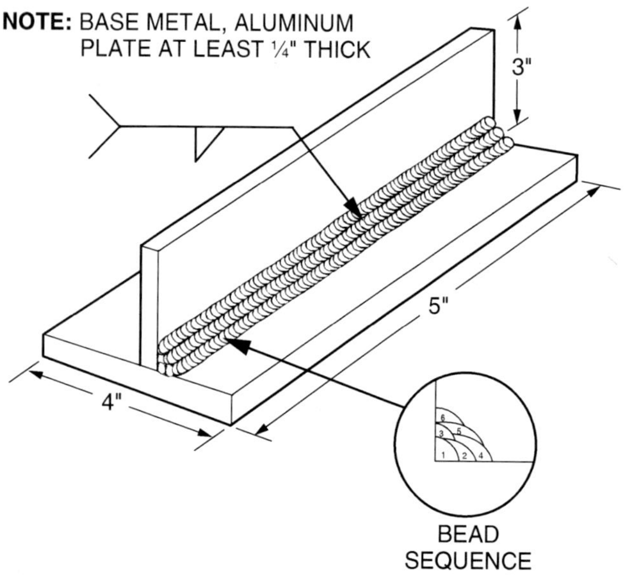

210A02.EPS

Criteria for Acceptance

- Uniform rippled appearance on the bead face ________
- Craters and restarts filled to the full cross section of the weld ________
- Uniform weld size ±1/16" ________
- Smooth flat transition with complete fusion at the toes of the welds ________
- Acceptable weld profile in accordance with the applicable code or standard ________
- No porosity ________
- No excessive undercut ________
- No overlap ________
- No inclusions ________

MAKE A GROOVE WELD IN THE (1G) FLAT POSITION

As directed by the instructor, use an appropriate aluminum filler metal to make a multi-pass groove weld using backing and stringer beads on aluminum plate, as shown.

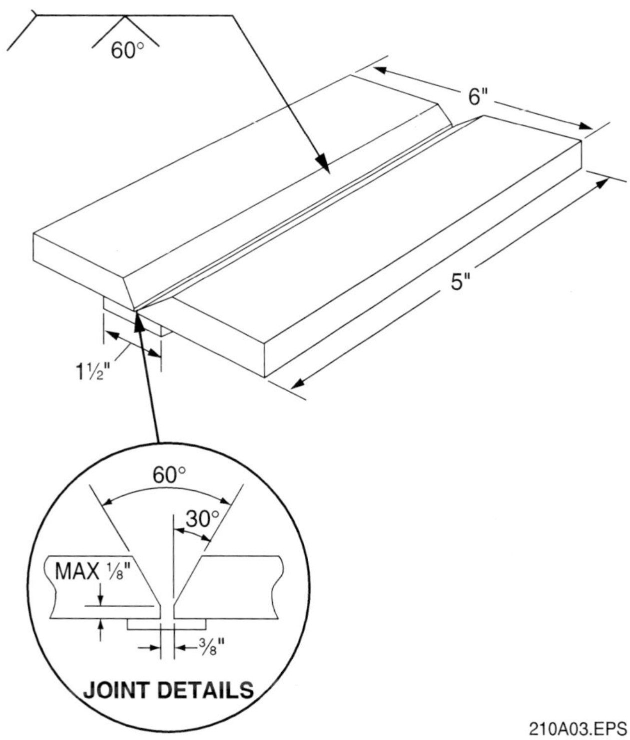

210A03.EPS

Criteria for Acceptance

- Uniform rippled appearance on the bead face __________
- Craters and restarts filled to the full cross section of the weld __________
- Uniform weld width ±1⁄16" __________
- Acceptable weld profile in accordance with the acceptable code or standard __________
- Smooth flat transition with complete fusion at the toes of the weld __________
- No porosity __________
- No excessive undercut __________
- No inclusions __________
- No cracks __________

Additional Resources

This module is intended to be a thorough resource for task training. The following reference works are suggested for further study. These are optional materials for continued education rather than for task training.

Gas Tungsten Arc Welding Handbook, Current Edition. William H. Minnick. Tinley Park, IL: The Goodheart-Willcox Company, Inc.

Modern Welding, 2000. A.D. Althouse, C.H. Turnquist, W.A. Bowditch, K.E. Bowditch. Tinley Park, IL: The Goodheart-Willcox Company, Inc.

The Procedure Handbook of Arc Welding, Current Edition. Cleveland, OH: The Lincoln Electric Company.

Welding Aluminum: Theory and Practice, Current Edition. Washington, DC: The Aluminum Association.

Welding Handbook, 2001. Miami, FL: The American Welding Society.

Welding Skills, Current Edition. R.T. Miller. Homewood, IL: American Technical Publishers, Inc.

Figure Credits

Terry Lowe	210F02, 210F03, 210F14, 210F15, 210F17
Miller Electric Manufacturing Company	210SA01, 210F13
Gerald Shannon	210F21
Charles Rogers	210F28, 210F29

Table 5 courtesy of American Welding Society

Adapted from AWS Committee on Filler Metals, 1999, *Specification for Bare Aluminum and Aluminum Alloy Welding Electrodes and Rods, AWS A5.10/A5.10M*: 1999, Miami: American Welding Society, Table A2.

CONTREN™ LEARNING SERIES — USER UPDATES

The NCCER makes every effort to keep these textbooks up-to-date and free of technical errors. We appreciate your help in this process. If you have an idea for improving this textbook, or if you find an error, a typographical mistake, or an inaccuracy in NCCER's Contren™ textbooks, please write us, using this form or a photocopy. Be sure to include the exact module number, page number, a detailed description, and the correction, if applicable. Your input will be brought to the attention of the Technical Review Committee. Thank you for your assistance.

Instructors – If you found that additional materials were necessary in order to teach this module effectively, please let us know so that we may include them in the Equipment/Materials list in the Instructor's Guide.

Write: Curriculum Revision and Development Department
National Center for Construction Education and Research
P.O. Box 141104, Gainesville, FL 32614-1104

Fax: 352-334-0932

E-mail: curriculum@nccer.org

Craft ____________ Module Name ____________

Copyright Date ____________ Module Number ____________ Page Number(s) ____________

Description

____

____

____

____

(Optional) Correction

____

____

____

(Optional) Your Name and Address

____

____

____

Index

Index